Prandtl Equations and Related Boundary Layer Equations

Yuming Qin • Xiaolei Dong • Xiuqing Wang

Prandtl Equations and Related Boundary Layer Equations

Springer

Yuming Qin
School of Mathematics and Statistics
Donghua University
Shanghai, Shanghai, China

Xiaolei Dong
School of Mathematics and Statistics
Zhoukou Normal University
Zhoukou, Henan, China

Xiuqing Wang
Faculty of Science
Kunming University of Science and Technology
Kunming, Yunnan, China

ISBN 978-981-97-4567-8 ISBN 978-981-97-4565-4 (eBook)
https://doi.org/10.1007/978-981-97-4565-4

This Springer imprint is published by the registered company Springer Nature Singapore Pte Ltd.
The registered company address is: 152 Beach Road, #21-01/04 Gateway East, Singapore 189721, Singapore

In memory of Yuming's father, Zhenrong QIN, and Yuming's mother, Xilan XIA
Dedicated to Xiaolei's parents, Anmin DONG and Xiurong SUN
Dedicated to Xiuqing's parents, Dianqin WANG and Jun LI
Dedicated to Yuming's wife and son, Yu YIN and Jia QIN
and Xiuqiung's husband Hongjun Liu

Preface

At the International Mathematical Congress held in Heidelberg in 1904, Prandtl [111], in his lecture "Fluid Motion with Very Small Friction," suggested an explanation of this phenomenon and described the main principles underlying his new theory, currently called the theory of boundary layer. He showed that the flow about a solid body can be divided into two regions: a very thin layer in the neighborhood of the body (the boundary) where viscous friction plays an essential part, and the remaining region outside this layer where friction may be neglected (the outer flow). Thus, for fluids whose viscosity is small, its influence is perceptible only in a very thin region adjacent to the walls of a body in the flow; the said region, according to Prandtl, is called the boundary layer. Prandtl derived the system of equations for the first approximation of the flow velocity in the boundary layer. This system served as a basis for the development of the boundary layer theory, which has now become one of the fundamental parts of fluid dynamics.

Normally, the boundary layer equation of Navier-Stokes equations is generally called the Prandtl equations. According to the Prandtl theory, the Prandtl equations can be derived from the Navier-Stokes equations. In the more than 100 years since the Prandtl equation was put forward, there have been many research results about the Prandtl equation. These results mainly focus on the well-posedness of local and global solutions. Since the Prandtl equations and the Navier-Stokes equations are closely related, it is also very important to study the relationship between the Navier-Stokes equations and the Prandtl equations, that is, the zero-viscosity limit of Navier-Stokes equations. The construction of the Navier-Stokes solution is performed as a composite asymptotic expansion involving an Euler solution, a Prandtl boundary layer solution, and a correction term. The zero-viscosity limit for the incompressible Navier-Stokes equations is a challenging problem due to the formation of a boundary layer whose thickness is proportional to the square root of the viscosity. The Navier-Stokes solution goes to an Euler solution outside a boundary layer, and it is close to a solution of the Prandtl equations within the boundary layer.

Of course, many mathematicians also study the boundary layer of other fluids such as boundary layer in Magnetohydrodynamics, boundary layer in non-Newtonian flows, Maxwell fluid, and other boundary layer equations.

This book aims to present some recent results on the Prandtl equations and Magnetohydrodynamics (for short, MHD) boundary layer equations. This book is essentially divided into two parts. The first part includes Chap. 1 in which we shall systematically survey the results till 2020 on the Prandtl equations and MHD boundary layer equations. The second part includes Chaps. 2– 6 where the local and the global well-posedness of solutions to the Prandtl equations and MHD boundary layer equations. In detail, Chap. 2 is concerned with global well-posedness of solutions to the 2D Prandtl-Hartmann equations in analytic framework. Chapter 3 investigates the local existence of solutions to the 2D Prandtl equations in a weighted Sobolev space. Chapter 4 studies the local well-posedness of solutions to the 2D mixed Prandtl equations in a Sobolev space without monotonicity and lower bound. Chapter 5 is concerned with local existence of solutions to the 2D magnetic Prandtl equations in the Prandtl-Hartmann regime. Chapter 6 proves the local existence of solutions to the 3D Prandtl equations with a special structure. The results in Chaps. 2–6 have been obtained recently by the authors and have never been published before, i.e., which are updated new results on the Prandtl equations and MHD boundary layer equations.

We sincerely wish that the reader will know the main ideas and essence of the basic theories and methods of the Prandtl equations and the MHD boundary layer equations in deriving the local and global well-posedness of solutions. We also wish that the reader can be stimulated by some ideas from this book and continue to undertake the further study and research after having read this book.

We also want to take this opportunity to thank all the people who concern about us, including our teachers, colleagues, collaborators, etc.

Yuming Qin also acknowledges the National Natural Science Foundation (NNSF) of China for its support. Currently, this book project is being supported by the NNSF of China with contract number 12171082, the Fundamental Research Funds for the Central Universities with contract numbers 2232022G-13 and 2232023G-13. Xiuqing Wang is being supported by the Yunnan Fundamental Research Projects of China with contract number 202401AT070411. Xiaolei Dong is being supported by the Natural Science Foundation of Henan with contract number 242300420673 and the Zhoukou Normal University high level talents research start funding project with contract number ZKNUC2022008.

Shanghai, P. R. China — Yuming Qin
Zhoukou, Henan, P. R. China — Xiaolei Dong
Kunming, Yunnan, P. R. China — Xiuqing Wang

Contents

Chapter 1
Survey on the Prandtl Equations and Related Boundary Layer Equations

The incompressible Navier-Stokes equations are the equations of motion describing the momentum conservation of viscous incompressible fluid. The motion equation of viscous fluid was first proposed by Navier in 1827. He only considered the motion of incompressible fluids. Poisson proposed the motion equation of compressible fluids in 1831, and Stokes proposed that the viscosity coefficient is a positive constant in 1845. Fluid equations are collectively called Navier-Stokes equations, and the 2D incompressible Navier-Stokes equations are as follows:

$$\begin{cases} \mathbf{u}_t + (\mathbf{u} \cdot \nabla)\mathbf{u} - \nu \Delta \mathbf{u} + \nabla p = 0, \\ \operatorname{div} \mathbf{u} = 0. \end{cases} \tag{1.0.1}$$

Here $\mathbf{u} = (u, v)$ is the velocity, $p(t, x, y)$ is the pressure, and $\nu > 0$ is the coefficient of kinematic viscosity.

At the Heidelberg International Mathematical Conference in 1904, Prandtl put forward a basic criterion for describing a phenomenon in his lecture "Fluid motion with very small friction." He pointed out that the flow motion of a solid can be divided into two areas: in a thin layer near the object, viscous friction plays an important role in it; in the area outside this thin layer, the friction effect can be ignored. Prandtl called this thin layer a boundary layer and can use the so-called Prandtl equations to describe. The Prandtl equations are the theoretical model of the Navier-Stokes equations in the thin layer, and the boundary layer equations are the theoretical model of the fluid equations in the thin layer. According to the Prandtl's lecture, let

$$t' = t, \quad x' = x, \quad y' = \varepsilon^{-1} y,$$

Y. Qin et al., *Prandtl Equations and Related Boundary Layer Equations*,
https://doi.org/10.1007/978-981-97-4565-4_1

and let $\overline{u}(t', x', y') = u(t', x', \varepsilon y')$, $\overline{v}(t', x', y') = \varepsilon^{-1} v(t', x', \varepsilon y')$, from (1.0.1), making ε tends to zero, he obtained

$$\begin{cases} u_t + uu_x + vu_y = -p_x + \nu u_{yy}, \\ u_x + v_y = 0, \end{cases} \tag{1.0.2}$$

and p_x satisfies the Bernoulli's law

$$U_t + UU_x + p_x = 0. \tag{1.0.3}$$

Before the twentieth century, there were very few results about the Prandtl equation, but until 1999, Oleinik [104, 105] and Oleinik and Samokhin [106] gave the analysis framework of boundary layer theory, for which there have been many results, for example, well-posedness, viscosity limit, blow up, etc.

Euler equations have a closed relationship with the Navier-Stokes equations, and the 2D incompressible Euler equations are as follows ($\nu = 0$ in (1.0.1)):

$$\begin{cases} \mathbf{u}_t + (\mathbf{u} \cdot \nabla)\mathbf{u} + \nabla p = 0, \\ \operatorname{div} \mathbf{u} = 0. \end{cases} \tag{1.0.4}$$

In recent decades, the relationship among the solutions of Navier-Stokes equations, Euler equations, and Prandtl equations has always been a hot research topic of many scholars. There have been some results, for example, well-posedness, viscosity limit, blow up, etc.

Kármán and Tsien [63] in 1938 concerned with the theory of the laminar boundary layer in compressible fluid:

$$\rho u u_x + \rho v u_y = \nu u_{yy}.$$

Narasimha and Vasantha [102] used numerical solution to boundary layer flows of fluids with a high Prandtl number in heat transfer to liquids and vapors. Nickel [103] reviewed the results about the two-dimensional, steady, and incompressible Prandtl's boundary layer theory.

There are currently four methods used for boundary layer equations. The first method is von Mises transformation and Crocco transformation. And the second method is to separate out the shear flow method, that is, $u(t, x, y) = u^s(t, y) + \widetilde{u}(t, x, y)$, where the shear flow satisfies the heat equation. The third method is to study the Prandtl equations in Gevrey space. The fourth method is the energy estimation method. In this chapter, we list some results and main theorems about the boundary layer equations. We also refer to Refs. [1, 4, 14, 18, 25, 92, 99, 112–114, 123, 124, 132, 153].

1.1 2D Prandtl Equations

Serrin [126] in 1968 illustrated approximation results by studying in detail a specific example of fluid flow where a boundary layer occurs, to some extent, obtained in this way a miniature version of the full theory.

1.1.1 2D Steady Prandtl Equations

Fife [31, 32] in 1965 and Serrin [125] in 1967 considered the approximation between Navier-Stokes equations and the steady two-dimensional Prandtl equations:

$$\begin{cases} uu_x + vu_y = UU_x + \nu u_{yy}, \\ u_x + v_y = 0. \end{cases} \tag{1.1.1}$$

Walter [131] considered the steady 2D (two-dimensional) Prandtl equations (1.1.1) and gave various estimates on the rate of convergence of $u(x, y) \to U(x)$ as $y \to \infty$ if $u(0, y) \to U(0)$. Let dimensionless parameters λ, ξ, and z be determined by the relations

$$\lambda = \frac{\delta}{2aP}, \ \delta = \inf_{x \in \mathbb{R}} U^2(x), \ W = \sup_{x \in \mathbb{R}} U^2(x),$$

$$P = \sup_{x \in \mathbb{R}} |2\frac{dp(x)}{dx}|, \ \xi = \frac{\delta}{W}, \ z = \frac{\psi^2 P}{\nu W^{3/2}}, \ \psi = \int_0^y u_0(\tau) d\tau.$$

Kuznetsov [68] in 1994 considered the boundary layer equations (1.1.1) in the domain $\Omega = \{(x, y) | x \in (0, a), y \in (0, \infty)\}$ and used von Mises transform to prove that there exists a unique global solution, i.e., the following theorem.

Theorem 1.1.1 ([68]) *Assume the following inequalities holds:*

$$\frac{(2+\lambda)^2}{\lambda^2(1+\lambda)} < z < \frac{3}{32}\Big(\frac{\xi(2+\lambda)}{2(1+\lambda)}\Big)^{3/2}.$$

Then there exists a unique solution u, v to problem (1.1.1) in Ω such that $u > 0$, the derivatives $\partial u/\partial y$ and $\partial^2 u/\partial y^2$ are continuous and bounded in Ω, $\partial u/\partial y \geq m > 0$ for $y = 0$, and the functions $\partial u/\partial x$, v, and $\partial v/\partial y$ are continuous and bounded in any finite part of $\overline{\Omega}$.

Liu and Wan [80] in 1985 considered a steady, two-dimensional, laminar, incompressible flow (1.1.1) and found an exact solution if the transverse component of the external flow velocity is allowed to be deflected upward, while the longitudinal component is assumed to be linearly retarded. It is that this solution is regular at separation and exhibits reverse flow thereafter.

Theorem 1.1.2 ([80]) *For the given positive pressure gradient (1.1.2), let g_0, g_1 be the solution of (1.1.3), with constants $\alpha \approx 9.45$, $\beta \approx 1.255$. Then,*

$$u = g_0' + x g_1', \quad v = -g_1$$

in an analytic solution of the problem (1.1.1) in $D_B = \{-1 \le x \le 1, 0 \le y \le \delta\}$, the boundary layer thickness to be $\delta = B/R^{\frac{1}{2}}$, R denotes the Reynolds number, $B = \beta/k^{1/2}$, possessing the following properties in the region

$$\{(x, y) \in D_B | y \le \beta^*/k^{\frac{1}{2}} R^{\frac{1}{2}}\}$$

for some β^*, $0 < \beta^* \le \beta$:

(i) $v \ge 0$, $v = O(R^{-\frac{1}{2}})$; moreover $u_y > 0$ on $y = 0$ if $x > 0$, and $u_y > 0$ on $y = 0$ if $x > 0$, and $u_y < 0$ on $y = 0$ if $x < 0$;
(ii) the separating streamline $\gamma : \varphi = 0$ (i.e., $x = -g_0(y)/g_1(y)$, $y > 0$), has slope

$$\frac{3\alpha k^{3/2}}{p_0 R^{1/2}}$$

at $(0, 0)$ and $x > 0$ on γ;
(iii) the curve T; $u = 0$ (i.e., $x = -g_0'(y)/g_1'(y)$, $y > 0$) starting from $(0, 0)$ lies to the right of γ. Thus $u < 0$ ($u > 0$) to the right (left) of T.

Here

$$p_x = p_0 - k^2 x, \quad p_y = -k^2 y, \quad -1 \le x \le 1, \quad 0 \le y < +\infty, \tag{1.1.2}$$

$$\begin{cases} \nu g_0''' + g_1 g_0'' - g_1' g_0' = p_0, \\ \nu g_1''' + g_1 g_1'' - g_1'^2 = -k^2. \end{cases} \tag{1.1.3}$$

Oleinik and Samokhin [106] in 1999 considered 2D stationary boundary layer based on the von Mises transformation which reduces the boundary layer system to a single quasilinear parabolic equation. Consider

$$\begin{cases} u u_x + v u_y + p_x = u_{yy}, \\ u_x + v_y = 0, \\ u(0, y) = u_0(y), \end{cases} \tag{1.1.4}$$

in the domain $D = \{0 < x < X, 0 < y < +\infty\}$ with the conditions

$$\begin{cases} u(0, y) = u_0(y), \quad u(x, 0) = 0, \quad v(x, 0) = v_0(x), \\ \lim_{y\to+\infty} u(x, y) = U(x). \end{cases} \tag{1.1.5}$$

Introduce von Mises transformation of the variables by

$$x = x, \quad \psi = \psi(x, y), \quad w(x, \psi) = u^2(x, y),$$

where

$$u = \frac{\partial \psi}{\partial y}, \quad v - v_0 = -\frac{\partial \psi}{\partial x}, \quad \psi(x, 0) = 0,$$

the domain D turns into $G = \{0 < x < X, 0 < \psi < \infty\}$. Then the system (1.1.4) reduces to the equation

$$\frac{\partial w}{\partial x} + v_0 \frac{\partial w}{\partial \psi} = \sqrt{w} \frac{\partial^2 w}{\partial \psi^2} - 2\frac{\partial p(x)}{\partial x}. \tag{1.1.6}$$

Theorem 1.1.3 ([106]) *Assume that $u_0(y)$ for $y > 0$; $u_0(0) = 0$, $u_0'(0) > 0$, $u_0(y) \to U(0) \neq 0$ as $y \to \infty$; p_x and $v_0(x)$ are continuously differentiable on [0,X]; $u_0(y), u_0'(y), u_0''(y)$ are bounded for $0 \leq y \leq \infty$ and satisfy the Hölder condition. Moreover, assume that for small y the following compatibility condition is satisfied at the point* $(0, 0)$:

$$\frac{d^2 u_0(y)}{dy^2} - \frac{dp(0)}{dx} - v_0 \frac{du_0(y)}{dy} = O(y^2). \tag{1.1.7}$$

Then, for some $X > 0$ there exists a solution $u(x, y), v(x, y)$ of problem (1.1.4)–(1.1.5) in D, which has the following properties: $u(x, y)$ is bounded and continuous in $\overline{D}$, $u > 0$ for $y > 0$; $u_y > m > 0$ for $0 \leq y \leq y_0$, where m and y_0 are constants; u_y and u_{yy} are bounded and continuous in D; u_x, v, v_y are bounded and continuous in any finite portion of $\overline{D}$.

Moreover, if $|u'_0(y)| \leq m_1 \exp(-m_2 y)$, m_1, m_2 =const.> 0, then u_x and v_y are bounded in D. If either $p_x \leq 0$ and $v_0(x) \leq 0$ or $p_x \leq 0$, then such a solution of problems (1.1.4)–(1.1.5) exists in D for any $X > 0$.

Theorem 1.1.4 ([106]) *Let u and v be two functions satisfying system (1.1.4) in D , continuous in $\overline{D}$, and satisfying the conditions (1.1.5). Moreover, assume the following estimates hold: $0 < u < c_1$ for $\psi > 0$,*

$$k_1 y \leq u \leq k_2 y, \quad for \;\; 0 < y < y_0, \tag{1.1.8}$$

$$u_{yy} \leq k_3 \;\; in \;\; D, \tag{1.1.9}$$

where $k_1, k_2, k_3, y_0,$ *are positive constants. Then* u, v *is the only solution of problems (1.1.4)–(1.1.5) with these properties. If* $p_x \le 0$ *and* $v_0(x) \le 0$, *then the conditions (1.1.8)–(1.1.9) can be dropped. If* $p_x \le 0$ *and* $v_0(x)$ *is allowed to change sign, then it is only the condition (1.1.9) that is required for the uniqueness.*

Theorem 1.1.5 ([106]) *Let all the assumptions of Theorem 1.1.3, except the compatibility condition (1.1.7), hold for* $u_0(y), v_0(x), p_x$. *Then, for some* $X > 0$, *problems (1.1.4)–(1.1.5) in the domain D admit a solution* u, v *with the following properties:* $u(x, y)$ *is continuous and bounded in* $\overline{D}$; $u > 0$ *for* $y > 0$; $u_y \ge m > 0$ *for* $0 \le y \le y_0$ *and* $x \ge x_0 > 0$, *where* m, y_0 *are constants depending on* x_0; u_y *is continuous and bounded in D;* u_{yy} *is continuous in D and bounded in* $D \backslash \Omega_\rho$, *where* Ω_ρ *is the neighborhood of* $(0, 0)$ *given by the inequality* $x^2 + y^2 \le \rho$; u_x, v, v_y *are continuous in D and bounded in any finite portion of the domain* $D \backslash \Omega_\rho$. *If* $|u\prime_0(y)| \le k_1 \exp(-k_2 y)$ *for* $y \to \infty$, *then* u_x, v_y *are bounded in* $D\ \Omega_\rho$. *If* $p_x \le 0$ *and* $v_0 \le 0$ *or* $p_x \le 0$, *then problems (1.1.4)–(1.1.5) admit a solution in D for any* $X > 0$. *If* $p_x \le 0$ *in a neighborhood of* $x = 0$, *this solution is unique.*

Theorem 1.1.6 ([106]) *Let* $U(x) = U_1(x + d)^{(\alpha+1)/2}$, U_1 *=const.*$> 0, \alpha \ge -1$, d *=const.*≥ 0, $v_0(x) \equiv 0$. *Let* $u(x, y)$ *be a solution of problems (1.1.4)–(1.1.5) corresponding to the initial velocity profile* $u_0(y)$, *and let* $\overline{u}(x, y), \overline{v}(x, y)$ *be a self-similar solution of system (1.1.4), i.e., a solution of the form*

$$\overline{u}(x, y) = U(x) f'(\eta), \quad \overline{v}(x, y) = \int_0^y \frac{\partial \overline{u}}{\partial x} dy + v_0(x).$$

Then

$$\left| \frac{u(x, y)}{U(x)} - \frac{\overline{u}(x, y)}{U(x)} \right| = o\left(\left[1 + (\frac{\alpha + 1}{2} \ln 2) \right] x^{-\frac{\alpha+1}{2}} \right) \quad \text{as } x \to \infty,$$

uniformly with respect to y.

Further, consider two solutions $u_1(x, y)$ *and* $u_2(x, y)$ *of problems (1.1.4)–(1.1.5) corresponding to two distinct initial velocity profiles* $u_{10}(y)$ *and* $u_{20}(y)$, *let* $v_0(x) \equiv 0$, *and assume that*

$$C_1(x + d)^{2m-1} \le U U_x \le C_2(x + d)^{2n-1}, \quad C_1, C_2 = \text{const.} > 0,$$

where $m \le n < 5m/3$. *Then*

$$\left| \frac{u_1}{U} - \frac{u_2}{U} \right| = o(1) \quad \text{as } x \to \infty,$$

uniformly with respect to y.

Theorem 1.1.7 ([106]) *Assume that*

$$\lim_{y\to+\infty} U(x) = U_\infty = const. > 0, \quad v_0(x) \equiv 0,$$
$$0 \le u_0(y) \le U(0), \quad u_0(0) = 0, \quad u'_0(0) > 0,$$
$$0 \le \frac{dU}{dx} \le \frac{M_0}{(x+1)^{\gamma_0+1}}, \quad \gamma_0 > 0.$$

Then

$$|u(x, y) - u_1(x, y)| \to 0, \quad as \ x \to \infty,$$

uniformly in $y \in [0, \infty)$, *where* $u(x, y)$ *is the solution of problems (1.1.4)–(1.1.5) corresponding to the initial velocity profile* $u_0(y)$;

$$u_1(x, y) = U_\infty f'(\eta), \quad when \ \eta = y\sqrt{\frac{U_\infty}{2(x+1)}}$$

and $f(\eta)$ *is the solution of the following boundary value problem:*

$$f''' + ff'' = 0, \quad 0 < \eta < \infty,$$
$$f(0) = 0, \quad f'(0) = 0, \quad f'(\infty) = 1.$$

Under the additional conditions

$$U(0)f'(y - N) \le u_0(y), \quad y \in [N, \infty),$$
$$|u_0(y) - U(0)| \le M_1 \exp(-\gamma_1 y^2), \quad y \in [0, \infty),$$

with some constants $N, M, \gamma_1 > 0$, *the following estimate holds:*

$$|u(x, y) - u_1(x, y)| \le \frac{M_2}{(x+1)^{\gamma_2}}$$

where $M_2, 0 \le \gamma_2, \ \gamma_0$ *are constants depending only on the data of the problem.*

Theorem 1.1.8 ([106]) *Assume that problems (1.1.4)–(1.1.5) admit a solution in a domain* D. *Then* $X < x_0$, *where* x_0 *is determined by the conditions*

$$\max u_0^2(y) - 2\int_0^{x_0} \frac{dp}{dx} dx = 0, \quad \frac{dp(x_0)}{dx} > 0.$$

Theorem 1.1.9 ([106]) *Assume that $dp/dx \geq 0$ and there exist x_1, x_2 such that $x_1 \leq x_2$ and*

$$\int_{x_1}^{x_2} v_0(t)dt > \sqrt{6M_0(x_2 - x_1)}.$$

Then separation of the boundary layer occurs at some points $x_1 < x_2$.

Theorem 1.1.10 ([106]) *Assume the following: $v_0(x) \leq 0$, $v\prime_0(x) \leq 0$ in the interval $0 \leq x \leq X$; dp/dx and $v_0(x)$ belong to the class $C^1([0, X])$; the initial velocity profile $u_0(y)$ is such that*

$$u_0(0) = 0, \quad u_0(y) > 0 \text{ for } y > 0, \quad \lim_{y\to\infty} u_0(y) = U(0),$$
$$|u_0(y)|, |u\prime_0(y)|, |u\prime\prime_0(y)| \leq M, \quad y \in [0, \infty);$$

the Hölder inequality with exponent $\alpha = \alpha(N) \in (0, 1)$ holds for $u\prime\prime_0(y)$ on the segment $N^{-1} \leq y \leq N$, for any $N > 1$; moreover,

$$l_0(u_0) = u\prime\prime_0(y) - u\prime_0(y) - \frac{dp(0)}{dx} = O(y^2), \quad \text{as } y \to 0.$$

Assume also that

$$\frac{dp}{dx} \leq (Cx+1)^{-3}\frac{dp(0)}{dx} \quad \text{for } 0 \leq x < \infty, \quad C = \sup_{y\geq 0}\left|\frac{l_0(u_0)}{u_0^2}\right|.$$

Then for any finite $X > 0$, problems (1.1.4)–(1.1.5) in D admits a unique solution u, v such that u, u_y, u_{yy} are bounded and continuous in D; v, u_x, v_y are continuous in D; and the velocity profile $u(x_0, y)$, for any $x_0 \in [0, X]$ has the properties of the initial profile $u_0(y) = u(0, y)$.

Assume that $dp/dx \leq 0$ and $v_0(x) \equiv 0$. Applying the von Mises transformation, then we obtain the equation

$$\frac{\partial w}{\partial x} = \sqrt{w}\frac{\partial^2 w}{\partial \psi^2} - 2\frac{\partial p(x)}{\partial x} \tag{1.1.10}$$

in $G = \{0 < x < X, 0 < \psi < \infty\}$, with the boundary conditions

$$\begin{cases} w(x, 0) = 0, \quad w(0, \psi) = w_0(\psi), \\ \lim_{y\to\infty} w(x, \psi) = U^2(x) \text{ uniformly on } [0, X]. \end{cases} \tag{1.1.11}$$

Theorem 1.1.11 ([106]) *Assume that $U(x)$ has a bounded second derivative on the segment $[0, X]$, $U(x) > 0$, $U'(x) \geq 0$; $u_0(y) \in C^2(0 \leq y \leq \infty)$, $u_0(0) =$*

$0, u'_0(0) > 0, u''_0(y) \le 0$, *and* $u''_0(y)$ *satisfies the local Hölder condition for* $y > 0$. *Moreover, let*

$$u_0(y) \to U(0), \quad as \quad y \to 0,$$
$$u''_0(y) + U(0)U'(0) = O(y^2) \; as \quad y \to 0.$$

Then problems (1.1.10)–(1.1.11) have a unique solution w *with the following properties:* $w, \partial w/\partial x, \partial w/\partial \psi$ *are bounded and continuous in* $\overline{G}$ *and satisfy the inequalities*

$$w > 0, \quad \left|\frac{\partial w}{\partial x}\right| \le C\psi,$$
$$\frac{\partial w}{\partial \psi}(x, 0) \ge \alpha > 0, \quad \frac{\partial w}{\partial x} \le 2UU' \;\; in \;\; G, \;\; C, \alpha = const.$$

Theorem 1.1.12 ([106]) *Let* $f(x, z) = (f_0(x, z_0, z_1, \ldots), f_1(x, z_0, z_1, \ldots), \ldots)$ *be a mapping* $[0, X] \times B \to B$ *which is continuous in* x *and satisfies the Lipschitz condition in* z*:*

$$\|f(x, z) - f(x, \overline{z}); B\| \le L\|z - \overline{z}; B\|, \quad z, \overline{z} \in B; \quad x \in [0, X],$$

where L *=const.*> 0. *Then the Cauchy problem*

$$\frac{du}{dx} = f(x, u), \quad x \in [0, X], \quad u(0) \in B, \tag{1.1.12}$$

has a unique solution $u(x) \in C^1(0, X; B)$. *Moreover, if* $u_0(x)$ *belongs to the space* $C^0(0, X; B)$, *then the successive approximations* $u_n(x)$ *given by*

$$u_{n+1} = u(0) + \int_0^x f(\tau, u_n(\tau))d\tau, \quad n = 0, 1, 2, \ldots,$$

are uniformly convergent to $u(x)$ *in the norm of* B, *as* $n \to \infty$.

Theorem 1.1.13 ([106]) *Assume that the conditions of Theorem 1.1.12 hold for* $f(x, z)$ *and* $f(x, z)$ *increases quasimonotonically in* z. *Let* $u \in C^1(0, X; B)$ *be a solution of problem* (1.1.12), *and let* $v \in C^1(0, X; B)$ *be a function such that*

$$\frac{dv}{dx} \le f(x, v), \quad x \in [0, X], \quad v(0) \le u(0).$$

Then, $v(x) \le u(x)$ *for* $x \in [0, X]$.

Theorem 1.1.14 ([106]) *Problems (1.1.10)–(1.1.11) with* $p_x(x) \leq 0$ *admit a unique solution such that* $w_x, w_\psi, w_{\psi\psi}$ *are continuous in* G*, and* $w \geq 0, w_{\psi\psi} \leq 0$ *in* G.

Oleinik and Samokhin [106] in 1999 considered the boundary layer system for an axially symmetric incompressible stationary flow that has the form

$$\begin{cases} u\frac{\partial u}{\partial x} + v\frac{\partial u}{\partial y} = \nu\frac{\partial^2 u}{\partial y^2} + U\frac{dU}{dx}, \\ \frac{\partial(ru)}{\partial x} + \frac{\partial(rv)}{\partial y} = 0, \end{cases} \tag{1.1.13}$$

in a domain $D = \{0 < x < X, 0 < y < \infty\}$, with the boundary conditions

$$\begin{cases} u(0, y) = 0, \quad u(x, 0) = 0, \quad v(x, 0) = v_0(x), \\ u(x, y) \to U(x) \quad \text{as} \quad y \to \infty. \end{cases} \tag{1.1.14}$$

The function $r(x)$ determines the surface of the body past which the fluid flows, $r(0) = 0, r_x(0) \neq 0, r(x) > 0$ for $x > 0$; $U(x)$ is a given longitudinal velocity component of the outer flow, $U(0) = 0, U(x) > 0$ for $x > 0$, $U_x > 0$ for $x \geq 0$; ν is the viscosity coefficient.

Introduce new independent Crocco variables by

$$\xi = x, \quad \eta = \frac{u(x, y)}{U(x)},$$

and a new unknown function

$$w(\xi, \eta) = \frac{u_y(x, y)}{U(x)}.$$

From (1.1.13), they obtained the following equation for $w(\xi, \eta)$:

$$\nu w^2 w_{\eta\eta} - \eta U w_\xi + A w_\eta + B w = 0 \tag{1.1.15}$$

in the domain $\Omega = \{0 < \xi < X, 0 < \eta < 1\}$, with the boundary conditions

$$w|_{\eta=1} = 0, \quad (\nu w^2 w_{\eta\eta} - v_0 w + C)|_{\eta=0} = 0, \tag{1.1.16}$$

where

$$A = (\eta^2 - 1)U_x(\xi), \quad B = -\eta U_x(\xi) + \frac{\eta r_x(\xi)U(\xi)}{r(\xi)}, \quad C = U_x(\xi).$$

Note that from the above assumptions on $U(x)$ and $r(x)$, one has

$$\lim_{\xi \to 0} B(\xi, \eta) = 0.$$

Theorem 1.1.15 ([106]) *Assume that the functions $r(x), U(x)$ are twice continuously differentiable, U_x, $U(0) = 0$, $|U_{xx}| \le N_1 x$, $v_0 \le N_2 x$, N_i =const.> 0, and v_0 has its first derivative bounded.*

Then problems (1.1.15)–(1.1.16) have a solution w in the domain $\Omega = \{0 < \xi < X, 0 < \eta < 1\}$, where $X > 0$ depends only on U, r, v_0. This solution has the following properties: $w(\xi, \eta)$ is continuous in $\overline{\Omega}$; $M_9(1-\eta)\sigma \le w \le M_2(1-\eta)\sigma$; w_η is continuous in η for $\eta < 1$; $-M_{22}\sigma \le w_\eta \le -M_{23}\sigma$ in Ω, $|w_\xi| \le M_{24}(1-\eta)\sigma$, $ww_{\eta\eta} \le -M_{25}$, and $ww_{\eta\eta}$ is bounded in Ω, $\sigma = (-\ln\mu(1-\eta))^{1/2}, 0 < \mu < 1$. In any closed interior subdomain of Ω, the function w and its derivatives in (1.1.15) satisfy the Hölder condition. Furthermore, w satisfies equation (1.1.15) in Ω and also the condition (1.1.16) for $0 \le \xi \le X$.

Theorem 1.1.16 ([106]) *Assume the assumptions of Theorem 1.1.15 hold for $U(x), r(x), v_0(x)$. Then problems (1.1.13)–(1.1.14) in the domain D admit a solution with the following properties: $u/U, u_y/U$ are continuous and bounded in $\overline{D}$; $u > 0$ for $y > 0$ and $x > 0$; $u \to U$ as $y \to \infty$, $u(x,0) = u(0,y) = 0$; $u_y/U > 0$ for $y > 0$; $u_y/U \to 0$ as $y \to \infty$; u_{yy}, u_x, u_y are bounded and continuous with respect to y in $\overline{D}$; v is bounded for bounded y; u_{yy} is bounded in $\overline{D}$; u_{xy} is bounded in D for bounded y; u_{xy} and u_{yy} are continuous in $\overline{D}$; u_{yy}/u_y is continuous in $\overline{D}$ with respect to y. Moreover,*

$$
\begin{aligned}
&M_9(U-u)\sigma \le u_y \le M_2(U-u)\sigma,\\
&-M_{22}\sigma \le \frac{u_{yy}}{u_y} \le -M_{23}\sigma, \quad -M_{26} \le \frac{u_{yyy}u_y - u_{yy}^2}{u_y^2} \le -M_{25},\\
&\left|\frac{1}{u_y}(u_{xy}u_y - u_x u_{yy}) + \frac{U_x}{Uu_y}(uu_{yy} - u_y^2)\right| \le M_{24}(U-u)\sigma,\\
&M_{32}U\exp(-M_{33}y^2) \le U - u \le M_{34}U\exp(-M_{35}y^2),
\end{aligned}
$$

where $\sigma = (-\ln\mu(1-u/U))^{1/2}$, M_i, μ are positive constants, $\mu < e^{-1/2}$.

Theorem 1.1.17 ([106]) *Assume (u, v) is a solution of problems (1.1.13)–(1.1.14) such that the derivatives $u_x, u_y, v_y, u_{yy}, u_{yyy}, u_{xy}$ are continuous in D; u/U and u_y/U are continuous in $\overline{D}$; $u_y > 0$ for $y \ge 0, x > 0$; $u_y/U > 0$ for $y = 0$; $u_y/U \to 0$ as $y \to \infty$; $u_{yy}/u_y, u_x$ are continuous with respect to y at $y = 0$; and $(u_{yyy}u_y - u_y^2)/u_y^2 \le 0$. Then u, v is the only solution of problems (1.1.13)–(1.1.14) with these properties.*

Oleinik and Samokhin [106] in 1999 considered the boundary layer system for the symmetric incompressible stationary flow:

$$
\begin{cases}
u\frac{\partial u}{\partial x} + v\frac{\partial u}{\partial y} = \nu\frac{\partial^2 u}{\partial y^2} + U\frac{dU}{dx},\\
\frac{\partial u}{\partial x} + \frac{\partial v}{\partial y} = 0,
\end{cases}
\tag{1.1.17}
$$

in a domain $D = \{0 < x < X, 0 < y < \infty\}$, with the boundary conditions

$$\begin{cases} u(0, y) = 0, \;\; u(x, 0) = 0, \;\; v(x, 0) = v_0(x), \\ u(x, y) \to U(x) \;\text{ as }\; y \to \infty. \end{cases} \tag{1.1.18}$$

Let

$$U(x) = \sum_{m=0}^{q} a_m x^{2m+1} + a_{q+1}(x), \tag{1.1.19}$$

where $a_0 = a, a_j$ =const., $j = 1, \dots, q$, and

$$|a_{q+1}(x)| \le C_1 x^{2(q+1)+1}, \;\; |a_{(q+1)x}(x)| \le C_2 x^{2(q+1)}, \;\; |a_{(q+1)xx}(x)| \le C_3 x^{2q+1};$$

and let

$$v_0(x) = \sum_{m=0}^{q} b_m x^{2m} + b_{q+1}(x), \tag{1.1.20}$$

where $b_0 = b, b_j$ =const., $j = 1, \dots, q$,

$$|b_{q+1}(x)| \le C_4 x^{2(q+1)}, \;\; |b_{(q+1)x}(x)| \le C_5 x^{2q+1},$$

$$C_j = \text{const.} > 0, \;\; j = 1, \dots, 5, \;\; q \ge 0.$$

Theorem 1.1.18 ([106]) *Suppose that conditions (1.1.19)–(1.1.20) with some $q \ge 0$ hold for $U(x)$, $v_0(x)$. Then problems (1.1.17)–(1.1.18) in the domain $D = \{0 < x < X, 0 < y < \infty\}$ (X depends on U, v_0) admit a solution with the following properties: u/U, u_y/U are continuous and bounded in $\overline{D}$; $u > 0$ for $y > 0, x > 0$; $u \to U(x)$ as $y \to \infty$ so that*

$$1 - \frac{u}{U} = \exp\left\{-\frac{U_x(0)}{2\nu}[y^2(1 + O(x^2)) + O(y^{1+\varepsilon}))]\right\}, \quad \text{as } y \to \infty,$$

where $\varepsilon > 0$ is arbitrarily small; $u_y/U > 0$ for $y \ge 0$; $u_y/U \to 0$ as $y \to \infty$; u_{yy}, u_x, u_y are bounded and continuous in $\overline{D}$ with respect to y and are continuous in D with respect to x, y; v is continuous in $\overline{D}$ with respect to y, and in D it is continuous with respect to x, y and bounded for bounded y; u_{xy} and u_{yyy} are continuous in D; u_{xy} is bounded in D for bounded y; u_{yy}/u_y is continuous in $\overline{D}$ with respect to y. Moreover, the following estimates hold:

$$U(x)Y(\frac{u}{U})(1 - M_{22}x^2) \le u_y \le U(x)Y(\frac{u}{U})(1 + M_{22}x^2),$$

$$Y_\eta(\frac{u}{U})(1+M_{14}x^2) \le \frac{u_{yy}}{u_y} \le Y_\eta(\frac{u}{U})(1-M_{15}x^2),$$

$$-M_{25} < \frac{u_{yyy}u_y - u_{yy}^2}{u_y^2} < -M_{26}.$$

Theorem 1.1.19 ([106]) *A solution u, v of problems (1.1.17)–(1.1.18) in D is unique if it possesses the following properties: the derivatives $u_x, u_y, v_y, u_{yy}, u_{yyy}$, u_{xy} are continuous in D; $u/U, u_y/U$ are continuous in $\overline{D}$; v is continuous in $\overline{D}$ with respect to y; $u_y > 0$ for $y \ge 0$, $x > 0$; $u_y/U > 0$ for $y = 0$; $u_y/U \to 0$ os $y \to \infty$; u_{yy}/u_y, u_x are continuous with respect to y at $y = 0$.*

Theorem 1.1.20 ([106]) *Assume that conditions (1.1.19)–(1.1.20) with some $q \ge 0$ hold for $U(x)$, $v_0(x)$. Then the following estimate holds for the solution u, v of problems (1.1.17)–(1.1.18):*

$$\left| \frac{u_y}{U} - \sum_{m=0}^{q} Y_m \frac{u}{U} x^{2m} \right| \le M_{24} Y_0 \frac{u}{U} x^{2q+2},$$

where M_{24} is a positive constant.

Theorem 1.1.21 ([106]) *Let $U(x) = x^m V(x)$, $v_0(x) = x^{(m-1)/2} v_1(x)$, where $V(x)$ and $v_1(x)$ satisfy*

$$V(x) = a + a_1(x), \quad a > 0, \quad v_1(x) = b + b_1(x),$$
$$|a_1(x)| \le M_{32}x^3, \quad |a_{1x}(x)| \le M_{33}x^2, \quad |a_{1xx}(x)| \le M_{34}x,$$
$$|b_1(x)| \le M_{35}x^2, \quad |b_{1x}(x)| \le M_{36}x.$$

Then problems (1.1.17)–(1.1.18) in the domain D, for some X depending on U and v_0, admit a solution $u(x, y)$, $v(x, y)$ with the following properties: $u_y > 0$ for $y \ge 0$, $x > 0$; u/U, $u_y/(x^{(m-1)/2}U(x))$ are bounded and continuous in $\overline{D}$; $u > 0$ for $y > 0$ and $x > 0$; $u(x, y) \to U(x)$, $u_y \to 0$ as $y \to \infty$; u_x, u_y, u_{yy} are bounded and continuous in $\overline{D}$ with respect to y, and continuous with respect to x, y in D; v is continuous in $\overline{D}$ with respect to y, and continuous with respect to x, y in D; $u_{yy}/(x^{(m-1)/2}u_y)$ is continuous in $\overline{D}$ with respect to y; u_{xy} and u_{yyy} are continuous in D. Moreover, the following inequalities hold:

$$M_{44}\exp(-M_{45}x^{m-1}y^2) \le 1 - \frac{u}{U} \le M_{44}\exp(-M_{46}x^{m-1}y^2),$$
$$x^{\frac{m-1}{2}} Y\left(\frac{u}{U}\right)(1 - M_{37}x^2) \le \frac{u_y}{U} \le x^{\frac{m-1}{2}} Y\left(\frac{u}{U}\right)(1 + M_{38}x^2),$$
$$Y_\eta\left(\frac{u}{U}\right)(1 + M_{39}x^2) \le \frac{u_{yy}}{u_y} \le Y_\eta\left(\frac{u}{U}\right)(1 - M_{40}x^2).$$

The solution with the above properties is unique.

Oleinik and Samokhin [106] in 1999 considered the boundary layer system (1.1.17) in a domain $D = \{0 < x < X, 0 < y < \infty\}$, with the boundary conditions

$$\begin{cases} u(0, y) = u_0(y), \quad u(x, 0) = 0, \quad v(x, 0) = v_0(x), \\ u(x, y) \to U(x) \text{ as } y \to \infty. \end{cases} \tag{1.1.21}$$

Theorem 1.1.22 ([106]) *Assume that $U(x) > 0$ for $0 \le x \le X$; $U_x(x)$ and $v_0(x)$ are continuously differentiable on the segment $0 \le x \le X$; $u_0(y)$ is continuous for $0 \le y < \infty$, $u_0(0) = 0$, $u_0(y) > 0$ for $y > 0$; $u_0(y) \to U(0)$ as $y \to \infty$; the derivatives $u_{0y}(y)$, $u_{0yy}(y)$, $u_{0yyy}(y)$ exist and satisfy the inequalities*

$$M_1\left(1 - \frac{u_0(y)}{U(0)}\right)\sqrt{-\ln\mu\left(1 - \frac{u_0}{U(0)}\right)} \le \frac{u_{0y}(y)}{U(0)}$$

$$\le M_2\left(1 - \frac{u_0(y)}{U(0)}\right)\sqrt{-\ln\mu\left(1 - \frac{u_0}{U(0)}\right)},$$

$$-M_3\sqrt{-\ln\mu\left(1 - \frac{u_0}{U(0)}\right)} \le \frac{u_{0yy}}{u_{0y}} \le -M_4\sqrt{-\ln\mu\left(1 - \frac{u_0}{U(0)}\right)},$$

$$\left|\frac{u_{0yyy}u_{0y} - u_{0yy}^2}{u_{0y}^2}\right| \le M_5,$$

where μ, $M_i (i = 1, 2, \cdots, 5)$ are positive constants and $0 < \mu < 1$. Moreover, let the following compatibility conditions hold:

$$v_0(0)u_{0y}(0) = U(0)U_x(0) + \nu u_{0yy}(0),$$

$$v_0(0)u_{0yy}(y) - \nu u_{0yyy}(y) = O(y), \quad \textit{as } y \to 0.$$

Then problems (1.1.17) and (1.1.21) in the domain D, with X depending on U, v_0, u_0, admit a unique solution with the following properties: u, u_y are bounded and continuous in $\overline{D}$; $u > 0$ for $y > 0$, $u_y > 0$ for $y \ge 0$; u_{yy}, u_x, v_y are bounded and continuous in $\overline{D}$ with respect to y; v is bounded for bounded y; u_{xy}, u_{yyy} are continuous in $\overline{D}$, u_{yyy} is bounded in D, u_{xy} is bounded in D for bounded y; u_{yy}/u_y is continuous in D with respect to y. Moreover,

$$M_6\left(1 - \frac{u}{U}\right)\sqrt{-\ln\mu\left(1 - \frac{u}{U}\right)} \le \frac{u_y(y)}{U} \le M_7\left(1 - \frac{u}{U}\right)\sqrt{-\ln\mu\left(1 - \frac{u}{U}\right)},$$

$$M_8\exp(-M_9y^2) \le 1 - \frac{u}{U} \le M_{10}\exp(-M_{11}y^2),$$

$$-M_{12}\sqrt{-\ln\mu\left(1-\frac{u}{U}\right)} \le \frac{u_{yy}}{u_y} \le -M_{13}\sqrt{-\ln\mu\left(1-\frac{u}{U}\right)},$$

$$\left|\frac{u_{yyy}u_y - u_{yy}^2}{u_y^2}\right| \le M_{14},$$

where $M_i (i = 1, 2, \dots, 5)$, μ *are positive constants,* $\mu < 1$.

Theorem 1.1.23 ([106]) *Assume that* $U_x(x)$, $v_0(x)$ *have piecewise continuous first derivatives, and* $u_0(y)$ *is continuous for* $0 < y < \infty$, $u_0(0) = 0$, $u_0(y) > 0$ *for* $y > 0$, $u_0(y) \to U(0)$ *as* $y \to \infty$, $U(x) > 0$ *for* $x \ge 0$,

$$M_{15}(U(0) - u_0(y)) \le u_{0y}(y) \le M_{16}(U(0) - u_0(y))$$

where M_i *are positive constants. Assume also that* u_0 *has bounded measurable derivatives up to the third order such that*

$$\frac{u_{0yy}}{u_{0y}}, \quad \frac{u_{0yyy}u_{0y} - u_{0yy}^2}{u_{0y}^2}$$

are bounded. Moreover, assume the following compatibility conditions hold:

$$v_0(0)u_{0y}(0) = -p_x(0) + \nu u_{0yy}(0),$$
$$v_0(0)u_{0yy}(y) - \nu u_{0yyy}(y) = O(y), \quad y \to 0.$$

Then problems (1.1.17) and (1.1.21) in D, *with* X *depending on the given functions, admit a unique solution with the following properties:* u, u_y *are continuous and bounded in* $\overline{D}$; $u_y > 0$ *for* $y \ge 0$; v *is continuous in* y *and bounded for bounded* y; *the weak derivatives* $u_x, u_{yx}, u_{yy}, u_{yyy}, v_y$ *exist and are bounded measurable functions in* D. *Moreover,*

$$\frac{u_{yy}}{u_y}, \quad \frac{u_{yyy}u_y - u_{yy}^2}{u_y^2}$$

are bounded, and

$$M_{17}\exp(-M_{18}y) \le 1 - \frac{u}{U} \le M_{19}\exp(-M_{20}y),$$
$$M_{21}(U(x) - u(x, y)) \le u_y(x, y) \le M_{22}(U(x) - u(x, y)) \le u_y(x, y),$$
$$\left|\frac{1}{u_y}(u_{yx}u_y - u_x u_{yy}) + \frac{U_x}{u_y U}(uu_{yy} - u_y^2)\right| \le M_{23}(U - u).$$

Theorem 1.1.24 ([106]) *Assume that the function $U(x)$ is continuous in the interval $0 \le x \le X$ and $U_x(x)$, $v_0(x)$ are piecewise continuous, $U_x(x) \ge const. > 0$. Assume also that $u_0(y)$ is continuous in the interval $0 \le y < \infty$, $u_0(0) = 0$, $u_0(y) > 0$ for $y > 0$, $u_0(y) \to U(0)$ as $y \to \infty$, the derivative $u_{0y}(y)$ is continuous, $u_{0y} > 0$; there exists the weak derivative $u_{0yy}(y)$ and the following inequalities hold:*

$$M_{21}\left(1-\frac{u_0(y)}{U(0)}\right)\left(-\ln\mu\left(1-\frac{u_0(y)}{U(0)}\right)\right)^{\frac{1}{2}} \le \frac{u_0(y)}{U(0)}$$

$$\le M_{22}\left(1-\frac{u_0(y)}{U(0)}\right)\left(-\ln\mu\left(1-\frac{u_0(y)}{U(0)}\right)\right)^{\frac{1}{2}},$$

$$\int_0^\infty \frac{u_{0yy}^2(y)}{u_{0y}(y)} \le \infty, \quad \mu = const., \quad 0 < \mu < 1.$$

Then problems (1.1.17) and (1.1.21) has a weak solution $u(x, y)$, $v(x, y)$ with the following properties: $u(x, y)$ is a bounded measurable function continuous in $\overline{D}$ with respect to y; $u > 0$ for $y > 0$, $u = 0$ for $y = 0$; $u_y(x, y)$ is bounded in D and

$$M_{23}\left(1-\frac{u}{U}\right)\left(-\ln\mu\left(1-\frac{u}{U}\right)\right)^{\frac{1}{2}} \le \frac{u_y}{U} \le M_{24}\left(1-\frac{u}{U}\right)\left(-\ln\mu\left(1-\frac{u}{U}\right)\right)^{\frac{1}{2}},$$

$$M_{25}\exp(-M_{26}y^2) \le 1-\frac{u}{U} \le M_{27}\exp(-M_{28}y^2);$$

there exist weak derivatives u_{yy} and u_{yyy} such that

$$\int_D \frac{u_{yy}^2}{u_y}dxdy < \infty, \quad \int_D \frac{(u_{yyy}u_y - u_y^2)^2}{u_y^4}u^2dxdy < \infty;$$

and $v(x, y)$ is a measurable function in D such that

$$\int_{D\cap\{(x,y):0\le x\le X-\varepsilon\}} v(x, y)\exp(-My^2)dxdy < \infty, \quad \forall M, \varepsilon = const. > 0.$$

Equations (1.1.17) and the boundary conditions (1.1.21) at $x = 0$, $y = 0$ hold in the sense of the following integral identities:

$$\int_D \left(\nu\frac{u_y\phi_{yy} - u_{yy}\phi_y}{u_y} - u\phi_x + (u^2 - U^2)\frac{U_x}{U}\frac{\phi_y}{u_y}\right)dxdy - \int_0^X v_0(x)\phi(x, 0)dx$$

$$-\int_0^\infty u_0(y)\phi(0, y)dy = 0,$$

$$\int_D \left(\nu u_{yy}\phi + \frac{u^2}{2}\phi_x - vu_y\phi + UU_x\phi\right)dxdy + \int_0^\infty \frac{u_0(y)^2}{2}\phi(0, y)dy = 0,$$

for any $\phi(x, y)$ such that $\phi_x, \phi_y, \phi_{yy}$ continuous in $\overline{D}$ and $\phi(X, y) = 0$, $\phi_y(x, 0) = 0$, $|\phi| \le M_{29} \exp(-M_{30}y^2)$ as $y \to \infty$. A solution (u, v) of problems (1.1.17) and (1.1.21) with the above properties is unique.

Oleinik and Samokhin [106] in 1999 reached boundary layer with unknown border between two media, i.e., they considered a stationary plane-parallel incompressible flow past a porous surface. Assume that another incompressible fluid, whose properties differ from those of the main stream, is injected into the boundary layer through the porous wall.

The velocity components of the injected fluid $u_1(x, y), v_1(x, y)$ satisfy the Prandtl system

$$\begin{cases} u_1\frac{\partial u_1}{\partial x} + v_1\frac{\partial u_1}{\partial y} = \alpha_1\frac{\partial^2 u_1}{\partial y^2} - \frac{1}{\rho_1}\frac{\partial p}{\partial x}, \\ \frac{\partial u_1}{\partial x} + \frac{\partial v_1}{\partial y} = 0 \end{cases} \tag{1.1.22}$$

in the domain $D_1 = \{0 < x < X, 0 < y < y_*(x)\}$, where α_1 is the viscosity of the injected fluid and ρ_1 is its density.

In the domain $D_2 = \{0 < x < X, y_*(x) < y < \infty\}$, another system is given, namely,

$$\begin{cases} u_2\frac{\partial u_2}{\partial x} + v_2\frac{\partial u_2}{\partial y} = \alpha_2\frac{\partial^2 u_2}{\partial y^2} - \frac{1}{\rho_2}\frac{\partial p}{\partial x}, \\ \frac{\partial u_1}{\partial x} + \frac{\partial v_1}{\partial y} = 0 \end{cases} \tag{1.1.23}$$

where $u_2(x, y), v_2(x, y)$ are the velocity components of the main fluid, α_2 is its viscosity, and ρ_2 is its density.

The boundary conditions are given in the form

$$\begin{aligned} &u_1(0, y) = u_{10}(y), \quad \text{for } 0 \le y \le y_*(0), \\ &u_2(0, y) = u_{20}(y), \quad \text{for } y_*(0) \le y \le \infty, \\ &u_1(x, 0) = 0, \quad v_1(x, 0) = v_0(x) > 0, \\ &u_2(x, y) \to U(x), \quad \text{as } y \to \infty, \end{aligned} \tag{1.1.24}$$

where $y_*(0)$ is regarded as given and $2p(x) + \rho_2 U^2(x)$ =const. Assume that the fluids do not penetrate the interface between the two media. In this case, the curve $y = y_*(x)$ coincides with a streamline and, therefore,

$$\frac{dy_*(x)}{dx} = \frac{v_1(x, y_*(x))}{u_1(x, y_*(x))} = \frac{v_2(x, y_*(x))}{u_2(x, y_*(x))}. \tag{1.1.25}$$

On the line $y = y_*(x)$, transmission conditions are imposed, which amount to the continuity of the velocity and of the viscous forces, namely,

$$u_1(x, y_*(x)) = u_2(x, y_*(x)), \tag{1.1.26}$$

$$\nu_1\rho_1\frac{\partial u_1(x, y_*(x))}{\partial y} = \nu_2\rho_2\frac{\partial u_2(x, y_*(x))}{\partial y}. \quad (1.1.27)$$

It follows from (1.1.25) and (1.1.26) that $v_1(x, y_*(x)) = v_2(x, y_*(x))$.

Theorem 1.1.25 ([106]) *For some $X > 0$, problems (1.1.22)–(1.1.27) have a solution $u_1, v_1, u_2, v_2, y_*(x)$ with the following properties: u_1 is continuous in $\overline{D}_1$, $u_1 > 0$ for $y > 0$; $\partial u_1/\partial y > m_1 > 0$ for $0 < y \le y_0$ with constant m_1 and y_0; $\partial u_1/\partial y, \partial^2 u_1/\partial y^2$ are continuous and bounded in D_1; $\partial u_1/\partial x, v_1, \partial v_1/\partial y$ are bounded in D_1 and continuous at $y = 0$; u_2 is continuous and bounded in $\overline{D}_2$; $u_2 > 0$ in $\overline{D}_2$; $\partial u_2/\partial y, \partial^2 u_2/\partial y^2$ are continuous and bounded in D_2; $\partial u_2/\partial x, v_2, \partial v_2/\partial y$ are continuous and bounded in any finite subdomain of D_2; $\partial u_i/\partial y, v_i$ are continuous in D_i $(i = 1, 2)$ up to the curve $y = y_*(x)$; if $u\prime_{20}(y) \to 0$ as $y \to \infty$ and this convergence is fast enough, so that $u\prime_{20}(y) \le m_2\exp(-m_3 y^2), m_2, m_3 > 0$, then $\partial u_2/\partial x, \partial v_2/\partial y$ are bounded in D_2; the function $y_*(x)$ is continuously differentiable on $[0, X]$. If $dp/dx < 0$, then such a solution of problems (1.1.22)–(1.1.27) exists in D for any $X > 0$.*

Theorem 1.1.26 ([106]) *The solution of problems (1.1.22)–(1.1.27) with the properties specified in Theorem 1.1.25 is unique.*

Oleinik and Samokhin [106] in 1999 considered the stationary mixing layer at the interface between two fluids with distinct properties. Assume that the interface coincides with the straight line $y = 0$. In this case, the fluid motion in the domain $D_1 = \{0 < x < X, 0 < y < +\infty\}$ is described by the system (1.1.22) with the boundary conditions

$$\begin{aligned} &u_1(0, y) = u_{10}(y), \quad v_1(x, 0) = 0,\\ &u_1(x, y) \to U_1(x), \quad \text{as } y \to +\infty, \end{aligned} \quad (1.1.28)$$

where $u_1(x, y), v_1(x, y)$ are the velocity components; ν_1 is the viscosity of the fluid; and ρ_1 is its density.

In the domain $D_2 = \{0 < x < X, -\infty < y < 0\}$, the velocity field is described by the system (1.1.23) with the boundary conditions

$$\begin{cases} u_2(0, y) = u_{20}(y), \quad v_2(x, 0) = 0,\\ u_2(x, y) \to U_2(x), \quad \text{as } y \to -\infty, \end{cases} \quad (1.1.29)$$

where $u_2(x, y), v_2(x, y)$ are the velocity components; ν_2 is the viscosity of the fluid; and ρ_2 is its density.

For $y = 0$, the following transmission conditions are imposed:

$$\begin{cases} u_1(x, 0) = u_2(x, 0),\\ \nu_1\rho_1 u_{1y}(x, 0) = \nu_2\rho_2 u_{2y}(x, 0). \end{cases} \quad (1.1.30)$$

Theorem 1.1.27 ([106]) *Under the compatibility assumptions, problems (1.1.22)–(1.1.23) and (1.1.28)–(1.1.30), for some* $X > 0$, *admit a solution* u_1, v_1, u_2, v_2 *with the following properties:* u_i *is continuous and bounded in* $\overline{D}_i$; $u_i > 0$; $\partial u_i/\partial y, \partial^2 u_i/\partial y^2$ *are continuous and bounded in* D_i; $\partial u_i/\partial x, v_i, \partial v_i/\partial y$ *are continuous and bounded in any finite subdomain of* D_i; $\partial u_i/\partial y, v_i$ *are continuous in* D_i *up to the line* $y = 0$; $i = 1, 2$. *If*

$$|u'_0(y)| \leq k_1 \exp(-k_2|y|) \quad \text{as} \quad |y| \to \infty$$

with constant $k_1, k_2 > 0$, *then* $\partial u_i/\partial x, \partial v_i/\partial y$ *are bounded in* D_i. *If* $dp/dx < 0$, *the solution of problems (1.1.22)–(1.1.23) and (1.1.28)–(1.1.30) exists for any* $X > 0$. *The solution of problems (1.1.22)–(1.1.23) and (1.1.28)–(1.1.30) with these properties is unique. s*

Oleinik and Samokhin [106] in 1999 considered the stationary Prandtl system for a planar incompressible flow by the von Mises, the stationary Prandtl system

$$\begin{cases} u\frac{\partial u}{\partial x} + v\frac{\partial u}{\partial y} = \nu\frac{\partial^2 u}{\partial y^2} + U(x)\frac{dU}{dx}, \\ \frac{\partial u}{\partial x} + \frac{\partial v}{\partial y} = 0, \end{cases} \tag{1.1.31}$$

in the domain $D = \{0 < x < X, 0 < y < \infty\}$, with the boundary conditions

$$\begin{cases} u(0, y) = u_1(y), \quad u(x, 0) = 0, \quad v(x, 0) = v_0(x), \\ u(x, y) \to U(x), \quad \text{as} \quad y \to \infty. \end{cases} \tag{1.1.32}$$

Theorem 1.1.28 ([106]) *Suppose that in (1.1.31)–(1.1.32), we have* $U_x(x) > 0$ *for* $0 \leq x \leq X$, *and* $U(x), u_1(y), v_0(x) \equiv v_0(\varepsilon^{-1}x)$ *(for any* $\varepsilon \in (0, \varepsilon_0]$*) satisfy the assumptions that* $u_0(y)$ *for* $y > 0$; $u_0(0) = 0$, $u_0'(0) > 0$, $u_0(y) \to U(0) \neq 0$ *as* $y \to \infty$; p_x *and* $v_0(x)$ *are continuously differentiable on [0,X];* $u_0(y), u_0'(y), u_0''(y)$ *are bounded for* $0 \leq y \leq \infty$ *and satisfy the Hölder condition. Moreover, assume that for small y the following compatibility condition is satisfied at the point* $(0, 0)$:

$$\frac{d^2 u_0(y)}{dy^2} - \frac{dp(0)}{dx} - v_0\frac{du_0(y)}{dy} = O(y^2).$$

Then the solution $u_\varepsilon, v_\varepsilon$ *of problems (1.1.31)–(1.1.32) converges, as* $\varepsilon \to 0$, *to a solution of the system*

$$\begin{cases} \overline{u}\frac{\partial \overline{u}}{\partial x} + \overline{v}\frac{\partial \overline{u}}{\partial y} = \nu\frac{\partial^2 \overline{u}}{\partial y^2} + U(x)\frac{dU}{dx}, \\ \frac{\partial u}{\partial x} + \frac{\partial v}{\partial y} = 0, \end{cases}$$

in the domain D, with the conditions

$$\begin{cases} \overline{u}(0, y) = u_1(y), \ \ \overline{u}(x, 0) = 0, \ \ \overline{v}(x, 0) = v_0(x), \\ \overline{u}(x, y) \to U(x), \ \ as \ \ y \to \infty, \end{cases}$$

so that

$$\begin{cases} u_\varepsilon \to \overline{u} \ \ uniformly\, for \ \ 0 \le x \le X, \ \ 0 \le y \le y_1 < \infty, \\ v_\varepsilon \to \overline{v} \ \ uniformly\, for \ \ x_0 \le x \le X, \ \ 0 \le y \le y_1, \\ grad\, u_\varepsilon \to grad\, \overline{u} \ \ weakly\ in \ \ L_2(\{0 \le x \le X, 0 \le y \le y_1\}), \end{cases}$$

for arbitrary fixed $y_1 \in (0, \infty)$, $x_0 \in (0, X)$.

In the case of favorable pressure gradient, Oleinik proved the global existence of classical solution for the 2D steady Prandtl equations for a class of positive data. In the case of adverse pressure gradient, an important physical phenomena is the boundary layer separation. Shen, Wang, and Zhang [127] in 2019 proved the boundary layer separation for a large class of Oleinik's data and confirmed Goldstein's hypothesis concerning the local behavior of the solution near the separation. The authors studied the 2D steady Prandtl equations

$$\begin{cases} u\partial_x u + v\partial_y u - \partial_y^2 u + \partial_x p = 0, \quad x \ge 0, y \ge 0, \\ \partial_x u + \partial_y v = 0, \\ u|_{y=0} = v|_{y=0} = 0, \quad \text{and} \quad \lim_{y \to +\infty} u(x, y) = U(x), \end{cases} \tag{1.1.33}$$

where the outer flow $(U(x), p(x))$ satisfies

$$U(x)U'(x) + p'(x) = 0.$$

This system derived by Prandtl is used to describe the behavior of the solution near $y = 0$ for the steady Navier-Stokes equations with non-slip boundary condition when the viscosity coefficient ν tends to zero:

$$\begin{cases} u^\nu \cdot \nabla u^\nu - \nu \Delta u^\nu + \nabla p = f^\nu, \\ \operatorname{div} u^\nu = 0, \\ u^\nu|_{y=0} = 0. \end{cases}$$

Roughly speaking, away from the boundary, the solution u^ν can be described by the Euler equations; near the boundary $y = 0$, u^ν behaves as

$$u^\nu(x, y) = (u(x, y/\sqrt{\nu}), \sqrt{\nu} v(x, y/\sqrt{\nu})),$$

where (u, v) satisfies the Prandtl type equation.

The authors of [127] first introduced a class of data denoted by $\mathcal{K}$, which satisfies

$$u \in C_b^{3,\alpha}([0,+\infty))\ (\alpha > 0), \quad u(0) = 0, u_y(0) > 0, u_y(y) \geq 0 \text{ for } y \in [0,+\infty),$$

$$\lim_{y\to+\infty} u(0,y) = U(0) > 0, \quad u_{yy}(y) - \partial_x p(0) = O\left(y^2\right).$$

The data $u_0 \in \mathcal{K}$.

Proposition 1.1.1 *If $u_0 \in \mathcal{K}$, then there exists $X > 0$ such that the steady Prandtl equation admits a solution $u \in C^1([0,X) \times \mathbb{R}_+)$ with the following properties:*

1. *Regularity: u is bounded and continuous in $[0,X] \times [0,+\infty)$; u_y, u_{yy} are bounded and continuous in $[0,X) \times \mathbb{R}_+$; and v, v_y, u_x are locally bounded and continuous in $[0,X) \times \mathbb{R}_+$.*
2. *Non-degeneracy: $u(x,y) > 0$ in $[0,X) \times (0,+\infty)$ and for all $\bar{x} < X$, there exists $y_0 > 0$, $m > 0$ so that $\partial_y u(x,y) \geq m$ in $[0,\bar{x}] \times [0,y_0]$.*
3. *Global existence: if $p'(x) \leq 0$, then the solution is global in x, together with*

$$w(0,\psi) = w_0(\psi) = u_0(y)^2, \quad w(x,0) = 0, \quad \lim_{\psi\to+\infty} w(x,\psi) = U(x)^2.$$

Theorem 1.1.29 ([127]) *Fix any $\mu \in (0,1)$. Let u be a solution constructed in Proposition 1.1.1 with $u_0 \in \mathcal{K}$ satisfying*

$$\left\|\partial_y u_0\right\|_{L^\infty(0,y_0)} \leq \frac{1}{2}\epsilon_0 x_0^{\frac{1}{4}},$$

where y_0 is determined by

$$Bx_0^{\frac{3}{4}} = \psi_0 = \int_0^{y_0} u_0(z)dz,$$

and ϵ_0, B are positive constants depending only on μ. Then there exists a separation point $x_s = X^*$ with $X^* < \frac{\mu}{2}x_0$.

Wang and Zhang [142] in 2019 proved the global C^∞ regularity of the Oleinik's solution for the steady Prandtl equation with favorable pressure. The authors studied the steady Prandtl equations:

$$\begin{cases} u\partial_x u + v\partial_y u - \partial_y^2 u = -\frac{dp}{dx}(x), & x \geq 0, y \geq 0, \\ \partial_x u + \partial_y v = 0, \\ u|_{y=0} = v|_{y=0} = 0, \quad \text{and} \quad \lim_{y\to+\infty} u(x,y) = U(x), \end{cases} \tag{1.1.34}$$

where the outer flow $(U(x), p(x))$ satisfies

$$U(x)U'(x) + p'(x) = 0.$$

This system derived by Prandtl could be used to describe the behavior of the solution near $y = 0$ for the steady Navier-Stokes equations when the viscosity coefficient is small.

Oleinik (see Theorem 2.1.1 in [106]) proved the following classical result:

If $u_0 \in \mathcal{K}$ and $\frac{dp}{dx}(x)$ is smooth, then there exists $X > 0$ such that the steady Prandtl equation (1.1.34) admits a solution $u \in C^1([0, X) \times \mathbb{R}_+)$ with the following properties:

1. Regularity: u is bounded and continuous in $[0, X] \times \mathbb{R}_+$; u_y, u_{yy} are bounded and continuous in $[0, X) \times \mathbb{R}_+$; and v, v_y, u_x are locally bounded and continuous in $[0, X) \times \mathbb{R}_+$.
2. Non-degeneracy: $u(x, y) > 0$ in $[0, X) \times (0, +\infty)$ and for all $\bar{x} < X$, there exists $y_0 > 0, m > 0$ so that

$$\partial_y u(x, y) \geq m \quad \text{in } [0, \bar{x}] \times [0, y_0] .$$

3. Global existence: if $p'(x) \leq 0$, then the solution is global in x.

Theorem 1.1.30 ([142]) *Let u be a global solution to the equation constructed by Oleinik with $u_0 \in \mathcal{K}$ and $\frac{dp(x)}{dx} \leq 0$ smooth. For any positive integers m, k and any positive constants X, Y with $\epsilon < X$, there exists a positive constant C depending only on $\epsilon, X, Y, u_0, p, k, m$ so that*

$$\left|\partial_x^k \partial_y^m u(x, y)\right| \leq C \quad \textit{for} \quad (x, y) \in [\epsilon, X] \times [0, Y].$$

Dalibard and Masmoudi [16] in 2019 were interested in the stationary version of the Prandtl equation, namely,

$$\begin{cases} u u_x + v u_y - u_{yy} = -\frac{dp_E(x)}{dx}, \quad x > 0, y > 0, \\ u_x + v_y = 0, \quad x > 0, y > 0, \\ u|_{x=0} = u_0, \quad u|_{y=0} = 0, \lim\limits_{y \to \infty} u(x, y) = u_E(x), \end{cases} \tag{1.1.35}$$

where $y = 0$ stands for the rigid wall, x (resp. y) is the tangential (resp. normal) variable to the wall. The functions u_E, p_E are given by the outer flow: more precisely u_E (resp. p_E) is the trace at the boundary of the tangential velocity (resp. of the pressure) of a flow satisfying the Euler equations. The functions u_E, p_E are linked by the relation

$$u_E u_E' = -\frac{dp_E(x)}{dx}.$$

Let $\alpha > 0, X \in (0, +\infty]$. Let $u_0 \in C_b^{2,\alpha}(\mathbb{R})$ such that $u_0(0) = 0$, $u_0'(0) > 0$, $\lim_{y \to \infty} u_0(y) = u_E(0) > 0$, and such that $u_0(y) > 0$ for $y > 0$. Assume that

$dp_E/dx \in C^1([0, X])$, and that for $y \ll 1$ the following compatibility condition is satisfied

$$u_0''(y) - \frac{dp_E(0)}{dx} = O\left(y^2\right).$$

The authors in [16] are interested in the case where the solution of (1.1.35) is not global: more precisely, they considered the equation with $dp_E/dx = 1$, i.e.,

$$\begin{cases} uu_x + vu_y - u_{yy} = -1, & x \in (0, x_0), y > 0, \\ u_x + v_y = 0, & x \in (0, x_0), y > 0, \\ u_{|x=0} = u_0, \quad u_{|y=0} = 0, \lim_{y\to\infty} u(x, y) = u_E(x), \end{cases}$$

with $u_E(x) = \sqrt{2(x_0 - x)}$, for some $x_0 > 0$, and u_0 satisfies the assumptions.
The authors' assumptions on the initial data u_0 are the following:

(H1) $u_0 \in C^7(\mathbb{R}_+)$, u_0 is increasing in y and $\lambda_0 := u_0'(0) > 0$,
(H2) There exists a constant $C_0 > 0$ such that

$$\begin{aligned} \forall y \geq 0, \quad -C_0 \inf\left(y^2, 1\right) \leq u_0''(y) - 1 \leq 0, \\ C_0^{-1} \leq -u_0^{(4)}(0) \leq C_0, \\ \|u_0\|_{W^{7,\infty}} \leq C_0. \end{aligned}$$

(H3) $u_0 = u_0^{\text{app}} + v_0$, where

$$\begin{aligned} u_0^{\text{app}} =& \lambda_0 y + \frac{y^2}{2} + u_0^{(4)}(0)\frac{y^4}{4!} \\ &- c_7\left(u_0^{(4)}(0)\right)^2 \frac{y^7}{\lambda_0} + c_{10}\left(u_0^{(4)}(0)\right)^3 \frac{y^{10}}{\lambda_0^2} \\ &+ c_{11}\left(u_0^{(4)}(0)\right)^3 \frac{y^{11}}{\lambda_0^3} \quad \text{for } y \leq \lambda_0^{3/7}, \\ \left|u_0^{\text{app}}\right| \leq& C_0 \text{ for } y \geq \lambda_0^{3/7}, \end{aligned}$$

and

$$|v_0| \leq C_0\left(\lambda_0^{-\frac{3}{2}}\left(\lambda_0 y^7 + c_8 y^8\right) + \lambda_0^{-2} y^{10} + \lambda_0^{-3} y^{11}\right) \quad \text{for } y \leq \lambda_0^{3/7}.$$

Theorem 1.1.31 ([16]) *Consider the Prandtl equation with adverse pressure gradient (P) and with an initial data $u_0 \in C^7(\mathbb{R}_+)$ satisfying (H1)–(H3). Then for any $\eta > 0$, $C_0 > 0$, there exists $\epsilon_0 > 0$ such that if $\lambda_0 < \epsilon_0$, the existence time x^* is* *finite, and* $x^* = O\left(\lambda_0^2\right)$. *Furthermore, setting* $\lambda(x) := \partial_y u|_{y=0}(x)$, *there exists a constant* $C > 0$, *depending on* u_0, *such that*

$$\lambda(x) \sim C\sqrt{x^* - x} \quad \text{as } x \to x^*.$$

Gao and Zhang [38] in 2020 considered the zero viscosity limit of the 2D steady Navier-Stokes equations in $(0, L) \times \mathbb{R}^+$ with non-slip boundary conditions. By estimating the stream function of the remainder, we justify the validity of the Prandtl boundary layer expansions. The authors considered the vanishing viscosity limit of steady Navier-Stokes equations

$$\begin{cases} U^\varepsilon U_X^\varepsilon + V^\varepsilon U_Y^\varepsilon - \varepsilon \Delta U^\varepsilon + P_X^\varepsilon = 0, \\ U^\varepsilon V_X^\varepsilon + V^\varepsilon V_Y^\varepsilon - \varepsilon \Delta V^\varepsilon + P_Y^\varepsilon = 0, \\ U_X^\varepsilon + V_Y^\varepsilon = 0, \\ U^\varepsilon|_{Y=0} = V^\varepsilon|_{Y=0} = 0, \end{cases} \tag{1.1.36}$$

in a two-dimensional domain $\Omega = \{(X, Y) : 0 \leqslant X \leqslant L, Y \geqslant 0\}$. A formal limit $\varepsilon \to 0$ should lead to the Euler flow $\left[U^0, V^0\right]$ inside Ω :

$$\begin{cases} U^0 U_X^0 + V^0 U_Y^0 + P_X^0 = 0, \\ U^0 V_X^0 + V^0 V_Y^0 + P_Y^0 = 0, \\ U_X^0 + V_Y^0 = 0, \\ V^0\big|_{Y=0} = 0. \end{cases} \tag{1.1.37}$$

In their first result, they assumed that the outside Euler flow $\left[U^0, V^0\right] \equiv \left(u_e^0(X, Y), v_e^0(X, Y)\right)$ satisfying the following hypothesis:

$$\begin{aligned} & 0 < c_0 \leqslant u_e^0 \leqslant C_0 < \infty, \\ & \left\| v_e^0 \right\|_{L^\infty} \ll 1, \\ & \left\| \langle Y \rangle^k \nabla^m v_e^0 \right\|_{L^\infty} < \infty \text{ for sufficiently large } k, m \geqslant 0, \\ & \left\| \langle Y \rangle^k \nabla^m u_e^0 \right\|_{L^\infty} < \infty \text{ for sufficiently large } k, m \geqslant 1. \end{aligned}$$

Here $\langle Y \rangle = Y + 1$. The authors in [38] considered the Prandtl equations with the positive data

$$\begin{cases} u_p^0 u_{px}^0 + v_p^0 u_{py}^0 - u_{pyy}^0 + p_{px}^0 = 0, \quad p_{py}^0 = 0, \quad u_{px}^0 + v_{py}^0 = 0, \\ \quad (x, y) \in (0, L) \times \mathbb{R}_+, \\ u_p^0\big|_{x=0} = U_P^0(y), \quad u_p^0\big|_{y=0} = v_p^0\big|_{y=0} = 0, \quad u_p^0\big|_{y\uparrow\infty} = u_e^0|_{Y=0}, \end{cases}$$

where U_P^0 is a prescribing smooth function such that

$$U_P^0 > 0 \text{ for } y > 0, \quad \partial_y U_P^0(0) > 0, \quad \partial_y^2 U_P^0 - u_e^0 u_{ex}^0(0) \sim y^2 \text{ near } y = 0,$$

$\partial_y^m \left\{U_P^0 - u_e^0(0)\right\}$ decay fast for any $m \geqslant 0$.

Theorem 1.1.32 ([38]) *Assume the Euler flows $\left[u_e^0, v_e^0\right]$ satisfy the former assumptions, U_P^0 is a smooth function satisfying and high-order compatibility conditions, L is a constant small enough, and then there exist $C(L)$, $\varepsilon_0(L) > 0$ depending on L, such that for $0 < \varepsilon \leqslant \varepsilon_0$, equation* (1.1.36) *admits a solution $[U^\varepsilon, V^\varepsilon] \in W^{2,2}(\Omega)$, satisfying*

$$\begin{cases} \|U^\varepsilon - u_e^0 + u_e^0|_{Y=0} - u_p^0\|_{L^\infty} \leqslant C\sqrt{\varepsilon}, \\ \|V^\varepsilon - v_e^0\|_{L^\infty} \leqslant C\sqrt{\varepsilon}, \end{cases} \tag{1.1.38}$$

with the following boundary conditions

$$\begin{cases} (U^\varepsilon, V^\varepsilon)|_{Y=0} = 0, \\ (U^\varepsilon, V^\varepsilon)|_{X=0} = \left(u_e^0(0, Y) - u_e^0(0, 0) + u_p^0(0, \frac{Y}{\sqrt{\varepsilon}}) + \sqrt{\varepsilon}a_0, v_e^0(0, Y) + \sqrt{\varepsilon}b_0\right), \\ (U^\varepsilon, V^\varepsilon)|_{X=L} = \left(u_e^0(L, Y) - u_e^0(L, 0) + u_p^0(L, \frac{Y}{\sqrt{\varepsilon}}) + \sqrt{\varepsilon}a_L, v_e^0(L, Y) + \sqrt{\varepsilon}b_L\right). \end{cases} \tag{1.1.39}$$

Here

$$\begin{cases} a_0(Y) = u_e^1(0, Y) + u_b^1\left(0, \frac{Y}{\sqrt{\varepsilon}}\right) + \sqrt{\varepsilon}u_e^2(0, Y) + \sqrt{\varepsilon}\hat{u}_b^2\left(0, \frac{Y}{\sqrt{\varepsilon}}\right), \\ a_L(Y) = u_e^1(L, Y) + u_b^1\left(L, \frac{Y}{\sqrt{\varepsilon}}\right) + \sqrt{\varepsilon}u_e^2(L, Y) + \sqrt{\varepsilon}\hat{u}_b^2\left(L, \frac{Y}{\sqrt{\varepsilon}}\right), \\ b_0(Y) = v_b^0\left(0, \frac{Y}{\sqrt{\varepsilon}}\right) + v_e^1(0, Y) + \sqrt{\varepsilon}v_b^1\left(0, \frac{Y}{\sqrt{\varepsilon}}\right) + \sqrt{\varepsilon}v_e^2(0, Y) + \varepsilon\hat{v}_b^2\left(0, \frac{Y}{\sqrt{\varepsilon}}\right), \\ b_L(Y) = v_b^0\left(L, \frac{Y}{\sqrt{\varepsilon}}\right) + v_e^1(L, Y) + \sqrt{\varepsilon}v_b^1\left(L, \frac{Y}{\sqrt{\varepsilon}}\right) + \sqrt{\varepsilon}v_e^2(L, Y) + \varepsilon\hat{v}_b^2\left(L, \frac{Y}{\sqrt{\varepsilon}}\right), \end{cases} \tag{1.1.40}$$

are smooth functions.

For the second result, they considered L is any given positive constant. The authors assumed the Euler flow $\left[U^0, V^0\right] \equiv \left[u_e^0(Y), 0\right]$ is a shear flow, that is, it satisfies the following hypothesis: $0 < c_0 \leqslant u_e^0 \leqslant C_0 < \infty$, $\left\|\langle Y\rangle^k \nabla^m u_e^0\right\|_{L^\infty} < \infty$, for sufficiently large $k, m \geqslant 1$. Here $\langle Y\rangle = Y + 1$.

While the authors assumed $\left[u_p^0, v_p^0\right]$ is a smooth solution of Prandtl equations satisfying the following hypothesis:

$$\begin{aligned} &u_p^0 > 0, -u_{pyy}^0 \geqslant 0, \text{ for } y > 0, \\ &u_{py}^0 > 0, \text{ for } y \geqslant 0, \end{aligned}$$

and $\nabla^m \left\{u_p^0 - u_e^0(0)\right\}$ decays fast for any $m \geqslant 0$.

Theorem 1.1.33 ([38]) *Assume the Euler flows* $\left[u_e^0, v_e^0\right]$ *satisfy the assumptions, the Prandtl profiles* $\left[u_p^0, v_p^0\right]$ *satisfy the assumptions, and* $L > 0$ *is any given constant, and then there exist* $C(L), \varepsilon_0(L) > 0$ *depending on* L*, such that for* $0 < \varepsilon \leqslant \varepsilon_0$*, equation* (1.1.36) *admits a solution* $[U^\varepsilon, V^\varepsilon] \in W^{2,2}(\Omega)$*, satisfying*

$$\left\| U^\varepsilon - u_e^0 + u_e^0\big|_{Y=0} - u_p^0 \right\|_{L^\infty} \leqslant C\sqrt{\varepsilon},$$
$$\|V^\varepsilon\|_{L^\infty} \leqslant C\sqrt{\varepsilon},$$

with the boundary conditions

$$\begin{aligned}
&[U^\varepsilon, V^\varepsilon]|_{Y=0} = 0,\\
&[U^\varepsilon, V^\varepsilon]|_{X=0} = \left[u_e^0(Y) - u_e^0(0) + u_p^0\left(0, \tfrac{Y}{\sqrt{\varepsilon}}\right) + \sqrt{\varepsilon}a_0, \sqrt{\varepsilon}b_0\right],\\
&[U^\varepsilon, V^\varepsilon]_{X=L} = \left[u_e^0(Y) - u_e^0(0) + u_p^0\left(L, \tfrac{Y}{\sqrt{\varepsilon}}\right) + \sqrt{\varepsilon}a_L, \sqrt{\varepsilon}b_L\right].
\end{aligned}$$

Here

$$\begin{aligned}
a_0(Y) &= u_e^1(0, Y) + u_b^1\left(0, \tfrac{Y}{\sqrt{\varepsilon}}\right) + \sqrt{\varepsilon}u_e^2(0, Y) + \sqrt{\varepsilon}\hat{u}_b^2\left(0, \tfrac{Y}{\sqrt{\varepsilon}}\right),\\
a_L(Y) &= u_e^1(L, Y) + u_b^1\left(L, \tfrac{Y}{\sqrt{\varepsilon}}\right) + \sqrt{\varepsilon}u_e^2(L, Y) + \sqrt{\varepsilon}\hat{u}_b^2\left(L, \tfrac{Y}{\sqrt{\varepsilon}}\right),\\
b_0(Y) &= v_b^0\left(0, \tfrac{Y}{\sqrt{\varepsilon}}\right) + v_e^1(0, Y) + \sqrt{\varepsilon}v_b^1\left(0, \tfrac{Y}{\sqrt{\varepsilon}}\right) + \sqrt{\varepsilon}v_e^2(0, Y) + \varepsilon\hat{v}_b^2\left(0, \tfrac{Y}{\sqrt{\varepsilon}}\right),\\
b_L(Y) &= v_b^0\left(L, \tfrac{Y}{\sqrt{\varepsilon}}\right) + v_e^1(L, Y) + \sqrt{\varepsilon}v_b^1\left(L, \tfrac{Y}{\sqrt{\varepsilon}}\right) + \sqrt{\varepsilon}v_e^2(L, Y) + \varepsilon\hat{v}_b^2\left(L, \tfrac{Y}{\sqrt{\varepsilon}}\right),
\end{aligned}$$

are smooth functions constructed.

Iyer and Masmoudi [60] in 2020 established the convergence of 2D, stationary Navier-Stokes flows, $(u^\varepsilon, v^\varepsilon)$ to the classical Prandtl boundary layer, $(\bar{u}_p, \bar{v}_p)$ posed on the domain $(0, \infty) \times (0, \infty)$:

$$\left\|u^\varepsilon - \bar{u}_p\right\|_{L_y^\infty} \lesssim \sqrt{\varepsilon}\langle x\rangle^{-\frac{1}{4}+\delta}, \quad \left\|v^\varepsilon - \sqrt{\varepsilon}\bar{v}_p\right\|_{L_y^\infty} \lesssim \sqrt{\varepsilon}\langle x\rangle^{-\frac{1}{2}}.$$

This validates Prandtl's boundary layer theory globally in the x-variable for a large class of boundary layers, including the entire one parameter family of the classical Blasius profiles, with sharp decay rates.

Consider the Navier-Stokes (NS) equations posed on the domain $Q := (0, \infty) \times (0, \infty)$:

$$\begin{cases}
u^\varepsilon u_x^\varepsilon + v^\varepsilon u_Y^\varepsilon + P_x^\varepsilon = \varepsilon\Delta u^\varepsilon,\\
u^\varepsilon v_x^\varepsilon + v^\varepsilon v_Y^\varepsilon + P_Y^\varepsilon = \varepsilon\Delta v^\varepsilon,\\
u_x^\varepsilon + v_Y^\varepsilon = 0.
\end{cases} \tag{1.1.41}$$

The authors in [60] took the following boundary conditions in the vertical direction:

$$\left[u^\varepsilon, v^\varepsilon\right]\big|_{Y=0} = [0,0], \quad \left[u^\varepsilon(x,Y), v^\varepsilon(x,Y)\right] \overset{Y\to\infty}{\longrightarrow} [u_E(x,\infty), v_E(x,\infty)],$$

which coincide with the classical no-slip boundary condition at $\{Y = 0\}$ and the Euler matching condition as $Y \to \infty$. The authors in [60] fixed the vector field

$$[u_E, v_E] := [1,0], \quad P_E = 0,$$

as a solution to the steady, Euler equation condition above reads $[u^\varepsilon, v^\varepsilon] \overset{Y\to\infty}{\longrightarrow} [1,0]$. They expanded the rescaled solution as

$$\begin{cases} U^\varepsilon := 1 + u_p^0 + \sum_{i=1}^{N_1} \varepsilon^{\frac{i}{2}} \left(u_E^i + u_P^i\right) + \varepsilon^{\frac{N_2}{2}} u =: \bar{u} + \varepsilon^{\frac{N_2}{2}} u, \\ V^\varepsilon := v_p^0 + v_E^1 + \sum_{i=1}^{N_1-1} \varepsilon^{\frac{i}{2}} \left(v_P^i + v_E^{i+1}\right) + \varepsilon^{\frac{N_1}{2}} v_p^{N_1} + \varepsilon^{\frac{N_2}{2}} v =: \bar{v} + \varepsilon^{\frac{N_2}{2}} v, \\ P^\varepsilon := \sum_{i=0}^{N_1+1} \varepsilon^{\frac{i}{2}} P_p^i + \sum_{i=1}^{N_1} \varepsilon^{\frac{i}{2}} P_E^i + \varepsilon^{\frac{N_2}{2}} P. \end{cases}$$

Above,

$$\begin{aligned} &\left[u_E^0, v_E^0\right] := [1,0], \quad \left[u_p^i, v_p^i\right] = \left[u_p^i(x,y), v_p^i(x,y)\right], \\ &\left[u_E^i, v_E^i\right] = \left[u_E^i(x,Y), v_E^i(x,Y)\right], \end{aligned}$$

and N_1, N_2 are the expansion parameters for the sake of precision.

Introduce the Blasius boundary layer, which takes the following form:

$$\begin{cases} \left[\bar{u}_*^{x_0}, \bar{v}_*^{x_0}\right] = \left[f'(z), \frac{1}{\sqrt{x+x_0}}\left(zf'(z) - f(z)\right)\right], \\ z := \frac{y}{\sqrt{x+x_0}}, \\ ff'' + f''' = 0, \quad f'(0) = 0, \quad f'(\infty) = 1, \quad \frac{f(z)}{z} \overset{\eta\to\infty}{\longrightarrow} 1, \end{cases} \tag{1.1.42}$$

where, above, $f' = \partial_z f(z)$ and x_0 is a free parameter. Physically, x_0 has the meaning that at $x = -x_0$, the fluid interacts with the leading edge of, say, a plate (hence the singularity at $x = -x_0$). The authors' complete norm will be

$$\begin{aligned} \|U,V\|_{\mathcal{X}} := &\sum_{n=0}^{10} \left(\|U,V\|_{X_n} + \|U,V\|_{X_{n+\frac{1}{2}}} + \|U,V\|_{Y_{n+\frac{1}{2}}} \right) \\ &+ \|U,V\|_{X_{11}} + \|U,V\|_{E}. \end{aligned}$$

Theorem 1.1.34 ([60]) *Fix $N_1 = 400$ and $N_2 = 200$. Fix the leading order Euler flow to be*

$$\left[u_E^0, v_E^0, P_E^0\right] := [1, 0, 0].$$

Assume the following pieces of initial data at $\{x = 0\}$ are given for $i = 0, \dots, N_1$, and $j = 1, \dots N_1$,

$$u_p^i\Big|_{x=0} =: U_p^i(y), \quad v_E^j\Big|_{x=0} =: V_E^j(Y),$$

where we make the following assumptions on the initial datum:

(1) For $i = 0$, the boundary layer datum $\bar{U}_p^0(y)$ is in a neighborhood of Blasius. More precisely, we will assume

$$\left\| \left(\bar{U}_p^0(y) - \bar{u}_*(0, y)\right) \langle y\rangle^{m_0} \right\|_{C^{\ell_0}} \leq \delta_*,$$

where $0 < \delta_* << 1$ is small relative to universal constants, where m_0, ℓ_0, are large, explicitly computable numbers. Assume also the difference $\bar{U}_p^0(y) - \bar{u}_*(0, y)$ satisfies generic parabolic compatibility conditions at $y = 0$.

(2) For $i = 1, \dots, N_1$, the boundary layer datum, $U_p^i(\cdot)$, is sufficiently smooth and decays rapidly

$$\left\| U_p^i \langle y\rangle^{m_i} \right\|_{C^{\ell_i}} \lesssim 1,$$

where $m_i and \ell_i$ are large, explicitly computable constants (for instance, we can take $m_0 = 10{,}000$, $\ell_0 = 10{,}000$ and $m_{i+1} = m_i - 5, \ell_{i+1} = \ell_i - 5$), and satisfy generic parabolic compatibility conditions at $y = 0$.

(3) The Euler datum $V_E^i(Y)$ satisfies generic elliptic compatibility conditions,

(4) Assume Dirichlet datum for the remainders, that is

$$[u, v]|_{x=0} = [u, v]|_{x=\infty} = 0.$$

Then there exists an $\varepsilon_0 \ll 1$ small relative to universal constants such that for every $0 \leq \varepsilon \leq \varepsilon_0$, there exists a unique solution $(u^\varepsilon, v^\varepsilon)$ to system which satisfies the expansion in the quadrant, Q. Each of the intermediate quantities in the expansion satisfies the following estimates for $i = 1, \dots, N_1$

$$\left\| \partial_x^k \partial_y^j u_p^i \langle z\rangle^M \right\|_{L_y^\infty} \leq C_{M,k,j} \langle x\rangle^{-\frac{1}{4}-k-\frac{j}{2}+\sigma_*},$$

$$\left\| v_p^i \right\|_{L_y^\infty} \leq C_{M,k,j} \langle x\rangle^{-\frac{3}{4}-k-\frac{j}{2}+\sigma_*},$$

$$\left\| \partial_x^k \partial_Y^j u_E^i \right\|_{L_y^\infty} \leq C_{k,j} \langle x\rangle^{-\frac{1}{2}-k-j}, \quad \left\| v_E^i \right\|_{L_y^\infty} \leq C_{k,j} \langle x\rangle^{-\frac{1}{2}-k-j},$$

where $\sigma_* := \frac{1}{10{,}000}$. *Finally, the remainder* (u, v) *exists globally in the quadrant,* Q, *and satisfies the following estimates:*

$$\|u, v\|_X \lesssim 1.$$

Li and Ding [69] in 2020 concerned with the validity of the Prandtl boundary layer theory in the inviscid limit of the steady incompressible Navier-Stokes equations. Under the symmetry assumption, they established the validity of the Prandtl boundary layer expansions and the error estimates. The convergence rate as $\varepsilon \to 0$ is also given. They considered the following steady incompressible Navier-Stokes equations:

$$\begin{cases} UU_X + VU_Y + P_X = \varepsilon U_{XX} + \varepsilon U_{YY}, \\ UV_X + VV_Y + P_Y = \varepsilon V_{XX} + \varepsilon V_{YY}, \\ U_X + V_Y = 0, \end{cases} \tag{1.1.43}$$

in the domain

$$\Omega := \{(X, Y) \mid 0 \le X \le L, 0 \le Y \le 2\},$$

with moving boundary conditions

$$U(X, 0) = U(X, 2) = u_b > 0, \; V(X, 0) = V(X, 2) = 0.$$

They further focused on the problem when $\varepsilon \to 0$. As $\varepsilon \to 0$, a formal limit of the solution of (1.1.43) should be the shear flow $[U_0, V_0] = \left[u_e^0(Y), 0\right]$, which satisfies the corresponding Euler equations. We assume that this smooth positive function $u_e^0(\cdot)$ satisfies $u_e^0(1 - Y) = u_e^0(1 + Y)$, for any $Y \in [0, 1]$ and $u_e^0(0) = u_e^0(2) = u_e \neq u_b$. Accordingly, they assumed that the solution $[U, V]$ to (1.1.43) satisfies the following symmetrical conditions with respect to $Y = 1$:

$$U(X, 1 - Y) = U(X, 1 + Y), \; V(X, 1 - Y) = -V(X, 1 + Y), \; Y \in [0, 1],$$

It should be noted that, due to this assumption, the pair $[U, V]_{1 \le Y \le 2}$ satisfies equation (1.1.43) as long as $[U, V]_{0 \le Y \le 1}$ does. Then their discussion can be restricted to the domain

$$\Omega_0 := \{(X, Y) \mid 0 \le X \le L, 0 \le Y \le 1\},$$

and the boundary conditions turn to

$$[U, V](X, 0) = [u_b, 0], \quad [U_Y, V](X, 1) = [0, 0],$$

introduce the Prandtl's scaling

$$x = X,\, y = \frac{Y}{\sqrt{\varepsilon}},$$

and the new unknown functions

$$U^\varepsilon(x, y) = U(X, Y),\, V^\varepsilon(x, y) = \frac{1}{\sqrt{\varepsilon}} V(X, Y).$$

Under this transformation, system (1.1.43) can be rewritten as

$$\begin{cases} U^\varepsilon U^\varepsilon_x + V^\varepsilon U^\varepsilon_y + P^\varepsilon_x = U^\varepsilon_{yy} + \varepsilon U^\varepsilon_{xx}, \\ U^\varepsilon V^\varepsilon_x + V^\varepsilon V^\varepsilon_y + P^\varepsilon_y/\varepsilon = V^\varepsilon_{yy} + \varepsilon V^\varepsilon_{xx}, \, , \\ U^\varepsilon_x + V^\varepsilon_y = 0 \end{cases} \tag{1.1.44}$$

in the domain

$$\Omega_\varepsilon := \left\{(x, y) \mid 0 \le x \le L, 0 \le y \le \frac{1}{\sqrt{\varepsilon}}\right\},$$

with the boundary conditions

$$\left[U^\varepsilon, V^\varepsilon\right](x, 0) = [u_b, 0], \quad \left[U^\varepsilon_y, V^\varepsilon\right]\left(x, \frac{1}{\sqrt{\varepsilon}}\right) = [0, 0].$$

In what follows, the authors in [60] intend to find the exact solutions $[U^\varepsilon, V^\varepsilon, P^\varepsilon]$ in form of

$$\begin{cases} U^\varepsilon(x, y) = u_{app}(x, y) + \varepsilon^{\gamma+\frac{1}{2}} u^\varepsilon(x, y) \\ V^\varepsilon(x, y) = v_{app}(x, y) + \varepsilon^{\gamma+\frac{1}{2}} v^\varepsilon(x, y) \\ P^\varepsilon(x, y) = p_{app}(x, y) + \varepsilon^{\gamma+\frac{1}{2}} p^\varepsilon(x, y) \end{cases} \tag{1.1.45}$$

where

$$\begin{cases} u_{app}(x, y) = u^0_e(\sqrt{\varepsilon}y) + u^0_p(x, y) + \sqrt{\varepsilon}u^1_e(x, \sqrt{\varepsilon}y) + \sqrt{\varepsilon}u^1_p(x, y), \\ v_{app}(x, y) = v^0_p(x, y) + v^1_e(x, \sqrt{\varepsilon}y) + \sqrt{\varepsilon}v^1_p(x, y), \\ p_{app}(x, y) = \sqrt{\varepsilon}p^1_e(x, \sqrt{\varepsilon}y) + \sqrt{\varepsilon}p^1_p(x, y) + \varepsilon p^2_p(x, y). \end{cases} \tag{1.1.46}$$

For convenience, denote $z := \sqrt{\varepsilon}y$. Boundary conditions on $\{y = 0\}$

$$u^0_e(0) + u^0_p(x, 0) = u_b, \quad u^1_e(x, 0) + u^1_p(x, 0) = 0, \quad u^\varepsilon(x, 0) = 0, \tag{1.1.47}$$

$$v^0_p(x, 0) + v^1_e(x, 0) = 0, \quad v^1_p(x, 0) = 0, \quad v^\varepsilon(x, 0) = 0. \tag{1.1.48}$$

Boundary conditions on $\left\{y = \frac{1}{\sqrt{\varepsilon}}\right\}$

$$u^0_{py}\left(x, \frac{1}{\sqrt{\varepsilon}}\right) = 0, \quad u^1_{ez}(x, 1) = 0, u^1_{py}\left(x, \frac{1}{\sqrt{\varepsilon}}\right) = 0, \quad u^\varepsilon_y\left(x, \frac{1}{\sqrt{\varepsilon}}\right) = 0; \tag{1.1.49}$$

$$v^0_p\left(x, \frac{1}{\sqrt{\varepsilon}}\right) = 0, \quad v^1_e(x, 1) = 0, v^1_p\left(x, \frac{1}{\sqrt{\varepsilon}}\right) = 0, \quad v^\varepsilon\left(x, \frac{1}{\sqrt{\varepsilon}}\right) = 0. \tag{1.1.50}$$

Boundary conditions on $\{x = 0\}$

$$u^0_p(0, y) = \bar{u}_0(y), \quad u^1_e(0, z) = u^1_b(z), \quad u^1_p(0, y) = \bar{u}_1(y), \quad u^\varepsilon(0, y) = 0; \tag{1.1.51}$$

$$v^1_e(0, z) = V_{b0}(z), \quad v^\varepsilon(0, y) = 0. \tag{1.1.52}$$

Boundary conditions on $\{x = L\}$

$$v^1_e(L, z) = V_{bL}(z), \quad \left[p^\varepsilon - 2\varepsilon u^\varepsilon_x, u^\varepsilon_y + \varepsilon v^\varepsilon_x\right](L, y) = 0. \tag{1.1.53}$$

Theorem 1.1.35 ([69]) *Let $u_b > 0$ be a constant tangential velocity of the Navier-Stokes flow on the boundary $\{Y = 0\}$, and let $u^0_e(Y)$ be a smooth positive Euler flow satisfies $u^0_{ez}(1) = 0$. Suppose that the boundary conditions prescribed in (1.1.47)–(1.1.53) hold. Suppose further that the positive condition $\min_y \left\{u^0_e(\sqrt{\varepsilon}y) + \bar{u}_0(y)\right\} > 0$ holds. Then there exists a constant $L_0 > 0$, which depends only on the prescribed data, such that for $0 < L \leq L_0$ and $\gamma \in \left(0, \frac{1}{5}\right)$, the asymptotic expansions stated in (1.1.45)–(1.1.46) is a solution to equation (1.1.44) on Ω_ε together with the corresponding boundary conditions. The remainder solutions $[u^\varepsilon, v^\varepsilon]$ satisfies the estimate*

$$\left\|\nabla_\varepsilon u^\varepsilon\right\|_{L^2(\Omega_\varepsilon)} + \left\|\nabla_\varepsilon v^\varepsilon\right\|_{L^2(\Omega_\varepsilon)} + \left\|u^\varepsilon\right\|_{L^\infty(\Omega_\varepsilon)} + \sqrt{\varepsilon}\left\|v^\varepsilon\right\|_{L^\infty(\Omega_\varepsilon)} \leq C_0.$$

1.1.2 2D Prandtl Equations-Local Existence

Usually von Mises transformation and Crocco transformation are used to prove the existence of the global solutions. Therefore, in this section, we list the papers that used the shear flow method to study of the Prandtl equations in Gevrey space and the energy estimation method to prove the existence of local solutions.

The authors in [3, 7, 8, 11, 12, 15, 135, 144, 145] studied the local well-posedness of the Prandtl equations around a shear flow.

Cannone, Lombardo, and Sammartino [7] in 2013 concerned with Prandtl equations with incompatible data, i.e., with initial data that, in general, do not fulfil the boundary conditions imposed on the solution. Under the hypothesis of analyticity in the streamwise variable, the authors proved that Prandtl equations, on the half-plane or on the half-space, are well-posed for a short time. Prandtl equations on the half-plane are written as

$$\begin{cases} \partial_t u^P + u^P \partial_x u^P + v^P \partial_Y u^P + \partial_x p^E = \partial_{YY} u^P, \quad (x, Y) \in \mathbb{R} \times \mathbb{R}^+, \\ \partial_x u^P + \partial_Y v^P = 0, \\ u^P(x, Y = 0, t) = 0, \\ u^P(x, Y, t = 0) = u_0(x, Y), \\ u^P(x, Y \to \infty, t) = U(x, t), \\ v^P(x, Y = 0, t) = 0 \end{cases} \tag{1.1.54}$$

where the Euler matching datum $U(x, t)$ and the pressure $p^E = p^E(x, t)$ satisfy Bernoulli's law

$$\partial_t U + U \partial_x U + \partial_x p^E \Big|_{y=0} = 0.$$

We first introduce the norms of the spaces $\mathcal{K}^{l,\rho}$, $\mathcal{K}^{l,\rho,\alpha}$ and $\mathcal{K}^{l,\rho,\alpha}_{\beta,T}$, respectively,

$$|f|_{l,\rho} = \sum_{j=0}^{l} \sup_{\lambda \in (-\rho,\rho)} \|\partial_x^j f(\cdot + i\lambda)\|_{L_x} < +\infty, \tag{1.1.55}$$

$$|f|_{l,\rho,\alpha} = \sum_{0 \le j \le 2} \sum_{i \le l-j} \sup_{Y \in \mathbb{R}^+} \langle Y \rangle |^{\alpha} \partial_Y^j \partial_x^i f(\cdot, Y)|_{0,\rho} < +\infty, \tag{1.1.56}$$

$$\begin{aligned} |f|_{l,\rho,\alpha,\beta,T} = & \sum_{0 \le j \le 2} \sum_{i \le l-j} \sup_{0 \le t \le T} |\partial_Y^j \partial_x^i f(\cdot, \cdot, t)|_{0,\rho-\beta t,\alpha} \\ & + \sum_{i \le l-2} \sup_{0 \le t \le T} |\partial_t \partial_x^i f(\cdot, \cdot, t)|_{0,\rho-\beta t,\alpha} < +\infty. \end{aligned} \tag{1.1.57}$$

Theorem 1.1.36 ([7]) *Suppose that* $U \in \mathcal{K}^{l,\rho_0}_{\beta_0,T_0}$ *and* $u_0 - U \in K^{l,\rho_0,\alpha}$. *Then there exist* $0 < \rho < \rho_0$, $\beta > \beta_0 > 0$ *and* $0 < T < T_0$ *such that Prandtl equation* (1.1.54) *admits a unique solution* u^P. *This solution can be written as*

$$u^P(x, Y, t) = u^S + u^R,$$

where (i) u^S *is an initial layer corrector and has the following form:*

$$u^S = -2u_0(x, Y = 0)\,\mathrm{erfc}\left(\frac{Y}{2\sqrt{t}}\right),$$

where $\mathrm{erfc}(y) = \frac{2}{\sqrt{\pi}} \int_0^y e^{-z^2} dz$. *(ii)* u^R *is the solution of the Prandtl equation with compatible data and with a source term that keeps into account the interaction with* u^S *and has following form:*

$$u^R(x, Y, t) = \tilde{u}(x, Y, t) + U(x, t),$$

where $\tilde{u} \in \mathcal{K}_{\beta,T}^{l,\rho,\alpha}$;

(iii) The solution $u^P(x, Y, t)$, *for* $0 < t < T$, *is analytic both in* x *and* Y.

Cannone, Lombardo, and Sammartino [8] in 2014 proved the well-posedness of the Prandtl boundary layer equations (1.1.58) on a periodic strip when the initial and the boundary data are not assigned to be compatible in the domain $(x, y) \in [0, 2\pi] \times \mathbb{R}_T$,

$$\begin{cases} u_t + uu_x + vu_y = \nu u_{yy} + p_x, \\ u_x + v_y = 0, \\ u(0, x, y) = u_0, \quad u(t, x, 0) = v(t, x, 0) = 0, \\ \lim\limits_{y \to +\infty} u(t, x, y) = U(t, x). \end{cases} \tag{1.1.58}$$

But, initial and boundary data are not compatible, i.e.,

$$u_0(x, 0) \neq 0 = u^P(0, x, 0).$$

The space $\mathcal{H}^\sigma$ is the space of the 2π periodic real functions $f(x)$ such that

$$|f|_\sigma := \sum_{k \in \mathbb{Z}} |\widehat{f}(k)| e^{|k|\sigma} < \infty,$$

where $\widehat{f}(k)$ are the Fourier coefficients of f.

The space $\mathcal{H}_{\beta,T}^{\sigma,\alpha}$ is the space of the functions $f(t, x, y)$, 2π periodic with respect to x, such that

$$f, \partial_t f, \partial_y^j f \in \mathcal{H}^{\sigma-\beta t,\alpha}, \forall 0 \le t \le T, \text{and } j \le 2.$$

Moreover,

$$|f|_{\sigma,\alpha,\beta,T} = \sum_{0 \le j \le 2} \sup_{0 \le t \le T} |\partial_y^j f(\cdot, \cdot, t)|_{\sigma-\beta t,\alpha} + \sup_{0 \le t \le T} |\partial_t f(\cdot, \cdot, t)|_{\sigma-\beta t,\alpha} < \infty.$$

Theorem 1.1.37 ([8]) *Suppose that the Euler datum* U *and the initial datum* u_0 *satisfy the following hypotheses:*

(i) U *satisfies the Bernoulli's law equation (1.0.3);*
(ii) $U(x, t) \in \mathcal{H}_{\beta_0,T_0}^{\sigma_0}$;
(iii) $u_0(x, y) - U(0, x) \in \mathcal{H}^{\sigma_0,\alpha}$.

Then there exist $0 < \sigma < \sigma_0$, $\beta > \beta_0 > 0$ *and* $0 < T < T_0$ *such that Prandtl equation (1.1.58) admits, in [0,T], a unique solution* u^P. *This solution can be written as*

$$u^P(t, x, y) = u^s + u^R,$$

where

1. u^s *is an initial layer corrector and has the following form:*

$$u^s = -2u_0(x, 0)\, \mathrm{erfc}\left(\frac{y}{2\sqrt{t}}\right).$$

2. u^R *is the solution of the Prandtl equation with compatible data and with a source term that keeps into account the interaction with* u^s; u^R *can be decomposed as follows:*

$$u^R(t, x, y) = \widetilde{u}(t, x, y) + U$$

where

$$\widetilde{u} \in \mathcal{H}^{\sigma,\alpha}_{\beta,T}.$$

Moreover, the solution $u^P(t, x, y)$, *for* $t > 0$, *is analytic both in* x *and* Y.

Alexander, Wang, Xu, and Yang [3] in 2015 studied the local well-posedness of the Prandtl equations (1.1.58) around a monotonic shear flow in the domain $\mathbb{R}^2_+ = \{(x, y)|x \in \mathbb{R}, y \in [0, +\infty)\}$. In this paper, assume that

$$u_0(x, y) = u_0^s(y) + \widehat{u}_0(x, y),$$

where u_0^s is monotonic in y,

$$\partial_y u_0^s(y) > 0, \quad \forall y \geq 0. \tag{1.1.59}$$

Set $\Omega_T = [0, T] \times \mathbb{R}^2_+$. For any nonnegative integer k and real number ℓ, define

$$\|u\|_{\mathcal{A}^k_\ell(\Omega_T)} = \left(\sum_{k_1+[\frac{k_2+1}{2}]\leq k} \|\langle y\rangle^\ell \partial^{k_1}_{t,x} \partial^{k_2}_y u\|^2_{L^2([0,T]\times\mathbb{R}^2_+)}\right)^{\frac{1}{2}},$$

and

$$\|u\|_{\mathcal{D}^k_\ell(\Omega_T)} = \sum_{k_1+[\frac{k_2+1}{2}]\le k} \|\langle y\rangle^\ell \partial^{k_1}_{t,x}\partial^{k_2}_y u\|_{L^\infty(L^2_{t,x})},$$

with $\langle y\rangle = (1+y^2)^{\frac{1}{2}}$.

Theorem 1.1.38 ([3]) *Concerning the problem (1.1.58) with U being a constant, we have the following existence and stability results:*

(1) Given any integer $k \ge 5$ and real number $\ell > \frac{1}{2}$, let the initial data $u_0(x, y) = u_0^s(y) + \tilde{u}_0(x, y)$ satisfy the compatibility conditions of the initial boundary value problem (1.1.58) up to order $k+4$. Assume the following two conditions:

(i) The monotonicity condition (1.1.59) holds for u_0^s, and

$$\begin{cases} \partial_y^{2j} u_0^s(0) = 0, \quad \forall 0 \le j \le k+4, \\ \|\langle y\rangle^\ell (u_0^s - 1)\|_{H^{2k+9}(\mathbb{R}^2_+)} + \|\frac{\partial_y^2 u_0^s}{\partial_y u_0^s}\|_{H^{2k+7}(\mathbb{R}^2_+)} \le C, \end{cases} \tag{1.1.60}$$

for a fixed constant $C > 0$.

(ii) There exists a small constant $\epsilon > 0$ depending only on u_0^s, such that

$$\|\tilde{u}_0\|_{\mathcal{A}^{2k+9}_\ell(\mathbb{R}^2_+)} + \|\frac{\partial_y \tilde{u}_0}{\partial_y u_0^s}\|_{\mathcal{A}^{2k+9}_\ell(\mathbb{R}^2_+)} \le \epsilon. \tag{1.1.61}$$

Then there is a $T > 0$, such that the problem (1.1.58) admits a classical solution u satisfying

$$u - u^s \in \mathcal{A}^k_\ell(\Omega_T), \ \frac{\partial_y(u - u^s)}{\partial_y u_0^s} \in \mathcal{A}^k_\ell(\Omega_T), \ v \in \mathcal{D}^{k-1}_0(\Omega_T). \tag{1.1.62}$$

Here, note that $(u^s(t, y), 0)$ is the shear flow of the Prandtl equation defined by the initial data $(u_0^s(y), 0)$.

(2) The solution is unique in the function space described in (1.1.62). Moreover, we have the stability with respect to the initial data in the following sense. For any given two initial data

$$u_0^1 = u_0^s + \tilde{u}_0^1, \ \ u_0^2 = u_0^s + \tilde{u}_0^2,$$

if u_0^s satisfy (1.1.59)–(1.1.60), and $\tilde{u}_0^1, \tilde{u}_0^2$ satisfy (1.1.61), then the corresponding solutions (u^1, v^1) and (u^2, v^2) of (1.1.58) satisfy

$$\|u^1 - u^2\|_{\mathcal{A}_\ell^p(\Omega_T)} + \|v^1 - v^2\|_{\mathcal{D}_0^{p-1}(\Omega_T)} \le C\|\frac{\partial}{\partial y}\left(\frac{u_0^1 - u_0^2}{\partial y u_0^s}\right)\|_{\mathcal{A}_\ell^p(\Omega_T)}$$

for all $p \le k - 1$, where the constant $C > 0$ depends only on T and the upper bounds of the norms of u_0^1, u_0^2.

Wu [145] in 2015 studied the local well-posedness of classical solutions to the unsteady Prandtl equations with Robin boundary condition in half space $\mathbb{R}_+$ in weighted Sobolev spaces with shear flow. His results are also valid for the Dirichlet boundary case. He studied the following Prandtl equations with Robin boundary condition:

$$\begin{cases} u_t + uu_x + vu_y = u_{yy}, \\ u_x + v_y = 0, \\ (u_y - \beta u)|_{y=0} = 0, \quad v_{y=0} = 0, \\ u(0, x, y) = u_0, \\ \lim_{y\to+\infty} u(t, x, y) = U = 1 \end{cases} \tag{1.1.63}$$

and defined the following functional spaces:

$$\|u\|_{\mathcal{A}_\ell^k(\Omega_T)} = \left(\sum_{k_1+\left[\frac{k_2+1}{2}\right]\le k} \|\langle y\rangle^\ell \partial_{t,x}^{k_1}\partial_y^{k_2} u\|_{L^2([0,T]\times\mathbb{R}_+^2)}^2\right)^{\frac{1}{2}},$$

$$\|u\|_{C_\ell^k(\Omega_T)} = \sum_{k_1+\left[\frac{k_2+1}{2}\right]\le k} \|\langle y\rangle^\ell \partial_{t,x}^{k_1}\partial_y^{k_2} u\|_{L_y^2(L_{t,x}^\infty)},$$

$$\|u\|_{\mathcal{D}_\ell^k(\Omega_T)} = \sum_{k_1+\left[\frac{k_2+1}{2}\right]\le k} \|\langle y\rangle^\ell \partial_{t,x}^{k_1}\partial_y^{k_2} u\|_{L_y^\infty(L_{t,x}^2)},$$

with $\langle y\rangle = (1 + y^2)^{\frac{1}{2}}$.

Theorem 1.1.39 ([145]) *Concerning the nonlinear unsteady Prandtl equations with Robin boundary condition (1.1.63), giving any integer $k \ge 5$ and real number $\ell > \frac{1}{2}$, we have the following existence, uniqueness, and stability results:*

(1) For any $\delta_\beta > 0$, $\delta_\beta \le \beta < +\infty$, assume the initial data $u_0(x,y) = u_0^s(y) + \tilde{u}_0(x,y)$ satisfies the following two conditions:

(i) u_0^s satisfies

$$\begin{cases} u_0^s(y) > 0, \ \ \partial_y u_0^s(y) > 0, \ \ \beta - \frac{\partial_{yy} u_0^s(y)}{\partial_y u_0^s(y)} \ge \delta_{s,0} > 0, \forall y \in [0,+\infty), \\ \partial_y^{2j}(\partial_y u_0^s(y) - \beta u_0^s(y))|_{y=0} = 0, \ \ \forall 0 \le j \le 2k+10, \\ \lim\limits_{y\to+\infty} u_0^s(y) = 1, \\ \|u_0^s - 1\|_{L^2} + \|u_0^s\|_\infty + \|u_0^s\|_{C_\ell^{2k+11}} + \|\frac{\partial_{yy} u_0^s(y)}{\partial_y u_0^s(y)}\|_{C_\ell^{2k+10}} \le C, \end{cases} \tag{1.1.64}$$

for a fixed constant $C > 0$.

(ii) There exists a small constant $\epsilon > 0$ such that $\tilde{u}_0 = u_0 - u_0^s$ satisfies

$$\begin{cases} \partial_y^{2j}(\partial_y \tilde{u}_0(x,y) - \beta \tilde{u}_0(x,y))|_{y=0} = 0, \ \ \forall 0 \le j \le 2k+8, \\ \lim\limits_{y\to+\infty} \tilde{u}_0(x,y) = 0, \\ \|\tilde{u}_0\|_{\mathcal{A}_\ell^{2k+9}(\mathbb{R}_+^2, t=0)} + \|\frac{\partial_y \tilde{u}_0}{\partial_y u_0^s}\|_{\mathcal{A}_\ell^{2k+9}(\mathbb{R}_+^2, t=0)} \le \epsilon. \end{cases} \tag{1.1.65}$$

Then there exists a time $T \in [0,+\infty)$, such that the Prandtl system (1.1.63) admits a unique classical solution (u,v) satisfying

$$u > 0, \ \ \partial_y u > 0, \ \ \beta - \frac{\partial_{yy} u}{\partial_y u} \ge \delta > 0, \tag{1.1.66}$$

and

$$\begin{cases} u - u^s \in \mathcal{A}_\ell^k([0,T]\times\mathbb{R}_+^2), \ \ \partial_y(u-u^s), \frac{\partial_y(u-u^s)}{\partial_y u^s} \in \mathcal{A}_\ell^k([0,T]\times\mathbb{R}_+^2), \\ v \in \mathcal{D}_0^{k-1}([0,T]\times\mathbb{R}_+^2), \ \ \partial_y v, \partial_{yy} v \in \mathcal{A}_\ell^{k-1}([0,T]\times\mathbb{R}_+^2), \\ \partial_y^j u|_{y=0} - \partial_y^j u^s|_{y=0} \in \mathcal{A}_\ell^{k-[\frac{j+1}{2}]}([0,T]\times\mathbb{R}), \ \ 0 \le j \le 2k, \\ \partial_y^{j+1} v|_{y=0} \in \mathcal{A}_\ell^{k-1-[\frac{j+1}{2}]}([0,T]\times\mathbb{R}), \ \ 0 \le j \le 2k-2. \end{cases} \tag{1.1.67}$$

(2) For any $\delta_\beta > 0$, $\delta_\beta \le \beta < +\infty$, the classical solution to the Prandtl system (1.1.63) is stable with respect to the initial data in the following sense: for any given two initial data

$$u_0^1 = u_0^s + \tilde{u}_0^1, \ \ u_0^2 = u_0^s + \tilde{u}_0^2,$$

if u_0^s satisfies (1.1.64) and u_0^1, u_0^2 satisfy (1.1.65) , then for all $p \leq k-2$, the corresponding solutions (u^1, v^1) and (u^2, v^2) of the Prandtl system (1.1.63) satisfy

$$
\begin{aligned}
&\|u^1 - u^2\|_{\mathcal{A}_\ell^p([0,T]\times\mathbb{R}_+^2)} + \|\frac{\partial_y(u^1-u^2)}{\partial_y u^s}\|_{\mathcal{A}_\ell^p([0,T]\times\mathbb{R}_+^2)} \\
&\quad + \|v^1 - v^2\|_{\mathcal{D}_\ell^{p-1}([0,T]\times\mathbb{R}_+^2)} + \sum_{j=0}^{2p} \|\partial_y^j u^1|_{y=0} - \partial_y^j u^2|_{y=0}\|_{\mathcal{A}_\ell^{p-[\frac{j+1}{2}]}([0,T]\times\mathbb{R})} \\
&\leq C(T,\epsilon,u_0^s)\|\partial_y\left(\frac{u_0^1-u_0^2}{\partial_y(u_0^1+u_0^2)}\right)\|_{\mathcal{A}_\ell^p(\mathbb{R}_+^2,t=0)} \\
&\quad + \frac{C(T,\epsilon,u_0^s)}{\max\{\sqrt{\beta - C_\eta}\}, \sqrt{\delta}}\|\partial_y\left(\frac{u_0^1-u_0^2}{\partial_y(u_0^1+u_0^2)}\right)|_{y=0}\|_{\mathcal{A}_\ell^p(\mathbb{R})}.
\end{aligned}
$$

(3) As $\beta \to +\infty$, $\partial_t^j u^s|_{y=0} = O(\frac{1}{\beta})$, $0 \leq j \leq k$, $\|u|_{y=0} - u^s|_{y=0}\|_{\mathcal{A}_\ell^k} = O(\frac{1}{\beta})$ and (u, v) satisfies (1.1.67) uniformly. When $\beta = +\infty$, (u, v) satisfies (1.1.67) and for $p \leq k-2$,

$$
\begin{aligned}
&\|u^1 - u^2\|_{\mathcal{A}_\ell^p([0,T]\times\mathbb{R}_+^2)} + \|\frac{\partial_y(u^1-u^2)}{\partial_y u^s}\|_{\mathcal{A}_\ell^p([0,T]\times\mathbb{R}_+^2)} \\
&+\|v^1 - v^2\|_{\mathcal{D}_\ell^{p-1}([0,T]\times\mathbb{R}_+^2)} \\
&\quad + \sum_{j=0}^{2p} \|\partial_y^j u^1|_{y=0} - \partial_y^j u^2|_{y=0}\|_{\mathcal{A}_\ell^{p-[\frac{j+1}{2}]}([0,T]\times\mathbb{R})} \\
&\leq C\|\partial_y(\frac{u_0^1-u_0^2}{\partial_y(u_0^1+u_0^2)})\|_{\mathcal{A}_\ell^p(\mathbb{R}_+^2,t=0)}.
\end{aligned}
$$

Authors of [15] in 2017 studied the linear stability of shear flow solutions to the interactive boundary layer with high-frequency perturbations. The general concern of [15] is the boundary layer behavior of high Reynolds number flows. The authors of [15] restricted to the two-dimensional case near a flat boundary and considered the Navier-Stokes system for $t > 0$, $\mathbf{x} = (x, Y) \in \mathbb{R} \times \mathbb{R}_+$,

$$
\partial_t u^\nu - \nu\Delta u^\nu + u^\nu \cdot \nabla u^\nu + \nabla P^\nu = 0, \quad div u^\nu = 0, \quad u^\nu(x, Y=0) = 0. \tag{1.1.68}
$$

The parameter $\nu \ll 1$ refers to the inverse Reynolds number. They wrote the linearized system as the following system, with unknowns U, V, and u_e:

$$\begin{cases} \partial_t U + U_s \partial_x U + V U'_s - \partial_y^2 U = \partial_t u_e + \partial_x u_e, \\ \partial_x U + \partial_y V = 0, \end{cases} \tag{1.1.69}$$

$$U|_{y=0} = V|_{y=0} = 0, \quad \lim_{y \to +\infty} = u_e, \; u_e = \sqrt{\nu} DN \int_0^{+\infty} (u_e - U) dy,$$

where we use the convention $DNf = -\partial_Y \psi|_{y=0}$, ψ is the solution of $\Delta \psi = 0$.

After linearization around the shear flow, we find

$$\partial_t U + U_s \partial_x U + V U'_s - \partial_y^2 U = \partial_t u_e + \partial_x u_e, \; \partial_x U + \partial_y V = 0, \tag{1.1.70}$$

$$U|_{y=0} = V|_{y=0} = 0, \quad \lim_{y \to +\infty} = u_e, \; \Delta_s u_e = \int_0^{+\infty} (u_e - U) dy,$$

with $\Delta_s u_e$ the displacement thickness associated to the shear flow.

Criterion 1. Assume that $\xi(\Delta_s) = 0$. Assume, moreover, that either $U_s''(0) > 0$ and $\xi_+(\Delta_s) \geq \xi_-(\Delta_s)$ or $U_s''(0) < 0$ and $\xi_+(\Delta_s) > \xi_-(\Delta_s)$.

Criterion 2. Assume that $\xi(0) = 0$. Assume, moreover, that either $U_s''(0) > 0$ and $\xi_+(0) \geq \xi_-(0)$ or $U_s''(0) < 0$ and $\xi_+(0) > \xi_-(0)$.

Criterion 3. Assume that there exists an arbitrary abscissa $\gamma > 0$ such that $\xi(\gamma) = 0$ and, moreover, either $U_s''(0) > 0$ and $\xi_+(\gamma) \geq \xi_-(\gamma)$ or $U_s''(0) < 0$ and $\xi_+(\gamma) > \xi_-(\gamma)$.

Theorem 1.1.40 ([15]) *Let U_s be a shear flow satisfying Criterion 1. There exist $K, \eta > 0$ such that, for $k \geq K$, there exists $\lambda_k \in \mathbb{C}$ with $Re\lambda_k \geq \eta k$ such that the system* (1.1.70) *displays* (λ_k, k) *instabilities.*

Theorem 1.1.41 ([15])

- *Let U_s be a shear flow satisfying Criterion 1. There exist $K, \eta, \gamma_+ > 0$ such that, for $k \geq K$ and $\nu > 0$ satisfying $(\sqrt{\nu} k)^{-1} \leq \gamma_+$, there exists $\lambda_{k,\nu} \in \mathbb{C}$ with $Re\lambda_{k,\nu} \geq \eta k$ such that the system* (1.1.69) *displays* (λ_k, k) *instabilities.*
- *Let U_s be a shear flow satisfying Criterion 1. There exist $K, \eta, \gamma_-, \gamma_+ > 0$ such that, for $k \geq K$ and $\nu > 0$ satisfying $\gamma_- \leq (\sqrt{\nu} k)^{-1} \leq \gamma_+$, there exists $\lambda_{k,\nu} \in \mathbb{C}$ with $Re\lambda_{k,\nu} \geq \eta k$ such that the system* (1.1.69) *displays* (λ_k, k) *instabilities.*
- *If U_s satisfies $U_s''(0) > 0$, there exist $\eta, \gamma_-, > 0$ such that, for $\nu k^3 \geq 1$, there exists $\lambda_{k,\nu}$ such that the system* (1.1.69) *displays* $(\lambda_{k,\nu}, k)$ *instabilities with $Re\lambda_{k,\nu} \geq \eta \nu k^3$.*

Theorem 1.1.42 ([15]) *Let U_s be an arbitrary monotone shear flow. There exist positive constants $\bar{\nu}, K, S, \eta$ such that for $\nu \leq \bar{\nu}$, $|\bar{k}| \geq K$ and $k\nu^{3/4} \geq S$, there* *exists* $\lambda_{k,\nu}$ *such that the system* (1.1.69) *displays* $(\lambda_{k,\nu}, k)$ *instabilities with* $Re\lambda_{k,\nu} \geq \eta\nu^{3/4}k^2$.

Theorem 1.1.43 ([15]) *Both for the* (1.1.69) *and* (1.1.70) *systems, we can build shear flows displaying instabilities with chosen behavior within a given spectral domain. More precisely,*

$\cdot$*System* (1.1.70)*: let* $\mu \in \Gamma_1 = \left\{a + ib \in \mathbb{C}, 0 < a < 1, 0 < b < \sqrt{\frac{3a(1-a)^2}{(4-3a)}}\right\}$. *There exist a shear flow* U_s *and* $K > 0$ *such that, for* $k \geq K$, *there exists a sequence* λ_k *with* $\lambda_k \sim -ik\mu$ *such that the system* (1.1.70) *displays* (λ_k, k) *instabilities.*

$\cdot$*System* (1.1.69)*: let* $\mu \in \Gamma_2 = \left\{a + ib \in \mathbb{C}, 0 < a < \frac{2}{3}, 0 < b < \sqrt{\frac{a(2-3a)}{(3)}}\right\}$. *For any* $\varepsilon > 0$, *there exist a shear flow* U_s *and* $K > 0$ *and* $\gamma_+ > 0$ *such that* $k \geq K$ *and* $(\sqrt{\nu}k)^{-1} \leq \gamma_+$ *there exists a sequence* $\lambda_{\nu,k}$ *with* $|\lambda_{\nu,k}/(k + i\mu)| \leq \varepsilon$ *such that the system* (1.1.69) *displays* $(\lambda_{\nu,k}, k)$ *instabilities.*

$\cdot$ *System*(1.1.69)*: let* $\mu \in \Gamma_3 = \left\{a + ib \in \mathbb{C}, 0 < a < 1, 0 < b < \sqrt{a(1-a)}\right\}$. *For any* $\gamma > 0$ *such that, for any* $\varepsilon > 0$, *there exist a shear flow* U_s *and* $K > 0$ *and* $\gamma_+ > \gamma > \gamma_-$ *such that* $k \geq K$ *and* $\gamma_- < (\sqrt{\nu}k)^{-1} \leq \gamma_+$, *there exists a sequence* $\lambda_{\nu,k}$ *with* $|\lambda_{\nu,k}/(k + i\mu)| \leq \varepsilon$ *such that the system* (1.1.69) *displays* $(\lambda_{\nu,k}, k)$ *instabilities.*

Using the paralinearization technique, Chen, Wang, and Zhang [12] in 2018 proved the local well-posedness of the Prandtl equation for monotonic data in anisotropic Sobolev space with an exponential weighted and low regularity and provided a new way for the zero viscosity limit problem of the Navier-Stokes equations with the non-slip boundary condition. They obtained the local well-posedness of Prandtl equations when weighted function μ is an exponential function in $H^{3,1}_\mu(\mathbb{R}^2_+) \cap H^{1,2}_\mu(\mathbb{R}^2_+)$. They studied the Prandtl equation in $\mathbb{R}_+ \times \mathbb{R}^2_+$:

$$\begin{cases} u_t + u\partial_x u + v\partial_y u - \partial_y^2 u + \partial_x p = 0, \\ \partial_x u + \partial_y v = 0, \\ (u, v)|_{y=0} = 0, \quad \lim\limits_{y\to+\infty} u = U, \\ u|_{t=0} = u_0. \end{cases} \tag{1.1.71}$$

For the simplicity, they considered in [12] the case of a uniform outflow $U(t, x) = 1$, which implies that p is a constant. Then the Prandtl equation (1.1.71) reduces to

$$\begin{cases} u_t + u\partial_x u + v\partial_y u - \partial_y^2 u = 0, \\ \partial_x u + \partial_y v = 0, \\ (u, v)|_{y=0} = 0, \quad \lim\limits_{y\to+\infty} u = 1, \\ u|_{t=0} = u_0. \end{cases} \tag{1.1.72}$$

They considered the data around a shear flow, i.e.,

$$\begin{cases} \partial_t u_s - \partial_y^2 u_s = 0, \\ u_s|_{y=0} = 0, \lim_{y\to+\infty} u = 1, \\ u_s|_{t=0} = u_{0s}(y). \end{cases}$$

Obviously, $(u^s(t, y), 0)$ is a shear flow solution of the Prandtl equation (1.1.71). Let $u(t, x, y) = u^s(t, y) + \tilde{u}(t, x, y)$. Then $\tilde{u}$ satisfies the following equations:

$$\begin{cases} \partial_t \tilde{u} + u\partial_x \tilde{u} + v\partial_y u - \partial_y^2 \tilde{u} = 0, \\ \partial_x \tilde{u} + \partial_y v = 0, \\ \tilde{u}|_{y=0} = 0, \lim_{y\to+\infty} u = 0, \\ \tilde{u}|_{t=0} = \tilde{u}_0. \end{cases}$$

Theorem 1.1.44 ([12]) *Let $\mu(y) = e^{\frac{y}{2}}$. Assume that $u_0^s - 1 \in H^2_{y,\mu}$, $\tilde{u_0} \in H_\mu^{3,1} \cap H_\mu^{1,2}$ with $u_0^s(0) = \tilde{u_0}|_{y=0} = 0$. Moreover, for some $c > 0$ and $k = 0, 1, 2$*

$$\partial_y u_0^s(y) \geq ce^{-y}, |\partial_y^k(u_0^s(y) - 1)| \leq c^{-1}e^{-y}, \text{ for } y \in [0, +\infty),$$
$$|\partial_y \tilde{u}_0^s(x, y)| + |\partial_x\partial_y \tilde{u}_0^s(x, y)| \leq \epsilon e^{-y} \text{ for } (x, y) \in \mathbb{R}_+^2,$$
$$\|\tilde{u}_0\|_{H_\mu{}^{3,1}} + \|\tilde{u}_0\|_{H_\mu{}^{1,2}} \leq \epsilon.$$

There exist $\epsilon_0 > 0$ and $T > 0$ such that for any $\epsilon \in (0, \epsilon_0)$, the Prandtl equation (1.1.71) has a unique solution in $[0, T]$, which satisfies

$$c_1 e^{-y} \leq \partial_y u(t, x, y) \leq C_1 e^{-y} \text{ for } (t, x, y) \in [0, T] \times \mathbb{R}_+^2.$$
$$|\partial_y^2 u(t, x, y)| + |\partial_x \partial_y u(t, x, y)| \leq C_1 e^{-y},$$
$$u - u^s \in L^\infty(0, T; H_\mu^{3,1} \cap H_\mu^{1,2}) \cap L^2(0, T; H_\mu^{1,3}).$$

for some $c_1 > 0$ and $C_1 > 0$. Here $H_\mu^{s,\sigma}$ is the weighted anisotropic Sobolev space.

When shear flow is non-monotonic, Chen, Wang, and Zhang [11] in 2018 proved the well-posedness of the linearized Prandtl equation in Gevrey class $2 - \theta$ for any $\theta > 0$ and constructed a class of solution with the growth like $e^{\sqrt{kt}}$ for the linearized Prandtl equation by energy method. The authors in [11] studied the Prandtl equation in $\mathbb{R}_+ \times \mathbb{R}_+^2$

$$\begin{cases} u_t + u\partial_x u + v\partial_y u - \partial_y^2 u + \partial_x p = 0, \\ \partial_x u + \partial_y v = 0, \\ (u, v)|_{y=0} = 0, \quad \lim\limits_{y\to+\infty} u = U, \\ u|_{t=0} = u_0. \end{cases} \tag{1.1.73}$$

where (u, v) denotes the tangential and normal velocity of the boundary layer flow and $(U(t,x), p(t,x))$ are the values on the boundary of the tangential velocity and pressure of the outflow, which satisfies Bernoulli's law:

$$\partial_t U + U\partial_x U + \partial_x p = 0.$$

The linearized Prandtl equation around $(u^s, 0)$ takes as follows:

$$\begin{cases} \partial_t u + u^s\partial_x u + v\partial_y u^s - \partial_y^2 u_s = 0, \\ u|_{y=0} = 0, \quad \lim\limits_{y\to+\infty} u = 0, \\ u|_{t=0} = u_0 \end{cases} \tag{1.1.74}$$

The main result of [11] is stated as follows.

Theorem 1.1.45 ([11]) *Let* $\theta \in (0, \frac{1}{2}]$. *Assume that* $e^{\langle D_x\rangle^{\frac{1}{2}+2\theta}} u_0 \in H_\mu^{\frac{1}{2},1}$ *with* $\partial_y^k u_0|_{y=0} = 0$ *for* $k = 0, 2$. *Then there exists* $T > 0$, *so that* (1.1.73) *has a unique solution* u *in* $[0, T]$. *In particular, we have*

$$u_\Phi \in L^\infty(0, T; H_\mu^{\frac{1}{4}+\theta,1}).$$

Here we denote

$$f_\Phi = \mathcal{F}^{-1}(e^{\Phi(t,\xi)}\hat{f}(\xi)),\ \Phi(t,\xi) = (1-\lambda t)\langle\xi\rangle^{\frac{1}{2}+2\theta},$$

and $H_\mu^{s,\sigma}$ *is the weighted Sobolev space with* $\mu = e^{\frac{y}{2}}$.

Yang and Zhu [144] in 2020 considered the 2D Prandtl equation with constant outer flow and monotonic data and proved that if the curvature of the velocity distribution (i.e., $\partial_y^2 u$) is bounded near the boundary, then the solution cannot develop the singularity. They studied the Prandtl equation in $\mathbb{R}_+ \times \mathbb{R}_+^2$

$$\begin{cases} \partial_t u + u\partial_x u + v\partial_y u - \partial_y^2 u + \partial_x p = 0, \\ \partial_x u + \partial_y v = 0, \\ u|_{y=0} = v|_{y=0} = 0, \qquad \lim\limits_{y\to+\infty} u(t,x,y) = U(t,x), \\ u|_{t=0} = u_0, \end{cases} \tag{1.1.75}$$

where (u, v) denotes the tangential and normal velocity of the boundary layer flow and $(U(t,x), p(t,x))$ is the values on the boundary of the tangential velocity and pressure of the outer flow, which satisfies the Bernoulli's law

$$\partial_t U + U\partial_x U + \partial_x p = 0.$$

They considered a class of monotonic data $u_0(x, y)$, which holds that there exist positive constants $c, C > 0$ such that

$$\partial_y u_0(x, y) \geq ce^{-y} \quad \text{for} \quad y \in \mathbb{R}_+,$$

and for $k = 0, 1, 2$,

$$\left|\partial_y^k \left(u_0(x, y) - u_0^s(y)\right)\right| + \left|\partial_x \partial_y u_0(x, y)\right| \leq Ce^{-y} \quad \text{for } (x, y) \in \mathbb{R}_+^2$$

where $u_0^s(y) \in C^2(\mathbb{R}_+)$ satisfies

$$u_0^s(0) = 0, \quad \left|\partial^k \left(u_0^s(y) - 1\right)\right| \leq Ce^{-y} \quad \text{for } y \in \mathbb{R}_+ \text{ and } k = 0, 1, 2.$$

Let $\omega(y)$ be a nonnegative function in $\mathbb{R}_+$. They introduced the weighted L^p norm:

$$\|f\|_{L^p_\omega} = \|\omega(y) f(x, y)\|_{L^p}, \quad \|f\|_{L^p_{y,\omega}} = \|\omega(y) f(y)\|_{L^p},$$

Let $k, \ell \in \mathbb{N}$, the weighted anisotropic Sobolev space $H^{k,\ell}_\omega$ consists of all functions $f \in L^2_\omega$ satisfying

$$\|f\|^2_{H^{k,\ell}_\omega} = \sum_{\alpha \leq k} \sum_{\beta \leq \ell} \left\|\partial_x^\alpha \partial_y^\beta f\right\|^2_{L^2_\omega} < +\infty.$$

They denoted by $H^\ell_{y,\omega}$ the weighted Sobolev space in $\mathbb{R}_+$, which consists of all functions $f \in L^2_{y,\omega}$ satisfying

$$\|f\|^2_{H^\ell_{y,\omega}} \stackrel{\text{def}}{=} \sum_{\beta \leq \ell} \left\|\partial_y^\beta f\right\|^2_{L^2_{y,\omega}} < +\infty.$$

When $\omega = 1$, they denoted $H^{k,\ell}_\omega$ by $H^{k,\ell}$, and $H^\ell_{y,\omega}$ by H^ℓ_y for the simplicity.

Theorem 1.1.46 ([144]) *Assume that the initial data satisfies the assumptions and* $\widetilde{u}_0 \in H^{3,1}_\mu \cap H^{1,2}_\mu$ *with* $\widetilde{u}_0(x, 0) = 0$. *There exists a constant* $T > 0$ *and a unique solution of the Prandtl equation in* $[0, T]$, *which satisfies*

$$c_1 e^{-y} \leq \partial_y u(t, x, y) \leq C_1 e^{-y} \quad \textit{for } (t, x, y) \in [0, T] \times \mathbb{R}^2_+,$$

$$\left|\partial_y^2 u(t, x, y)\right| + \left|\partial_x \partial_y u(t, x, y)\right| \leq C_1 e^{-y} \quad \textit{for } (t, x, y) \in [0, T] \times \mathbb{R}^2_+,$$

$\widetilde{u} \in L^\infty\left(0, T; H_\mu^{3,1} \cap H_\mu^{1,2}\right) \cap L^2\left(0, T; H_\mu^{1,3}\right)$ *for some* $c_1, C_1 > 0$. *Let* T^* *be the maximal existence time of the solution. If the solution satisfies the following conditions:*

$$ce^{-y} \le \partial_y u(t, x, y) \le Ce^{-y} \quad \text{for} \quad (t, x, y) \in \left[0, T^*\right) \times \mathbb{R}_+^2,$$
$$\sup_{t \in [0,T^*)} A(t) \le C,$$

for some constants $c, C > 0$, *then it can be extended after* $t = T^*$. *Here*

$$A(t) = \sup_{(s,x,y) \in [0,t] \times \mathbb{R}_+^2} e^y \left(\sum_{k=0}^{2} \left| \partial_y^k (u-1) \right| + (1+y)^{-1} \left(|\partial_x u| + \left| \partial_x \partial_y u \right| \right) \right),$$

and $H_\mu^{k,\ell}$ *is the weighted anisotropic Sobolev space.*

Theorem 1.1.47 ([144]) *Let* u *be a solution in* $\left[0, T^*\right)$ *constructed in the Theorem 1.1.46. Then there exists* c_0, C_0 *such that*

$$c_0 e^{-y} \le \partial_y u(t, x, y) \le C_0 e^{-y} \quad \text{for } (t, x, y) \in \left[0, T^*\right) \times \mathbb{R}_+^2.$$

Moreover, if

$$\left| \partial_y^2 u(t, x, y) \right| \le C \quad \text{for } (t, x) \in \left[0, T^*\right) \times \mathbb{R},\ y \in [0, \delta],$$

for some constants $C > 0$, $\delta > 0$, *then*

$$A(t) \le C_* \quad \text{for any } t \in \left[0, T^*\right),$$

where $C_* > 0$ *is a constant depending only on* C_0, c_0, T^*, δ.

Wang and Wang [135] in 2020 under the assumption that the initial velocity and outer flow velocity are analytic in the horizontal variable obtained the local well-posedness of the geophysical boundary layer problem by using energy method in the weighted Chemin-Lerner spaces. Moreover, when the initial velocity and outer flow velocity satisfy certain condition on a transversal plane, for any smooth solution decaying exponentially in the normal variable to the geophysical boundary layer problem, it is proved that its $W^{1,\infty}$-norm blows up in a finite time.

The authors in [135] considered the following initial boundary value problem in the domain $Q_T = \{0 < t < T, x \in \mathbb{R}, y > 0\}$,:

$$\begin{cases} \partial_t u + u\partial_x u + v\partial_y u - \int_y^{+\infty} (u - U) dy' - \partial_y^2 u = \partial_t U + U \partial_x U, \\ \partial_x u + \partial_y v = 0, \\ u|_{t=0} = u_0(x, y), \\ (u, v)|_{y=0} = (0, 0), \quad \lim_{y \to +\infty} u(t, x, y) = U(t, x), \end{cases} \tag{1.1.76}$$

where (u, v) is the velocity field and $U(t, x)$ is the tangential velocity of the outer flow.

Let $u^s(t, x, y) = U(t, x)\phi(t, y)$ and $w = u - u^s$. From (1.1.76), we know that w satisfies the following problem in $\{t > 0, x \in \mathbb{R}, y > 0\}$,:

$$\begin{cases} \partial_t w+(w+u^s)\,\partial_x w+w\partial_x u^s-\int_0^y \partial_x\,(w+u^s)\,dy'\partial_y\,(w+u^s)-\int_y^{+\infty} w dy'-\partial_y^2 w, \\ = (1-\phi)\,(\partial_t U+(1+\phi)U\partial_x U)-\int_y^{+\infty} U(1-\phi)dy' \\ w|_{y=0}=0, \ \lim\limits_{y\to+\infty} w=0, \\ w|_{t=0}=w_0(x, y) \triangleq u_0(x, y)-U(0, x)\,\mathrm{erfc}\left(\frac{y}{2}\right). \end{cases} \tag{1.1.77}$$

Introduce the following function spaces with parameters $s > 0, l \in \mathbb{N}_+$ and $p \in [1, +\infty)$:

(i) The space B^s is the set of functions $u \in \mathcal{S}'(\mathbb{R})$ such that

$$\|u\|_{B^s} := \sum_{k\in\mathbb{Z}} 2^{ks}\,\|\Delta_k u\|_{L^2(\mathbb{R})} < +\infty.$$

(ii) The space $B_\psi^{s,l}$, with a positive function $\psi(y)$, is the set of functions $u \in \mathcal{S}'\left(\mathbb{R}_+^2\right)$ such that

$$\|u\|_{B_\psi^{s,l}} := \sum_{j=0}^{l}\sum_{k\in\mathbb{Z}} 2^{ks}\left\|e^{\psi(y)}\Delta_k\partial_y^j u\right\|_{L^2(\mathbb{R}_x\times\mathbb{R}^+)} < +\infty.$$

(iii) The space $\tilde{L}_t^p\,(B^s)$ is defined as the completion of $C([0, t]; \mathcal{S}(\mathbb{R}))$ with the norm

$$\|u\|_{\tilde{L}_t^p(B^s)} := \sum_{k\in\mathbb{Z}} 2^{ks}\left(\int_0^t \|\Delta_k u\,(t', \cdot)\|_{L^2(\mathbb{R})}^p\,dt'\right)^{\frac{1}{p}}.$$

(iv) For any positive function $\psi\left(t', y\right)$ and nonnegative $f\left(t'\right) \in L_{loc}^1\left(\mathbb{R}^+\right)$, the space $\tilde{L}_{t,f}^p\left(B_\psi^{s,l}\right)$ is defined as the completion of $C\left([0, t]; \mathcal{S}\left(\mathbb{R}_+^2\right)\right)$ with the norm

$$\|u\|_{\tilde{L}_{t,f}^p\left(B_\psi^{s,l}\right)} := \sum_{j=0}^{l}\sum_{k\in\mathbb{Z}} 2^{ks}\left(\int_0^t f\left(t'\right)\left\|e^{\psi(t',y)}\Delta_k\partial_y^j u\left(t', \cdot\right)\right\|_{L^2(\mathbb{R}_x\times\mathbb{R}^+)}^p dt'\right)^{\frac{1}{p}}.$$

Denote $\tilde{L}^p_{t,1}\left(B^{s,l}_\psi\right)$ by $\tilde{L}^p_t\left(B^{s,l}_\psi\right)$ for simplicity when $f(t') \equiv 1$. The above notations can be properly modified to the case $p = +\infty$.

To obtain energy estimates for solutions, we introduce the weights

$$\psi(t, y) = \frac{1 + y^2}{16(1 + t)^\gamma} \quad \text{and} \quad \psi_0(y) = \frac{1 + y^2}{16},$$

with $\gamma \geq 2$, then there holds

$$\sqrt{-\left(\psi_t + 2\psi_y^2\right)} \geq (1 + y)\sqrt{\frac{\gamma - 1}{32(1 + t)^{\gamma+1}}}.$$

Theorem 1.1.48 ([135]) *For a given $T_0 > 0$, assume that the initial velocity $w_0(x, y)$ and the outer flow velocity $U(t, x)$ are analytic in $x \in \mathbb{R}$, and*

$$e^{\langle D\rangle} w_0 \in B^{\frac{1}{2},0}_{\psi_0},$$

and

$$e^{\langle D\rangle} U \in \tilde{L}^\infty_{T_0}\left(B^{\frac{1}{2}}\right) \cap \tilde{L}^\infty_{T_0}\left(B^1\right) \cap \tilde{L}^\infty_{T_0}\left(B^{\frac{3}{2}}\right), \quad e^{\langle D\rangle} U_t \in \tilde{L}^\infty_{T_0}\left(B^{\frac{1}{2}}\right).$$

Then, there exists $0 < T^* \leq T_0$ such that the problem has a unique solution $e^{\Phi(t,D)} w \in \tilde{L}^\infty_{T^*}\left(B^{\frac{1}{2},0}_\psi\right)$ and $\Phi(t, D)$ is the operator associated with the symbol $\Phi(t, \xi)$. D denotes the Fourier multiplier.

Theorem 1.1.49 ([135]) *For a given $T > 0$, assume that the initial data and outer flow satisfy*

$$u_0(0, y) = U(t, 0) = 0,\, U_x(t, 0) \geq 0,\, u_{0x}(0, y) \leq U_x(0, 0),$$

and there is $M \geq 0$ depending on T and $\|U_x(\cdot, 0)\|_{L^\infty([0,T])}$ such that

$$\int_0^{+\infty} \rho(y)\,(U_x(0, 0) - u_{0x}(0, y))\,dy > M,$$

for a weight function $\rho(y)$ if the smooth solution $u \in C^3(Q_T)$ of the first equation (1.1.76) *satisfies*

$$\sup_{0 \leq t \leq T} \int_0^{+\infty} (u(t, 0, y))^2 e^{2\mu y}\,dy < +\infty,$$

and

$$\lim_{y\to+\infty}\sup_{0\le t\le T}\left(|u_x(t,0,y)-U_x(t,0)|\,e^{\mu y}\right)=0,$$

for some $\mu>0$, *then the* $W^{1,\infty}$- *norm of* u *blows up in a finite time.*

The authors in [36, 65, 98] studied the local well-posedness of the Prandtl equations by the energy method. In the 2D setting, it is convenient to study the evolution of the vorticity $w=\partial_y u$. The nonlinear transport equation is as follows:

$$\partial_t w+u\partial_x w+v\partial_y w=0, \tag{1.1.78}$$

where one may compute (u,v) from w via

$$u=-\partial_y\mathcal{A}(w)\text{ and }v=\partial_x\mathcal{A}(w), \tag{1.1.79}$$

where the stream function $\mathcal{A}(w)$ solves

$$-\partial_{yy}\mathcal{A}(w)=w$$

with the boundary condition $\mathcal{A}(w)|_{y=0,1}=0$.

Kukavica, Masmoudi, Vicol, and Wong [65] in 2014 found a new class of data for which the Prandtl boundary layer equations and the hydrostatic Euler equations are locally in time well-posed. They showed that the local existence and uniqueness of solution hold.

Theorem 1.1.50 ([65]) *Let* $I\subset\mathbb{R}$ *be an open interval,* $s\ge 4$ *be an even integer,* $\gamma\ge 1$, $\sigma>\gamma+\frac{1}{2}$, *and* $\delta\in(0,\frac{1}{4})$. *Assume that the initial velocity obeys* $u_0-U_0\in H^{s,\gamma-1}(I)$, *the initial vorticity satisfies* $w_0\in H^{s,\gamma}_{\sigma,2\delta}(I)$, *and the outer Euler flow* U *is sufficiently smooth (e.g., the following equation will suffice):*

$$\|U\|_{W^{s+9,\infty}_{t,x}}=\sup_t\sum_{0\le 2l\le s+9}\|\partial_t^l U\|_{W^{s-2l+9,\infty}(I)}<\infty.$$

Then there exists a $T>0$ *and a smooth solution* u *of (1.1.58) such that the velocity obeys* $u-U\in L^\infty([0,T];H^{s,\gamma-1}(I))\cap C_w([0,T];H^s_{loc}(I))$ *and the vorticity obeys* $w\in L^\infty([0,T];H^{s,\gamma}_{\sigma,\delta}(I))\cap C_w([0,T],H^s_{loc}(I))$. *When* $I=\mathbb{R}$ *or* $I=[0,T]$, *the solution constructed above is the unique solution in this regularity class. When* $\partial I\ne\emptyset$, *e.g., if* $I=(a,b)$, *there exists a positive* $M\le\infty$ *such that* $u(t)$ *is the unique solution with* $u(t)-U(t)\in H^{s,\gamma-1}(I_t)$ *and* $w(t)\in H^{s,\gamma}_{\sigma,\delta}(I_t)$, *where* $I_t=\{x\in\mathbb{R}:(x-Mt,x+Mt)\subseteq I\}$.

Theorem 1.1.51 ([65]) *Let* $\sigma,\tau_0>0$ *and* $s\ge 4$. *Assume that the initial vorticity satisfies a uniform Rayleigh condition on* $\mathfrak{D}_m$, *we have* $\partial_y u_0\in H^s_{2\sigma,loc}(\mathfrak{D}_m)$, *and* $\partial_y u_0$ *is uniformly real-analytic on* $\mathfrak{D}_a$ *with radius of analyticity at least* τ_0, *i.e.,* $u_0\in$

$Y_{\tau_0}(\mathfrak{D}_a)$. *Also assume that the initial velocity* u_0 *satisfies the compatibility condition. Then there exist* $T > 0$ *and a unique smooth solution* $\partial_y u \in C(0, T; X_{\tau_0/2}(\mathfrak{D}_a)) \cap H^s_{\sigma,loc}(\mathfrak{D}_m)$ *of (1.1.78)–*(1.1.79) *on* $[0, T]$, *which has zero vertical mean.*

Here

$$H^{s,\gamma}_{\sigma,\delta}(I) = \Big\{ w \in H^{s,\gamma}(I) : (1+y)^\sigma w(x,y) \geq \delta, \sum_{|\alpha|\leq 2} |(1+y)^{\sigma+\alpha_2} D^\alpha w|$$

$$\leq \delta^{-2} \text{ for all } (x,y) \in I \times \mathbb{R}_+ \}$$

$$\mathfrak{D}_m = (-\infty, 1) \times (0, 1), \quad \mathfrak{D}_a = (0, \infty) \times (0, 1).$$

Masmoudi and Wong [98] in 2015 proved local existence and uniqueness for the 2D Prandtl system in weighted Sobolev spaces under the Oleinik's monotonicity assumption. This new energy estimate is based on a cancellation property which is valid under the monotonicity assumption, but not using the Crocco transform.

They considered the Prandtl equation (1.1.58) in a periodic domain $\mathbb{T} \times \mathbb{R}_+ = \{(x,y) | x \in \mathbb{R}/\mathbb{Z}, y \in [0, +\infty)\}$; $U(t,x)$ satisfies Bernoulli's law (1.0.3). In [98], let $w = \partial_y u$, and denote

$$H^{s,\gamma}_{\sigma,\delta} := \big\{ w : \mathbb{T} \times \mathbb{R}_+ \to \mathbb{R}, \|w\|_{H^{s,\gamma}} \leq +\infty, (1+y)^\sigma |w|$$

$$\geq \delta, \sum_{|\alpha|\leq 2} |(1+y)^{\sigma+\alpha_2} D^\alpha w|^2 \leq \frac{1}{\delta^2} \big\}.$$

Theorem 1.1.52 ([98] Local $H^{s,\gamma}_{\sigma,\delta}$ Existence and Uniqueness to the Prandtl Equations) *Let* $s \geq 4$ *be an even integer,* $\gamma \geq 1, \sigma > \gamma + \frac{1}{2}$ *and* $\delta \in (0, \frac{1}{2})$. *For simplicity, we suppose that the outer flow* U *satisfies*

$$\sup_t \| |U| \|_{s+9,\infty} := \sup_t \sup_{l=0}^{[\frac{s+9}{2}]} \|\partial_t^l U\|_{W^{s-2l+9,\infty}(\mathbb{T})} < +\infty.$$

Assume that $u_0 - U \in H^{s,\gamma-1}$ *and the initial vorticity* $w_0 = \partial_y u_0 \in H^{s,\gamma}_{\sigma,2\delta}$. *In addition, when* $s = 4$, *we further assume that* $\delta > 0$ *is chosen small enough such that*

$$\|w_0\|_{H^{s,\gamma}_g} \leq C\delta^{-1},$$

where the norm $\| \cdot \|_{H^{s,\gamma}_g}$ *will be defined by (1.1.80) and* C *is a universal constant. Then there exist a time* $T := T(s, \gamma, \sigma, \delta, \|w_0\|_{H^{s,\gamma}}, U) > 0$ *and a unique classical solution* (u, v) *to the Prandtl equations (1.1.58) such that* $u - U \in L^\infty([0,T], H^{s,\gamma-1}) \cap C([0,T]; H^s - w)$ *and the vorticity* $w := \partial_y u \in$

$L^\infty([0,T]; H^{s,\gamma}_{\sigma,\delta}) \cap C([0,T]; H^s - w)$, where $H^s - w$ is the space H^s endowed with its weak topology.

Here

$$\|w\|^2_{H^{s,\gamma}_g} := \|(1+y)^\gamma g_s\|^2_{L^2} + \sum_{\substack{|\alpha|\le s\\ \alpha_1\le s-1}} \|(1+y)^{\gamma+\alpha_2} D^\alpha w\|^2_{L^2}. \tag{1.1.80}$$

Gao, Huang, and Yao [36] in 2018 investigated the local-in-time well-posedness for the two-dimensional Prandtl equations in weighted Sobolev spaces under the Oleinik monotonicity condition. Throughout this paper, they are concerned with the two-dimensional Prandtl equations in a periodic domain $\mathbb{T}\times\mathbb{R}_+$:

$$\begin{cases} u_t + u\partial_x u + v\partial_y u - \partial_y^2 u + \partial_x p = 0,\\ \partial_x u + \partial_y v = 0,\\ (u,v)|_{y=0} = 0, \quad \lim\limits_{y\to+\infty} u = U,\\ u|_{t=0} = u_0. \end{cases} \tag{1.1.81}$$

Define the weighted Sobolev space

$$H^{s,\gamma}_{\sigma,\delta} := \left\{ w : \|w\|_{H^{s,\gamma}} < +\infty, (1+y)^\sigma w \ge \delta, \sum_{|\alpha|\le 2} |(1+y)^{\sigma+\alpha_2} D^\alpha w|^2 \le \frac{1}{\delta^2} \right\},$$

where $s\ge 4, \gamma\ge 1, \sigma > \gamma+\frac{1}{2}, \delta\in(0,\frac{1}{2})$.

Theorem 1.1.53 ([36]) *Let $s\ge 4$ be an even integer, $\gamma\ge 1, \sigma>\gamma+\frac{1}{2}, \delta\in(0,\frac{1}{2})$. Suppose the outer flow U satisfies*

$$M_U := \sup_t \sum_{t=0}^{\frac{s+1}{2}} \|\partial_t^l U\|_{W^{s-2l+2}(\mathbb{R})} < +\infty.$$

Assume that the initial tangential velocity, $u_0 - U|_{t=0} \in H^{s,\gamma-1}$ and the initial vorticity $w_0 := \partial_y u_0 \in H^{s,\gamma}_{\sigma,2\delta}$. Then there exist a time $T = T(s,\gamma,\delta,\delta_0,M_U) > 0$ and a unique classical solution (u,v) to the Prandtl equation (1.1.81) *such that*

$$\sup_{0\le t\le T} \|w\|^2_{H^{s,\gamma}_{\sigma,\delta}} \le C_{s,\gamma,\delta}(1 + \|w_0\|^8_{H^{s,\gamma}_{\sigma,\delta}} + M_U^4) < +\infty.$$

Theorem 1.1.54 ([36]) *Let $s\ge 4$ be an even integer, $\gamma\ge 1, \sigma>\gamma+\frac{1}{2}, \delta\in(0,\frac{1}{2})$, and $\varepsilon\in(0,1)$, the smooth solution $(u^\varepsilon, v^\varepsilon, w^\varepsilon)$ to the regularized Prandtl equations* (1.1.82)–(1.1.84). *Under the assumptions of the above theorem, there* *exists a time $T_a = T_a(s, \gamma, \delta, \delta_0, M_U) > 0$, independent of ε; such the following estimates hold on:*

$$\sup_{0\le t\le T} \|w^\varepsilon\|^2_{H^{s,\gamma}_{\sigma,\delta}} \le C_{s,\gamma,\delta}(1 + \|w^\varepsilon_0\|^8_{H^{s,\gamma}_{\sigma,\delta}} + M_U^4) < +\infty.$$

Denote vorticity $w^\varepsilon = \partial_y u^\varepsilon$, using the regularized Prandtl equations, we find that this vorticity satisfies the following evolution equations:

$$\begin{cases} w_t^\varepsilon + u^\varepsilon \partial_x w^\varepsilon + v \partial_y w^\varepsilon - \partial_y^2 w^\varepsilon = 0, \\ w^\varepsilon|_{t=0} = w_0 = \partial_y u_0, \\ \partial_y w^\varepsilon = \partial_x p, \end{cases} \tag{1.1.82}$$

where the velocity field $(u^\varepsilon, v^\varepsilon)$ is given by

$$u^\varepsilon(t, x, y) = U(t, x) - \int_y^{+\infty} w^\varepsilon dy \tag{1.1.83}$$

and

$$v^\varepsilon(t, x, y) = -\int_0^y \partial_x u^\varepsilon dy. \tag{1.1.84}$$

The authors in [19, 42, 43, 66, 72] studied the local well-posedness of the Prandtl equations in Gevrey space.

Kukavica and Vicol [66] in 2013 addressed the local well-posedness of the Prandtl boundary layer equations. Using a new change of variables, the authors in [66] allowed for more general data than previously considered, that is, they required the matching at the top of the boundary layer to be at a polynomial rather than exponential rate. The proof is direct, via analytic energy estimates in the tangential variables.

The foundations for the boundary layer theory were laid by Prandtl, who made the ansatz $u^{NS}(x, \tilde{y}, t) = (u(x, \tilde{y}/\sqrt{\nu}, t), \sqrt{\nu} w(x, \tilde{y}/\sqrt{\nu}, t))$. Inserting this velocity field in the Navier-Stokes equations and sending the kinematic viscosity ν to zero, one formally obtains the Prandtl boundary layer equations for the unknown velocity field $(u, \sqrt{\nu} w)$

$$\begin{cases} \partial_t u - \partial_{YY} u + u \partial_x u + w \partial_Y u + \partial_x P = 0, \\ \partial_x u + \partial_Y w = 0, \\ \partial_Y P = 0, \end{cases} \tag{1.1.85}$$

in $\mathbb{H} = \{(x, Y) \in \mathbb{R}^2 : Y > 0\}$, where $Y = \tilde{y}/\sqrt{\nu}$ is the normal variable in the boundary layer. For simplicity of the presentation, they in [66] considered the two-dimensional setting, but all the methods and results presented here extend to the three-dimensional case as well. The system (1.1.85) is supplemented with the no-slip and the no-influx boundary conditions

$$\begin{cases} u(x, Y, t)|_{Y=0} = 0, \\ w(x, Y, t)|_{Y=0} = 0, \end{cases}$$

for $t > 0$, and the matching conditions with the Euler flow as $Y \to \infty$, via the Bernoulli's law

$$\begin{cases} \lim_{Y\to\infty} u(x, Y, t) = U(x, t), \\ \partial_x P(x, t) = -(\partial_t + U(x, t)\partial_x) U(x, t), \end{cases}$$

for $x \in \mathbb{R}^2, t > 0$, where $U(x, t)$ is the trace at $\tilde{y} = 0$ of the tangential component of the Euler flow u^E.

Under the change of variables, the Prandtl system reads

$$\partial_t v - A^2\partial_{yy}v + N(v) + L(v) = F$$

where

$$\begin{aligned} N(v) &= v\partial_x v - \partial_x W(v)\partial_y v + \partial_x a W(v)\partial_y v, \quad W(v)(x, y) = \int_0^y v(x, \zeta)d\zeta, \\ L(v) &= \partial_x W(v)\partial_y \phi U + \partial_x v(1-\phi)U + \partial_y v\,(\Phi\partial_x U - \partial_x a\Phi U) \\ &\quad - W(v)\partial_x a\partial_y\phi U + v(1-\phi)\partial_x U, \\ F &= \left(\phi(1-\phi) + \Phi\partial_y\phi\right) U\partial_x U - \partial_x a\partial_y\phi\Phi U^2 - A^2\partial_{yy}\phi U - \phi\partial_x P. \end{aligned}$$

The system (1.1.85) is supplemented with the boundary conditions

$$\begin{aligned} &v(x, y, t)|_{y=0} = u(x, Y, t)|_{Y=0} - (1-\phi(0))U(x, t) = 0, \\ &\lim_{y\to\infty} v(x, y, t) = \lim_{Y\to\infty} u(x, Y, t) - U(x, t) = 0, \end{aligned}$$

for all $(x, t) \in \mathbb{R} \times [0, \infty)$, and initial condition

$$v(x, y, t)|_{t=0} = v_0(x, y) = u_0(x, Y) - (1-\phi(y))U_0(x).$$

The authors of [66] considered the x -analytic norm with y -weight given by

$$\rho(y) = \langle y\rangle^\alpha$$

for some $\alpha > 0$ to be fixed later. Namely, for a function $V(x, y)$ and a number $\tau_0 > 0$, denote

$$\|V\|^2_{X_{\tau_0}} = \sum_{m\geq 0} \left\|\rho(y)\partial_x^m V(x, y)\right\|^2_{L^2(\mathbb{H})} \tau_0^{2m} M_m^2,$$

where the analytic weights M_m are defined as

$$M_m = \frac{(m+1)^r}{m!},$$

for some $r > 0$ to be determined. If $v(x, y, t)$ and $\tau(t)$ have t -dependence, similarly denote

$$\|v(t)\|^2_{X_{\tau(t)}} = \sum_{m\geq 0} \left\|\rho(y)\partial_x^m v(x, y, t)\right\|^2_{L^2(\mathbb{H})} \tau(t)^{2m} M_m^2.$$

If the t dependence is clear from the context, we will omit it. Since the weight $\rho(y)$ does not depend on x, the analytic norm may also be written as

$$\|v\|^2_{X_\tau} = \sum_{m\geq 0} \|\rho v\|^2_{\dot{H}_x^m} \tau^{2m} M_m^2.$$

For a positive number $\tau > 0$, we write $v \in X_\tau$ if $\|v\|_{X_\tau} < \infty$.

Theorem 1.1.55 ([66]) *Fix real numbers $\alpha > 1/2, \theta > \alpha + 1/2$, and $r > 1$. Assume that the initial data for the underlying Euler flow is uniformly real analytic, with radius of analyticity at least $\tau_E > 0$ and analytic norm bounded by $G_E > 0$. There exists $\tau_0 = \tau_0(r, \tau_E, G_E) > 0$ such that for all $v_0 \in X_{\tau_0}$ there is $T_* = T_*\left(r, \alpha, \theta, \tau_E, G_E, \tau_0, \|v_0\|_{X_{\tau_0}}\right) > 0$ such that the initial value problem has a unique real-analytic solution on $[0, T_*]$.*

Gérard-Varet and Masmoudi [43] in 2015 showed that the Prandtl system (1.1.58) is actually locally well-posed for data that belong to the Gevrey class 7/4 in the horizontal variable x. Their result improved the classical local well-posedness result for data that are analytic in x (that is Gevrey class 1).

In [43], let $m \leq 1$, they defined the Gevrey space $G^m(\mathbb{T})$ which is the set of functions f satisfying as follows: there exist $C, \tau > 0$ such that

$$|f^j(x)| \leq C\tau^{-j}(j!)^m, \quad \forall j \in \mathbb{N}, \quad x \in \mathbb{T},$$

$G^m(\mathbb{T}) = \cup_{\tau>0} G^m_\tau(\mathbb{T})$, where

$$G^m_\tau(\mathbb{T}) := \left\{ f \in C^\Omega(\mathbb{T}), \sup_j \tau^j (j!)^{-m} (j+1)^{10} \|f^{(j)}\|_{L^2} < +\infty \right\}$$

is a Banach space. For $s \in \mathbb{N}$, $\gamma \geq 0$, define the spaces

$$H_\gamma^s := \left\{ g \in H^s(\mathbb{R}_+),\ \ (1+y)^{\gamma+k} g^{(k)} \leq L^2(\mathbb{R}_+), k = 0, \dots s \right\}, \|g\|_{H_\gamma^s}$$

$$:= \sum_{k=0}^{s} \|(1+y)^{\gamma+k} g^{(k)}\|_{L^2}^2.$$

They assumed the following assumptions:

(H) $w_0(x,y) = 0$ if $y = a_0(x)$, for some curve $a_0 > 0$, with $\partial_y w_0(x, a_0(x)) > 0, \forall x$.

(H') there exist $\sigma, \delta > 0$ such that for all $y > 3$, for all x, for all $\alpha \in \mathbb{N}^1$, $|\alpha| \leq 2$,

$$|w_0(x,y)| \geq \frac{2\delta}{(1+y)^\sigma}, \quad |\partial^\alpha w_0(x,y)| \leq \frac{1}{2\delta(1+y)^{\sigma+\alpha_2}}.$$

Theorem 1.1.56 ([43]) *Let $\tau_0 > 0, s > 8$ even, $\gamma \geq 1, \sigma \geq \gamma + \frac{1}{2}, \delta > 0$. Let U be a constant and u_0 satisfy*

$$u_0 \in G_{\tau_0}^{7/4}(\mathbb{T}, H_{\gamma-1}^{s+1}), \quad w_0 := \partial_y u_0 \in G_{\tau_0}^{7/4}(\mathbb{T}, H_\gamma^s).$$

Assume the compatibility condition: $u_0|_{y=0} = 0$, and assumptions (H), (H') above hold. Then there exists $T > 0, 0 < \tau < \tau_0$ and a unique solution

$$u \in L^\infty(0,T; G_\tau^{7/4}(\mathbb{T}, H_{\gamma-1}^{s+1}), w \in L^\infty(0,T; G_\tau^{7/4}(\mathbb{T}, H_\gamma^s),$$

of the Prandtl equation (1.1.58) with initial data u_0.

Gérard-Varet, Maekawa, and Masmoudi [42] in 2016 investigated the stability of boundary layer solutions of the two-dimensional incompressible Navier-Stokes equation (1.0.1). They considered shear flow solutions of Prandtl type :

$$u^\nu(t,x,y) = (U^E(t,y) + U^{BL}(t, \frac{y}{\sqrt{\nu}}), 0), \quad 0 \leq \nu \ll 1.$$

They showed that if U^{BL} is monotonic, then u^ν is stable over some time interval $(0,T)$, T independent of ν, under perturbations with Gevrey regularity in x and Sobolev regularity in y. They improved the classical stability results of Sammartino and Caflisch [121, 122] in analytic class (both in x and y), where $u^\nu(t,x,y)$ is the solution of (1.0.1), (U^E, V^E) is the solution of (1.0.4), (U^{BL}, V^{BL}) is a corrector localized near the boundary, and (U^P, V^P) is the solution of Prandtl equations;

they emphasized that the condition is always satisfied when U^P is the solution to the problem

$$\begin{cases} \partial_t U^P - \partial_Y^2 U^P = 0, \quad t > 0, Y > 0, \\ U^P|_{Y=0} = 0, \quad \lim_{Y\to\infty} U^P = U^E|_{y=0}, \quad t \geq 0, \\ U^P|_{t=0} = U. \end{cases} \tag{1.1.86}$$

They assumed that

$$\begin{aligned} &(i)\ U|_{Y=0} = 0,\ \lim_{Y\to\infty} U = U^E|_{y=0},\ \ U^E, U \in BC^2(\mathbb{R}_+),\\ &(ii)\ \|U\| := \sum_{k=0,1,2} \sup_{Y\geq 0}(1+Y)^k|\partial_Y^k U(Y)| < \infty,\\ &(iii)\ \text{For each } \sigma \in (0,1] \text{ there exists } M_\sigma > 0 \text{ such that } \ -M_\sigma \partial_Y^2 U \geq (\partial_Y U)^2\\ &\text{for } Y \geq \sigma, \end{aligned} \tag{1.1.87}$$

and assumed

$$\begin{aligned} &(i)\ U|_{Y=0} = 0,\ \lim_{Y\to\infty} U = U^E|_{y=0},\ \ U^E, U \in BC^2(\mathbb{R}_+),\\ &(ii)\ \|U\| := \sum_{k=0,1,2} \sup_{Y\geq 0}(1+Y)^k|\partial_Y^k U(Y)| < \infty,\\ &(iii)\ \text{There exists an } M > 0 \text{ such that } \ -M\partial_Y^2 U \geq (\partial_Y U)^2 \text{ for } Y \geq 0. \end{aligned} \tag{1.1.88}$$

To state our main theorem, we need to introduce a few notations. Let

$$(\mathcal{P}_n f)(y) = f_n(y)e^{inx},\ f_n(y) = \frac{1}{2\pi}\int_0^{2\pi} f(x,y)e^{-inx}dx,\ n \in \mathbb{Z},$$

the projection on the Fourier mode n in x. We then introduce, for $\gamma \in (0,1]$, $d \geq 0$ and $K \geq 0$, the Banach space $X_{d,\gamma,K}$ as

$$X_{d,\gamma,K} = \{f \in L^2_\sigma(\Omega) | \|f\|_{X_{d,\gamma,K}} = \sup_{n\in\mathbb{Z}}(1+|n|^d)e^{K\theta_{\gamma,n}}\|\mathcal{P}_n f\|_{L^2(\Omega)} < \infty\}, \tag{1.1.89}$$

$$\theta_{\gamma,n} = |n|^\gamma(1+(1-\gamma)log(1+|n|)).$$

Theorem 1.1.57 ([42]) *Assume that (1.1.86) holds for some $T > 0$. For any $\gamma \in [\frac{7}{9}, 1)$, $d > \frac{9}{2} - \frac{7}{2}\gamma$, and $K > 0$, there exist $C, T', K' > 0$ such that the following statement holds for any sufficiently small $\nu > 0$. If $\|a\|_{X_{d,\gamma,K}} \leq \nu^{\frac{1}{2}+\beta}$ with $\beta =$*

$\frac{2(1-\gamma)}{\gamma}$, *then the system (1.0.1) admits a unique solution* $u \in C([0, T']; L^2_\sigma(\Omega)) \cap L^2([0, T']; W_0^{1,2}(\Omega))$ *satisfying the estimate*

$$\sup_{0<t\le T'} (\|u(t)\|_{X_{d,\gamma,K}} + (\nu t)^{\frac{1}{4}} \|u(t)\|_{L^\infty(\Omega)} + (\nu t)^{\frac{1}{2}} \|\nabla u(t)\|_{L^2(\Omega)}) \le C\|a\|_{X_{d,\gamma,K}}.$$

If $U^E(t, y) = U^E(y)$ *and* $U^P(t, Y) = U(Y)$ *are steady and satisfy (1.1.87) instead of (1.1.86), the above result holds for any* $\gamma \in [\frac{5}{7}, 1]$.

If $U^E(t, y) = U^E(y)$ *and* $U^P(t, Y) = U(Y)$ *are steady and satisfy (1.1.88) instead of (1.1.86), the above result holds for any* $\gamma \in [\frac{2}{3}, 1]$.

Li, Wu, and Xu [72] in 2016 studied the Gevrey smoothing effects of the local solution for the Prandtl boundary layer equation (1.1.58) with U being a constant.

Theorem 1.1.58 ([72]) *Let* $u(t, x, y)$ *be a classical local in time solution to Prandtl equation (1.1.58) which* U *is a constant on* $[0, T]$ *with the properties subsequently listed below:*

(i) There exist two constants $C_* > 1, \sigma > \frac{1}{2}$ *such that for any* $(t, x, y) \in [0, T] \times \mathbb{R}^2_+$,

$$C_*\langle y\rangle^{-\sigma} \le u_y(t, x, y) \le C_*\langle y\rangle^{-\sigma},$$
$$|u_{yy}(t, x, y)| + |u_{yyy}(t, x, y)| \le C_*\langle y\rangle^{-\sigma},$$

where $\langle y\rangle = (1 + |y|^2)^{\frac{1}{2}}$.

(ii) There exist constants $c > 0$, $C_0 > 0$ *and integer* $N_0 \ge 7$ *such that*

$$\|e^{2cy}u_x\|_{L^\infty([0,T];H^{N_0})} + \|e^{2cy}u_{xy}\|_{L^\infty([0,T];H^{N_0})} \le C_0.$$

Then for any $0 < T_1 < T$, *there exists a constant* $L > 0$, *such that for any* $0 < t \in (0, T_1]$,

$$\forall m > 1 + N_0,\ \|e^{\bar{c}y}\partial_x^m u\|_{L^2(\mathbb{R}^2_+)} \le t^{-3(m-N_0-1)L^m(m!)^{3(1+\sigma)}},$$

where $0 < \bar{c} < c$. *The constant* L *depends only on* $C_0, T_1, C_*, c, \bar{c}$, *and* σ. *Therefore, the solution* u *belongs to the Gevrey class of index* $3(1+\sigma)$ *with respect to* $x \in \mathbb{R}$ *for any* $0 < t \le T_1$.

Dietert and Gérard-Varet [19] in 2019 showed the local in time well-posedness of the following Prandtl equations for data with Gevrey 2 regularity in x and Sobolev regularity in y:

$$\begin{cases} \partial_t U^P + U^P\partial_x U^P + V^P\partial_y U^P - \partial_y^2 U^P = \partial_t U^E + U^E\partial_x U^E, \\ \partial_x U^P + \partial_y V^P = 0, \end{cases} \tag{1.1.90}$$

in the domain $\Omega = \mathbb{T} \times \mathbb{R}_+$, completed with boundary conditions

$$U^P\Big|_{y=0} = V^P\Big|_{y=0} = 0, \qquad \lim_{y\to+\infty} U^P = U^E. \tag{1.1.91}$$

Let $\gamma \geq 1$, $\tau > 0$, $r \in \mathbb{R}$. For functions $f = f(x)$ of one variable, the authors in [19] defined the Gevrey norm

$$|f|^2_{\gamma,\tau,r} = \sum_{j\in\mathbb{N}} \left(\frac{\tau^{j+1}(j+1)^r}{(j!)^\gamma}\right)^2 \left\|f^{(j)}\right\|^2_{L^2(\mathbb{T})},$$

and for functions $f = f(x, y)$ of two variables, the norm

$$\|f\|^2_{\gamma,\tau,r} = \sum_{j\in\mathbb{N}} \left(\frac{\tau^{j+1}(j+1)^r}{(j!)^\gamma}\right)^2 \left\|\partial_x^j f\right\|^2_j,$$

where $\|\cdot\|_j$, $j \geq 0$, denotes a family of weighted L^2 norms. Namely,

$$\|f\|_j^2 = \int_{\mathbb{T}\times\mathbb{R}^+} |f(x,y)|^2 \rho_j(y)\mathrm{d}x\,\mathrm{d}y,$$

where ρ_j, $j \geq 0$, is the family of weights given by

$$\rho_0(y) = (1+y)^{2m}, \quad \rho_j(y) = \frac{\rho_{j-1}(y)}{\left(1+\frac{y}{j^\alpha}\right)^2} = \rho_0(y)\prod_{k=1}^{j}\left(1+\frac{y}{k^\alpha}\right)^{-2}, \quad j \geq 1,$$

for fixed constants $\alpha \geq 0$ and $m \geq 0$ chosen later (m large enough and α matching the constraints found from the estimates). The need for this family of weights will be clarified later. Let us note that locally in y, this family of norms is comparable to more classical families such as

$$|||f|||^2_{\gamma,\tau,r} = \sum_{j\in\mathbb{N}} \left(\frac{\tau^{j+1}(j+1)^r}{(j!)^\gamma}\right)^2 \left\|\partial_x^j f\right\|^2_{L^2}.$$

For instance, for functions f which are zero for $|y| \geq M$, one has

$$\|f\|_{\gamma,\tau,r} \leq C_M |||f|||_{\gamma,\tau,r}, \quad \| |f| \|_{\gamma,\tau,r} \leq C_{M,\tau'} \|f\|_{\gamma,\tau',r} \text{ for any } \tau' > \tau.$$

Theorem 1.1.59 ([19]) *There exists m and α such that for all $0 < \tau_1 < \tau_0, r \in \mathbb{R}$, for all $T_0 > 0$, for all U^E satisfying*

$$\begin{aligned}&\sup_{[0,T_0]} \left|\partial_t U^E\right|_{2,\tau_0,r} + \left|U^E\right|_{2,\tau_0,r} < +\infty,\\ &\sup_{[0,T_0]} \max_{l=0,\dots,3} \left\|\partial_t^l \left(\partial_t + U^E \partial_x\right) U^E\right\|_{H^{6-2l}(\mathbb{T})} < +\infty,\end{aligned}$$

for all U_{in}^P satisfying

$$\begin{aligned}\left\|U_{in}^P - U^E|_{t=0}\right\|_{2,\tau_0,r} < +\infty, \quad \left\|(1+y)\partial_y U_{in}^P\right\|_{2,\tau_0,r} < +\infty,\\ \left\|(1+y)^{m+6}\partial_y U_{in}^P\right\|_{H^6(\mathbb{T}\times\mathbb{R}_+)} < +\infty,\end{aligned}$$

and under usual compatibility conditions (see the last remark below), there exists $0 < T \le T_0$ and a unique solution U^P of (1.1.90)–(1.1.91) over $(0, T)$ with initial datum U_{in}^P that satisfies

$$\begin{aligned}&\sup_{t\in[0,T]} \left\|U^P(t) - U^E(t)\right\|_{2,\tau_1,r}^2 + \sup_{t\in[0,T]} \left\|(1+y)\partial_y U^P(t)\right\|_{2,\tau_1,r}^2\\ &+ \int_0^T \left\|(1+y)\partial_y^2 U^P(t)\right\|_{2,\tau_1,r}^2 \mathrm{d}t < +\infty.\end{aligned}$$

The authors in [55, 95, 143] studied the local well-posedness of the Prandtl equations around others.

Hong and Hunter [55] in 2003 used the method of characteristics to prove the short-time existence of smooth solutions of the unsteady inviscid Prandtl equations and presented a simple explicit solution that forms a singularity in a finite time.

Lombardo, Cannone, and Sammartino [95] in 2003 considered the mild solutions of the following Prandtl equations (1.1.92)–(1.1.93) on the half space and proved the short time existence and the uniqueness of the solution in the proper function space under the analyticity of the tangential variable by the abstract Cauchy-Kovalevskaya theorem.

$$\begin{cases}(\partial_t - \partial_{YY})u^P + u^P\partial_x u^P + v^P\partial_Y u^P + \partial_x p^P = 0,\\ \partial_Y p^P = 0,\\ \partial_x u^P + \partial_Y v^P = 0,\\ u^P(x, Y = 0, t) = v^P(x, Y = 0, t) = 0,\\ u^P(x, Y \to \infty, t) \to U(x, t),\\ p^P(x, Y \to \infty, t) \to p^E(x, y = 0, t),\\ u^P(x, Y, t = 0) = u_{in}^P,\end{cases} \tag{1.1.92}$$

$$u + F(t, u) = 0. \tag{1.1.93}$$

Theorem 1.1.60 ([95]) *Suppose that there exist $R > 0$, $\rho_0 > 0$, and $\beta_0 > 0$ such that if $0 < t \le \rho_0/\beta_0$, the following properties hold:*

(1) $\forall 0 < \rho' < \rho \le \rho_0$ and $\forall u$ such that $\{u \in X_\rho : \sup_{0\le t\le T} |u(t)|_\rho \le R\}$, the map $F(t,u) : [0,T] \mapsto X_{\rho'}$ is continuous.
(2) $\forall 0 < \rho \le \rho_0$, the function $F(t,0) : [0, \rho_0/\beta_0] \mapsto \{u \in X_\rho : \sup_{0\le t\le T} |u(t)|_\rho \le R\}$ is continuous and

$$|F(t,0)|_\rho \le R_0 < R. \tag{1.1.94}$$

(3) $\forall 0 < \rho' < \rho(s) \le \rho_0$ and $\forall u^1$ and $u^2 \in \{u \in X_\rho : \sup_{0\le t\le T} |u(t)|_{\rho-\beta_0 t} \le R\}$,

$$|F(t,u^1) - F(t,u^2)|_{\rho'} \le C \int_0^t ds (\frac{|u^1 - u^1|_{\rho(s)}}{\rho(s) = \rho'} + \frac{|u^1 - u^2|_{\rho'}}{\sqrt{t-s}}). \tag{1.1.95}$$

Then there exists $\beta > \beta_0$ such that $\forall 0 < \rho < \rho_0$, (1.1.93) *has a unique solution $u(t) \in X_\rho$ with $t \in [0, (\rho_0 - \rho)/\beta]$. Moreover, $\sup_{\rho<\rho_0-\beta t} |u(t)|_\rho \le R$. Here $\{X_\rho : 0 < \rho \le \rho_0\}$ be a Banach scale with norms $\|.\cdot\|_\rho$.*

Theorem 1.1.61 ([95]) *Suppose that u_1 and u_2 are in $K^{l,\rho_0,\mu_0}_{\beta_0,T}$ (see* (1.1.57)*). Suppose $0 < \rho' < \rho(s) < \rho_0 s$ and $0 < \mu' < \mu(s) < \mu_0$. Then the following estimate holds:*

$$|F(u^1,t) - F(u^2,t)|_{l,\rho',\mu'} \le c \int_0^t ds \left(\frac{|u^1-u^2|_{l,\rho(s),\mu}}{\rho(s)-\rho'} + \frac{|u^1-u^2|_{l,\rho,\mu(s)}}{\mu(s)-\mu'} \right.$$
$$\left. + \frac{|u^1-u^2|_{l,\rho',\mu'}}{\sqrt{t-s}} \right). \tag{1.1.96}$$

Theorem 1.1.62 ([95]) *Suppose $U \in K^{l,\rho_0}_{\beta_0,T}$ and $u^P_{in} - U \in K^{l,\rho_0,\mu_0}$. Moreover, let the compatibility conditions*

$$u^P_{in}(x, Y = 0) = 0, \tag{1.1.97}$$

$$u^P_{in}(x, Y \to \infty) - U \to 0, \tag{1.1.98}$$

hold. Then there exist $0 < \rho_1 < \rho_0$, $0 < \mu_1 < \mu_0$, $\beta_1 > \beta_0 > 0$, and $0 < T < T_1$ such that (1.1.92) *admits a unique mild solution u^p. This solution can be written as*

$$u^P(x,Y,t) = u(x,Y,t) + U, \tag{1.1.99}$$

where $u \in K^{l,\rho_1,\mu_1}_{\beta_1,T_1}$. Here

$$|f|_{K^{l,\rho}_{\beta,T}} = \sum_{0\le j\le 1} \sum_{i\le l-j} \sup_{0\le t\le T} |\partial_t^j \partial_x^i f(\cdot,t)|_{0,\rho-\beta t} < +\infty.$$

Wang and Zhu [143] in 2020 studied of the existence of a backflow point in the two-dimensional unsteady boundary layers under an adverse pressure gradient and considered the following problem for the Prandtl boundary layer equation with non-slip boundary condition for an unsteady incompressible flow in a domain $Q_T = \{(t, x, y) \mid 0 \le t < T, 0 \le x \le L, 0 \le y < +\infty\}$:

$$\begin{cases} \partial_t u + u\partial_x u + v\partial_y u = \partial_y^2 u - \partial_x P, \\ \partial_x u + \partial_y v = 0, \\ u|_{y=0} = v|_{y=0} = 0, \quad \lim_{y\to+\infty} u = U_e(t, x), \\ u|_{t=0} = u_0(x, y), \ u|_{x=0} = u_1(t, y), \end{cases} \tag{1.1.100}$$

where $(u(t, x, y), v(t, x, y))$ is the velocity field in the boundary layer and $U_e(t, x)$ and $P(t, x)$ are traces at the boundary $\{y = 0\}$ of the tangential velocity and pressure of the Euler outer flow, respectively, interrelated through Bernoulli's law

$$\partial_t U_e + U_e \partial_x U_e = -\partial_x P.$$

Assume that the data satisfy the following properties:

$$u_0(x, y) > 0, u_1(t, y) > 0, \forall t \in [0, T), x \in [0, L], y \in [0, +\infty),$$

and the monotonicity assumption

$$\partial_y u_0(x, y) > 0, \partial_y u_1(t, y) > 0, \forall t \in [0, T), x \in [0, L], y \in [0, +\infty).$$

Theorem 1.1.63 ([143])

(1) Assume that the trace at the boundary of the outer Euler flow satisfies

$$U_e \in C^1([0, T] \times [0, L]) \textit{ and } U_e(t, x) > 0 \quad \forall t \in [0, T], x \in [0, L],$$

and the uniformly adverse pressure gradient in the sense that

$$\partial_x P(t, x) > 0 \quad \forall t \in [0, T], x \in [0, L].$$

Assume that $(u, v) \in C^2(Q_T)$ is the solution to the problem (1.1.100) corresponding to the data u_0, u_1 satisfying the conditions above. Then, the first zero point of $\partial_y u(t, x, y)$, when the time evolves, should be at the boundary $\{y = 0\}$ if it exists for some time $t \in (0, T)$.

(2) Moreover, when the initial velocity $u_0(x, y)$ satisfies

$$\int_0^{+\infty}\int_0^L \frac{(L - x)^{\frac{3}{2}} \partial_y u_0}{\sqrt{(\partial_y u_0)^2 + u_0^2}} dxdy \ge C_*,$$

for a positive constant C_* depending only on L, T, U_e, and $\partial_x P$, then there is a backflow point $(t^*, x^*) \in (0, T) \times [0, L]$ such that

$$\begin{cases} \partial_y u\,(t^*, x^*, 0) = 0, \\ \partial_y u(t, x, y) > 0 \quad \forall 0 < t < t^*, x \in [0, L], y \geq 0. \end{cases}$$

Moreover, they have $\partial_y^2 u\,(t^*, x^*, 0) \neq 0$.

Li and Yang [76] in 2020 studied the well-posedness of the Prandtl system without monotonicity and analyticity assumption. Precisely, for any index $\sigma \in [3/2, 2]$, they obtained the local in time well-posedness in the space of Gevrey class G^σ in the tangential variable and Sobolev class in the normal variable so that the monotonicity condition on the tangential velocity is not needed to overcome the loss of tangential derivative.

The Prandtl equations introduced by Prandtl in 1904 described the behavior of the incompressible flow near a rigid wall at high Reynolds number

$$\begin{cases} \partial_t u^P + u^P \partial_x u^P + v^P \partial_y u^P - \partial_y^2 u^P + \partial_x p = 0, \quad t > 0, \quad x \in \mathbb{R}, \quad y > 0, \\ \partial_x u^P + \partial_y v^P = 0, \\ u^P\big|_{y=0} = v^P\big|_{y=0} = 0, \quad \lim\limits_{y \to +\infty} u = U(t, x), \\ u^P\big|_{t=0} = u_0^P(x, y), \end{cases} \tag{1.1.101}$$

where $u^P(t, x, y)$ and $v^P(t, x, y)$ represent the tangential and normal velocities of the boundary layer, with y being the scaled normal variable to the boundary, while $U(t, x)$ and $p(t, x)$ are the values on the boundary of the tangential velocity and pressure of the outflow satisfying the Bernoulli's law

$$\partial_t U + U \partial_x U + \partial_x p = 0.$$

To have a clear presentation, the authors of [76] constructed a solutions u^P that is a small perturbation around a shear flow, that is, $u^P(t, x, y) = u^s(t, y) + u(t, x, y)$. For this, they supposed that the initial data u_0^P can be written as

$$u_0^P(x, y) = u_0^s(y) + u_0(x, y),$$

with u_0^s being independent of x variable. Then one reduces the original Prandtl equation to the following two time evolutional equations, one of which is the equation for the shear flow $(u^s, 0)$ with u^s solving

$$\begin{cases} \partial_t u^s - \partial_y^2 u^s = 0, \\ u^s|_{y=0} = 0, \quad \lim\limits_{y \to +\infty} u^s = 1, \\ u^s|_{t=0} = u_0^s, \end{cases} \tag{1.1.102}$$

and another reads

$$\begin{cases} \partial_t u + (u^s + u)\,\partial_x u + v\partial_y\,(u^s + u) - \partial_y^2 u = 0, \\ u|_{y=0} = 0, \quad \lim_{y\to+\infty} u = 0, \\ u|_{t=0} = u_0, \end{cases} \tag{1.1.103}$$

where

$$v = -\int_0^y \partial_x u(x, \tilde{y})d\tilde{y}.$$

There exists a $y_0 > 0$ such that $u_0^s \in C^6\,(\mathbb{R}_+)$ satisfies the following properties:

(i) $\frac{du_0^s}{dy}\,(y_0) = 0$ and $\frac{d^2u_0^s}{dy^2}\,(y_0) \neq 0$. Moreover, there exist $0 < \delta < y_0/2$ and a constant $c_0 > 0$ such that

$$\forall y \in [y_0 - 2\delta, y_0 + 2\delta]\,, \qquad \left|\frac{d^2u_0^s}{dy^2}(y)\right| \geq c_0.$$

(ii) There exists a constant $0 < c_1 < 1$ such that

$$\forall y \in [0, y_0 - \delta] \cup \Big[y_0 + \delta, +\infty\Big[, \quad c_1\langle y\rangle^{-\alpha} \leq \left|\frac{du_0^s(y)}{dy}\right| \leq c_1^{-1}\langle y\rangle^{-\alpha},$$

for some $\alpha > 1$, and that

$$\forall y \geq 0, \qquad \left|\frac{d^j u_0^s(y)}{dy^j}\right| \leq c_1^{-1}\langle y\rangle^{-\alpha-1} \text{ for } 2 \leq j \leq 6.$$

(iii) The compatibility condition holds, that is, $u_0^s\big|_{y=0} = \partial_y^2 u_0^s\big|_{y=0} = 0$ and $u_0^s(y) \to 1$ as $y \to +\infty$.

Let α be the number given in the assumption and let ℓ be a fixed number satisfying that

$$\ell > 3/2, \quad \alpha \leq \ell < \alpha + \frac{1}{2}.$$

With each pair (ρ, σ), $\rho > 0, \sigma \geq 1$, we associate a Banach space $X_{\rho,\sigma}$, equipped with the norm $\|\cdot\|_{\rho,\sigma}$ that consists of all the smooth functions f such that $\|f\|_{\rho,\sigma} < +\infty$, where

$$\begin{aligned}
\|f\|_{\rho,\sigma} \stackrel{\text{def}}{=} & \sup_{m\geq 6} \frac{\rho^{m-5}}{[(m-6)!]^\sigma} \left\| \langle y\rangle^{\ell-1} \partial_x^m f \right\|_{L^2} \\
& + \sup_{m\geq 6} \frac{\rho^{m-5}}{[(m-6)!]^\sigma} \left\| \langle y\rangle^{\ell} \partial_x^m \left(\partial_y f\right) \right\|_{L^2} \\
& + \sup_{0\leq m\leq 5} \left(\left\| \langle y\rangle^{\ell-1} \partial_x^m f \right\|_{L^2} + \left\| \langle y\rangle^{\ell} \partial_x^m \left(\partial_y f\right) \right\|_{L^2} \right) \\
& + \sup_{\substack{1\leq j\leq 4 \\ i+j\geq 6}} \frac{\rho^{i+j-5}}{[(i+j-6)!]^\sigma} \left\| \langle y\rangle^{\ell+1} \partial_x^i \partial_y^j \left(\partial_y f\right) \right\|_{L^2} \\
& + \sup_{\substack{1\leq j\leq 4 \\ i+j\leq 5}} \left\| \langle y\rangle^{\ell+1} \partial_x^i \partial_y^j \left(\partial_y f\right) \right\|_{L^2}.
\end{aligned}$$

Theorem 1.1.64 ([76]) *For $\sigma \in [3/2, 2]$, let the initial datum u_0 belong to $X_{2\rho_0,\sigma}$ for some $\rho_0 > 0$ and moreover*

$$\|u_0\|_{2\rho_0,\sigma} \leq \eta_0,$$

for some $\eta_0 > 0$. Suppose that the compatibility condition holds for u_0. Then equation (1.1.101) *admits a unique solution $u \in L^\infty\left([0, T]; X_{\rho,\sigma}\right)$ for some $T > 0$ and some $0 < \rho < 2\rho_0$, provided η_0 is sufficiently small.*

1.1.3 2D Prandtl Equations-Global Existence

Oleinik and Samokhin [106] in 1999 considered the following Prandtl system for a nonstationary boundary layer in an axially symmetric incompressible flow past a solid body:

$$\begin{cases} \frac{\partial u}{\partial t} + u\frac{\partial u}{\partial x} + v\frac{\partial u}{\partial y} = \nu\frac{\partial^2 u}{\partial y^2} + U\frac{\partial U}{\partial x} + U\frac{dU}{dx}, \\ \frac{\partial (ru)}{\partial x} + \frac{\partial (rv)}{\partial y} = 0, \end{cases} \tag{1.1.104}$$

in a domain $D = \{0 < t < T, 0 < x < X, 0 < y < \infty\}$, with the boundary conditions

$$\begin{cases} u(0, x, y) = u_0(x, y), \;\; u(t, 0, y) = 0, \;\; u(t, x, 0) = 0, \;\; v(t, x, 0) = v_0(t, x), \\ u(t, x, y) \to U(t, x) \;\text{ as }\; y \to \infty. \end{cases} \tag{1.1.105}$$

The function $U(t, x)$ and $r(x)$ are given functions such that $U(t, 0) = 0$, $U(t, x) > 0$ for $x \geq 0$, $r(0) = 0$, $r(x) > 0$ for $x > 0$.

Introduce Crocco variables by

$$\tau = t, \quad \xi = x, \quad \eta = \frac{u(t, x, y)}{U(t, x)},$$

and a new unknown function

$$w(\tau, \xi, \eta) = \frac{u_y(t, x, y)}{U(t, x)}.$$

From (1.1.104), we obtain the following equation for $w(\tau, \xi, \eta)$:

$$\nu w^2 w_{\eta\eta} - w_\tau - \eta U w_\xi + A w_\eta + B w = 0 \tag{1.1.106}$$

in the domain $\Omega = \{0 < \tau < T, 0 < \xi < X, 0 < \eta < 1\}$, with the boundary conditions

$$w|_{\tau=0} = \frac{u_{0y}}{U} \equiv w_0(\xi, \eta), \quad w|_{\eta=1} = 0, \quad (\nu w^2 w_{\eta\eta} - v_0 w + C)_{\eta=0} = 0, \tag{1.1.107}$$

where

$$A = (\eta^2 - 1)U_x + (\eta - 1)\frac{U_t}{U}, \quad B = -\eta U_x(\xi) + \frac{\eta r_x(\xi) U(\xi)}{r(\xi)} - \frac{U_t}{U}, \quad C = U_x + \frac{U_t}{U}.$$

Theorem 1.1.65 ([106]) *Assume that $U_x, U_t/U, U r_x/r, v_0$ are bounded functions having bounded derivatives with respect to t, x in D; $u_0(x, y) \to \infty$ as $y \to \infty$, $u_0 = 0$ for $y = 0$; $u_0/U, u_{0y}/U$ are continuous in $\overline{D}$; $u_{0y} > 0$ for $y \geq 0, x > 0$,*

$$K_1(U(0, x) - u_0(x, y)) \leq u_{0y}(x, y) \leq K_2(U(0, x) - u_0(x, y)),$$

with positive constants K_1 and K_2. Assume also that there exist bounded derivatives $u_{0y}, u_{0yy}, u_{0yyy}, u_{0x}, u_{0xy}$, and the ratios

$$\frac{u_{0yy}}{u_{0y}}, \quad \frac{u_{0yyy}u_{0y} - u_{0yy}^2}{u_{0y}^2}$$

are bounded for $0 \leq x \leq X, 0 \leq y < \infty$. Let the following compatibility condition be satisfied:

$$v_0(0, x)u_{0y}(x, 0) = -p_x(0, x) + \nu u_{0yy}(x, 0),$$

and let

$$\left| \frac{u_{0yx} - u_{0x}u_{0yy}}{u_{0y}} + U_x \frac{u_0 u_{0yy} - u_{0y}^2}{U u_{0y}} \right| \le K_6(U - u_0(x, y)).$$

Then, problems (1.1.104)–(1.1.105) in D have a unique solution u, v with the following properties: $u/U, u_y/U$ are continuous and bounded in $\overline{D}$; $u_y/U > 0$ for $y \ge 0$; $u_y/U \to 0$ as $y \to \infty$; $u = 0$ for $y = 0$; v is continuous in D and bounded for bounded y; the weak derivatives $u_t, u_x, u_{yt}, u_{yx}, u_{yy}, u_{yyy}, u_y$ are bounded measurable functions in D; the equations of system (1.1.104) hold almost everywhere; in D, the functions u_t, u_x, v_y, u_{yy} are continuous with respect to y. Moreover,

$$\frac{u_{yy}}{u_y}, \quad \frac{u_{yyy}u_y - u_{yy}^2}{u_y^2}$$

are bounded and the following inequalities hold:

$$C_1(U(t,x) - u(t,x,y)) \le u_y(t,x,y) \le C_2(U(t,x) - u(t,x,y)),$$

$$\exp(-C_2 y) \le 1 - \frac{u(t,x,y)}{U(t,x)} \le \exp(-C_1 y),$$

$$\left| \frac{u_{yt}u_y - u_t u_{yy}}{u_y} + U_t \frac{u u_{yy} - u_y^2}{u_y U} \right| \le C_4(U(t,x) - u(t,x,y)),$$

$$\left| \frac{u_{yx}u_y - u_x u_{yy}}{u_y} + U_x \frac{u u_{yy} - u_y^2}{u_y U} \right| \le C_3(U(t,x) - u(t,x,y)).$$

Oleinik and Samokhin [106] in 1999 also considered system (1.1.104) in the domain $D = \{0 < t < T, 0 < x < X, 0 < y < \infty\}$, with the boundary and the initial conditions of the form:

$$\begin{cases} u(0,x,y) = u_0(x,y), \quad u(t,0,y) = u_1(t,y), \quad u(t,x,0) = 0, \\ v(t,x,0) = v_0(t,x), \\ u(t,x,y) \to U(t,x) \text{ as } y \to \infty. \end{cases} \tag{1.1.108}$$

Theorem 1.1.66 ([106]) *Suppose that $U_x, U_t, v_0, U, r_x/r$ are bounded and have bounded first derivatives in D with respect to t and x;*

$$u_0(x,y) \to U(0,x), \quad u_1(t,y) \to U(t,0), \quad \textit{as } y \to \infty,$$

and $U(t,x) > 0, u_{0y} > 0, u_{1y} > 0$ *for* $y \geq 0$,

$$K_1(U(0,x) - u_0(x,y)) \leq u_{0y}(x,y) \leq K_2(U(0,x) - u_0(x,y)),$$
$$K_1^1(U(t,0) - u_1(t,y)) \leq u_{1y}(x,y) \leq K_2^1(U(t,0) - u_1(x,y)).$$

Suppose also that there exist bounded weak derivatives $u_{0y}, u_{0yy}, u_{0yyy}, u_{0x}, u_{0xy}, u_{1y}, u_{1yy}, u_{1yyy}, u_{1t}, u_{1ty}$ *and the functions*

$$\frac{u_{iyy}}{u_{iy}}, \quad \frac{u_{iyyy}u_{iy} - u_{iyy}^2}{u_{iy}^2}, \quad i = 0, 1,$$

are bounded; moreover,

$$\left| \frac{u_{0yx}u_{0y} - u_{0x}u_{0yy}}{u_{0y}} + U_x \frac{u_0 u_{0yy} - u_{0y}^2}{U u_{0y}} \right| \leq K_5(U - u_0),$$

$$\left| \frac{u_{1yx}u_{1y} - u_{1x}u_{1yy}}{u_{1y}} + U_t \frac{u_1 u_{1yy} - u_{1y}^2}{U u_{1y}} \right| \leq K_5^1(U - u_1);$$

the function whose absolute value is estimated in the last inequality is continuous in t *and* y *for small* y. *Let the following compatibility conditions hold:*

$$v_0(0,x)u_{0y}(x,0) = -p_x(0,x) + \nu u_{0yy}(x,0),$$
$$v_0(t,0)u_{1y}(t,0) = -p_x(t,0) + \nu u_{1yy}(t,0),$$
$$u_0 = u_1 \text{ for } t = 0, x = 0; \quad u_1 = 0, u_0 = 0 \text{ for } y = 0;$$
$$u_{1ty}(t,y) + v_0(t,0)u_{1yy}(t,y) - \nu u_{1yyy}(t,y) = O(y) \text{ for small } y.$$

Then problems (1.1.104) and (1.1.108) in D *have a unique solution* u, v *with the following properties:* u, u_y *are continuous and bounded in* $\overline{D}$; $u_y > 0$ *for* $y \geq 0$; $u_y \to 0$ *as* $y \to \infty$; $u = 0$ *for* $y = 0$; v *is continuous in* $\overline{D}$ *with respect to* y *and bounded for bounded* y; *there exist bounded weak derivatives* $u_t, u_x, u_{yt}, u_{yx}, u_{yy}, u_{yyy}, v_y$, *and equation (1.1.104) holds almost everywhere in* D. *Moreover,* u_t, u_x, v_y, u_{yy} *are continuous in* y; $u_{yy}/u_y, (u_{yyy}u_y - u_{yy}^2)/u_y^2$ *are bounded, and*

$$C_1(U(t,x) - u(t,x,y)) \leq u_y(t,x,y) \leq C_2(U(t,x) - u(t,x,y)),$$
$$\exp(-C_2 y) \leq 1 - \frac{u(t,x,y)}{U(t,x)} \leq \exp(-C_1 y),$$

$$\left|\frac{u_{yt}u_y - u_t u_{yy}}{u_y} + U_t \frac{u u_{yy} - u_y^2}{u_y U}\right| \le C_3(U-u),$$

$$\left|\frac{u_{yx}u_y - u_x u_{yy}}{u_y} + U_x \frac{u u_{yy} - u_y^2}{u_y U}\right| \le C_4(U-u).$$

where C_i $(i = 1, 2, 3, 4)$ are positive constants.

Theorem 1.1.67 ([106]) *Let the assumptions of Theorem 1.1.66 hold in the domain $D = \{0 < t < \infty, 0 < x < X, 0 < y < \infty\}$. Then the solution u, v of problems (1.1.104) and (1.1.108) exists and is unique in the domain D with $X > 0$ depending on the data of problems (1.1.104) and (1.1.108). This solution has the same properties as the solution u, v constructed in Theorem 1.1.66.*

Oleinik and Samokhin [106] in 1999 also considered the nonstationary system of the two-dimensional boundary layer

$$\begin{cases} \frac{\partial u}{\partial t} + u\frac{\partial u}{\partial x} + v\frac{\partial u}{\partial y} = \nu\frac{\partial^2 u}{\partial y^2} - \frac{dp}{dx}, \\ \frac{\partial u}{\partial x} + \frac{\partial v}{\partial y} = 0, \end{cases} \tag{1.1.109}$$

in a domain $D = \{0 < t < T, 0 < x < X, 0 < y < \infty\}$, with the boundary conditions

$$\begin{cases} u(0,x,y) = u_0(x,y), \quad u(t,0,y) = u_1(t,y), \quad u(t,x,0) = 0, \\ v(t,x,0) = v_0(t,x), \\ u(t,x,y) \to U(t,x) \text{ as } y \to \infty. \end{cases} \tag{1.1.110}$$

The density of the fluid ρ is assumed constant and equal to 1. Therefore, $U_t + UU_x = -p_x(t,x)$. It is also assumed that $U(t,x) > 0, u_0 > 0, u_1 > 0$ for $y > 0$, $u_{0y} > 0, u_{1y} > 0$ for $y \ge 0$.

Theorem 1.1.68 ([106]) *Assume that $p(t,x), v_0(t,x), u_0(x,y), u_1(t,y), u_{0y}$ are sufficiently smooth and satisfy the compatibility conditions which amount to the existence of the function to w^* mentioned earlier. Then there is one and only one solution of problems (1.1.109), (1.1.110) in the domain D, with X being arbitrary and T depending on the data of problems (1.1.109) and (1.1.110) or T being arbitrary and X depending on the data. This solution has the following properties: $u > 0$ for $y > 0$, $u_y > 0$ for $y \ge 0$; the derivatives $u_t, u_x, u_y, u_{yy}, v_y$ are continuous and bounded in $\overline{D}$; moreover, the ratios*

$$\frac{u_{yy}}{u_y}, \quad \frac{u_{yyy}u_y - u_{yy}^2}{u_y^2}$$

are bounded in D.

Theorem 1.1.69 ([106] Global Solutions of the Prandtl System for Axially Symmetric Flows) *Let $U(t,x) = ax + b(t,x)x^3$, where a is a positive constant and $b(x,t)$ has bounded second-order derivatives. Assume also that*

$$v_0 \le M_1 x, \quad v_{0t} \ge -M_2 x, \quad \left|\left(U\frac{r_x}{r} - U_x\right)_t\right| \le M_3 x.$$

Assume that $u_0(x,y)$ satisfies the conditions: u_0/U is continuous in $\overline{D}$, $u_0 > 0$ for $y > 0, x > 0$; $u_0(x,0) = 0, u_0/U \to 1$ as $y \to \infty$; $u_{0y}/U > 0$ for $y \ge 0$; $u_{0y}/U \to 0$ as $y \to \infty$; moreover,

$$M_4(U - u_0)\sigma_0 \le u_{0y} \le M_5(U - u_0)\sigma_0,$$

$$-M_6\sigma_0 \le \frac{u_{0yy}}{u_{0y}} \le -M_7\sigma_0,$$

$$\left|\frac{u_{0yx}u_{0y} - u_{0x}u_{0yy}}{u_{0y}} + \frac{U_x}{U}\frac{u_0u_{0yy} - u_{0y}^2}{u_{0y}}\right| \le M_8\sigma_0(U - u_0),$$

$$\left|\frac{u_{0yyy}u_{0y} - u_{0yy}^2}{u_{0y}^2}\right| \le M_9,$$

$$\nu\frac{u_{0yyy}u_{0y} - u_{0yy}^2}{u_{0y}} + \left[u_0^2 - U^2\frac{U_x}{U}\frac{u_{0yy}}{u_{0y}}\right] \le M_{10}xu_{0y},$$

where $\sigma_0 = \sqrt{-\ln\mu(1 - u_0/U)}$, M_i $(i = 1,2,3,4,5,6,7,8,9,10)$, μ are positive constants. Finally, let the following compatibility conditions hold:

$$v_0(0,x)u_{0y}(x,0) = U_t(0,x) + U(0,x)U_x(0,x) + \nu u_{0yy}(x,0).$$

Then in the domain $D = \{0 < t < \infty, 0 < x < X, 0 < y < \infty\}$, problems (1.1.104)–(1.1.105) have one and only one solution u, v with the following properties: u/U, u_y/U are bounded and continuous in $\overline{D}$, $u > 0$ for $y > 0, x > 0$; $u \to U$ as $y \to \infty$; $u|_{y=0} = 0, u|_{x=0} = 0$, $u_y/U > 0$ for $y \ge 0$; $u_y/U \to 0$ as $y \to \infty$; $u_y, u_x, u_{yy}, u_t, v_y$ are bounded and continuous in $\overline{D}$ with respect to y; u_{yy}/u_y, v are continuous with respect to y in $\overline{D}$ and are bounded for bounded y; $v|_{y=0} = v_0(t,x)$, u_{yyy} is bounded in D; u_{yx}, u_{yt} are bounded for bounded y; equation (1.1.104) holds almost everywhere in D. Moreover,

$$M_{12}(U - u)\sigma \le u_y \le M_{11}(U - u)\sigma,$$

$$-M_{14}\sigma \le \frac{u_{yy}}{u_y} \le -M_{13}\sigma, \quad \left|\frac{u_{yyy}u_y - u_{yy}^2}{u_y^2}\right| \le M_{15},$$

$$\left|\frac{u_{yx}u_y - u_x u_{yy}}{u_y} + \frac{U_x}{U}\frac{uu_{yy} - u_y^2}{u_y}\right| \leq M_{16}\sigma(U-u),$$

$$- \leq M_{18}\sigma(U-u) \leq \frac{u_{yt}u_y - u_t u_{yy}}{u_y} + \frac{U_t}{U}\frac{uu_{yy} - u_y^2}{u_y} \leq M_{17}x\sigma(U-u),$$

$$M_{21}U\exp(-M_{22}y^2) \leq U - u \leq M_{19}U\exp(-M_{20}y^2),$$

where $\sigma = \sqrt{-\ln\mu(1-u/U)}$, M_i $(i = 1, 2, 3 \cdots 21)$, μ *are positive constants;* $\mu < 1$, $X > 0$ *depend on* U, r, v_0, u_0.

They have ensured the existence of a solution for the boundary layer system for t in the interval $0 \leq t < \infty$ (Theorem 1.1.69). It is natural to examine the question of stability and asymptotic stability under certain perturbations of the data and the parameters involved in problems (1.1.104)–(1.1.105) in the domain $D = \{0 < t < \infty, 0 < x < X, 0 < y < \infty\}$. And let u, v correspond to the given functions U, r, v_0, u_0; let $\widetilde{u}(t,x,y), \widetilde{v}(t,x,y)$ be a solution of the same problem corresponding to $\widetilde{U}, \widetilde{r}, \widetilde{v}_0, \widetilde{u}_0$.

Theorem 1.1.70 ([106]) *Assume that the functions*

$$U - \widetilde{U},\ U_x - \widetilde{U}_x,\ \frac{U_t}{U} - \frac{\widetilde{U}_t}{\widetilde{U}},\ v_0 - \widetilde{v}_0,\ \frac{r_x U}{r} - \frac{\widetilde{r}_x \widetilde{U}}{\widetilde{r}}, \tag{1.1.111}$$

identically vanish for $t \geq t_0 > 0$, *where* $t_0 =$*const.*$< \infty$. *Then, in the domain* D *for* $0 \leq y \leq y_0 < \infty$, *the following estimate holds:*

$$\left|\frac{u}{U} - \frac{\widetilde{u}}{\widetilde{U}}\right| \leq K_3 e^{-\alpha_0 t},$$

where $\alpha =$*const.*> 0, $K_3 =$*const.*> 0; K_3 *depends on* y_0.

Theorem 1.1.71 ([106]) *Assume that the functions (1.1.111) uniformly converge to* 0 *in the domain* D *as* $t \to \infty$. *Then, for any* $\delta > 0$, *there is a constant* $K(\delta) > 0$ *such that*

$$\left|\frac{u}{U} - \frac{\widetilde{u}}{\widetilde{U}}\right| \leq \delta + K(\delta)e^{-\alpha_0 t} \quad \text{in } D.$$

Theorem 1.1.72 ([106]) *Assume that the functions (1.1.111) have their absolute values* $\leq \varepsilon$, *and* $u_0, \widetilde{u}_0$ *such that* $|w - \widetilde{w}| \leq \varepsilon$, *where* w *and* $\widetilde{w}$ *are solutions of Crocco variables corresponding to* U, r, v_0, u_0, *and* $\widetilde{U}, \widetilde{r}, \widetilde{v}_0, \widetilde{u}_0$, *respectively. Then*

$$\left|\frac{u}{U} - \frac{\widetilde{u}}{\widetilde{U}}\right| \leq K_4\varepsilon \quad \text{in } D,$$

for any t, *where* $K_4 > 0$ *is a constant depending on* $u_0, \widetilde{u}_0$ *and their derivatives.*

Oleinik and Samokhin [106] in 1999 examined the boundary layer system for a nonstationary axially symmetric flow (1.1.104) in the case of time-periodic outer flow in the domain $D = \{-\infty < t < \infty, 0 < x < X, 0 < y < \infty\}$, with the boundary conditions

$$\begin{cases} u(t,0,y) = 0, \quad u(t,x,0) = 0, \quad v(t,x,0) = v_0(t,x), \\ u(t,x,y) \to U(t,x) \text{ as } y \to \infty, \end{cases} \tag{1.1.112}$$

and the condition of periodicity

$$u(t+T,x,y) = u(t,x,y), \tag{1.1.113}$$

where T is a given positive constant.

Theorem 1.1.73 ([106]) *Let $U(t,x) = ax + b(t,x)x^2$, where a is a positive constant and $b(x,t)$ has bounded second-order derivatives. Assume also that*

$$v_0 \le K_1 x, \quad v_{0t} \ge -K_2 x, \quad \left|\left(U\frac{r_x}{r} - U_x\right)_t\right| \le K_3 x, \quad r_x|_{x=0} > 0;$$

b, v_0, r are periodic in t with period T. Then problems (1.1.104) and (1.1.112) in $D = \{0 < t < T, 0 < x < X, 0 < y < \infty\}$ have one and only one solution u, v with the following properties: $u/U, u_y/U$ are bounded and continuous in $\overline{D}$, $u > 0$ for $y > 0, x > 0$; $u \to U$ as $y \to \infty$; $u|_{y=0} = 0, u|_{x=0} = 0$; $u_y/U > 0$ for $y \ge 0$; $u_y/U \to 0$ as $y \to \infty$; $u_y, u_x, u_{yy}, u_t, v_y$ are continuous and bounded in $\overline{D}$ with respect to y; $u_{yy}/u_y, v$ are continuous in $\overline{D}$ with respect to y and bounded for bounded y; $v|_{y=0} = v_0(t,x), u_{yyy}$ is bounded in D; u_{yx}, u_{yt} are bounded for bounded y; equation (1.1.104) hold almost everywhere in D, and the following inequalities are valid:

$$K_4(U-u)\sigma \le u_y \le K_5(U-u)\sigma,$$

$$-K_6\sigma \le \frac{u_{yy}}{u_y} \le -K_7\sigma, \quad \left|\frac{u_{yyy}u_y - u_{yy}^2}{u_y^2}\right| \le K_8,$$

$$\left|\frac{u_{yx}u_y - u_x u_{yy}}{u_y} + \frac{U_x}{U}\frac{uu_{yy} - u_y^2}{u_y}\right| \le K_9\sigma(U-u),$$

$$- \le K_{10}\sigma(U-u) \le \frac{u_{yt}u_y - u_t u_{yy}}{u_y} + \frac{U_t}{U}\frac{uu_{yy} - u_y^2}{u_y} \le K_{11}x\sigma(U-u),$$

where $\sigma = \sqrt{-\ln \mu(1-u/U)}$, K_i $(i = 1,2,3,4,5,6,7,8,9,10,11)$, μ are positive constants; $0 < \mu < e^{-1/2}$, X depend on U, r, v_0.

Oleinik and Samokhin [106] in 1999 considered the nonstationary system of the two-dimensional boundary layer

$$\begin{cases} \frac{\partial u}{\partial t} + u\frac{\partial u}{\partial x} + v\frac{\partial u}{\partial y} = \nu\frac{\partial^2 u}{\partial y^2} + U\frac{\partial U}{\partial x} + \frac{\partial U}{\partial t}, \\ \frac{\partial u}{\partial x} + \frac{\partial v}{\partial y} = 0 \end{cases} \tag{1.1.114}$$

in a domain $D = \{0 < t < T, 0 < x < X, 0 < y < \infty\}$, with the boundary conditions

$$\begin{cases} u(0, x, y) = 0, \quad u(t, 0, y) = 0, \quad u(t, x, 0) = 0, \quad v(t, x, 0) = v_0(t, x), \\ u(t, x, y) \to U(t, x) \text{ as } y \to \infty. \end{cases} \tag{1.1.115}$$

It is also assumed that $U(t, 0) = 0$, $U(0, x) = 0$, $U > 0$ for $t, x > 0$,

$$U(t, x) = t^n U_1(t, x), \quad n \geq 1,$$

and U_{1t}/U_1 is a bounded function.

Introduce new independent variables in (1.1.114)–(1.1.115), setting

$$\tau = t^{n-\frac{1}{2}}, \quad \xi = x, \quad \eta = \frac{u(t, x, y)}{U(t, x)},$$

and a new unknown function

$$w(\tau, \xi, \eta) = t^n \frac{u_y(t, x, y)}{U(t, x)}.$$

From (1.1.114), we obtain the following equation for $w(\tau, \xi, \eta)$:

$$\nu w^2 w_{\eta\eta} - \tau^3 (n - \frac{1}{2}) w_\tau - \eta U \tau^{2N} w_\xi + n(n-1)\tau^2 w_\eta + A_1 \tau^{2N} w_\eta + B_1 \tau^{2N} w$$
$$= 0, \tag{1.1.116}$$

where $N = 1 + \frac{1}{2n-1}$,

$$A_1 = (\eta^2 - 1)U_x + (\eta - 1)\frac{U_{1t}}{U_1}, \quad B_1 = -\eta U_x(\xi) - \frac{U_{1t}}{U_1},$$

in the domain $\Omega = \{0 < \tau < T^{n-\frac{1}{2}}, 0 < \xi < X, 0 < \eta < 1\}$, with the boundary conditions

$$w|_{\tau=0} = 0, \quad w|_{\eta=1} = 0, \quad (\nu w w_\eta - v_0 w \tau^N + n\tau^2 + C_1 \tau^{2N})_{\eta=0} = 0, \tag{1.1.117}$$

where

$$C_1 = U_x + \frac{U_{1t}}{U_1}.$$

Theorem 1.1.74 ([106]) *Let*

$$U(t,x) = t^n U_1(t,x), \quad n \geq 1,$$
$$U(t,0) = 0, \quad U_1(t,x) > 0, \quad x > 0.$$

Assume that U_{1x}, u_{1t}/U_1, v_0 have continuous first-order derivatives in t and x.

Then problems (1.1.114), (1.1.115) in $D = \{0 < t < T, 0 < x < X, 0 < y < \infty\}$ admit a solution u, v with the following properties: u/U, $u_y t^n/U$ are bounded and continuous in $\overline{D}$; $u(t,x,y) > 0$ for $t, x > 0$; $u_y t^n/U > 0$ for $t > 0$, $u_y t^n/U \to 0$ as $y \to 0$; the derivatives $u_y, u_x, u_{yy}, u_t, v_y$ are bounded and continuous in D with respect to y;

$$|u_y| \leq E_1 t^{n-\frac{1}{2}}, \quad |u_{yy}| \leq E_2 t^{n-1}, \quad |u_t| \leq E_3 t^{n-1}, \quad |u_x| \leq E_4 t^n;$$

the function v is continuous in $\overline{D}$ with respect to y and is bounded for bounded y; $t^{-n+\frac{1}{2}} u_{yx}, t^{-n+\frac{3}{2}} u_{yt}$ are bounded for bounded y,

$$|u_{yyy}| \leq E_5 t^{n-\frac{3}{2}}, \quad E_i = const. > 0, \ (i = 1,2,3,4,5);$$

equation (1.1.114) holds almost everywhere in D. Moreover, the following estimates hold for this solution:

$$\Phi^{-1}\left(yt^{-\frac{1}{2}}(1-\alpha t^{\frac{1}{2}})\right) U \leq u \leq \Phi^{-1}\left(yt^{-\frac{1}{2}}(1+\beta t^{\frac{1}{2}})\right) U, \quad \alpha, \beta = const. > 0,$$

$$\Phi(\zeta) \equiv \int_0^\zeta (Y_0(s))^{-1} ds \ \ (\Phi^{-1}) \text{ is the inverse of } \Phi,$$

$$U(1-e^{-\nu_1}) \leq u \leq U(1-e^{-\nu_2}),$$

$$\nu_2 = \left[\frac{M_1 y}{2t^{1/2}(1-\beta t^{1/2})}\right]^2 + \frac{M_1 y\sqrt{-\ln \mu}}{t^{1/2}(1-\beta t^{1/2})}, \quad \nu_1 = \left[\frac{M_2 y}{2t^{1/2}(1+\alpha t^{1/2})}\right]^2$$
$$+ \frac{M_2 y\sqrt{-\ln \mu}}{t^{1/2}(1+\alpha t^{1/2})},$$

$$\nu M_1^2 = 1, \quad \nu M_2^2 = \frac{1}{2} - \delta, \quad \delta, \varepsilon = const. > 0,$$

$$1 - \frac{u}{U} = \exp\left(-\frac{1}{4\nu t}\left[y^2\left(1 + O(\sqrt{t}) + O(y^{1+\varepsilon} t^{\frac{1-\varepsilon}{2}})\right)\right]\right), \quad as \ y \to \infty, t \to 0,$$

$$\left|t^n \frac{u_y}{U} - t^{n-\frac{1}{2}} Y_0\left(\frac{u}{U}\right)\right| \leq E_6 t^n Y_0\left(\frac{u}{U}\right),$$

$$\left|t^n \frac{u_{yy}}{u_y} - t^{n-\frac{1}{2}} Y_{0\eta}\left(\frac{u}{U}\right)\right| \leq E_7 t^n Y_{0\eta}\left(\frac{u}{U}\right),$$

$$t(u_{yyy}u_y - u_{yy}^2)u_y^{-2} \leq -E_8, \quad E_i = const. > 0, \ (i = 1, 2, 3, \cdots 8).$$

The solution u, v of problems (1.1.114)–(1.1.115) is unique in the class of functions.

Oleinik and Samokhin [106] in 1999 considered the nonstationary system of the two-dimensional boundary layer

$$\begin{cases} \frac{\partial u}{\partial t} + u\frac{\partial u}{\partial x} + v\frac{\partial u}{\partial y} = \nu\frac{\partial^2 u}{\partial y^2} + U\frac{\partial U}{\partial x} + \frac{\partial U}{\partial t}, \\ \frac{\partial u}{\partial x} + \frac{\partial v}{\partial y} = 0, \end{cases} \tag{1.1.118}$$

in a domain $D = \{0 < t < T, 0 < x < X, 0 < y < \infty\}$, with the boundary conditions

$$\begin{cases} u(0, x, y) = U(0, x), \ \ u(t, 0, y) = 0, \ \ u(t, x, 0) = 0, \ \ v(t, x, 0) = v_0(t, x), \\ u(t, x, y) \to U(t, x) \ \text{ as } \ y \to \infty. \end{cases} \tag{1.1.119}$$

It is also assumed that $U(t, 0) = 0, U(t, x) > 0$ for $x > 0, t \geq 0$; $v_0(t, x), U(t, x), U_x, U_t/U$ are bounded.

Introduce new independent variables in (1.1.118)–(1.1.119), setting

$$\tau = \sqrt{t}, \ \ \xi = x, \ \ \eta = \frac{u(t, x, y)}{U(t, x)},$$

and a new unknown function

$$w(\tau, \xi, \eta) = \frac{\sqrt{t}u_y(t, x, y)}{U(t, x)}.$$

From (1.1.118), we obtain the following equation for $w(\tau, \xi, \eta)$:

$$\nu w^2 w_{\eta\eta} - \frac{\tau}{2}w_\tau - \tau^2\eta U w_\xi + Aw_\eta + Bw = 0, \tag{1.1.120}$$

where

$$A = \tau^2(\eta^2 - 1)U_x + \tau^2(\eta - 1)\frac{U_t}{U}, \ \ B = -\eta\tau^2 U_x(\xi) - \tau^2\frac{U_t}{U} + \frac{1}{2},$$

in the domain $\Omega = \{0 < \tau < \sqrt{T}, 0 < \xi < X, 0 < \eta < 1\}$, with the boundary conditions

$$w|_{\eta=1} = 0, \quad (\nu w w_\eta - v_0 w \tau + C)_{\eta=0} = 0, \tag{1.1.121}$$

where

$$C = \tau^2 U_x + \tau^2 \frac{U_t}{U}.$$

Theorem 1.1.75 ([106]) *Assume that the functions $U, U_x, U_{xx}, U_t/U, (U_t/U)_x, \sqrt{t}U_{xt}, \sqrt{t}(U_t/U)_x, v_0$ are bounded, and*

$$U_x(t,0) > 0, \quad U_x + \frac{U_t}{U} > 0, \quad -M_1 U < \frac{U_t}{U}$$

for small x. Assume also that

$$v_0 \le M_2 t^{\frac{1}{2}+\varepsilon}, \quad |v_{0x}| \le M_3 t^{\frac{1}{2}}, \quad t^{\frac{1}{2}} |v_{0t}| \le M_4,$$

where ε is a positive constant. Then problems (1.1.118)–(1.1.119), in the domain D with small enough T, have one and only one solution u, v with the following properties: u/U is continuous for $t = 0$, $y > 0$ and for $t > 0$, $y \ge 0$; $\sqrt{t}u_y/U$ is bounded and continuous in $\overline{D}$; $u \to U$ as $y \to \infty$; $u_y, u_x, u_{yy}, u_t, u_y$ bounded for $t \ge t_0 > 0$; conditions (1.1.119) are satisfied for u, v, and equation (1.1.118) holds almost everywhere in D. Moreover, the following inequalities are valid:

$$C_8\left(1 - \frac{u}{U}\right)\sigma_\mu\left(\frac{u}{U}\right) \le \frac{\sqrt{t}u_y}{U} \le C_7\left(1 - \frac{u}{U}\right)\sigma_{\mu_1}\left(\frac{u}{U}\right),$$

$$-C_1\sigma_{\mu_1}\left(\frac{u}{U}\right) \le \frac{\sqrt{t}u_{yy}}{u_y} \le C_{12}\Phi\left(\frac{u}{U}\right),$$

$$\left|\frac{2tu_{yy}}{u_y}\frac{uU_t - u_tU}{U^2} + \frac{u_y}{U} + 2t\left(\frac{u_{yt}U - u_yU_t}{U^2}\right)\right| \le C_{10}\left(1 - \frac{u}{U}\right)\sigma_{\mu_1}\left(\frac{u}{U}\right),$$

$$U(t,x)\exp\left(-\frac{C_7^2 y^2}{4t} - \frac{C_7 y\sqrt{-\ln\mu_1}}{\sqrt{t}}\right) \le U(t,x)$$

$$-u \le U(t,x)\exp\left(-\frac{C_8^2 y^2}{4t} - \frac{C_8 y\sqrt{-\ln\mu_1}}{\sqrt{t}}\right),$$

where $\sigma_\gamma = \sqrt{-\ln\gamma(1-\eta)}$, K_i, $0 < \mu < e^{-1/2} < \mu_1 < 1$; $\Phi(s) = -s^2$ for $0 \le s \le 1 - \delta_0$; $\Phi(s) = -\sigma_\mu(s)$ for $1 - \delta_0 < s < 1$, $\delta_0 > 0$; $C_7, C_8, C_9, C_{10}, C_{11}, C_{12}$ are positive constants.

Oleinik and Samokhin [106] in 1999 considered the boundary layer theory for the numerical solution of these problems by finite differences method, the nonstationary system of the two-dimensional boundary layer

$$\begin{cases} \frac{\partial u}{\partial t} + u\frac{\partial u}{\partial x} + v\frac{\partial u}{\partial y} = \nu\frac{\partial^2 u}{\partial y^2} + U\frac{\partial U}{\partial x} + \frac{\partial U}{\partial t}, \\ \frac{\partial u}{\partial x} + \frac{\partial v}{\partial y} = 0, \end{cases} \tag{1.1.122}$$

in a domain $D = \{0 < t < T, 0 < x < X, 0 < y < \infty\}$, with the boundary conditions

$$\begin{cases} u(0, x, y) = u_0(x, y), \quad u(t, 0, y) = u_1(t, y), \quad u(t, x, 0) = 0, \\ \quad v(t, x, 0) = v_0(t, x), \\ u(t, x, y) \to U(t, x) \text{ as } y \to \infty. \end{cases} \tag{1.1.123}$$

Introduce new independent variables in (1.1.122)–(1.1.123), setting

$$\tau = t, \quad \xi = x, \quad \eta = u(t, x, y),$$

and a new unknown function

$$w(\tau, \xi, \eta) = u_y(t, x, y),$$

we obtain the following equation :

$$\nu w^2 w_{\eta\eta} - w_\tau - \eta U w_\xi + p_x w_\eta = 0, \tag{1.1.124}$$

in the domain $\Omega = \{0 < \tau < T, 0 < \xi < X, 0 < \eta < U(\tau, \xi)\}$, with the boundary conditions

$$\begin{aligned} &w|_{\tau=0} = u_{0y} = w_0(\xi, \eta), \quad w|_{\xi=0} = u_{1y} = w_1(\tau, \eta), \quad w|_{\eta=U\tau,\xi} = 0, \\ &(\nu w w_\eta - v_0 w - p_x)|_{\eta=0} = 0. \end{aligned} \tag{1.1.125}$$

Denote by Ω' the union of the domain Ω and the part of its boundary on which the initial and the boundary conditions are prescribed. A node is said to be internal if its neighboring nodes belong to Ω'. The remaining nodes belonging to Ω' are called boundary nodes.

For any function $f(\tau, \xi, \eta)$, denoted by $f^{m,l,k}$ its value at the node $(mh, l\sigma, k\sigma)$. They associated a finite difference equation for w which approximates equation (1.1.124), namely,

$$\begin{aligned} &\left(\nu(w^{m,l,k})^2 + M\sigma\right)\frac{w^{m,l,k+1} - 2w^{m,l,k} + w^{m,l,k-1}}{\sigma^2} - \frac{w^{m+1,l,k} - w^{m,l,k}}{h} \\ &- k\sigma\frac{w^{m,l,k} - w^{m,l-1,k}}{\sigma} + p_x^{m,l,k}\frac{w^{m,l,k} - w^{m,l,k-1}}{\sigma} = 0, \end{aligned} \tag{1.1.126}$$

where M is a positive constant such that $M > \max |p_x|$. At the boundary nodes of O' belonging to the planes $\tau = 0, \xi = 0, \eta = 0$, we set

$$\begin{cases} w_0^{0,l,k} = w_0(l\sigma, k\sigma), \quad w^{m,0,k} = w_1(mh, k\sigma), \\ \nu w^{m,l,0} \frac{w^{m+1,l,1} - w^{m+1,l,0}}{\sigma} - p_x^{m,l,0} - v_0^{m,l,0} w^{m,l,0} = 0. \end{cases} \tag{1.1.127}$$

At every boundary node of Ω' that does not belong to any of the planes $\tau = 0, \xi = 0, \eta = 0$ (the set of all such points is denoted by $\Gamma_{h\sigma}$), we set

$$w^{m,l,k} = 0. \tag{1.1.128}$$

Theorem 1.1.76 ([106]) *Let W be a solution of problem (1.1.124)–(1.1.125) with bounded second derivatives in Ω', and let w be the solution of finite difference equations (1.1.126)–(1.1.128). Then, for small enough h and σ, the following inequality holds:*

$$|W - w| \leq M_3(h + \sigma),$$

if either $\tau \leq T$ and $h/\sigma^2 < 1/(2\nu b_1^2)$ or $\xi \leq X$ and $h/\sigma^2 < 1/(2\nu b_2^2)$.

By $f^{m,l,k}$ they denoted the value of a function f at the point $(mh, l\sigma, k\sigma)$. To every internal node of Ω' with the coordinates $((m+1)h, l\sigma, k\sigma)$, they associated a difference equation for w which approximates equation (1.1.124):

$$\begin{aligned} &\left(\nu (w^{m,l,k})^2 + Mh\right) \frac{w^{m+1,l,k+1} - 2w^{m+1,l,k} + w^{m+1,l,k-1}}{h^2} - \frac{w^{m+1,l,k} - w^{m,l,k}}{h} \\ &- kh \frac{w^{m+1,l,k} - w^{m+1,l,k-1}}{h} + p_x^{m,l,k} \frac{w^{m+1,l,k} - w^{m+1,l,k-1}}{h} = 0, \end{aligned} \tag{1.1.129}$$

where M =const.> $\max |p_x|$. To the boundary conditions (1.1.125), we associate the following equations at boundary nodes:

$$\begin{cases} w^{0,l,k} = w_0(lh, kh), \quad w^{m,0,k} = w_1(mh, kh), \quad \text{on } \Gamma_h, \\ \nu w^{m,l,0} \frac{w^{m+1,l,1} - w^{m+1,l,0}}{h} - p_x^{m,l,0} - v_0^{m,l,0} w^{m,l,0} = 0. \end{cases} \tag{1.1.130}$$

By Γ_h, we denote the boundary nodes of Ω' that do not belong to the planes $\tau = 0, \xi = 0, \eta = 0$.

Theorem 1.1.77 ([106]) *Let W be the solution of problems (1.1.124)–(1.1.125) with bounded second derivatives in Ω', and let w be the solution of finite difference equations (1.1.129)–(1.1.130). Then, at the nodes of Ω', for small enough h, the inequality*

$$|W - w| \leq k_1 h$$

holds either for $\tau \le T$ or for $\xi \le X$.

Oleinik and Samokhin [106] in 1999 applied the finite difference method to construct a solution of the boundary layer system in the case of a nonstationary axially symmetric flow (1.1.104)–(1.1.105) in a domain $D = \{0 < t < \infty, 0 < x < X, 0 < y < \infty\}$. They replaced equation (1.1.106) and conditions (1.1.107) by the following linear system of algebraic equations with respect to $w^{m+1,l,k}$:

$$\begin{aligned}&\left(\nu(w^{m,l,k})^2\right)\frac{w^{m,l,k+1}-2w^{m,l,k}+w^{m,l,k-1}}{\delta^2}-\frac{w^{m+1,l,k}-w^{m,l,k}}{h}\\&-k\delta U^{m,l,k}\frac{w^{m,l,k}-w^{m,l-1,k-1}}{d}\\&+A^{m,l,k}\frac{w^{m,l,k}-w^{m,l,k-1}}{\delta}+B^{m,l,k}w^{m,l,k}=0,\end{aligned} \tag{1.1.131}$$

with $m = 0, 1 \dots$; $l = 0, 1, \dots [X/\varepsilon]$; $k = 1, \dots, J-1$, and the following conditions:

$$\begin{cases} w^{m+1,l,0} = \frac{1}{2}\Big[w^{m+1,l,1} - \delta\nu^{-1}v_0^{m+1,l,0} + \\ \qquad \sqrt{(w^{m+1,l,0} - \delta\nu^{-1}v_0^{m+1,l,0})^2 + 4 - \delta\nu^{-1}C^{m+1,l,0}}\Big], \\ w^{0,l,k} = w_0(ld, k\delta), \quad l = 0, 1, \dots [X/d]; \quad k = 0, \dots, J; \\ w^{m+1,l,J} = 0, \quad l = 0, 1, \dots [X/d]; \quad m = 0, 1, \dots. \end{cases} \tag{1.1.132}$$

Theorem 1.1.78 ([106]) *Let A, B, C, v_0, U be bounded functions in Ω. Assume that $U_x > 0$, $C > 0$ in Ω and*

$$E_1(1-\eta)\sigma \le w_0 \le E_2(1-\eta)\sigma,$$

where $\sigma = \sqrt{-\ln\mu(1-\eta)}$, $\mu < 1$, and let the function $W(\tau, \xi, \eta)$ be a solution of problems (1.1.106)–(1.1.107) such that

$$\begin{aligned}&M_{12}(1-\eta)\sigma \le W \le M_{13}(1-\eta)\sigma,\\&|W_\xi| \le M_{14}(1-\eta)\sigma, \quad -M_{15}\sigma \le W_\eta \le -M_{16}\sigma,\\&-M_{17}(1-\eta)\sigma \le W_\tau \le M_{18}(1-\eta)\sigma,\\&-M_{19} \le WW_{\eta\eta} \le -M_{20}.\end{aligned}$$

Assume, in addition, that W has bounded weak derivatives $W_{\tau\tau}$, $W_{\xi\xi}$ in Ω, and $W_{\eta\eta}$ satisfies the inequality

$$|W_{\eta\eta}(\tau,\xi,\eta_1)-W_{\eta\eta}(\tau,\xi,\eta_2)|\le\frac{M_{17}|\eta_1-\eta_2|}{(1-\eta_2)^2\sigma(\eta_2)}$$

for any η_1,η_2 such that $0<\eta_1<\eta_2<1$. Then for small enough δ and any $m,l,k,mh\le T$ (T is an arbitrary positive constant), we have

$$|W^{m+1,l,k}-w^{m+1,l,k}|\le M_{18}(h+d+\delta\sigma^{J-1}),$$

provided that $h\delta^{-2}\le(2\nu a^2+b\delta^2d^{-1})^{-1}$, where $w^{m+1,l,k}$ is a solution of (1.1.131)–(1.1.132), and the constant M_{18} is independent of h,d,δ, but may depend on T.

Xin and Zhang [149] in 2004 obtained the existence of global weak solutions to the Prandtl equation (1.1.134) when the pressure is favorable ($\partial_x p\le 0$), and the following conditions are satisfied:

$$\begin{cases}U(x,t)>0,\ u_0(x,t)>0,\ u_1(y,t)>0,\ v_0(x,t)\le 0,\\ \partial_y u_0(x,y)>0,\ \partial_y u_1(x,y)>0,\\ U(x,t)\equiv d=constant.\end{cases}\tag{1.1.133}$$

The considered problem in [149] is the following initial-boundary value problem:

$$\begin{cases}\partial_\tau w^{-1}+\eta U\partial_\xi w^{-1}+A\partial_\eta w^{-1}-Bw^{-1}=-v\partial_\eta^2 w\\ on\ Q=\{(\xi,\eta,\tau)|0<\tau<\infty,\ 0<\xi<L,\ 0<\eta<1\},\\ w|_{\tau=0}=\frac{\partial_y u_0}{U}\equiv w_0,\quad w|_{\eta=1}=0,\\ w|_{\xi=0}=w_1,\quad and\ (vw\partial_\eta w-v_0w)|_{\eta=0}=\frac{\partial_y U_1(0,y,t)}{U(0,t)},\end{cases}\tag{1.1.134}$$

where $A=(1-\eta^2)\partial_x U+(1-\eta)\frac{\partial_t U}{U}$, $B=\eta\partial_x U+\frac{\partial_t U}{U}$ and $w_1(\tau,\eta)=\frac{\partial_y U_1(0,y,t)}{U(0,t)}$.

To study problem (1.1.134), we need the following Lemmas 1.1.79–1.1.88. To this end, we consider the problem

$$\begin{cases}u_t-u^2u_{yy}=0,\ 0<y<d,\\ u|_{t=0}=u_0>0,\\ \frac{\partial u}{\partial y}|_{y=0}=v_0,\ u|_{y=d}=0.\end{cases}\tag{1.1.135}$$

Lemma 1.1.79 ([149]) *Let $\phi=e^{\alpha y}\sin\alpha(d-y)$, where $\alpha=\frac{\pi}{2d}$: Suppose $v_0\le 0$. Then there exist constants β and C_0 which depend only on d $|u_0|_{L^\infty}$ and $|v_0|_{L^\infty}$ such that the solution of* (1.1.135) *satisfies*

$$|u|\le C_0(d-y),\tag{1.1.136}$$

$$u\ge\theta_0e^{-\beta t}\phi,\tag{1.1.137}$$

provided that (1.1.136) *and* (1.1.137) *hold true initially, where* $\theta_0 = \min \frac{u_0}{\phi}$.

Lemma 1.1.80 ([149])

$$|u_y^2|_{L^\infty} \le a_0^2. \tag{1.1.138}$$

We estimate $w = u_x$ which solves the following problem:

$$\begin{cases} w_t - (u^2+\varepsilon)w_{yy} = 2uu_{yy}w + 2uu_{yy}v_{0,x}(y-d) - v_{0,tx}(y-d), \\ w|_{t=0} = u_{0,x} - v_{0,x}(y-d) \equiv w_0, \\ \frac{\partial w}{\partial y} = 0, \ w|_{y=d} = 0. \end{cases} \tag{1.1.139}$$

Lemma 1.1.81 ([149]) *Suppose that* $v_0 \le 0$. *Then there exists a constant* C_6 *depending only on T, d,* $|u_{0,x}|_{L^\infty}$, $|u_0|_{L^\infty}$, $|u_y|_{L^\infty}$, $|v_0|_{C^2}$ *and* $\int_0^d \frac{u_{0,x}^2}{(y-d)^2}dy$, *such that the solution to problem* (1.1.139) *satisfies*

$$|u_x(t,x,y)| \le C_6. \tag{1.1.140}$$

For the initial-boundary value problem,

$$\begin{cases} u_t + yu_x - u^2u_{yy} = 0, \ (0,T)\times\Omega, \\ u|_{t=0} = u_{0,y}, \ u|_{x=0} = u_{1,y}, \\ \frac{\partial u}{\partial y}|_{y=0} = v_0, \ u|_{y=1} = 0, \end{cases} \tag{1.1.141}$$

we have the following three lemmas

Lemma 1.1.82 ([149]) *For* $0 \le t \le T, (x,y) \in \Omega$, *it holds for solution u to problem* (1.1.141) *that*

$$|u_t(t,x,y)| \le |\overline{u}_0|_{L^\infty} + 2|v_0|_{L^\infty} + |\overline{u}_1|_{L^\infty}. \tag{1.1.142}$$

$$|u(t,x,y)| \le |\overline{u}_0|_{L^\infty} + 2|v_0|_{L^\infty} + |\overline{u}_1|_{L^\infty}. \tag{1.1.143}$$

Lemma 1.1.83 ([149]) *Suppose* $v_0 \le 0$. *Then there exist constants* C_0 *and* β *depending only on* $|v_0|_{L^\infty}$, $|\overline{u}_0|_{L^\infty}$ *and* $|\overline{u}_1|_{C^1}$, *such that the solution u to problem* (1.1.141)

$$u_t(t,x,y) \le C_0(1-y), \tag{1.1.144}$$

$$u(t,x,y) \ge \theta_0\varepsilon^{-\beta t}\phi. \tag{1.1.145}$$

Lemma 1.1.84 ([149]) *There exists a constant* C_3 *depending on*$|\overline{u}_0|_{L^\infty}$, $|v_0|_{L^\infty}$, *and* $|\overline{u}_1|_{C^1}$ *such that the solution u to problem* (1.1.141)

$$\begin{aligned}\int\int_\Omega |u_x(t,\cdot)|dxdy \le C_3 e^{C_3 t}[\int\int_\Omega (C_3^2|\overline{u}_{0,x}|)dxdy + \tfrac{L}{2}|v_{0,x}|_{L^\infty} \\ +\tfrac{1}{2}\int_0^1 \tfrac{|\overline{u}_{1,t}|}{\overline{u}_1^2}(1-y)^2dy].\end{aligned} \tag{1.1.146}$$

Now we consider the following problem:

$$\begin{cases} u_t = u^2 u_{yy} + b(t,x,y)u, & (t,x,y) \in (0,T)\times\Omega, \\ u|_{t=0} = u_{0,y} > 0, \\ \frac{\partial u}{\partial y}|_{y=0} = v_0, \quad u|_{y=1} = 0, \end{cases} \tag{1.1.147}$$

which is approximated by

$$\begin{cases} u_t = (u^2+\varepsilon) u_{yy} + b(t,x,y)u, & (t,x,y) \in (0,T)\times\Omega, \\ u|_{t=0} = u_{0,y} > 0, \\ \frac{\partial u}{\partial y}|_{y=0} = v_0, \quad u|_{y=1} = 0. \end{cases} \tag{1.1.148}$$

The following Lemmas 1.1.85–1.1.88 are concerned with problem (1.1.148).

Lemma 1.1.85 ([149]) *The solution of problem* (1.1.147) *satisfies*

$$\begin{cases} |v_0|_{L^\infty}(1-y) \mp ue^{-Bt} \le \max_{x,y}(|v_0|(1-y) \mp ue^{-Bt_0})|_{t=t_0}, \\ |u| \le C_0(1-y)e^{Bt}, \\ u \ge \theta_0 e^{-\beta t}\phi. \end{cases} \tag{1.1.149}$$

Lemma 1.1.86 ([149]) *There exists a constant* C_4, *independent of E, such that the solution u to* (1.1.148) *as well as* (1.1.147) *satisfies*

$$|u_y(t,x,y)| \le C_4. \tag{1.1.150}$$

Lemma 1.1.87 ([149]) *There exists a constant* C_5, *independent of E, such that the solution u to* (1.1.148) *as well as* (1.1.147) *satisfies*

$$|u_x(t,x,y)| + |u_t(t,x,y)| \le C_5. \tag{1.1.151}$$

Lemma 1.1.88 ([149]) *There exists a constant* C_9, *depending on* T, Ω, B, $|\overline{u}_0|_{L^\infty}$, $|v_0|_{L^\infty}$, *and* $|\overline{u}_1|_{C^1}$ *such that the solution u of* (1.1.148) *as well as* (1.1.147) *satisfies*

$$\int\int_\Omega |u_x(t,\cdot)|dxdy \le e^{C_9 t}[\int\int_\Omega |\overline{u}_{0x}| + C_9]dxdy, \tag{1.1.152}$$

for $t \le T$.

Based on Lemmas 1.1.79–1.1.88, we have the following two theorems.

Theorem 1.1.89 ([149]) *Assume that the data satisfy conditions* (1.1.133). *Then there exists a weak solution* $w \in BV(Q_T) \cap L^\infty(Q_T)$ *to the initial value problem* (1.1.134) *provided that the pressure is favorable, i.e.,* $\partial_x p(x,t) \le 0$ *holds for* $t > 0$, $0 < x < L$.

Theorem 1.1.90 ([149]) *Assume that* $v_0 \le 0$ *and the data are smooth and compatible. Then problem* (1.1.135) *has a unique positive bounded smooth solution in* $(0, T] \times \Omega$. *Moreover*

$$u \in L_p([0, T] \times \overline{\Omega}) \tag{1.1.153}$$

and u satisfies estimates (1.1.136)–(1.1.139) *and* (1.1.142).

From above estimates on problem (1.1.141), we have the following theorem.

Theorem 1.1.91 ([149]) *Assume that* (1.1.133) *hold and* $b(t, x, y) = 0$. *Then problem* (1.1.141) *has a weak solution*

$$u \in L^\infty(0, T, BV(\Omega)) \cap BV((0, t) \times \Omega), \tag{1.1.154}$$

which satisfies estimates (1.1.143)–(1.1.145). *Furthermore, its derivatives satisfy* (1.1.146) *and*

$$\begin{aligned}\int\int_\Omega |u_y(t,\cdot)|dxdy \le \int_0^t\int_0^1 |\overline{u}_{1,y}|dyds + 2L|u|_{L^\infty} + \int\int_\Omega |\overline{u}_{0,y}|dxdy \\ + \int_0^t\int\int_\Omega |u_x(s,\cdot)|dxdyds,\end{aligned} \tag{1.1.155}$$

$$\begin{aligned}\int\int_\Omega |u_t(t,\cdot)|dxdy \le \int_0^1\left[\tfrac{y(1-y)}{u(t,0,y)} + \tfrac{y(1-y)}{u(t,L,y)}\right]dy + \int_0^1 |v_0(t,\cdot)|dx \\ + \int\int_\Omega |u_y(t,\cdot)|dxdy.\end{aligned} \tag{1.1.156}$$

Theorem 1.1.92 ([149]) *Suppose that* $v_0 \le 0$ *and the data are smooth and compatible. Then problem* (1.1.147) *has a unique positive bounded smooth solution and*

$$u \in L_p([0, T] \times \overline{\Omega}), \tag{1.1.157}$$

Moreover, the solution satisfies (1.1.149)–(1.1.151).

Lemma 1.1.93 ([149]) *There exists a constant* C_{11} *depending on* $T, \Omega, B, |\overline{u}_0|_{L^\infty}$, $|v_0|_{L^\infty}$, *and* $|\overline{u}_1|_{C^1}$ *such that the solution u of problem* (1.1.148) *satisfies*

$$\int\int_\Omega |u_y(t,\cdot)|dxdy \le e^{Bt}\left[\int\int_\Omega\left[|\overline{u}_{0y}| + \tfrac{n}{T}\Sigma_{j=1}^i |u_x(t_{j-1},\cdot)|\right]dxdy + C_{11}.\right. \tag{1.1.158}$$

From Theorems 1.1.91–1.1.92 and Lemma 1.1.93, we conclude the following theorem.

Theorem 1.1.94 ([149]) *Under assumptions* (1.1.133), *problem* (1.1.141) *has a solution*

$$u \in L^\infty(0, T; BV(\Omega)) \cap BV((0, T) \times \Omega) \tag{1.1.159}$$

satisfying (1.1.149). *Furthermore, its derivatives are estimated by* (1.1.158) *and* (1.1.160) *and*

$$\int\int_\Omega |u_y(t,\cdot)|dxdy \le e^{Bt} \int\int_\Omega [|\overline{u}_{0y}| + \int_0^t |u_x|ds]dxdy + C_{11}. \tag{1.1.160}$$

Xu and Zhao [152] in 2009 studied the global existence and uniqueness of weak solutions to the nonstationary boundary layer (1.1.58) by Crocco transformation. The initial boundary value problem is

$$w_\tau + \eta U w_\xi + A w_\eta + B w - w^2 w_{\eta\eta} = 0 \quad in\ \Omega_T, \tag{1.1.161}$$

$$w(0, \xi, \eta) = w_0(\xi, \eta) = \frac{u_{0y}}{U(0,x)},\ w(\tau, \xi, 1) = 0, \tag{1.1.162}$$

$$w w_\eta - v_0 w + (U_\xi + \frac{U_\tau}{U}) = 0 \quad on\ \eta = 0, \tag{1.1.163}$$

where $\Omega_T = \{(\tau, \xi, \eta)|0 < \tau < T, 0 < \xi < L, 0 < \eta < 1\}$ and

$$A = (1 - \eta^2)U_\xi + (1 - \eta)\frac{U_\tau}{U},\ B = \eta U_\xi + \frac{U_\tau}{U}. \tag{1.1.164}$$

Theorem 1.1.95 ([152]) *Assume that* $U = \xi^m U_1(\tau, \xi)$ $m \ge 1$, $U_1(\tau, \xi)$, $w_0(\xi, \eta)$, $v_0(\tau, \xi)$ *are appropriately smooth and*

$$U_1(\tau, \xi) > 0,\ U_\tau + U U_\xi > 0,\ v_0(\tau, \xi) \le 0,\ w_0(\tau, \xi) \ge 0, \tag{1.1.165}$$

and there exists a constant C_0 *such that*

$$C_0^{-1}(1 - \eta) \le w_0(\xi, \eta) \le C_0(1 - \eta), \tag{1.1.166}$$

where w_0 *are the functions in* (1.1.162).

Theorem 1.1.96 ([152]) *The solution of* (1.1.161)–(1.1.163) *is unique.*

Xu and Zhang [150] in 2015 proved the local-in-time existence, uniqueness, and stability of solutions for the nonlinear Prandtl equation (1.1.58) where U is a constant in weighted Sobolev space with shear flow without monotonicity in the domain $\mathbb{R}_+^2$ and proved the global-in-time existence of solutions to the Prandtl equation (1.1.58) .

They assumed that $u_0(x, y) = u_0^s(y) + \tilde{u}_0(x, y)$, where u_0^s satisfies the following conditions: for some even integer $m \geq 6$, and $k > 1$, $u_0^s \in C^{m+4}([0, +\infty))$, $\lim_{y\to+\infty} u_0^s(y) = 1$ with the compatibility conditions $(\partial_y^{2p} u_0^s)(0) = 0$, $0 \leq 2p \leq m + 4$, and that there exists $0 < a < +\infty$, $c_1, c_2 > 0$ such that

$$\begin{cases} \partial_y u_0^s(a) = 0,\ \partial_y^2 u_0^s(a) \neq 0, \text{ and } \partial_y u_0^s(y) \neq 0, \text{ for any } y \in \bar{\mathbb{R}}_+ \ \{a\}, \\ c_1\langle y\rangle^{-k} \leq |\partial_y u_0^s(y)| \leq c_2\langle y\rangle^{-k}, \forall y \geq a + 1, \\ |\partial_y^p u_0^s(y)| \leq c_2\langle y\rangle^{-k-p+1}, \forall y \geq 0, 0 \leq p \leq m + 4. \end{cases} \tag{1.1.167}$$

The weighted Sobolev spaces, for $\lambda \in \mathbb{R}$, $m \in \mathbb{N}$,

$$\|u\|^2_{H^m_\lambda(\mathbb{R}^2_+)} = \sum_{|\alpha_1+\alpha_2|\leq m} \int_{\mathbb{R}^2_+} \langle y\rangle^{2\lambda+2\alpha_2} |\partial_x^{\alpha_1}\partial_y^{\alpha_2} u|^2 dxdy,$$

$$\|u\|^2_{L^2_\lambda(\mathbb{R}^2_+)} = \int_{\mathbb{R}^2_+} \langle y\rangle^{2\lambda} |u|^2 dxdy.$$

Theorem 1.1.97 ([150]) *Let $m \geq 6$ be an even integer, $k \geq 1$, $0 \leq \ell \leq \frac{1}{2}$, $k + \ell > \frac{1}{2}$. Assume that u_0^s satisfies (1.1.167), the initial data $u_0 - u_0^s \in H^{m+3}_{2k+\ell-1}(\mathbb{R}^2_+)$ and $u_0 - u_0^s$ satisfies the compatibility condition up to order $m + 2$. Then there exists a $T > 0$ such that if*

$$\|(u_0 - u_0^s)\|_{H^{m+3}_{2k+\ell-1}(\mathbb{R}^2_+)} \leq \delta_0, \tag{1.1.168}$$

for some $\delta_0 > 0$ small enough, the initial-boundary value problem (1.1.58) admits a unique solution (u, v) with

$$u - u^s \in L^\infty([0, T]; H^m_{k+\ell-1}(\mathbb{R}^2_+)),\ \partial_y(u - u^s) \in L^\infty([0, T]; H^m_{k+\ell}(\mathbb{R}^2_+)),$$

and

$$v \in L^\infty([0, T]; L^\infty(\mathbb{R}_{y,+}; H^{m-1}(\mathbb{R}_x))),\ \partial_y v \in L^\infty([0, T]; H^{m-1}_{k+\ell-1}(\mathbb{R}^2_+)).$$

Moreover, we have the stability with respect to the initial data in the following sense: given any two initial data

$$u_0^1 = u_0^s + \tilde{u}_0^1, \quad u_0^2 = u_0^s + \tilde{u}_0^2,$$

if u_0^s satisfies problem (1.1.167) and $\tilde{u}_0^1$, $\tilde{u}_0^2$ satisfies (1.1.168), then the solutions u^1 and u^2 of problem (1.1.58) with initial data u_0^1 and u_0^2, respectively, satisfy,

$$\|u^1 - u^2\|_{L^\infty([0,T]; H^{m-2}_{k+\ell-1}(\mathbb{R}^2_+))} \leq \|u_0^1 - u_0^2\|_{H^m_{2k+\ell-1}(\mathbb{R}^2_+)}.$$

Assume the initial data of the shear flow u_0^s is uniformly monotonic on $\mathbb{R}_+$ in the following sense:

$$\begin{cases} c_1\langle y\rangle^{-k} \le |\partial_y u_0^s(y)| \le c_2\langle y\rangle^{-k}, \forall y \ge 0, \\ |\partial_y^p u_0^s(y)| \le c_2\langle y\rangle^{-k-p+1}, \forall y \ge 0, 1 \le p \le m+2, \end{cases} \tag{1.1.169}$$

for certain $c_1, c_2 > 0$, they obtained the following existence of almost global-in-time solution.

Theorem 1.1.98 ([150]) *Let $m \ge 6$ be an even integer, $k \ge 1, 0 \le \ell \le \frac{1}{2}, k+\ell > \frac{3}{2}$. Assume that u_0^s satisfies (1.1.169), the initial data $u_0 - u_0^s \in H^{m+3}_{2k+\ell-1}(\mathbb{R}^2_+)$ and $u_0 - u_0^s$ satisfies the compatibility condition up to order $m+2$. Then for any $T > 0$, there exists $\delta_0 > 0$ small enough such that if*

$$\|(u_0 - u_0^s)\|_{H^{m+3}_{2k+\ell-1}(\mathbb{R}^2_+)} \le \delta_0,$$

then the initial-boundary value problem (1.1.58) admits a unique solution (u, v) with

$$u - u^s \in L^\infty([0,T]; H^m_{k+\ell-1}(\mathbb{R}^2_+)), \partial_y(u-u^s) \in L^\infty([0,T]; H^m_{k+\ell}(\mathbb{R}^2_+)),$$

and

$$v \in L^\infty([0,T]; L^\infty(\mathbb{R}_{y,+}; H^{m-1}(\mathbb{R}_x))), \partial_y v \in L^\infty([0,T]; H^{m-1}_{k+\ell-1}(\mathbb{R}^2_+)).$$

It also has the stability with respect to the initial data as in Theorem 1.1.97.

Wu [146] in 2016 studied the regularity and global well-posedness of classical solutions to the following nonlinear unsteady Prandtl equation (1.1.170) with Robin or Dirichlet boundary condition in half space. Under Oleinik's monotonicity assumption, Wu [145] in 2015 proved the local well-posedness of the following Prandtl equations with Robin boundary condition:

$$\begin{cases} u_t + uu_x + vu_y = u_{yy}, \\ u_x + v_y = 0, \\ (u_y - \beta u)|_{y=0} = 0, \quad v|_{y=0} = 0, \\ u(0,x,y) = u_0, \\ \lim\limits_{y\to+\infty} u(t,x,y) = U(t,x). \end{cases} \tag{1.1.170}$$

The following Prandtl equations with Dirichlet boundary condition takes the form:

$$\begin{cases} u_t + uu_x + vu_y = u_{yy}, \\ u_x + v_y = 0, \\ u|_{y=0} = 0, \quad v|_{y=0} = 0, \\ u(0, x, y) = u_0, \\ \lim_{y \to +\infty} u(t, x, y) = U(t, x). \end{cases} \tag{1.1.171}$$

The functional norms

$$\|f\|_{\mathcal{H}^k_\ell(\mathbb{R}^2_+)} = \sum_{|\alpha|+\sigma \leq k} \|(1+y)^{\ell+\sigma} \partial^\alpha_{t,x} \partial^\alpha_y f\|_{L^2_{\ell+\sigma}(\mathbb{R}^2_+)}.$$

Theorem 1.1.99 ([146]) *Considering the nonlinear unsteady Prandtl equations with Robin boundary condition (1.1.170) under Oleinik's monotonicity assumption* $w = u_y > 0$. *Giving any integer* $k \geq 4$, $U(t, x) \in C^{k+1}([0, +\infty) \times \mathbb{R})$ *and* $U(t, x) > 0$, *we have the following existence, uniqueness, and stability results:*

(1) For any fixed finite number $T \in (0, +\infty)$, *there exist sufficiently small real number* $0 < \varepsilon_1 = o(T^{-1})$ *and suitably large real numbers* $\ell_0 > 1$, $\delta_\beta > 0$ *such that if* $\ell \geq \ell_0$, $\beta \in [\delta_\beta, +\infty)$, *the compatible initial data satisfies*

$$\begin{cases} w_0 > 0, \ u_0|_{y=0} > 0, \ (\partial_y u_0 - \beta u_0)|_{y=0} = 0, \quad \lim_{y \to +\infty} u_0 = U|_{t=0}, \\ \|w_0\|_{\mathcal{H}^k_\ell(\mathbb{R}^2_+)} + \frac{1}{\sqrt{\beta}} \|w_0|_{y=0}\|_{\mathcal{H}^k_\ell(\mathbb{R})} \leq \varepsilon_1, \quad \|U(t, x)\|_{H^{k+1}([0,T]\times\mathbb{R})} \leq C_0 \varepsilon_1, \\ 0 < c_1(1+y)^{-\theta} \leq w_0 \leq c_2(1+y)^{-\theta}, \theta > \frac{\ell+1}{2}, \end{cases} \tag{1.1.172}$$

then the Prandtl system (1.1.170) admits a unique classical solution (w, u, v) *in* $[0, T]$ *satisfying*

$$\begin{aligned} & w \in \mathcal{H}^k_\ell(\mathbb{R}^2_+), \quad w, w_y \in \mathcal{H}^k_\ell([0, T] \times \mathbb{R}^2_+), \\ & u - U \in \mathcal{H}^k_{\ell-1}(\mathbb{R}^2_+) \cap \mathcal{H}^k_{\ell-1}([0, T] \times \mathbb{R}^2_+), \\ & \partial^j_y u|_{y=0} \in H^{k-j}(\mathbb{R}) \cap H^{k-j}([0, T] \times \mathbb{R}), \quad 0 \leq j \leq k, \\ & \partial^\alpha_{t,x} v + y \cdot \partial^\alpha_{t,x} \partial_x U \in L^\infty_{y,\ell-1}(L^2_{t,x}), \quad |\alpha| \leq k-1. \end{aligned} \tag{1.1.173}$$

(2) The classical solution to (1.1.170) is stable with respect to the initial data in the following sense: for any given two initial data satisfying (1.1.172), then for all $p \leq k-1$, *the corresponding solutions of the Prandtl system (1.1.171) satisfy*

$$\|u^1 - u^2\|_{\mathcal{H}^p_{\ell-1}(\mathbb{R}^2_+)} + \sum_{j=0}^{p} \|\partial^j_y u^1|_{y=0} - \partial^j_y u^2|_{y=0}\|_{H^{p-j}(\mathbb{R})}$$

$$+\|w^1 - w^2\|_{\mathcal{H}_\ell^p(\mathbb{R}_+^2)} + \sum_{|\alpha|\le p-1} \|\partial_{t,x}^\alpha v^1 - \partial_{t,x}^\alpha v^2\|_{L^\infty_{y,\ell-1}(L^2_{t,x})}$$

$$\le C(\varepsilon_1, T)[\|w_0^1 - w_0^2\|_{\mathcal{H}_\ell^p(\mathbb{R}_+^2)} + \frac{1}{\beta-\delta_\beta}\|w_0^1|_{y=0} - w_0^2|_{y=0}\|^2_{H^p(\mathbb{R})}].$$

(3) As $\beta \to +\infty$, $\|u|_{y=0}\|_{H^k(\mathbb{R})} = O(\frac{1}{\sqrt{\beta}})$, $\|w|_{y=0}\|_{H^k(\mathbb{R})} = O(\sqrt{\beta})$ and (w, u, v) satisfies the regularities (1.1.173) uniformly.

Theorem 1.1.100 ([146]) *Considering the nonlinear unsteady Prandtl equations with Dirichlet boundary condition (1.1.171) under Oleinik's monotonicity assumption $w = u_y > 0$. Giving any integer $k \ge 6$, $U(t, x) \in C^{k+1}([0, +\infty) \times \mathbb{R})$ and $U(t, x) > 0$, we have the following existence, uniqueness, and stability results:*

(1) For any fixed finite number $T \in (0, +\infty)$, there exist sufficiently small real number $0 < \varepsilon_1 = o(T^{-1})$ and suitably large real numbers $\ell_0 > 1$ such that if $\ell \ge \ell_0$, the compatible initial data and $U(t, x)$ satisfies

$$\begin{cases} w_0 > 0, \ \ u_0|_{y=0} > 0, \ \ \lim\limits_{y\to+\infty} u_0 = U|_{t=0}, \\ \|w_0\|_{\mathcal{H}_\ell^k(\mathbb{R}_+^2)} \le \varepsilon_2, \ \ \|U(t,x)\|_{H^{k+1}([0,T]\times\mathbb{R})} \le C_0\varepsilon_2, \\ 0 < c_1(1+y)^{-\theta} \le w_0 \le c_2(1+y)^{-\theta}, \theta > \frac{\ell+1}{2}, \end{cases} \tag{1.1.174}$$

then the Prandtl system (1.1.171) admits a unique classical solution (w, u, v) in $[0, T]$ satisfying

$$\begin{aligned} & w \in \mathcal{H}_\ell^k(\mathbb{R}_+^2), \ \ w, w_y \in \mathcal{H}_\ell^k([0,T]\times\mathbb{R}_+^2), \\ & u - U \in \mathcal{H}_{\ell-1}^k(\mathbb{R}_+^2) \cap \mathcal{H}_{\ell-1}^k([0,T]\times\mathbb{R}_+^2), \\ & \partial_y^j u|_{y=0} \in H^{k-j}(\mathbb{R}) \cap H^{k-j}([0,T]\times\mathbb{R}), \ \ 0 \le j \le k, \\ & \partial_{t,x}^\alpha v + y \cdot \partial_{t,x}^\alpha \partial_x U \in L^\infty_{y,\ell-1}(L^2_{t,x}), \ \ |\alpha| \le k-1. \end{aligned} \tag{1.1.175}$$

(2) The classical solution to (1.1.170) is stable with respect to the initial data in the following sense: for any given two initial data satisfy (1.1.174), then for all $p \le k-1$, the corresponding solutions of the Prandtl system (1.1.171) satisfy

$$\|u^1 - u^2\|_{\mathcal{H}_{\ell-1}^p(\mathbb{R}_+^2)} + \sum_{j=0}^p \|\partial_y^j u^1|_{y=0} - \partial_y^j u^2|_{y=0}\|_{H^{p-j}(\mathbb{R})}$$

$$+\|w^1 - w^2\|_{\mathcal{H}_\ell^p(\mathbb{R}_+^2)} + \sum_{|\alpha|\le p-1} \|\partial_{t,x}^\alpha v^1 - \partial_{t,x}^\alpha v^2\|_{L^\infty_{y,\ell-1}(L^2_{t,x})}$$

$$\le C(\varepsilon_2, T)\|w_0^1 - w_0^2\|_{\mathcal{H}_\ell^p(\mathbb{R}_+^2)}.$$

Ignatova and Vicol [59] in 2016 proved almost global existence of solutions to the following problem (1.1.176) if the Prandtl datum lies within ε of the error function

$\mathrm{erf}(\frac{y}{2})$ (in a suitable topology); then the Prandtl equations have a unique (classical in x weak in y) solution on $[0, T_\varepsilon]$, where $T_\varepsilon \geq \exp(\varepsilon^{-1}/\log(\varepsilon^{-1}))$.

The two-dimensional Prandtl boundary layer equations in the domain $\mathbb{H} = \{(x, y) \in \mathbb{R}^2_+ : y > 0\}$ for the velocity field (u^P, v^P) takes the form

$$\begin{cases} \partial_t u^P - \partial_y^2 u^P + u^P \partial_x u^P + v^P \partial_y u^P = -\partial_x p^E, \\ \partial_x u^P + \partial_y v^P = 0, \end{cases} \tag{1.1.176}$$

with the boundary conditions

$$\begin{cases} u^P\big|_{y=0} = v^P\big|_{y=0} = 0, \\ u^P\big|_{t=0} = u_0^P, \\ u^P\big|_{y=\infty} = u^E, \end{cases}$$

which are obtained by matching the Navier-Stokes no-slip boundary condition u^{NS} on $\partial\mathbb{H}$, with the Euler slip boundary condition at $y = \infty$. The trace at $\partial\mathbb{H}$ of the Euler tangential velocity u^E obeys Bernoulli's law.

Theorem 1.1.101 ([59] Almost Global Existence) *Let the Euler data be given by $u^E = \kappa$ and $\partial_x p^E = 0$. Define*

$$u_0(x, y) = u_0^P(x, y) - \kappa\,\mathrm{erfc}\left(\frac{y}{2}\right), \quad \mathrm{erf}\left(\frac{y}{2}\right) = \frac{1}{\sqrt{\pi}} \int_0^y \exp\left(-\frac{z^2}{4}\right) dz,$$

where erfc *is the Gauss error function. There exists a sufficiently large universal constant $C_* > 0$ and a sufficiently small universal constant $\varepsilon_* > 0$ such that the following holds. For any given $\varepsilon \in (0, \varepsilon_*]$, assume that there exists an analyticity radius $\tau_0 > 0$ such that*

$$\frac{C_*}{\log\frac{1}{\varepsilon}} \leq \tau_0^{3/2} \leq \frac{1}{C_* \varepsilon^3},$$

and such that the function

$$g_0(x, y) = \partial_y u_0(x, y) + \frac{y}{2} u_0(x, y),$$

obeys

$$\|g_0\|_{X_{2\tau_0,1/2}} := \sum_{m \geq 0} \left\| \exp\left(\frac{y^2}{8}\right) \partial_x^m g_0(x, y) \right\|_{L^2(\mathbb{H})} (2\tau_0)^m \frac{\sqrt{m+1}}{m!} \leq \varepsilon.$$

Here

$$X_{2\tau_0,1/2} = \{\|g\|_{X_{2\tau_0,1/2}} < +\infty\}.$$

Then there exists a unique solution u^p of the Prandtl boundary layer equations on $[0, T_\varepsilon]$, where

$$T_\varepsilon \geqq \exp\left(\frac{\varepsilon^{-1}}{\log(\varepsilon^{-1})}\right).$$

The solution u^p is real analytic in x, with analyticity radius larger than $\tau_0/2$, and lies in a weighted H^2 space with respect to y. We emphasize that $\varepsilon \ll 1$ and $\tau_0 = O(1)$ are independent of κ.

Greniner, Guo, and Nguyen [50] in 2016 constructed the linearized Navier-Stokes equations about generic stationary shear flows of the boundary layer type in a regime of sufficiently large Reynolds number : $R \to \infty$. Greniner, Guo, and Nguyen [49] also recalled the stability of Prandtl boundary layers of [50] and considered the linearization of the incompressible Navier-Stokes equations in the domain $\mathbb{R} \times \mathbb{R}_+$:

$$\begin{cases} v_t + u_0 \cdot \nabla v + v \cdot \nabla u_0 + \nabla p = \frac{1}{R}\Delta v, \\ \nabla \cdot v = 0, \\ v|_{z=0} = 0, \\ u_0 = (U(z), 0)^T. \end{cases} \tag{1.1.177}$$

The asymptotic behavior of these branches α_{low} and α_{up} depends on the profile:

(i) for plane Poiseuille flow in a channel: $U(z) = 1 - z^2$ for $-1 \le z \le 1$,

$$\alpha_{\text{low}}(R) \equiv A_{1c}R^{-1/7}, \quad \alpha_{\text{up}}(R) \equiv A_{2c}R^{-1/11}, \tag{1.1.178}$$

(ii) for boundary layer profiles,

$$\alpha_{\text{low}}(R) \equiv A_{1c}R^{-1/4}, \quad \alpha_{\text{up}}(R) \equiv A_{2c}R^{-1/6}, \tag{1.1.179}$$

(iii) for Blasius (a particular boundary layer) profile,

$$\alpha_{\text{low}}(R) = A_{1c}R^{-1/4}, \quad \alpha_{\text{up}}(R) = A_{2c}R^{-1/10}, \tag{1.1.180}$$

where A_{1c}, A_{2c} are constants> 0.

Theorem 1.1.102 ([50]) *Let $U(z)$ be an arbitrary shear profile with $U'(0) > 0$ and satisfy*

$$\sup_{z\ge 0} |\partial_z^k (U(z) - U_+)e^{\eta_0 z}| \le +\infty, \quad k = 0, \cdots, 4,$$

for some constants U_+ *and* $\eta_0 > 0$. *Let* $\alpha_{\text{low}}(R)$ *and* $\alpha_{\text{up}}(R)$ *be defined as in (1.1.179) for general boundary layer profiles or defined as in (1.1.180) for the Blasius profiles: those with additional assumptions,* $U''(0) = U'''(0) = 0$. *Then, there is a critical Reynolds number* R_c *so that for all* $R \geq R_c$ *and all* $\alpha \in (\alpha_{\text{low}}(R), \alpha_{\text{up}}(R))$, *there exist a nontrivial triple* $c(R), \widehat{v}(z, R), \widehat{p}(z, R)$, *with* $\operatorname{Im} c(R) > 0$, *such that* $v_R := e^{i\alpha(y-ct)}\hat{v}(z; R)$ *and* $p_R := e^{i\alpha(y-ct)}\hat{p}(z; R)$ *solve the problem (1.1.177) with the no-slip boundary conditions. In the case of instability, there holds the following estimate for the growth rate of the unstable solutions:*

$$\alpha \operatorname{Im} c(R) \approx R^{-1/2},$$

as $R \to \infty$.

Xu and Zhang [151] in 2017 discussed the global well-posedness of solutions for the nonlinear Prandtl boundary layer equations on the half plane. More specifically, they used energy method to obtain the existence, uniqueness, and stability of solutions in weighted Sobolev space $H^m_\mu(\mathbb{R}^2_+)$ ($m \geq 6$) when weighted functions μ is a polynomial function and proved the lifespan T of solutions could be any large when its initial datum is a perturbation around the monotonic shear profile of small size e^{-T}, which is the first result of global existence of solutions of 2D Prandtl equation in a polynomial weighted Sobolev space. In [151], they studied the initial-boundary value problem for the Prandtl boundary layer equation in two dimension, which reads

$$\begin{cases} \partial_t u + u\partial_x u + v\partial_y u + \partial_x p = \partial_y^2 u, \quad t > 0, (x, y) \in \mathbb{R}^2_+, \\ \partial_x u + \partial_y v = 0, \\ u|_{y=0} = v|_{y=0} = 0, \ \lim\limits_{y\to+\infty} u = U(t, x) = 1, \\ u|_{t=0} = u_0(x, y), \end{cases} \tag{1.1.181}$$

where $\mathbb{R}^2_+ = \{(x, y) \in \mathbb{R}^2; y > 0\}$, $u(t, x, y)$ represents the tangential velocity, $v(t, x, y)$ normal velocity.

Assume that u_0^s (initial datum of shear flow) satisfies the following conditions:

$$\begin{cases} u_0^s \in C^{m+4}([0, +\infty]), \quad \lim\limits_{y\to+\infty} u_0^s(y) = 1, \\ (\partial_y^{2p} u_0^s)(0) = 0, \quad 0 \leq 2p \leq m + 4, \\ c_1 < y >^{-k} \leq (\partial_y u_0^s)(y) \leq c_2 < y >^{-k}, \quad \forall y \geq 0, \\ |(\partial_y^p u_0^s)(y)| \leq c_2 < y >^{-k-p+1}, \quad \forall y \geq 0, 1 \leq p \leq m + 4. \end{cases} \tag{1.1.182}$$

Theorem 1.1.103 ([151]) *Let* $m \geq 6$ *be an even integer,* $k > 1$ *and* $-\frac{1}{2} < \nu < 0$. *Assume that* u_0^s, *the initial data* $\bar{u}_0 = u_0 - u_0^s \in H^{m+3}_{k+\nu}(\mathbb{R}^2_+)$, *and* $\bar{u}_0$ *satisfies the* *compatibility condition up to order* $m+2$. *Then for any* $T > 0$, *there exists a* $\delta_0 > 0$ *small enough such that if*

$$\|\bar{u}_0\|_{H^{m+1}_{k+\nu}(\mathbb{R}^2_+)} \le \delta_0, \tag{1.1.183}$$

then the initial-boundary value problem (1.1.181) *admits a unique solution* (u, v) *with*

$$(u - u^s) \in L^\infty([0, T]; H^m_{k+\nu-\delta'}(\mathbb{R}^2_+)),\ v \in L^\infty([0, T]; L^\infty(\mathbb{R}_{y,+}; H^{m-1}(\mathbb{R}_x))), \tag{1.1.184}$$

where $\delta' > 0$ *satisfies* $\nu + \frac{1}{2} < \delta' < \nu + 1$ *and* $k + \nu - \delta' > \frac{1}{2}$.

Moreover, we have the stability with respect to the initial data in the following sense: given any two initial data

$$u_0^1 = u_0^s + \bar{u}_0^1, \quad u_0^2 = u_0^s + \bar{u}_0^2.$$

If u_0^s *satisfies* (1.1.182) *and* $\bar{u}_0^1, \bar{u}_0^2$ *satisfy* (1.1.183), *then the solutions* u^1 *and* u^2 *to problem* (1.1.181) *satisfies*

$$\|u^1 - u^2\|_{L^\infty([0,T]; H^{m-3}_{k+\nu-\delta'}(\mathbb{R}^2_+))} \le C\|u_0^1 - u_0^2\|_{H^{m+1}_{k+\nu}(\mathbb{R}^2_+)},$$

where the constant C *depends on the norm of* $\partial_y u^1, \partial_y u^2$ *in* $L^\infty([0, T]; H^m_{k+\nu-\delta'+1}(\mathbb{R}^2_+))$.

Paicu and Zhang [107] in 2019 proved the global well-posedness of Prandtl system with small initial data, which is analytical in the tangential variable. The key ingredient used in the proof is to derive sufficiently fast decay-in-time estimate of some weighted analytic energy estimate of the tangent velocity, which is based on a Poincaré type inequality and a subtle interplay between the tangential velocity equation and its primitive one. Their result can be viewed as a global-in-time Cauchy-Kovalevskaya result for Prandtl system with small analytical data.

One of the key step to rigorously justify this inviscid limit of Navier-Stokes system with Dirichlet boundary condition is to deal with the well-posedness of the following Prandtl system:

$$\begin{cases} \partial_t U + U\partial_x U + V\partial_y U - \partial_y^2 U + \partial_x p = 0, \quad (t, x, y) \in \mathbb{R}_+ \times \mathbb{R} \times \mathbb{R}_+, \\ \partial_x U + \partial_y V = 0, \\ U|_{y=0} = V|_{y=0} = 0, \quad \text{and} \quad \lim\limits_{y\to+\infty} U(t, x, y) = w(t, x), \\ U|_{t=0} = U_0, \end{cases} \tag{1.1.185}$$

where U and V represent the tangential and normal velocities of the boundary layer flow. $(w(t,x), p(t,x))$ are the traces of the tangential velocity and pressure of the outflow on the boundary, which satisfy Bernoulli's law:

$$\partial_t w + w\partial_x w + \partial_x p = 0.$$

The authors of [107] investigated the global existence of the solutions to the Prandtl system (1.1.185) with small data which is analytic in the tangential variable. For simplicity, they took $w(t,x)$ in (1.1.185) to be $\varepsilon f(t)$ with $f(0)=0$. Let $R>0$ and $\chi \in C^\infty[0,\infty)$ with $\chi(y) = \begin{cases} 1 \text{ if } y \geq 2R, \\ 0 \text{ if } y \leq R, \end{cases}$ they denoted $W \stackrel{\text{def}}{=} U - \varepsilon f(t)\chi(y)$. Then W solves

$$\begin{cases} \partial_t W + (W + \varepsilon f(t)\chi(y))\partial_x W + V\partial_y (W + \varepsilon f(t)\chi(y)) - \partial_y^2 W = \varepsilon g, \\ \partial_x W + \partial_y V = 0, \quad (t,x,y) \in \mathbb{R}_+ \times \mathbb{R}_+^2, \\ W|_{y=0} = V|_{y=0} = 0, \quad \text{and} \quad \lim_{y\to+\infty} W(t,x,y) = 0, \\ W|_{t=0} = U_0, \end{cases}$$

where $\mathbb{R}_+^2 = \mathbb{R} \times \mathbb{R}_+$ and $g(t,y) \stackrel{\text{def}}{=} (1-\chi(y))f'(t) + f(t)\chi''(y)$.

In order to get rid of the source term in the equation of W, the authors of [107] introduced u^s via

$$\begin{cases} \partial_t u^s - \partial_y^2 u^s = \varepsilon g(t,y), \quad (t,y) \in \mathbb{R}_+ \times \mathbb{R}_+, \\ u^s|_{y=0} = 0, \quad \text{and} \quad \lim_{y\to+\infty} u^s = 0, \\ u^s|_{t=0} = 0. \end{cases} \tag{1.1.186}$$

With u^s being determined by the last equation, the authors set $u \stackrel{\text{def}}{=} W - u^s$ and $v \stackrel{\text{def}}{=} V$. Then (u,v) verifies

$$\begin{cases} \partial_t u + (u + u^s + \varepsilon f(t)\chi(y))\, \partial_x u + v\partial_y\, (u + u^s + \varepsilon f(t)\chi(y)) - \partial_y^2 u = 0, \\ \partial_x u + \partial_y v = 0, \quad (t,x,y) \in \mathbb{R}_+ \times \mathbb{R}_+^2, \\ u|_{y=0} = v|_{y=0} = 0, \quad \text{and} \quad \lim_{y\to\infty} u(t,x,y) = 0, \\ u|_{t=0} = u_0 \stackrel{\text{def}}{=} U_0. \end{cases} \tag{1.1.187}$$

Let s in $\mathbb{R}$. For u in $\mathcal{S}'_h(\mathbb{R}_+^2)$, which means that u is in $\mathcal{S}'(\mathbb{R}_+^2)$ and satisfies $\lim_{k\to-\infty} \left\| S_k^{\mathrm{h}} u \right\|_{L^\infty} = 0$, the authors set

$$\|u\|_{\mathcal{B}^{s,0}} \stackrel{\text{def}}{=} \left\| \left(2^{ks} \left\| \Delta_k^{\mathrm{h}} u \right\|_{L_+^2} \right)_{k\in\mathbb{Z}} \right\|_{\ell^1(\mathbb{Z})}.$$

For $s \leq \frac{1}{2}$, they defined $\mathcal{B}^{s,0}\left(\mathbb{R}_+^2\right) \stackrel{\text{def}}{=} \left\{u \in \mathcal{S}_h'\left(\mathbb{R}_+^2\right) \mid \|u\|_{\mathcal{B}^{s,0}} < \infty\right\}$.

If ℓ is a positive integer and if $\ell - \frac{1}{2} < s \leq \ell + \frac{1}{2}$, then they defined $\mathcal{B}^{s,0}\left(\mathbb{R}_+^2\right)$ as the subset of distributions u in $\mathcal{S}_h'\left(\mathbb{R}_+^2\right)$ such that $\partial_x^\ell u$ belongs to $\mathcal{B}^{s-\ell,0}\left(\mathbb{R}_+^2\right)$. In order to obtain a better description of the regularizing effect of the transport-diffusion equation, they needed to use Chemin-Lerner type spaces $\widetilde{L}_T^p\left(\mathcal{B}^{s,0}\left(\mathbb{R}_+^2\right)\right)$.

Let $p \in [1, +\infty]$ and $T_0, T \in [0, +\infty]$. The authors defined $\widetilde{L}^p\left(T_0, T; \mathcal{B}^{s,0}\left(\mathbb{R}_+^2\right)\right)$ as the completion of $C\left([T_0, T]; \mathcal{S}\left(\mathbb{R}_+^2\right)\right)$ by the norm

$$\|a\|_{\widetilde{L}^p(T_0,T;\mathcal{B}^{s,0})} \stackrel{\text{def}}{=} \sum_{k\in\mathbb{Z}} 2^{ks} \left(\int_{T_0}^T \left\|\Delta_k^{\mathrm{h}} a(t)\right\|_{L_+^2}^p dt\right)^{\frac{1}{p}}, \quad \Delta_k^{\mathrm{h}} a = \mathcal{F}^{-1}(\varphi(2^{-k}|\xi|)\widehat{a}),$$

φ decays to zero sufficiently fast as y approaching to $+\infty$, $\widehat{a}$ denotes the partial Fourier transform of the distribution a, with the usual change if $p = \infty$. In particular, when $T_0 = 0$, the authors denoted $\|a\|_{\widetilde{L}_T^p(\mathcal{B}^{s,0})} \stackrel{\text{def}}{=} \|a\|_{\widetilde{L}^p(0,T;\mathcal{B}^{s,0})}$ for simplicity.

Theorem 1.1.104 ([107]) *Let $\ell > 3$ be an integer, $\delta > 0$, and $f \in H^1(\mathbb{R}_+)$ which verifies*

$$C_{f,\ell}^2 \stackrel{def}{=} \int_0^\infty \langle t\rangle^{\frac{3}{2}}(1+\log\langle t\rangle)^\ell \left(f^2(t) + \left(f'(t)\right)^2\right) dt < \infty.$$

Let $u_0 = \partial_y \varphi_0$ satisfy $u_0(x, 0) = 0$ and

$$\mathfrak{C}(u_0) \stackrel{def}{=} \left\|e^{\frac{y^2}{8}} e^{\delta|D_x|}(\varphi_0, u_0)\right\|_{\mathcal{B}^{\frac{1}{2},0}} \leq c_0,$$

for some c_0 sufficiently small. Then there exists $\varepsilon_0 > 0$ so that for $\varepsilon \leq \varepsilon_0$, the system (1.1.187) has a unique global solution u which satisfies

$$\left\|e^{\frac{y^2}{8(t)}} e^{\frac{\delta}{2}|D_x|} u\right\|_{\widetilde{L}^\infty\left(\mathbb{R}_+;\mathcal{B}^{\frac{1}{2},0}\right)} + \left\|e^{\frac{y^2}{8(t)}} e^{\frac{\delta}{2}|D_x|} \partial_y u\right\|_{\widetilde{L}^2\left(\mathbb{R}_+;\mathcal{B}^{\frac{1}{2},0}\right)} \leq C \left\|e^{\delta|D_x|} u_0\right\|_{\mathcal{B}^{\frac{1}{2},0}}.$$

Moreover, for u^s being determined by (1.1.186) and for any integer j, there exists a positive constant C_j so that

$$\left\|e^{\frac{y^2}{8(t)}} e^{\frac{\delta}{2}|D_x|} u(t)\right\|_{\mathcal{B}^{\frac{1}{2},0}} \leq C_j \mathfrak{C}(u_0) \langle t\rangle^{-\frac{3}{4}} \log^{-\frac{j}{2}}\langle t\rangle \text{ and}$$

$$\int_0^\infty \langle t\rangle^{\frac{1}{4}} \left(\left\|e^{\frac{y^2}{8(t)}} \partial_y u^s(t)\right\|_{L_\mathrm{v}^2} + \left\|e^{\frac{y^2}{8(t)}} e^{\frac{\delta}{2}|D_x|} \partial_y u(t)\right\|_{\mathcal{B}^{\frac{1}{2},0}}\right) dt \leq C\left(C_{f,\ell}\varepsilon + \mathfrak{C}(u_0)\right).$$

Here

$$e^{\frac{\delta}{2}|D_x|}u(t) = \mathcal{F}^{-1}_{\xi\to x}\left(e^{(\delta-\lambda\theta(t))|\xi|}\hat{u}(t,\xi,y)\right).$$

Petrov and Suslina [110] in 2020 obtained theorem about existence and uniqueness of the solution in the classes $\widetilde{H}^s(-1,1)$ with $0 \leqslant s \leqslant 1$. In particular, for $s = 1$ the result is as follows: if $r^{1/2}f \in L^2$, then $r^{-1/2}u, r^{1/2}u' \in L_2$, where $r(x) = 1 - x^2$.

The Prandtl equation

$$u(x) - p(x)\frac{1}{2\pi}\int_{-1}^{1}\frac{u'(t)}{t-x}dt = p(x)f(x), \quad u(-1) = u(1) = 0, \tag{1.1.188}$$

is one of the universal equations of mathematical physics. It is convenient to put $V(x) := p(x)^{-1}$ and rewrite the first Prandtl equation (1.1.188) with the Dirichlet conditions as

$$V(x)u(x) - \frac{1}{2\pi}\int_{-1}^{1}\frac{u'(t)}{t-x}dt = f(x), \quad u(-1) = u(1) = 0. \tag{1.1.189}$$

Here and below, singular integrals are understood in the mean value sense. The authors of [110] assumed that $V(x)$ is a measurable function on $(-1,1)$ satisfying the following conditions:

$$V(x) \geqslant 0 \text{ for a. e. } x \in (-1,1); \quad \left(1-x^2\right)V(x) \leqslant M < \infty.$$

Any function $u \in H^{1/2}_{00}(-1,1)$ satisfies

$$\int_{-1}^{1}\frac{|u(x)|^2}{1-x^2}dx \leqslant \frac{1}{2}\int_{-\infty}^{\infty}|\zeta\,||\widehat{u}(\zeta)|^2 d\zeta,$$

where the norm of $H^{1/2}_{00}(-1,1)$ is defined as the standard norm in $H^{1/2}(\mathbb{R})$:

$$\|u\|^2_{H^{1/2}_{00}} := \|u\|^2_{H^{1/2}_{00}(-1,1)} = \frac{1}{2}\int_{-\infty}^{+\infty}(1+|\xi|^2)^{\frac{1}{2}}|\widehat{u}(\zeta)|^s d\zeta.$$

Theorem 1.1.105 ([110]) *Suppose that $V(x)$ satisfies the assumptions. Then for any $f \in \left(H^{1/2}_{00}\right)^*$, there exists a unique weak solution $u \in H^{1/2}_{00}$ of problem (1.1.188). The solution satisfies the following estimate:*

$$\|u\|_{H^{1/2}_{00}} \leqslant (2\pi+2)\|f\|_{\left(H^{1/2}_{00}\right)^*}.$$

Dalibard and Paddick [17] in 2021 proved the global existence of solutions of the following stationary Prandtl-like equations by using the von Mises variables:

$$
\begin{cases}
\lambda_0 \nu(y)(u\partial_\xi v + v\partial_y v) + \psi - \nu(y)^2\partial_\xi^2 v = \psi^0(y), \\
(u, v) = \nabla^\perp \psi := (-\partial_y \psi, \partial_\xi \psi), \\
v|_{y=0} = v_0(\xi), \\
(u, v)|_{\xi=0} = 0, \\
\lim\limits_{\xi\to+\infty} \psi(\xi, y) = \psi^0(y),
\end{cases}
\tag{1.1.190}
$$

in the domain $\Omega_Y = \{(\xi, y) \in \mathbb{R}^2 | \xi > 0, 0 < y < Y\}$. The function ψ^0 is given and smooth, λ_0 is a strictly positive parameter, and the function ν is a smooth function with $\nu \le 1$.

The von Mises change of variables

$$
(\xi, y) \mapsto (\psi(\xi, y), y),
$$

and define a new unknown function, $w = w(\psi, y)$ such that $w(\psi, y) = v^2(\psi(\xi, y), y)$. The chain rule allows us to express the derivatives:

$$
\frac{\partial v}{\partial \xi} = \frac{1}{2}\frac{\partial w}{\partial \psi}, \quad \frac{\partial^2 v}{\partial \xi^2} = \frac{\sqrt{w}}{2}\frac{\partial^2 w}{\partial \psi^2},
$$

$$
\frac{\partial v}{\partial y} = \frac{1}{2\sqrt{w}}\frac{\partial w}{\partial y} + \frac{1}{2\sqrt{w}}\frac{\partial w}{\partial \psi}\frac{\partial \psi}{\partial y}.
$$

Then (1.1.190) turns to

$$
\begin{cases}
\lambda_0 \nu(y)\partial_y w - \nu(y)^2\sqrt{w}\partial_\psi^2 w = 2(\psi^0(y) - \psi), \\
w|_{y=0} = w_0(\psi), \\
w|_{\psi=0} = w|_{\psi=\psi^0(y)} = 0,
\end{cases}
\tag{1.1.191}
$$

in the domain $\Omega = \{(\psi, y) \in \mathbb{R}^2 | 0 < \psi < \psi^0(y), 0 < y < Y\}$.

Theorem 1.1.106 ([17]) *Let* $\psi^0 \in C^2([0, Y])$ *such that* $\inf \psi^0 > 0$, $\inf \psi_0' > 0$. *Assume that* $\lambda_0(\psi^0)'(y)/\nu(y)^{1/3} > \eta + 9/2 > \sqrt[3]{\frac{27}{4}}$ *for every* $y \ge 0$ *and for some* $\eta > 0$. *Let* $v_0 \in W^{2,\infty}(\mathbb{R}^+)$, *positive on* $(0, +\infty)$, *with* $v(0) = v_0$, $v_0' > 0$, *such that* $\int_0^{+\infty} v_0(\xi)d\xi = \psi^0(0)$. *We assume that* v_0 *satisfies the corner compatibility condition*

$$
\nu(0)^2 v_0''(\xi) + \psi^0(0) = O(\xi^2)
$$

as $\xi \to 0$, *and the decay constraint*

$$v_0(\xi) \sim a(0)(\psi^0(0) - \psi_0(\xi))$$

as $\xi \to +\infty$, *in which* $\psi_0(\xi) = \int_0^\xi v_0(s)ds$ *is the initial stream function.*

Then, for every $Y > 0$, *there exists a unique classical Lipschitz solution to the steady rotating Prandtl equation (1.1.190) with* $v > 0$ *in* Ω_Y. *Furthermore,* v *satisfies the following properties:*

(i) *Behavior close to* $\xi = 0$: *for some* $\xi_0, m > 0$, *depending on* Y, *we have* $\partial_\xi v > m$ *for* $\xi \le \xi_0$.
(ii) *Behavior close to* $\xi = \infty$: *for every* y, $v(\xi, y) \sim a(y)(\psi^0(y) - \psi(\xi, y))$ *as* $\xi \to \infty$.

Wang, Wang, and Zhang [139] in 2021 proved the global existence and large time behavior of small solutions to 2D Prandtl system (1.1.58) which U is a constant for data with Gevrey 2 regularity in the x variable and Sobolev regularity in the y variable. They extended the global well-posedness result in [107] for 2D Prandtl system with analytic data to data with optimal Gevrey regularity in the sense of [40].

In [139], there exists a potential function φ such that $u = \partial_y\varphi$ and $v = -\partial_x\varphi$. Recalled the Gevrey form denoted by

$$G := u + \frac{y}{2\langle t\rangle}\varphi, \quad g := \partial_y G = \partial_y g + \frac{y}{2\langle t\rangle}u + \frac{\varphi}{2\langle t\rangle},$$

here $\langle t\rangle = (1+t^2)^{\frac{1}{2}}$, and denoted by

$$e^{\Phi(t,D_x)} := e^{\delta(t)\langle D_x\rangle^{\frac{1}{2}}} \text{ with } \delta(t) = \delta - \lambda\theta(t) \text{ and } f_\Phi := e^{\Phi(t,D_x)}f, \tag{1.1.192}$$

$$\|f\|_{H^{s,k}_\Psi} := \sum_{0\le\ell\le k}\left(\int_0^\infty e^{2\Phi(t,x)}\|\partial_y^\ell f(\cdot,y)\|^2_{H^s_h}dy\right)^{\frac{1}{2}} \text{ with } \Phi(t,y) = \frac{y^2}{8\langle t\rangle}. \tag{1.1.193}$$

Theorem 1.1.107 ([139]) *Let* $u_0 = \partial_y\varphi_0$ *satisfy* $u_0(x,0) = 0$ *and* $\int_0^\infty u_0 dy = 0$. *Let* $G_0 = u_0 + \frac{y}{2\langle t\rangle}\varphi_0$. *For some sufficiently small but fixed* $\eta \in (0, \frac{1}{6})$, *we denote*

$$\begin{aligned}E(t) := &\|\langle t\rangle^{\frac{1-\eta}{4}}u_\Phi\|^2_{H^{\frac{11}{2},0}_\Psi} + \|\langle t\rangle^{\frac{3-\eta}{4}}\partial_y u_\Phi\|^2_{H^{5,0}_\Psi} + \|\langle t\rangle^{\frac{5-\eta}{4}}G_\Phi\|^2_{H^{4,0}_\Psi}\\ &+\|\langle t\rangle^{\frac{7-\eta}{4}}\partial_y G_\Phi\|^2_{H^{3,0}_\Psi} + \|\langle t\rangle^{\frac{9-\eta}{4}}\partial_y^2 G\|^2_{H^{3,0}_\Psi} + \|\langle t\rangle^{\frac{11-\eta}{4}}\partial_y^3 G\|^2_{H^{2,0}_\Psi},\end{aligned}$$

and

$$D(t) := \|\langle t\rangle^{\frac{1-\eta}{4}}\partial_y u_\Phi\|^2_{H_\Psi^{\frac{11}{2},0}} + \|\langle t\rangle^{\frac{3-\eta}{4}}\partial_y^2 u_\Phi\|^2_{H_\Psi^{5,0}} + \|\langle t\rangle^{\frac{5-\eta}{4}}\partial_y G_\Phi\|^2_{H_\Psi^{4,0}}$$
$$+\|\langle t\rangle^{\frac{7-\eta}{4}}\partial_y^2 G_\Phi\|^2_{H_\Psi^{3,0}} + \|\langle t\rangle^{\frac{9-\eta}{4}}\partial_y^3 G\|^2_{H_\Psi^{3,0}} + \|\langle t\rangle^{\frac{11-\eta}{4}}\partial_y^4 G\|^2_{H_\Psi^{2,0}}.$$

Moreover, assume that

$$\|u_\Phi(0)\|^2_{H_\Psi^{\frac{25}{4},0}} + E(0) \le \varepsilon^2.$$

Then there exists $\varepsilon_0, \lambda_0 > 0$ so that for $\varepsilon \le \varepsilon_0$ and $\lambda \ge \lambda_0, \theta(t)$ in (1.1.192) satisfies $\sup\limits_{t\in[0,\infty)} \theta(t) \le \frac{\delta}{4\lambda}$, and the system (1.1.58) has a unique global solution u which satisfies

$$E(t) + c\eta\int_0^t D(t')dt' \le C\varepsilon^2, \quad \forall t \in \mathbb{R}.$$

1.1.4 Compressible 2D Prandtl Equations

Liu and Wang [82] in 2014 studied the asymptotic behavior of circularly symmetric solutions to the following initial boundary value problem (1.1.194) of the nonisentropic compressible Navier-Stokes equations in a two-dimensional exterior domain with impermeable boundary conditions when the viscosities and the heat conduction coefficient tend to zero at the same rate. The compressible viscous flow in two space variables:

$$\begin{cases} \partial_t\rho^\epsilon + \nabla\cdot(\rho^\epsilon \mathbf{u}^\epsilon) = 0, \\ \rho^\epsilon\{\partial_t \mathbf{u}^\epsilon + (\mathbf{u}^\epsilon\cdot\nabla)\mathbf{u}^\epsilon\} + \nabla p(\rho^\epsilon,\theta^\epsilon) = \epsilon\{\mu\Delta\mathbf{u}^\epsilon + (\lambda+\mu)\nabla(\nabla\cdot\mathbf{u}^\epsilon)\}, \\ c_V\rho^\epsilon\{\partial_t\theta^\epsilon + (\mathbf{u}^\epsilon\cdot\nabla)\theta^\epsilon\} + p(\rho^\epsilon,\theta^\epsilon)\nabla\cdot\mathbf{u}^\epsilon = \epsilon\{k\Delta\theta^\epsilon + \lambda(\nabla\cdot\mathbf{u}^\epsilon)^2 + 2\mu D\cdot D\}. \end{cases} \tag{1.1.194}$$

where $\rho^\epsilon, \theta^\epsilon, \mathbf{u}^\epsilon = (u^\epsilon, v^\epsilon)^T, p(\rho^\epsilon,\theta^\epsilon)$ are the density, the absolute temperature, velocity, and the pressure, respectively,

$$D_{ij} = \frac{1}{2}\left(\frac{\partial_{u_i^\epsilon}}{\partial_{x_j}} + \frac{\partial_{u_j^\epsilon}}{\partial_{x_i}}\right), D\cdot D = \sum_{i,j=2} D_{ij}^2.$$

Take the circularly symmetric solutions of the form

$$\rho^\epsilon(t,x) = \widetilde{\rho}^\epsilon(t,r), \quad \mathbf{u}^\epsilon(t,x) = (\widetilde{u}^\epsilon, \widetilde{v}^\epsilon)(t,r), \quad \theta^\epsilon(t,x) = \widetilde{\theta}^\epsilon(t,r),$$

and let

$$U^\epsilon(t,x) = (\frac{1}{\widetilde{\rho}^\epsilon}, \widetilde{u}^\epsilon, \widetilde{v}^\epsilon, \widetilde{\theta}^\epsilon)^T(t, r(t,x)),$$

then $U^\epsilon(t,x)$ satisfies the following problem:

$$\begin{cases} \partial_t U^\epsilon + A(U^\epsilon)\partial_x U^\epsilon + Q_1(U^\epsilon)(U^\epsilon, U^\epsilon) - \epsilon B(U^\epsilon)\partial_x^2 U^\epsilon \\ -\epsilon Q_2(U^\epsilon)(\partial_x U^\epsilon, \partial_x U^\epsilon) - \epsilon Q_3(U^\epsilon)(U^\epsilon, U^\epsilon) - \epsilon V(\partial_x U^\epsilon, U^\epsilon) = 0, \\ U^\epsilon(0,x) = U_0(x) =)(\widetilde{\tau}_0^\epsilon, \widetilde{u}_0^\epsilon, \widetilde{v}_0^\epsilon, \widetilde{\theta}_0^\epsilon)^T(x), \quad \tau = \frac{1}{\rho}, \\ U_1^\epsilon(t,0) = (0, v^0(t), \theta^0)^T. \end{cases} \tag{1.1.195}$$

Theorem 1.1.108 ([82]) *Let the initial data* $U_0(x) = (\tau_0, u_0, v_0, \theta_0)^T(x)$ *satisfy Assumption 2, and* $U_0 - (\overline{\tau}, 0, 0, \overline{\theta})^T \in H^s(\mathbb{R}_+)$ *with* $s > 11$*. Let* $U^a(t,x)$ *be an approximate solution to problem (1.1.195). Then there exists* $\delta > 0$ *and* $C > 0$*, independent of* ϵ*, such that when*

$$\sup_{t\in[0,T*],\beta\le 2} \|\partial_\xi^\beta U^{B,0}(t,\cdot)\|_{L^\infty} + \sup_{t\in[0,T*]} \|\partial_\xi U^{B,0}(t,\cdot)\|_{L^1} \le \delta,$$

holds, there exists a unique solution $U^\epsilon(t,x)$ *to (1.1.195) such that* $U^\epsilon - U^a \in C([0,T*]; H^1)$ *and*

$$\|U^\epsilon - U^a\|^2_{L^\infty([0,T*];L^2)} + \epsilon^2 \|\partial_x U^\epsilon - \partial_x U^a\|^2_{L^\infty([0,T*];L^2)} \le C\epsilon^2.$$

Moreover, we have

$$U^\epsilon(t,x) - \left(U^{I,0}(t,x) + U^{B,0}(t, \frac{x}{\sqrt{\epsilon}}) = \mathbb{O}(\sqrt{\epsilon})\right).$$

Wang, Xie, and Yang [140] in 2015 obtained the well-posedness of the compressible Prandtl equations locally in time in the domain $\{(x,y)|x \in \mathbb{T} = \mathbb{R}/\mathbb{Z}, y \in [0,+\infty)\}$

$$\begin{cases} u_t + uu_x + vu_y = \frac{1}{\rho(t,x)}\nu u_{yy} + p_x, \\ \partial_x(\rho u) + \partial_y(\rho v) = -\rho_t, \\ u(0,x,y) = u_0, \quad u(t,x,0) = v(t,x,0) = 0, \\ \lim_{y\to+\infty} u(t,x,y) = U(t,x), \end{cases} \tag{1.1.196}$$

where $U(t,x)$ satisfies Bernoulli's law (1.0.3).

Main assumptions (H) on the initial data are the following:

(H1) For a fixed integer $k_0 > 9$, the initial data $u_0(x,\eta)$ satisfy the compatibility condition of the problem (1.1.196) up to order $4k_0 + 2$;

(H2) monotone condition $\partial_\eta u_0(x,\eta) \geq \frac{\sigma_0}{(1+\eta)^{\gamma+2}} > 0$ holds for all $x \in \mathbb{T}$ and $\eta \geq 0$ with some positive constant σ_0 and a positive integer $\gamma \geq 2$;

(H3) $\|(1+\eta)^{\gamma+\alpha_2} D^\alpha (u_0(x,\eta) - U(0.x))\|_{L^2(\mathbb{T}\times\mathbb{R}^+)} \leq C_0$, where $D^\alpha = \partial_x^{\alpha_1}\partial_y^{\alpha_2}$ with $\alpha = (\alpha_1, \alpha_2)$ and $|\alpha| = \alpha_1 + \alpha_2 \leq 4k_0 + 2$;

(H4) $\|(1+\eta)^{\gamma+2+\alpha_2} D^\alpha \partial_\eta u_0\|_{L^\infty(\mathbb{T}\times\mathbb{R}^+)} \leq \frac{1}{\sigma_0}$ for $|\alpha| \leq 3k_0$.

Theorem 1.1.109 ([140]) *Suppose that the outer Euler flow is smooth for* $0 \leq t \leq T_0$*, the density* $\rho(t,x)$ *has both positive lower and upper bounds, and the Sobolev norm* $H^s([0, T_0 \times \mathbb{R}_+])$ *of* (ρ, U, V) *is bounded for a suitably large integer* s*. Moreover, the main assumptions (H) on the initial data* $u_0(x,y)$ *are satisfied. Then there exists a* $0 < T \leq T_0$*, such that the initial boundary-value problem (1.1.196) has a unique classical solution* (u, v) *satisfying*

$$\sum_{|m_1|+[(m_2+1)/2]\leq k_0} \|\langle\eta\rangle^l \partial_{t,x}^{m_1}\partial_\eta^{m_2}(u-U)\|_{L^2([0,T]\times\mathbb{T}\times\mathbb{R}_+)} < +\infty$$

for a fixed $l > \frac{1}{2}$ *depending only on* γ *given in (H) with* $\langle\eta\rangle = (1+\eta)$*, and*

$$\sum_{|m_1|+[(m_2+1)/2]\leq k_0-1} \sup_{\eta\in\mathbb{R}_+} \|\partial_{t,x}^{m_1}\partial_\eta^{m_2}(v - Vy)(\cdot, y)\|_{L^2([0,T]\times\mathbb{T})} < +\infty.$$

Here $V(t,x)$ *denotes the trace of* $\partial_y u_2^E$ *on* $y = 0$ *for the normal velocity* u_2^E *of outer Euler flow.*

Gong, Guo, and Wang [46] in 2016 studied the well-posedness for compressible steady boundary layer equations in the domain $D = \{0 < t < T, 0 < x < X, y > 0\}$ by the Crocco transformation or von Mises transformation.

The unsteady boundary layer equations are governed by

$$\begin{cases} \partial_x(\rho u) + \partial_y(\rho v) = -\partial_t \rho, \\ \partial_t u + u\partial_x u + v\partial_y u = \frac{1}{\rho}\partial_y^2 u - \frac{\partial_x p(\rho)}{\rho}, \end{cases} \tag{1.1.197}$$

associated steady boundary layer equations are governed by

$$\begin{cases} \partial_x(\rho u) + \partial_y(\rho v) = 0, \\ u\partial_x u + v\partial_y u = \frac{1}{\rho}\partial_y^2 u - \frac{\partial_x p(\rho)}{\rho}, \end{cases} \tag{1.1.198}$$

with the following boundary and initial conditions:

$$\begin{cases} u_{x=0} = u_1(t,y), \\ u|_{y=0} = 0, \quad v_{y=0} = v_0(t,x), \\ \lim\limits_{y\to+\infty} u(t,x,y) = U(t,x), \end{cases} \tag{1.1.199}$$

and

$$u(0, x, y) = u_0(x, y). \tag{1.1.200}$$

Theorem 1.1.110 ([46]) *For the steady boundary value problems (1.1.198)–(1.1.199), assume that* $\rho(x)$ *and* $v_0(x)$ *have second- and first-order bounded derivatives, respectively;* $u_1(0) = 0, u_1'(0) > 0, u_1(y) > 0$ *for* $y > 0$ *and* $u_1 \to U(0) > 0$ *as* $y \to \infty$*;* u_1, u_1' *and* u_1'' *are bounded and Hölder continuous. Moreover, the compatibility condition holds at the origin, i.e.,*

$$\frac{1}{\rho(0)}u_1''(y) - \frac{\partial_x p(0)}{\rho(0)} - v_0 u_1'(y) = O(y^2)$$

in the neighborhood of $y = 0$*. Then, for some* $X > 0$ *there exists a solution* (u, v) *of problems (1.1.198)–(1.1.199) in* $D = \{0 < x < X, y > 0\}$*, satisfying that* u *is bounded and continuous in* $\overline{D}$*,* $u > 0$ *for* $y > 0$*,* $u_y > m_1$ *in* $0 \le y \le y_0$ *for some positive constants* m_1 *and* y_0*,* u_y *and* u_{yy} *are bounded and continuous in* D*;* u_x*,* v *and* v_y *are bounded locally in* D*.*

In particular, if $|u_1'(y)| \le m_2 \exp(-m_3 y)$ *for two positive constants* m_2 *and* m_3*, then* u_x *and* v_y *are bounded in* D*. Moreover, if either* $\rho_x \le 0, v_0 \le 0$ *or* $\rho_x \le -m_4$ *for a positive constant* m_4*, then the solution exists globally.*

Theorem 1.1.111 ([46]) *For the initial boundary value problems (1.1.197), (1.1.199), and (1.1.200), let* $\rho, v_0, u_0, u_1, \frac{u_{0y}}{U(0,x)}$ *and* $\frac{u_{1y}}{U(t,0)}$ *be smooth enough and the data is sixth order compatible at the intersection of the parabolic boundary. In addition,* $\partial_y u_i > 0, i = 0, 1,$ *for* $y \ge 0$*, and satisfy*

$$K_1(U - u_i)^k \le \frac{\partial u_i}{\partial_y} \le K_2(U - u_i)^k, \quad i = 0, 1$$

for some $k \ge 1$ *and* $K_j > 0\,(j = 1, 2)$*. Then, there exists a unique solution* (u, v) *to problems (1.1.197), (1.1.199), and (1.1.200) in* $D = \{0 < t < T, 0 < x < X, y > 0\}$ *, with either arbitrary large* X *and small* T *or arbitrary large* T *and small* X*. In addition, the solution satisfies that* $u_y > 0$ *for* $y \ge 0$*, the derivatives appearing in (1.1.197) are bounded and continuous in* $\overline{D}$*;* $\frac{u_{yy}}{u_y}$ *and* $\frac{u_{yyy}u_y - u_{yy}^2}{u_y^3}$ *are bounded in* D*. Furthermore,*

$$K_1(U - u)^k \le u_y \le K_2(U - u)^k.$$

Assuming that $\rho = 1 + \varepsilon\rho_1$, then (1.1.197)–(1.1.198) formally can be regarded as the variation of the following steady and unsteady Prandtl equations for small ε, respectively,

$$\begin{cases} u\partial_x u + v\partial_y u = \partial_y^2 u, \\ \partial_x u + \partial_y v = 0, \end{cases} \tag{1.1.201}$$

and

$$\begin{cases} \partial_t u + u\partial_x u + v\partial_y u = \partial_y^2 u, \\ \partial_x u + \partial_y v = 0. \end{cases} \tag{1.1.202}$$

Theorem 1.1.112 ([46])

(1) Assume that $\rho_1(x)$ *and* $\rho_1'(x)$ *are bounded,* $|v_0^1 - v_0| \le \lambda_1 \varepsilon^{\frac{1}{2}}$, $|u_1^1(y) - u_1(y)| \le \lambda_2 \varepsilon^{\frac{1}{2}} \min\{y, 1\}$, $|u_{1y}^1| + |u_{1y}| \le \lambda_3 e^{-\lambda_4 y}$ *hold for some* $\lambda_j (1 \le j \le 4)$*. Then, the solution* u *of (1.1.198)–(1.1.199) converges to the solution* u^1 *of (1.1.201) uniformly in* D *and that*

$$|u^1 - u| \le \lambda_5 \varepsilon^{\frac{1}{2}}$$

for some $\lambda_5 > 0$.

(2) Assume that $\rho_1(x)$ *and* $\rho_1'(x)$ *are bounded,* $|v_0^1 - v_0| + |U_0^1 - U_0| \le \lambda_6 \varepsilon$, $\frac{u_0}{u_0^1} = \gamma(t, x, y)\frac{U_0}{U_0^1}$ *with* $|\gamma - 1| + |\gamma_x| + |\gamma_y| \le \lambda_7 \varepsilon$, $U_0^1(1 - e^{-\lambda_8 y})$ *and* $|u_{0y}^1| \le \lambda_{10} e^{-\lambda_{11} y}$ *hold for some* $\lambda_j > 0$ $(6 \le j \le 11)$*, then the solution* u *of (1.1.197), (1.1.199), and (1.1.200) converges to the solution* u^1 *of the corresponding problem of (1.1.202) uniformly and that*

$$|u - u^1| \le \lambda_{12} \varepsilon$$

for a positive constant λ_{12}.

Ding and Gong [21] in 2017 used the viscous splitting method to establish global existence of weak solutions to the two-dimensional compressible Prandtl equations in $\Omega = \{0 < t < \infty,\ 0 < x < L,\ 0 < y < +\infty\}$ when the density is favorable:

$$\begin{cases} \partial_t u + u\partial_x u + v\partial_y u + \frac{\partial_x p(\rho)}{\rho} = \frac{1}{\rho}\partial_y^2 u, \\ \partial_x(\rho u) + \partial_y(\rho v) = -\partial_t \rho, \\ u|_{t=0} = u_0(x, y), \\ (u, v)|_{y=0} = (0, v_0(t, x)), \ \lim\limits_{y \to +\infty} u = U(t, x) = 1, \\ u|_{x=0} = u_1(x, y). \end{cases} \tag{1.1.203}$$

Assume that

$$u_0(x, y) \ge 0, \ \ v_0(t, x) \le 0, \ \ u_1(t, y) \ge 0, \ \ U(t, x) > 0. \tag{1.1.204}$$

In [21], they initially gave a positive answer to this problem with the favorable pressure P, which implies that

$$\partial_x \rho \le 0. \tag{1.1.205}$$

We firstly introduce the following Crocco transformation:

$$\tau = t, \quad \xi = x, \quad \eta = \frac{u}{U}, \quad w = \frac{u_y}{U}, \tag{1.1.206}$$

then the problem (1.1.203) can be transformed into the following scalar degenerated parabolic equation in $\Omega_1 = \{0 < \tau < \infty, 0 < \xi < L, 0 < \eta < l\}$,

$$\begin{cases} w_\tau + \eta U w_\xi + A w_\eta + B w = \frac{1}{\rho} w^2 w_{\eta\eta}, \\ w|_{\tau=0} = w_0, \quad w|_{\xi=0} = w_1, \\ w|_{\eta=1} = 0, \quad \frac{1}{\rho} w_\eta - v_0 + \frac{C}{w}|_{\eta=0} = 0. \end{cases} \tag{1.1.207}$$

Here

$$\begin{cases} A = (1-\eta)\frac{U_t}{U} + (1-\eta^2)U_x, \quad w_i = \frac{u_{iy}}{U}, \quad i = 0, 1, \\ B = \frac{U_t}{U} + \eta U_x - \frac{\rho_t}{\rho} - \frac{\rho_t}{\rho}\eta U, \quad C = U_x + \frac{U_t}{U}. \end{cases}$$

Theorem 1.1.113 ([21]) *Suppose the data w_i $(i = 0, 1)$ satisfy*

$$c_1(1-\eta) \le w_i \le c_1(1-\eta). \tag{1.1.208}$$

Assumption condition (1.1.204) *and the favorable condition* (1.1.205) *hold for* $(\tau, \xi) \in (0, \infty) \times (0, L)$, *then problem* (1.1.207) *admits a global weak solution* w.

Fan, Ruan, and Yang [29] in 2021 proved by the energy method for the local well-posedness of the two-dimensional compressible boundary layer equations (1.1.196) by energy method in a periodic domain $\mathbb{T} \times \mathbb{R}_+ = \{(x, y)|x \in \mathbb{R}/\mathbb{Z}, y \in [0, +\infty)\}$ with the same method as in Masmoudi and Wong [98] in 2015. The space $H^{s,\gamma}_{\sigma,2\delta}$ for $w : \Omega \to \mathbb{R}$ is defined by

$$H^{s,\gamma}_{\sigma,2\delta} := \left\{ w | \|w\|_{H^{s,\gamma}} \le \infty, (1+y)^\sigma w \ge \delta, \sum_{|\alpha|\le 2} |(1+y)^{\sigma+\alpha_2} D^\alpha w|^2 \le \frac{1}{\delta^2}, \right\}$$

where $s \ge 4, \gamma \ge 1, \sigma > \gamma + \frac{1}{2}, \delta \in (0, 1)$, $D^\alpha = \partial_x^{\alpha_1}\partial_y^{\alpha_2}$ with the index $\alpha = (\alpha_1, \alpha_2)$, and the weighted H^s norm $\|\cdot\|_{H^{s,\gamma}(\Omega)}$ is defined by

$$\|w\|^2_{H^{s,\gamma}} := \sum_{|\alpha|\le s} \|(1+y)^{\gamma+\alpha_2} D^\alpha w\|^2_{L^2(\Omega)}.$$

Theorem 1.1.114 ([29]) *Given any even integer* $s \geq 4$ *and real numbers* γ, σ, δ *satisfying* $\gamma \geq 1, \sigma > \gamma + \frac{1}{2}$, *and* $\delta \in (0, 1)$, *assume the following conditions on the initial data and the outer flow* (ρ, U):

(i) Suppose that the initial data $u_0 - U(0, x) \in H^{s,\gamma-1}$ *and* $\partial_y u_0 = \partial_y u_0 \in H^{s,\gamma}_{\sigma,2\delta}$ *satisfy the compatibility conditions* $u_0|_y = 0, \lim_{y\to+\infty} u_0|_{y=0} = U|_{t=0}$. *In addition, when* $s = 4$, *it is further assumed that* $\delta > 0$ *is chosen small enough such that* $\|\partial_y u_0\|_{H^{s,\gamma}_g} \leq C\delta^{-1}$ *with a generic constant* $C > 0$.

(ii) The outer flow (ρ, U) *is supposed to satisfy that the density* $\rho(t, x)$ *has both positive lower and upper bounds, and*

$$\sup_t \sum_{l=0}^{\frac{s}{2}+1} \|\partial_t^l U\|_{H^{s-2l+2}(\mathbb{T})} + \sup_t \sum_{l=0}^{\frac{s}{2}+1} \|\partial_t^l \rho\|_{H^{s-2l+2}(\mathbb{T})} < +\infty.$$

Then there exist a time $T := T(s, \gamma, \sigma, \delta, \|w_0\|_{H^{s,\gamma}_g} \rho, , U) > 0$ *such that the initial-boundary value problem (1.1.196) has a unique classical solution* (u, v) *satisfying*

$$u - U \in L^\infty([0, T], H^{s,\gamma-1}) \cap C([0, T]; H^s - w)$$

and

$$\partial_y u \in L^\infty([0, T]; H^{s,\gamma}_{\sigma,\delta}) \cap C([0, T]; H^s - w),$$

where $H^s - w$ *is the space* H^s *endowed with its weak topology.*

1.2 Viscosity Limit of the Navier-Stokes Equations

Sammartino and Caflisch [121] in 1998 proved the existence of solutions to the Euler and Prandtl equations with analytic initial data. The solution will be found as a composite expansion, using the Prandtl solution near the boundary and the Euler solution far from the boundary. These results can be used in [122] as the leading order terms in an asymptotic expansion for the solution of the Navier-Stokes equations with small viscosity.

Theorem 1.2.1 ([121]) *Suppose that* $u^E(t, x, y)$ *and* $u^P(t, x, Y)$ *are solutions of the Euler and Prandtl equations (1.0.4) and (1.0.2), respectively, which are analytic in the spatial variables* x, y, Y. *Then for a short time* T, *independent of* ϵ, *there is a solution* $u^{NS}(t, x, y)$ *of the Navier-Stokes equations (1.0.1) with*

$$u^{NS} = \begin{cases} u^E + O(\epsilon), & \text{outside boundary layer,} \\ u^P + O(\epsilon), & \text{inside boundary layer.} \end{cases}$$

Authors of [26] in 2000 reviewed recent progresses on the analysis of Prandtl equation and the related issue of the zero viscosity limit for the solutions of the Navier-Stokes equations and considered the following equations:

$$\begin{cases} u_t^\varepsilon + (u^\varepsilon + \nabla)u^\varepsilon + \nabla p^\varepsilon = \varepsilon \Delta u^\varepsilon, \\ \nabla \cdot u^\varepsilon = 0 \; on \; \Omega, \\ u^\varepsilon \cdot n|\partial\Omega = 0, \quad u^\varepsilon \cdot \tau|\partial\Omega = 0, \\ u^\varepsilon(x, 0) = u_0(x). \end{cases} \tag{1.2.1}$$

Here n and τ are the unit normal and tangent vectors at the boundary $\partial\Omega$. They restricted their attention to two space dimensions.

The steady Prandtl equations to (1.2.1) are

$$\begin{cases} u_t + uu_x + vu_y + p_x = u_{yy}, \\ p_y = 0, u_x + v_y = 0. \end{cases} \tag{1.2.2}$$

The steady-state version of (1.2.2) has also been of considerable interest in the engineering literature, which reads

$$\begin{cases} uu_x + vu_y + p_x = u_{yy}, \\ u_x + v_y = 0, \\ u(0, y) = u_0(y). \end{cases} \tag{1.2.3}$$

Theorem 1.2.2 ([26]) *Assume that u_0, u_{0y}, u_{0yy} are bounded and Hölder continuous. Assume also that u_0 satisfies the compatibility condition*

$$u_{0yy} - P_x = O(y^2) \tag{1.2.4}$$

as $y \to 0$. Then there exists an X such that (1.2.3) *has a unique strong solution u in the domain $D_x = \{(x, y), 0 \le x \le X, y \ge 0\}$. Moreover, u, u_x, u_y, u_{yy} are continuous and bounded in D_X. If $P_x \le 0$, then $X = +\infty$.*

Theorem 1.2.3 ([26]) *Assume that u_0 satisfies*

$$u_0^2(y) - \frac{3}{2}u_{0y}(y)\int_0^y u_0(z)dz \ge 0 \tag{1.2.5}$$

for $y \ge 0$. Then

(1) There exists an x^*, such that (1.2.3) *has a unique strong solution on $D_{x^*} = \{(x, y), 0 \le x \le x^*, y > 0\}$, but this strong solution cannot be extended to $x > x^*$.*

(2) The sequences of functions $\{u_\lambda\}$ defined by

$$u_\lambda(x, y) = \frac{1}{\lambda^{\frac{1}{2}}} u(x^* - x\lambda, \lambda^{\frac{1}{4}} y) \tag{1.2.6}$$

is compact in $C^0(D)$, $D = \{(x, y) : x, y \geq 0\}$.

Theorem 1.2.4 ([26]) *Assume that u_0 satisfies*

$$u_{0y}(x, y) > 0 \ for \ y \geq 0. \tag{1.2.7}$$

Then (1.2.2) *has a unique global strong solution.*

Theorem 1.2.5 ([26]) *Assume that*

$$E(a_0) = \int_0^\infty (\frac{1}{2} a_{0y}^2 + \frac{1}{4} a_0^3) dy < 0, \tag{1.2.8}$$

where $a_0(y) = a(y, 0)$. Then global strong solutions of (1.2.2) *do not exist.*

Theorem 1.2.6 ([26]) *For any $N > 0$, there exists a profile $u_* \in C^\infty(\mathbb{R}_+)$, such that the solution of the Navier-Stokes equation u_*^ε with initial data*

$$u_*^\varepsilon(x, y, 0) = (u_*(\frac{y}{\sqrt{\varepsilon}}), 0) \tag{1.2.9}$$

has the following property: for any $s > 0$, and sufficiently small $\varepsilon > 0$, there exists a solution to the Navier-Stokes equation u^ε, such that

$$\|u^\varepsilon(\cdot, 0) - u_*^\varepsilon(\cdot, 0)\|_{H^s(\Omega)} \leq \varepsilon^N \tag{1.2.10}$$

and

$$\|u^\varepsilon(\cdot, T_\varepsilon) - u_*^\varepsilon(\cdot, T_\varepsilon)\|_{H^s(\Omega)} \to +\infty \tag{1.2.11}$$

as $\varepsilon \to 0$, where

$$T_\varepsilon = O(\sqrt{\varepsilon} \log \frac{1}{\varepsilon}). \tag{1.2.12}$$

Theorem 1.2.7 ([26]) *Fix $T > 0$. Then $u^\varepsilon(\cdot, t) \to u^0(\cdot, 0)$ in $L^2(\Omega)$, uniformly for $t \in [0, T]$ if*

$$\varepsilon \int_0^T \int_\Omega |\nabla u^\varepsilon|^2 d^2x dt \to 0 \tag{1.2.13}$$

or

$$\varepsilon \int_0^T \int_{T_\varepsilon} |\nabla u^\varepsilon|^2 d^2x dt \to 0 \tag{1.2.14}$$

where T_ε is a strip of width $O(\varepsilon)$ around $\partial\Omega$.

Theorem 1.2.8 ([26]) *Fix any* $T > 0$*. Then* $\max_{0\le t\le T} \|u^\varepsilon(\cdot,t) - u^0(\cdot,t)\|_{L^2(\Omega)} \to 0$ *as* $\varepsilon_\parallel \to 0$ *if* $\varepsilon_\perp/\varepsilon_\parallel \to 0$*, here* $\delta = \sqrt{\varepsilon_\perp}$*.*

Greniner [48] in 2000, by giving examples of nonlinearly unstable solutions of Euler equations (1.0.4), proved an instability result for Prandtl-type boundary layers (1.0.1).

Theorem 1.2.9 ([48] Nonlinear Instability of Euler Equations) *Let* $u^s(y) \in C^\infty(\Omega)$ *be a shear flow. If* $u^s(y)$ *is linearly unstable (in the sense that there exists an eigenvalue of a linearized Euler operator with a nonnegative real part), then it is also nonlinearly unstable in the following sense: There exist positive constants* $\delta_0 > 0$ *and* $C_s > 0$ *(for* $s \in \mathbb{N}$*) such that for every s arbitrarily large and every* $\delta > 0$ *arbitrarily small, there exists a smooth solution* $u(t, x, y)$ *of problem (1.0.4), such that*

$$\|u(0, x, y) - u^s(y)\|_{H^s} \le \delta$$

and a time

$$T_\delta \le C_s \log \delta^{-1} + C_s$$

such that

$$\|u(T_\delta, x, y) - u^s(y)\|_{L^\infty} \ge \delta_0$$

and

$$\|u(T_\delta, x, y) - u^s(y)\|_{L^2} \ge \delta_{\delta_0}.$$

Theorem 1.2.10 ([48] Nonlinear Instability of Prandtl Boundary Layers) *Let* $\Omega = \mathbb{R}^{d-1} \times \mathbb{R}^+$ *and* $u^s(y) \in C^\Omega$ *with* $u^s(y) = 0$ *be a smooth shear layer profile, linearly unstable for Euler equations. Let n be an integer, arbitrarily large. Then there exists a positive constant* $\delta > 0$ *such that for every s arbitrarily large and for* ν *small enough, there exists a sequence of solutions* $u^\nu(t, x, y)$ *of (1.0.1) with*

$$\|u^\nu(0, x, y) - u^s(\frac{y}{\sqrt{\nu}})\|_{H^n} \le \nu^n$$

and times T^ν *with* $\lim\limits_{\nu\to 0} T^\nu = 0$ *and*

$$\|curl\ u^\nu(T^\nu, x, y) - curl\ u^s(T^\nu, \frac{y}{\sqrt{\nu}})\|_{L^\infty} \to +\infty,$$

$$\|u^\nu(T^\nu, x, y) - u^s(T^\nu, \frac{y}{\sqrt{\nu}})\|_{L^\infty} \geq \delta_0 \nu^{\frac{1}{4}}$$

where $u^s(y)$ *is the solution of*

$$\partial_t u^s(y) - \partial^2_{yy} u^s(y) = 0$$

with initial data $u_0^s(y)$.

Teman [129] in 2002 studied the local well-posed for the boundary layer of wall bounded flows in a channel at small viscosity when the boundaries are uniformly noncharacteristic. Denoting now v^ϵ the perturbation, i.e., $v^\epsilon = u^\epsilon - (0, 0, -U)$, the author found after dropping the nonlinear terms:

$$\begin{cases} \frac{\partial v^\varepsilon}{\partial t} - \varepsilon\Delta v^\varepsilon - UD_3 v^\varepsilon + \nabla p^\varepsilon = f, \\ div v^\varepsilon = 0, \\ v^\varepsilon = 0 \quad at\ the\ wall, i.e.\ at \quad z = 0, h. \end{cases} \tag{1.2.15}$$

Setting formally $\epsilon = 0$ in (1.2.15), the author found

$$\begin{cases} \frac{\partial v^0}{\partial t} - UD_3 v^0 + \nabla p^0 = f, \\ div v^0 = 0, \\ v^0 = 0 \quad at \quad z = h, v_3^0 = 0 \quad at \quad z = 0, \end{cases} \tag{1.2.16}$$

where

$$\varphi^\varepsilon = curl\ \psi^\varepsilon = (-\frac{\partial \psi_2^\varepsilon}{\partial z}, -\frac{\partial \psi_1^\varepsilon}{\partial z}, -\frac{\partial \psi_1^\varepsilon}{\partial y} + \frac{\partial \psi_2^\varepsilon}{\partial x}), \tag{1.2.17}$$

$$\begin{cases} \|\varphi^{\varepsilon,2}\|_{L^\infty(0;T;H^k)} \leq k\varepsilon, \quad \forall k, \\ \|\varphi^{\varepsilon,1}\|_{L^\infty(0;T;H^k)} \leq k\varepsilon^{1/2-k}, \\ \|\varphi^{\varepsilon,1}\|_{L^\infty(0;T;L^\infty)} \leq k, \\ \|z\varphi^{\varepsilon,1}\|_{L^\infty(0;T;L^\infty)} \leq k\varepsilon, \\ \|z\varphi^{\varepsilon,1}\|_{L^\infty(0;T;L^2)} \leq k\varepsilon^{\frac{3}{2}}, \\ \|z\nabla\varphi^{\varepsilon,1}\|_{L^\infty(0;T;L^2)} \leq k\varepsilon^{\frac{3}{2}}, \end{cases} \tag{1.2.18}$$

where k is a generic constant independent of the kinematic viscosity ε.

We deduce that the adjusted difference satisfies the equation

$$\begin{cases} \frac{\partial w^\varepsilon}{\partial t} - \varepsilon \Delta w^\varepsilon - U D_3 w^\varepsilon + \nabla(p^\varepsilon - p^0) = -\frac{\partial \varphi^\varepsilon}{\partial t} + \varepsilon \Delta v^0 - \varepsilon \Delta \varphi^\varepsilon - U D_3 \varphi^\varepsilon, \\ div w^\varepsilon = 0, \\ w^\varepsilon = 0 \quad at \quad z = 0, h, \\ w^\varepsilon = 0 \quad at \quad t = 0. \end{cases} \tag{1.2.19}$$

Recall the Navier-Stokes equations

$$\begin{cases} \frac{\partial u^\varepsilon}{\partial t} - \varepsilon \Delta u^\varepsilon + (u \cdot \nabla) u^\varepsilon + \nabla p^\varepsilon = f, \\ div u^\varepsilon = 0. \end{cases} \tag{1.2.20}$$

The corresponding "inviscid" Euler equation ($\epsilon = 0$) is

$$\begin{cases} \frac{\partial u^0}{\partial t} - (u \cdot \nabla) u^0 + \nabla p^0 = f, \\ div u^0 = 0, \\ u^0 = (0, 0, -U) \quad at \quad z = h, \\ u_3^0 = -U \quad at \quad z = 0, \end{cases} \tag{1.2.21}$$

and

$$\frac{\|v^0(t)\|_{L^\infty(z=0)}}{U} \le \frac{e}{32} \quad for\ t \le T_*. \tag{1.2.22}$$

Theorem 1.2.11 ([129]) *For f an v_0 given, let v^ε and v^0 be the solutions of* (1.2.15) *and* (1.2.16) *and assume that v^0 is sufficiently regular. Then, as $\varepsilon \to 0$, $v^\varepsilon \to v^0$ is estimated by* (1.2.19) *and*

$$\begin{cases} \|w^\varepsilon\|_{L^\infty([0,T]\times\Omega)} = \|u^\varepsilon - u^0 - \varphi^\varepsilon\|_{L^\infty([0,T]\times\Omega)} \le k\varepsilon^{\frac{3}{4}}, \\ \|w^\varepsilon\|_{L^\infty(0;T;H)} \le k\varepsilon, \\ \|w^\varepsilon\|_{L^2(0;T;(H^1)^3)} \le k\varepsilon^{\frac{1}{2}}. \end{cases}$$

The difference $w^\varepsilon = v^\varepsilon - v^0 - \varphi^\varepsilon$, the corrector φ^ε being given by (1.2.17)*, is estimated by* (1.2.18).

Theorem 1.2.12 ([129]) *Let f and u_0 be smooth functions satisfying certain compatibility conditions so that the inviscid problem* (1.2.21) *is well-posed (at least locally in time). Let u^ε and u^0 be the solution of* (1.2.20) *and* (1.2.21)*. Then there exists a time $T_* > 0$ so that* (1.2.22) *is satisfied. Moreover, as $\varepsilon \to 0$, $v^\varepsilon \to v^0$ is estimated as*

$$\begin{cases} \|u^\varepsilon - u^0\|_{L^\infty(0;T_*;H)} = \|v^\varepsilon - v^0\|_{L^\infty(0;T_*;H)} \le k\varepsilon^{\frac{1}{2}}, \\ \|u^\varepsilon - u^0 + \varphi^\varepsilon\|_{L^2(0;T_*;V)} = \|v^\varepsilon - v^0 + \varphi^\varepsilon\|_{L^2(0;T_*;V)} \le k\varepsilon^{\frac{1}{2}}. \end{cases}$$

The adjusted difference $w^\varepsilon = u^\varepsilon - u^0 + \varphi^\varepsilon = v^\varepsilon - v^0 + \varphi^\varepsilon$, *the corrector* φ^ε *being given by* (1.2.17), *is estimated by* (1.2.18).

Theorem 1.2.13 ([129]) *Under the assumptions of Theorem 1.2.12, we have also*

$$\|u^\varepsilon - u^0 + \varphi^\varepsilon\|_{L^\infty((0,T_*)\times\overline{\Omega})} = \|v^\varepsilon - v^0 + \varphi^\varepsilon\|_{L^\infty((0,T_*)\times\overline{\Omega})} \le k\varepsilon^{\frac{1}{4}}.$$

Filho [33] in 2007 made a survey on results related to boundary layers and the vanishing viscosity limit for incompressible flow. The main topics covered are a derivation of Prandtl's boundary layer equation; an outline of the rigorous theory of Prandtl's equation, without proofs; Kato's criterion for the vanishing viscosity limit; the vanishing viscosity limit with Navier friction condition; and rigorous boundary layer theory for the Navier friction condition and boundary layers for flows in a rotating cylinder. For the problem

$$\begin{cases} \partial_t v^1 + v \cdot \nabla v^1 = U_t + UU_{x_1} + \partial_{x_2}^2 v^1, \\ div v = 0, \\ v(x_1, 0, t) = 0 \text{ and } \lim\limits_{x_2\to\infty} v^1(x,t) = U(x_1, t), \\ v(x, 0) = v_0(x), \end{cases} \tag{1.2.23}$$

the author in [33] obtained the following Theorems 1.2.14–1.2.18.

Theorem 1.2.14 ([33]) *Assume that both* U *and* v_0^1 *are positive and that, in addition,* $\partial_{x_2} v_0^1 \ge 0$. *Then there exists a unique global strong solution of* (1.2.23).

Let u^ν be a sequence of solutions of the incompressible Navier-Stokes equations in two space dimensions.

Theorem 1.2.15 ([33]) *Fix* $T > 0$. *There exists a constant* $c > 0$ *such that* $u^\nu \to u$ *strongly in* $L^\infty((0,T); L^2(\Omega))$; *if and only if* $\nu \int_0^T \|\nabla u^\nu(\cdot,t)\|^2_{L^2(\Gamma_{c\nu})} dt \to 0$ *as* $\nu \to 0$, *where* $\Gamma_{c\nu} \equiv \{1 - c\nu < |x| < 1\}$.

Let $\omega^\nu = \omega^\nu(\cdot, t)$ to be the unique solution of the IBVP:

$$\begin{cases} \omega_t^\nu + u^\nu \cdot \nabla \omega^\nu = \nu \Delta \omega^\nu, \\ u^\nu = K_\Omega[\omega^\nu], \\ \omega^\nu(x,t) = (2k - \alpha)(u(x,t)\cdot\tau) \text{ for } x \in \partial\Omega, \\ \omega(x,0) = \omega_0. \end{cases}$$

Theorem 1.2.16 ([33]) *For each* $p > 2$, *there exists a constant* $C > 0$, *independent of* ν, *such that*

$$\|\omega^\nu(\cdot,t)\|_{L^p(\Omega)} \le C(\|\omega_0\|_{L^p(\Omega)} + \|u_0\|_{L^2(\Omega)}),$$

with $u_0 = K_\Omega[\omega_0]$.

Theorem 1.2.17 ([33]) *Let u^υ be the solution of the 2D Navier-Stokes equations in the unit disk, with no slip boundary data with respect to boundary rotation with prescribed angular velocity $\alpha \in BV(\mathbb{R})$. Assume that the initial velocity $u_0 \in L^2(D)$ has circular symmetry, $D \equiv \{|x| \leq 1\} \subset \mathbb{R}^2$, i.e., $u_0 = v_0(|x|)x^\perp$. Then, u_0 is a steady solution of the 2D Euler equation and u^υ converges strongly in $L^\infty([0,T], L^2(D))$ to u_0 as $\upsilon \to 0$. Here*

$$x^\perp = (-x_2, x_1), \quad \nabla^\perp = (-\partial_{x_2}, \partial_{x_1}).$$

Theorem 1.2.18 ([33]) *Let $\alpha \in BV(\mathbb{R})$ and assume that α is compactly supported in $(0,\infty)$. Let $u_0 \in L^2(D,\mathbb{R}^2)$ be covariant under rotations and assume that $\omega_0 = \nabla^\perp \cdot u_0 \in L^1(D)$, $D \equiv \{|x| \leq 1\} \subset \mathbb{R}^2$. Then we have:*

(1) There exists a constant $C > 0$ such that

$$\int_D |\omega^\upsilon(x,t)|dx \leq C(\|\omega_0\|_{L^1} + \|\alpha\|_{BV}),$$

for almost all time.

(2) For almost every $t \in \mathbb{R}$,

$$\alpha(t) = \int_D \omega^\upsilon(x,t)dx.$$

(3) We also have, for any $0 < a < 1$,

$$\|\omega^\upsilon - \omega_0\|_{L^1(|x|\leq a)} \to 0 \text{ as } \upsilon \to 0,$$

a.e. in time. If $\omega_0 \in C^0(D)$, then the convergence is uniform in time.

Wang, Wang, and Xin [137] in 2010 studied the vanishing viscosity limit for the incompressible Navier-Stokes equations with the Navier friction boundary condition. To simplify the expansion of solutions in terms of the viscosity, they only considered the case that the slip length α in the Navier boundary condition is a power of the viscosity ε. They considered the following problem:

$$\begin{cases} u_t^\varepsilon + u^\varepsilon \nabla u^\varepsilon + \nabla p^\varepsilon = \varepsilon^{\frac{1}{2}}\partial_x^2 u^\varepsilon + \varepsilon \partial_y^2 u^\varepsilon, \\ \partial_x u^\varepsilon + \partial_y v^\varepsilon = 0, \\ u_2^\varepsilon = 0, \ \beta u_1^\varepsilon - \varepsilon^{\frac{1}{4}} \frac{\partial u_1^\varepsilon}{\partial y} = 0, \ on \ y = 0, \\ u^\varepsilon|_{t=0} = u_0(x,y), \end{cases} \tag{1.2.24}$$

where β is a positive constant.

Theorem 1.2.19 ([137]) *For the problem* (1.2.24)*, suppose that the initial data* u_0 *of* (1.2.24) *belongs to* $H^s(\Omega)$ *for a fixed* $s > 18$*, with* $div\, u_0 = 0$ *and* $u_{0,2} = 0$ *on* $y = 0$*. Then, in* $L^\infty((0, T] \times \mathbb{R}^3_+)$*, we have*

$$\begin{cases} u_1^\varepsilon(t,x,y) = u_1^{I,0}(t,x,y) + \sum_{j=1}^3 \varepsilon^{j/4}(u_1^{I,j}(t,x,y) + u_1^{B,j}(t,x,\frac{y}{\sqrt{\varepsilon}})) + O(\varepsilon), \\ u_2^\varepsilon(t,x,y) = \sum_{j=1}^3 \varepsilon^{j/4}(u_2^{I,j}(t,x,y) + u_2^{B,3}(t,x,\frac{y}{\sqrt{\varepsilon}})) + O(\varepsilon), \\ p^\varepsilon(t,x,y) = \sum_{j=1}^3 \varepsilon^{j/4}(u_1^{I,j}(t,x,y) + O(\varepsilon), \end{cases}$$

for rapidly decreasing $(u_1^{B,j}, u_2^{B,3})$ *in* $z \to +\infty$*, where* $u_1^{I,0}$ *is a solution for the Euler equations,* $(u_1^{I,j}, u_2^{I,j})$ *are solutions for the linearized Euler equations.*

The authors of [137] further considered the following problem in $\{t, y > 0,\ x \in \mathbb{R}\}$,

$$\begin{cases} u_t^\varepsilon + u^\varepsilon \nabla u^\varepsilon + \nabla p^\varepsilon = \varepsilon^{1-\frac{1}{2}\gamma}\partial_x^2 u^\varepsilon + \varepsilon \partial_y^2 u^\varepsilon, \\ \partial_x u^\varepsilon + \partial_y v^\varepsilon = 0, \\ u_2^\varepsilon = 0,\ \beta u_1^\varepsilon - \varepsilon^\gamma \frac{\partial u_1^\varepsilon}{\partial y} = 0,\ on\ y = 0, \\ u^\varepsilon|_{t=0} = u_0(x,y). \end{cases} \tag{1.2.25}$$

Theorem 1.2.20 ([137]) *For problem* (1.2.25)*, under certain assumptions on the regularity and compatibility conditions on the initial data* u_0*, the solutions to* (1.2.25) *have the following expansions:*

$$\begin{cases} u_1^\varepsilon(t,x,y) = \sum_{k=0}^{b-a-1} \varepsilon^{rk} u_1^{I,k}(t,x,y) + \sum_{j=b-a}^{J} \varepsilon^{rj}(u_1^{I,j}(t,x,y) \\ \qquad + u_1^{B,j}(t,x,\frac{y}{\sqrt{\varepsilon}})) + O(\varepsilon^{rJ}), \\ u_2^\varepsilon(t,x,y) = \sum_{k=0}^{2b-a-1} \varepsilon^{rk} u_2^{I,k}(t,x,y) + \sum_{j\ge 2b-a}^{J} \varepsilon^{rj}(u_2^{I,j}(t,x,y) \\ \qquad + u_2^{B,j}(t,x,\frac{y}{\sqrt{\varepsilon}})) + O(\varepsilon^{rJ}), \\ p^\varepsilon(t,x,y) = \sum_{k=0}^{3b-a-1} \varepsilon^{rk} p^{I,k}(t,x,y) + \sum_{j\ge 3b-a}^{J} \varepsilon^{rj}(p^{I,j}(t,x,y) \\ \qquad + p^{B,j}(t,x,\frac{y}{\sqrt{\varepsilon}})) + O(\varepsilon^{rJ}), \end{cases} \tag{1.2.26}$$

in $L^\infty((0, T] \times \mathbb{R}^2)$ *for a fixed* $J \ge 1$.

Liu and Wang [81] in 2014 using curvilinear coordinate, proved that boundary layers of the Navier-Stokes equations in a bounded domain, still satisfy the classical Prandtl equations when the viscosity goes to zero.

Theorem 1.2.21 ([81]) *The solutions* $(\mathbf{u}^\nu, p^\nu)$ *to the problem (1.0.1) formally admit the following asymptotic expansions:*

$$\begin{cases} \mathbf{u}^\nu(t,x) = \mathbf{u}^{I,0} + u^{B,0}(t,s(x),\frac{d(x)}{\sqrt{\nu}}) + \sqrt{\nu} v^p(t,s(x),\frac{d(x)}{\sqrt{\nu}})\vec{n} + \mathbb{O}(\sqrt{\nu}), \\ p^\nu = p^{I,0}(t,x) + \mathbb{O}(\sqrt{\nu}), \end{cases}$$

where $v^p(t,s,\eta)$ is a scalar function, and $u^{B,0}(t,s,\eta)$, $v^p(t,s,\eta)$ are rapidly decreasing as $\eta \to +\infty$. Also, $(\mathbf{u}^{I,0}, p^{I,0})(t,x)$ satisfies the boundary value problem for the incompressible Euler system (1.0.4); the term $u^{B,0}(t,s,\eta)$ satisfies

$$u^{B,0}(t,s,\eta)\cdot\vec{n}=0,$$

and $((u^{B,0}+\overline{u^{I,0}})\cdot\vec{\tau}, v^p)(t,s,\eta)$ satisfies the boundary value problem for the Prandtl system (1.1.58), where $\overline{u^{I,0}}$ satisfies the Bernoulli's law.

Guo and Nguyen [53] in 2017 studied the validity of the Prandtl boundary layer theory in the inviscid limit for steady incompressible Navier-Stokes flows. To this end, they considered the stationary incompressible Navier-Stokes equations,

$$\begin{cases} UU_X+VU_Y+P_X=\varepsilon U_{XX}+\varepsilon U_{YY},\\ UV_X+VV_Y+P_Y=\varepsilon V_{XX}+\varepsilon V_{YY},\\ U_X+V_Y=0, \end{cases} \tag{1.2.27}$$

posed in a two-dimensional domain $\Omega=\{(X,Y):0\le X\le L, 0\le Y<+\infty\}$, with a no-slip boundary condition

$$[U(X,0),V(X,0)]\equiv[u_e,0].$$

Throughout [53], they assumed that the outside Euler flow is a shear flow

$$[U_0,V_0]\equiv\left[u_e^0(Y),0\right].$$

They worked with the scaled boundary layer, or Prandtl's variables:

$$x=X,\quad y=\frac{Y}{\sqrt{\varepsilon}}.$$

In these new variables, the Navier-Stokes equations (1.2.27) now read as

$$\begin{cases} U^\varepsilon U^\varepsilon_x+V^\varepsilon U^\varepsilon_y+P^\varepsilon_x=U^\varepsilon_{yy}+\varepsilon U^\varepsilon_{xx},\\ U^\varepsilon V^\varepsilon_x+V^\varepsilon V^\varepsilon_y+\frac{P^\varepsilon_y}{\varepsilon}=V^\varepsilon_{yy}+\varepsilon V^\varepsilon_{xx},\\ U^\varepsilon_x+V^\varepsilon_y=0. \end{cases} \tag{1.2.28}$$

Precisely, we search for asymptotic expansions of the scaled Navier-Stokes solutions $(U^\varepsilon, V^\varepsilon, P^\varepsilon)$ in the following form:

$$\begin{cases} U^\varepsilon(x,y)=u_e^0(\sqrt{\varepsilon}y)+u_p^0(x,y)+\sqrt{\varepsilon}u_e^1(x,\sqrt{\varepsilon}y)+\sqrt{\varepsilon}u_p^1(x,y)+\varepsilon^{\gamma+\frac{1}{2}}u^\varepsilon(x,y),\\ V^\varepsilon(x,y)=v_p^0(x,y)+v_e^1(x,\sqrt{\varepsilon}y)+\sqrt{\varepsilon}v_p^1(x,y)+\varepsilon^{\gamma+\frac{1}{2}}v^\varepsilon(x,y),\\ P^\varepsilon(x,y)=\sqrt{\varepsilon}p_e^1(x,\sqrt{\varepsilon}y)+\sqrt{\varepsilon}p_p^1(x,y)+\varepsilon p_p^2(x,y)+\varepsilon^{\gamma+\frac{1}{2}}p^\varepsilon(x,y), \end{cases} \tag{1.2.29}$$

for some $\gamma > 0$, $\left[u_e^j, v_e^j, p_e^j\right]$ and $\left[u_p^j, v_p^j, p_p^j\right]$, $j = 0, 1$, denote the Euler and Prandtl profiles, respectively, and $[u^\varepsilon, v^\varepsilon, p^\varepsilon]$ collects all the remainder solutions.

Theorem 1.2.22 ([53]) *Let $u_b > 0$ be a constant tangential velocity of the Navier-Stokes flow on the boundary $Y = 0$ and let be a given smooth and positive Euler flow so that $\partial_Y u_e^0$ and its derivatives decay exponentially fast to zero at infinity. We take Euler data $u_b^1(Y)$, $V_{b0}(Y)$, $V_{bL}(Y)$, and Prandtl data $\bar{u}_0(y)$ and $\bar{u}_1(y)$ to be sufficiently smooth and decay exponentially fast at infinity in their arguments. We assume that $|V_{bL}(Y) - V_{b0}(Y)| \lesssim L$ for all L, and*

$$\min_{0\leq y\leq\infty} \left\{u_e^0(\sqrt{\varepsilon}y) + \bar{u}_0(y)\right\} > 0.$$

Then, there exists a positive number L_0 that depends only on the given data so that the boundary layer expansions (1.2.29), *with the profiles and the remainder satisfying the boundary conditions for $\gamma \in \left(0, \frac{1}{4}\right)$ and $0 < L \leq L_0$. Precisely, $[U^\varepsilon, V^\varepsilon, P^\varepsilon]$ as defined in* (1.2.29) *is the unique solution to the Navier-Stokes equations* (1.2.28), *so that the remainder solutions $[u^\varepsilon, v^\varepsilon]$ satisfy*

$$\left\|\nabla_\varepsilon u^\varepsilon\right\|_{L^2} + \varepsilon^{\frac{1}{2}} \left\|\nabla_\varepsilon v^\varepsilon\right\|_{L^2} + \varepsilon^{\frac{\gamma}{2}} \left\|u^\varepsilon\right\|_{L^\infty} + \varepsilon^{\frac{1}{2}+\frac{\gamma}{2}} \left\|v^\varepsilon\right\|_{L^\infty} \leq C_0,$$

for some constant C_0 that depends only on the given data. Here, $\nabla_\varepsilon = \left(\sqrt{\varepsilon}\partial_x, \partial_y\right)$, $\|\cdot\|_{L^p}$ denotes the usual L^p norm over $[0, L] \times \mathbb{R}_+$.

Wang and Liu [133] in 2017 used λ-weighted energy method and the matched asymptotic expansion method to prove the well-posedness of solutions to the boundary layer problem. They studied the boundary layer problem and the quasineutral limit of the isentropic Euler-Poisson system in plasma for $t > 0$, $x = (x_1, x_2, x_3) = (y, x_3) \in \mathbb{R}_+^3 = \mathbb{R}^2 \times \mathbb{R}_+$,

$$\begin{cases} \partial_t n^\lambda + div(n^\lambda u^\lambda) = 0, \\ \partial_t u^\lambda + u^\lambda \cdot \nabla u^\lambda + \nabla h(u^\lambda) = -\nabla\phi^\lambda, \\ -\lambda^2 \Delta\phi^\lambda = n^\lambda - 1, \end{cases} \tag{1.2.30}$$

where λ is the (scaled) Debye length, $(n^\lambda, u^\lambda) = (u_1^\lambda, u_2^\lambda, u_3^\lambda) = (u_y^\lambda, u_3^\lambda)$, ϕ^λ denote the electron density, the electron velocity, and the electric potential, respectively. The function $h = h(n)$ is the enthalpy for the system and satisfies

$$h'(n) = \frac{p'(n)}{n}, \quad n > 0$$

where $p = p(n)$ is the pressure density function, and we assume

$$p(n) = a^2 n^\gamma, \; n > 0, a \neq 0, \; \gamma \geq 1.$$

From (1.2.30) in the limit $\lambda \to 0$ and the fact that $n = 1$, we have the following incompressible Euler system:

$$\begin{cases} divu = 0, \\ \partial_t u + u \cdot \nabla u = -\phi. \end{cases} \tag{1.2.31}$$

Theorem 1.2.23 ([133]) *Let* $(u(0), \phi(0))$ *be a solution to* (1.2.31) *such that* $u(0) \in C^0([0, T], H^s(R^3_+))$ *with* s *large enough. There is* $\lambda_0 > 0$ *such that for any* $\lambda \in (0, \lambda_0]$, *there exists* $(n^\lambda, u^\lambda, \phi^\lambda)$, *a unique solution to* (1.2.30) *also defined on* $[0, T]$ *such that*

$$\lim_{\lambda \to 0} \sup_{[0,T]} (\|n^\lambda - 1\|_{L^2} + \|u^\lambda - u(0)\|_{L^2} + \|\lambda \nabla \phi^\lambda - \lambda \nabla \phi(0)\|_{L^2}) \to 0.$$

Guo and Lyer [51] in 2018 investigated the asymptotic behavior of solutions to incompressible Navier-Stokes equations on the two dimensional as the viscosity vanishes, that is, as $\varepsilon \to 0$.

They considered the steady, incompressible Navier-Stokes equations on the two-dimensional domain $(X, L) \in \Omega \times (0, \infty)$. Denoting the velocity $\mathbf{U}^{N,S} = (U^{N,S}, V^{N,S})$, the equations read:

$$\begin{cases} \mathbf{U}^{N,S} \cdot \nabla \mathbf{U}^{N,S} + \nabla P^{N,S} = \varepsilon \mathbf{U}^{N,S}, \\ \nabla \cdot \mathbf{U}^{N,S} = 0. \end{cases} \tag{1.2.32}$$

The system above is taken with the no-slip boundary condition on

$$(U^{N,S}, V^{N,S})|_{Y=0} = 0.$$

In [51], they addressed Prandtl's classical setup and worked with the scaled boundary layer variable:

$$x = X, \quad y = \frac{Y}{\sqrt{\varepsilon}}.$$

In these new variables, the Navier-Stokes equations (1.2.32) now read

$$\begin{cases} U^\varepsilon U^\varepsilon_x + V^\varepsilon U^\varepsilon_y + P^\varepsilon_x = \Delta_\varepsilon U^\varepsilon, \\ U^\varepsilon V^\varepsilon_x + V^\varepsilon V^\varepsilon_y + \frac{P^\varepsilon_y}{\varepsilon} = \Delta_\varepsilon V^\varepsilon, \\ U^\varepsilon_x + V^\varepsilon_y = 0. \end{cases}$$

They expanded the solution in ε as

$$\begin{cases} U^\varepsilon = u_e^0 + u_p^0 + \sum_{i=1}^n \sqrt{\varepsilon}^i u_e^i + \sqrt{\varepsilon} u_p^i + \varepsilon^{N_0} u^\varepsilon := u_s + \varepsilon^{N_0} u^\varepsilon, \\ V^\varepsilon = v_p^0 + v_e^1 + \sum_{i=1}^{n-1} \sqrt{\varepsilon}^i (v_p^i + v_e^{i+1}) + \varepsilon^{n/2} v_p^n + \varepsilon^{N_0} v^\varepsilon := v_s + \varepsilon^{N_0} v^\varepsilon, \\ P^\varepsilon = P_e^0 + P_p^0 + \sum_{i=1}^n \sqrt{\varepsilon}^i P_e^i + \sqrt{\varepsilon} P_p^i + \varepsilon^{N_0} P^\varepsilon := P_s + \varepsilon^{N_0} P^\varepsilon. \end{cases} \tag{1.2.33}$$

At the leading order, $[u_p^0, v_p^0]$ solves the nonlinear Prandtl equations:

$$\begin{cases} \bar{u}_p^0 u_{px}^0 + \bar{v}_p^0 u_{px}^0 - u_{pyy}^0 + P_{px}^0 = 0, \\ u_{px}^0 + u_{py}^0 = 0, \ P_{py}^0 = 0, \ u_p^0|_{x=0} = U_p^0, \ u_p^0|_{y=0} = -u_e^0|_{Y=0}. \end{cases} \tag{1.2.34}$$

It is well-known that Prandtl equation (1.2.34) admits the two parameter scaling invariance

$$[\bar{u}^{\lambda,\delta}, \bar{v}^{\lambda,\delta}] = [\frac{\lambda^2}{\delta} \bar{u}_p^0(\delta x, \lambda y), \lambda \bar{v}_p^0(\delta x, \lambda y)], \tag{1.2.35}$$

which means that $[u_p^0, v_p^0]$ solves (1.2.34), and then so do $[\bar{u}^{\lambda,\delta}, \bar{v}^{\lambda,\delta}]$ (with appropriately modified initial data).

The boundary condition is taken as the following:

$$\begin{cases} v_x^{(\varepsilon)}|_{x=L} = a_1^\varepsilon(y), \ v_{xx}^{(\varepsilon)}|_{x=L} = a_2^\varepsilon(y), \ v_{xxx}^{(\varepsilon)}|_{x=L} = a_3^\varepsilon(y), \\ v^{(\varepsilon)}|_{x=0} = v_0(y), \ v_y^0| + u^0 = h(y) \in C^\infty, \ h(0) = 0, \\ v^{(\varepsilon)}|_{y=0} = v_y^{(\varepsilon)}|_{y=0} = u^{(\varepsilon)}|_{y=0} = 0, \ v_y^{(\varepsilon)}|_{y=\infty} = 0. \end{cases} \tag{1.2.36}$$

Here, the $a_i^\varepsilon(y)$ are prescribed boundary data which are assumed to satisfy

$$\|\partial_y^j a_i^\varepsilon \{\frac{1}{\varepsilon^{1/2}}(y)(Y)^m\}\| \le o(1) \ for \ j = 0, 1, 2, 3, 4, and \ m \ lager. \tag{1.2.37}$$

Theorem 1.2.24 ([51]) *Assume boundary data are prescribed as in* (1.2.36)–(1.2.37). *Assume* $0 < \delta \ll 1$ *in* (1.2.35). *Then let* $0 < \varepsilon \ll L \ll 1$. *Take* $N_0 = 1+$ *and* $n = 4$ *in* (1.2.33). *Then all terms in the expansion* (1.2.33) *exist and are regular,* $\|u_s, v_s\| \le 1$. *The remainders,* $[u^{(\varepsilon)}, v^{(\varepsilon)}]$ *exists uniquely in the space X and satisfy*

$$\|\boldsymbol{u}^{(\varepsilon)}\|_X \le 1.$$

The Navier-Stokes solutions satisfy

$$\|U^{N,S} - u_e^0 - u_p^0\|_\infty \le \sqrt{\varepsilon} \ and \ \|V^{N,S} - \sqrt{\varepsilon} v_e^1 - \sqrt{\varepsilon} v_p^0\|_\infty \le \varepsilon,$$

where

$$X = \{(v, u^0, v^0) \in L^2(\Omega) \times L^2(\mathbb{R}_+) \times L^2(\mathbb{R}_+) : \|(v, u^0, v^0)\|_X < +\infty;$$
$$v|_{y=0} = v_y|_{y=0} = v_x|_{x=L} = v_{xx}|_{x=0} = v_{xxx}|_{x=L} = v|_{y=\infty} = 0,$$
$$v^0(0) = v_y^0(0) = \partial_y^k v(\infty) = 0 \, for \, k \geq 1, u^0 + v_y^0 = h(y), u^0(0) = 0\}. \quad (1.2.38)$$

We first record the properties of the leading order ($i = 0$) layers. For the outer Euler flow, we will take a shear flow, $[u_e^0(Y), 0, 0]$. The derivatives of u_e^0 decay rapidly in Y and that is bounded below, $|u_e^0| \geq 1$,

For the leading order Prandtl boundary layer, the equations are the following:

$$\begin{cases} \bar{u}_p^0 u_{px}^0 + \bar{v}_p^0 u_{px}^0 - u_{pyy}^0 + P_{px}^0 = 0, \\ u_{px}^0 + u_{py}^0 = 0, \ P_{py}^0 = 0, \ u_p^0|_{x=0} = U_p^0, \ u_p^0|_{y=0} = -u_e^0|_{Y=0}. \end{cases} \quad (1.2.39)$$

Theorem 1.2.25 ([51]) *Assume boundary data is prescribed satisfying* $U_P^0 \in C^\infty$ *and exponentially decaying* $|\partial_y^j (U_P^0 - u_e^0(0))|$ *for* $j \geq 0$ *satisfying*

$$\bar{U}_P^0 > 0 \ for \ y > 0, \ \partial_y \bar{U}_P^0(0) > 0, \ \partial_y^2 \bar{U}_P^0(0) \sim y^2 \ near \ y = 0. \quad (1.2.40)$$

Then for some $L > 0$, *there exists a solution,* $[\bar{u}_p^0, \bar{v}_p^0]$ *to problem* (1.2.39) *satisfying, for some* $y_0, m_0 > 0$,

$$\sup_{x\in[0,L]} \sup_{y\in(0,y_0)} |\bar{u}_p^0, \bar{v}_p^0, \partial_y \bar{u}_p^0, \partial_{yy} \bar{u}_p^0, \partial_x \bar{u}_p^0| \leq 1,$$

$$\sup_{x\in[0,L]} \sup_{y\in(0,y_0)} \partial_y \bar{u}_p^0 > m_0 > 0.$$

By evaluating the system (1.2.39) *and* ∂_y *of* (1.2.39) *at* $y = 0$, *we conclude*

$$\partial_y^2 \bar{u}_p^0|_{y=0} = \partial_y^3 \bar{u}_p^0|_{y=0} = 0.$$

Theorem 1.2.26 ([51]) *Assume the shear flow* $u_e^0(Y) \in C^\infty$, *whose derivatives decay rapidly. Assume* (1.2.40) *regarding* $\bar{u}_p^0|_{x=0}$, *and the conditions*

$$\begin{cases} \partial_y^3 \bar{v}_p^i|_{x=0}(0) = \partial_x g_1|_{x=0,y=0} = 0, \\ \bar{v}_p^i|''''_{x=0}(0) = \partial_{xy} g_1|_{y=0}(x = 0), \\ \bar{u}_{py}^0|_{x=0}(0) u_e^i|_{x=0}(0) - \int_0^\infty \bar{u}_p^0 e^{-\int_1^y \bar{v}_p^0} \{f^i(y) - r^i(y)\} dy = 0, \end{cases} \quad (1.2.41)$$

where $r^i(y) = \bar{v}_p^i \bar{u}_{py}^0 - \bar{u}_p^0 \bar{v}_{py}^i$. *We assume also standard higher-order versions of the parabolic compatibility conditions* (1.2.41). *Let* $v_e^i|_{x=0}$, $v_e^i|_{x=L}$, $u_e^i|_{x=0}$, *be prescribed smooth and rapidly decaying Euler data. We assume on the data standard elliptic compatibility conditions at the corners* $(0, 0)$ *and* $(L, 0)$ *obtained by evaluating the equation at the corners. In addition, assume*

$$\begin{cases} v_e^1|_{x=0} \sim Y^{-m_1} \text{ or for some } 0 < m_1 < \infty, \\ \|\partial_Y^k (v_e^i|_{x=0} - v_e^i|_{x=L})(Y)^M\|_\infty \leq L. \end{cases}$$

Then all profiles in $[u_s, v_s]$ *exist and are smooth on* Ω. *The following estimates hold:*

$$\begin{cases} \bar{u}_p^0 > 0,\ \bar{u}_{py}^0|_{y=0} > 0,\ \bar{u}_{pyy}^0|_{y=0} = \bar{u}_{pyyy}^0|_{y=0} = 0, \\ \|\nabla^K (u_p^0, v_p^0) e^{My}\|_\infty \lesssim 1 \text{ for any } K \geq 0, \\ \|u_p^1\|_\infty + \|\nabla^K (u_p^i e^{My}\|_\infty + \|\nabla^j (v_p^i e^{My}\|_\infty \lesssim 1 \text{ for any } K \geq 1,\ M \geq 0, \\ \|\nabla^K (u_e^1, v_e^1) w_{m_1}\|_\infty \lesssim 1 \text{ for some fixed } m_1 > 1, \\ \|\nabla^K (u_e^i, v_e^i) w_{m_i}\|_\infty \lesssim 1 \text{ for some fixed } m_1 > 1, \end{cases}$$

where $w_{m_i} \sim e^{m_i Y}$ *or* $(1+Y)^{m_i}$.

Gérard-Varet and Maekawa [41] in 2019 show the H^1 stability of shear flows of Prandtl type: $U^\upsilon = \left(U_s(y/\sqrt{\upsilon}), 0\right)$, in the steady two-dimensional Navier-Stokes equations, under the natural assumptions that $U_s(Y) > 0$ for $Y > 0$, $U_s(0) = 0$, and $U_s'(0) > 0$. Their result is in sharp contrast with the unsteady ones, in which at most Gevrey stability can be obtained, even under global monotonicity and concavity hypotheses. The authors of [41] also considered the vanishing viscosity limit of the two-dimensional steady Navier-Stokes equations:

$$\begin{cases} v^\upsilon \cdot \nabla v^\upsilon - \upsilon \Delta v^\upsilon + \nabla q^\upsilon = g^\upsilon, & (x, y) \in \mathbb{T}_\kappa \times \mathbb{R}_+, \\ \operatorname{div} v^\upsilon = 0, & (x, y) \in \mathbb{T}_\kappa \times \mathbb{R}_+, \\ v^\upsilon|_{y=0} = 0. \end{cases}$$

Here $\mathbb{T}_\kappa = \mathbb{R}/(2\pi\kappa)\mathbb{Z}$, $\kappa > 0$, is a torus with periodicity $2\pi\kappa$, $\mathbb{R}_+ = \{y \in \mathbb{R} \mid y > 0\}$, while $v^\upsilon = \left(v_1^\upsilon, v_2^\upsilon\right)$ and q^υ are, respectively, the unknown velocity field and pressure field of the fluid. The positive constant υ is the viscosity coefficient. The vector field g^υ is an external force, decaying fast enough at infinity. The usual no-slip condition is prescribed at $y = 0$.

Let $U_s = U_s(Y) \in C^2\left(\overline{\mathbb{R}_+}\right)$ such that

$$\begin{gathered} U_s(0) = 0, \quad U_s > 0 \quad \text{in } Y > 0, \quad \lim_{Y \to \infty} U_s(Y) = U_E > 0, \\ \partial_Y U_s(0) > 0, \\ \sum_{k=1,2} \sup_{Y \geq 0} (1+Y)^3 \left|\partial_Y^k U_s(Y)\right| < \infty. \end{gathered}$$

Denoting $u^\nu = v^\nu - U^\nu$ as the perturbation induced by $f^\nu = g^\nu + \nu\partial_y^2 U^\nu$, we get

$$\begin{cases} U_s^\nu \partial_x u^\nu + u_2^\nu \partial_y U_s^\nu \mathbf{e}_1 - \nu\Delta u^\nu + \nabla p^\nu = -u^\nu \cdot \nabla u^\nu + f^\nu, & (x, y) \in \mathbb{T}_\kappa \times \mathbb{R}_+, \\ \operatorname{div} u^\nu = 0, & (x, y) \in \mathbb{T}_\kappa \times \mathbb{R}_+, \\ u^\nu|_{y=0} = 0. \end{cases} \tag{1.2.42}$$

Here $\mathbf{e}_1 = (1, 0)$.

Theorem 1.2.27 ([41]) *There exist positive numbers $\kappa_0, \nu_0, \varepsilon$ such that the following statement holds for $0 < \kappa \leqq \kappa_0$ and $0 < \nu \leqq \nu_0$: if $f^\nu = Q_0 f^\nu$ and $\|f^\nu\|_{L^2} \leqq \varepsilon \nu^{\frac{3}{4}} |\log \nu|^{-1}$, then there exists a unique solution $(u^\nu, \nabla p^\nu) \in \left(X \cap W_{loc}^{2,2} (\mathbb{T}_\kappa \times \mathbb{R}_+)^2\right) \times L^2 (\mathbb{T}_\kappa \times \mathbb{R}_+)^2$ to the problem (1.2.42) such that*

$$\begin{aligned} &\left\| u_{0,1}^\nu \right\|_{L^\infty} + \nu^{\frac{1}{4}} \left\| \partial_y u_{0,1}^\nu \right\|_{L^2} \\ &\quad + \sum_{n \neq 0} \left\| u_n^\nu \right\|_{L^\infty} + \nu^{-\frac{1}{4}} \|Q_0 u^\nu\|_{L^2} + \nu^{\frac{1}{4}} \|\nabla Q_0 u^\nu\|_{L^2} \leqq \frac{C|\log \nu|^{\frac{1}{2}}}{\nu^{\frac{1}{4}}} \|f^\nu\|_{L^2}, \end{aligned}$$

where $Q_0 = (I - \mathcal{P}_0)$, P_n is the orthogonal projection on the n-th Fourier mode in variable x and I is the identity operator.

Ding and Li [23] in 2020 were concerned with the validity of the Prandtl boundary layer theory in the inviscid limit of the steady incompressible Navier-Stokes equations. Under the symmetry assumption, they established the validity of the Prandtl boundary layer expansions and the error estimates. The convergence rate as $\varepsilon \to 0$ is also given.

In [23], they considered the following steady incompressible Navier-Stokes equations

$$\begin{cases} UU_X + VU_Y + P_X = \varepsilon U_{XX} + \varepsilon U_{YY}, \\ UV_X + VV_Y + P_Y = \varepsilon V_{XX} + \varepsilon V_{YY}, \\ U_X + V_Y = 0, \end{cases} \tag{1.2.43}$$

in the domain

$$\Omega := \{(X, Y) \mid 0 \le X \le L, 0 \le Y \le 2\},$$

with moving boundary conditions

$$U(X, 0) = U(X, 2) = u_b > 0,\ V(X, 0) = V(X, 2) = 0.$$

Under the transformation, the system (1.2.43) can be rewritten as

$$\begin{cases} U^\varepsilon U_x^\varepsilon + V^\varepsilon U_y^\varepsilon + P_x^\varepsilon = U_{yy}^\varepsilon + \varepsilon U_{xx}^\varepsilon, \\ U^\varepsilon V_x^\varepsilon + V^\varepsilon V_y^\varepsilon + P_y^\varepsilon/\varepsilon = V_{yy}^\varepsilon + \varepsilon V_{xx}^\varepsilon, \\ U_x^\varepsilon + V_y^\varepsilon = 0, \end{cases} \tag{1.2.44}$$

in the domain

$$\Omega_\varepsilon := \left\{(x, y) \mid 0 \le x \le L, 0 \le y \le \frac{1}{\sqrt{\varepsilon}}\right\},$$

with the boundary conditions

$$\left[U^\varepsilon, V^\varepsilon\right](x, 0) = [u_b, 0], \quad \left[U_y^\varepsilon, V^\varepsilon\right]\left(x, \frac{1}{\sqrt{\varepsilon}}\right) = [0, 0].$$

In what follows, they intended to find the exact solutions $[U^\varepsilon, V^\varepsilon, P^\varepsilon]$ in form of

$$\begin{cases} U^\varepsilon(x, y) = u_{app}(x, y) + \varepsilon^{\gamma+\frac{1}{2}} u^\varepsilon(x, y), \\ V^\varepsilon(x, y) = v_{app}(x, y) + \varepsilon^{\gamma+\frac{1}{2}} v^\varepsilon(x, y), \\ P^\varepsilon(x, y) = p_{app}(x, y) + \varepsilon^{\gamma+\frac{1}{2}} p^\varepsilon(x, y), \end{cases}$$

where

$$\begin{cases} u_{app}(x, y) = u_e^0(\sqrt{\varepsilon} y) + u_p^0(x, y) + \sqrt{\varepsilon} u_e^1(x, \sqrt{\varepsilon} y) + \sqrt{\varepsilon} u_p^1(x, y), \\ v_{app}(x, y) = v_p^0(x, y) + v_e^1(x, \sqrt{\varepsilon} y) + \sqrt{\varepsilon} v_p^1(x, y), \\ p_{app}(x, y) = \sqrt{\varepsilon} p_e^1(x, \sqrt{\varepsilon} y) + \sqrt{\varepsilon} p_p^1(x, y) + \varepsilon p_p^2(x, y). \end{cases}$$

For convenience, denote $z := \sqrt{\varepsilon} y$ boundary conditions on $\{y = 0\}$:

$$\begin{cases} u_e^0(0) + u_p^0(x, 0) = u_b, \quad u_e^1(x, 0) + u_p^1(x, 0) = 0, \ u^\varepsilon(x, 0) = 0, \\ v_p^0(x, 0) + v_e^1(x, 0) = 0, \ v_p^1(x, 0) = 0, \qquad\qquad v^\varepsilon(x, 0) = 0. \end{cases}$$

Boundary conditions on $\left\{y = \frac{1}{\sqrt{\varepsilon}}\right\}$ are as follows:

$$\begin{cases} u_{py}^0\left(x, \frac{1}{\sqrt{\varepsilon}}\right) = 0, \ u_{ez}^1(x, 1) = 0, u_{py}^1\left(x, \frac{1}{\sqrt{\varepsilon}}\right) = 0, \ u_y^\varepsilon\left(x, \frac{1}{\sqrt{\varepsilon}}\right) = 0, \\ v_p^0\left(x, \frac{1}{\sqrt{\varepsilon}}\right) = 0, \quad v_e^1(x, 1) = 0, v_p^1\left(x, \frac{1}{\sqrt{\varepsilon}}\right) = 0, \quad v^\varepsilon\left(x, \frac{1}{\sqrt{\varepsilon}}\right) = 0. \end{cases}$$

Boundary conditions on $\{x = 0\}$ are the following:

$$\begin{cases} u_p^0(0, y) = \bar{u}_0(y), \quad u_e^1(0, z) = u_b^1(z), \quad u_p^1(0, y) = \bar{u}_1(y), \quad u^\varepsilon(0, y) = 0, \\ v_e^1(0, z) = V_{b0}(z), \quad v^\varepsilon(0, y) = 0. \end{cases}$$

Boundary conditions on $\{x = L\}$ are

$$v_e^1(L, z) = V_{bL}(z), \quad \left[p^\varepsilon - 2\varepsilon u_x^\varepsilon, u_y^\varepsilon + \varepsilon v_x^\varepsilon\right](L, y) = 0.$$

For the existence of the Euler corrector $\left[u_e^1, v_e^1, p_e^1\right]$, it is necessary to impose the following compatibility conditions:

$$V_{b0}(0) = -v_p^0(0,0), \quad V_{bL}(0) = -v_p^0(L,0), \quad V_{b0}(1) = V_{bL}(1) = 0.$$

Theorem 1.2.28 ([23]) *Let $u_b > 0$ be a constant tangential velocity of the Navier-Stokes flow on the boundary $\{Y = 0\}$, and let $u_e^0(Y)$ be a smooth positive Euler flow satisfies $u_{ez}^0(1) = 0$. Both the boundary condition and the compatibility condition are assumed to be true. Suppose further that the positive condition $\min_y \left\{u_e^0(\sqrt{\varepsilon}y) + \bar{u}_0(y)\right\} > 0$ holds. Then there exists a constant $L_0 > 0$, which depends only on the prescribed data, such that for $0 < L \le L_0$ and $\gamma \in \left(0, \frac{1}{5}\right)$, the asymptotic expansion stated before is a solution to equations on Ω_ε together with the corresponding boundary conditions. The approximate solutions appearing in the expansions are constructed in sections 2–5 of [23], in which the remainder solutions $[u^\varepsilon, v^\varepsilon]$ satisfy the estimate*

$$\left\|\nabla_\varepsilon u^\varepsilon\right\|_{L^2(\Omega_\varepsilon)} + \left\|\nabla_\varepsilon v^\varepsilon\right\|_{L^2(\Omega_\varepsilon)} + \left\|u^\varepsilon\right\|_{L^\infty(\Omega_\varepsilon)} + \sqrt{\varepsilon}\left\|v^\varepsilon\right\|_{L^\infty(\Omega_\varepsilon)} \le C_0,$$

where $\Omega_\varepsilon = \{(x, y) | 0 \le x \le L,\ 0 \le y \le \frac{1}{\sqrt{\varepsilon}}\}$.

Wang, Yang, and Zhang [138] in 2020 considered 2D incompressible Navier-Stokes equations in a thin domain when the depth of the domain and the viscosity coefficient converges to zero simultaneously in a related way

$$\begin{cases} \partial_t U + U \cdot \nabla U - \varepsilon^2 \left(\partial_x^2 + \eta \partial_y^2\right) U + \nabla P = 0 & \text{in } \mathcal{S}^\varepsilon \times (0, \infty), \\ \operatorname{div} U = 0, \\ U|_{y=0} = U|_{y=\varepsilon} = 0, \end{cases} \tag{1.2.45}$$

where $\mathcal{S}^\varepsilon = \{(x, y) \in \mathbb{T} \times \mathbb{R} : 0 < y < \varepsilon\}$, and $U(t, x, y)$, $P(t, x, y)$ stand for the velocity and pressure function, respectively, and η is a positive constant independent of ε. The system (1.2.45) is prescribed with the initial data of the form

$$U|_{t=0} = \left(u_0\left(x, \frac{y}{\varepsilon}\right), \varepsilon v_0\left(x, \frac{y}{\varepsilon}\right)\right) = U_0^\varepsilon$$

Now rescale (U, P) as follows:

$$U(t, x, y) = \left(u^\varepsilon\left(t, x, \frac{y}{\varepsilon}\right), \varepsilon v^\varepsilon\left(t, x, \frac{y}{\varepsilon}\right)\right) \quad \text{and} \quad P(t, x, y) = p^\varepsilon\left(t, x, \frac{y}{\varepsilon}\right).$$

Then the system (1.2.45) reduces to the following scaled anisotropic Navier-Stokes system:

$$\begin{cases} \partial_t u^\varepsilon + u^\varepsilon \partial_x u^\varepsilon + v^\varepsilon \partial_y u^\varepsilon - \varepsilon^2 \partial_x^2 u^\varepsilon - \eta \partial_y^2 u^\varepsilon + \partial_x p^\varepsilon = 0, & \text{in } \mathcal{S} \times (0, \infty), \\ \varepsilon^2 \left(\partial_t v^\varepsilon + u^\varepsilon \partial_x v^\varepsilon + v^\varepsilon \partial_y v^\varepsilon - \varepsilon^2 \partial_x^2 v^\varepsilon - \eta \partial_y^2 v^\varepsilon \right) + \partial_y p^\varepsilon = 0, & \text{in } \mathcal{S} \times (0, \infty), \\ \partial_x u^\varepsilon + \partial_y v^\varepsilon = 0, \quad \text{in } \mathcal{S} \times (0, \infty), \\ (u^\varepsilon, v^\varepsilon)|_{y=0,1} = 0. \\ (u^\varepsilon, v^\varepsilon)|_{t=0} = (u_0, v_0), \quad \text{in } \mathcal{S}, \end{cases} \tag{1.2.46}$$

where $\mathcal{S} = \{(x, y) \in \mathbb{T} \times (0, 1)\}$. This is a classical model in geophysical fluid, where the vertical dimension of the domain is very small compared with the horizontal dimension of the domain. For simplicity, they took $\eta = 1$ in the sequel and denoted $\Delta_\varepsilon = \varepsilon^2 \partial_x^2 + \partial_y^2$.

Formally, taking $\varepsilon \to 0$, we derive the hydrostatic Navier-Stokes/Prandtl system

$$\begin{cases} \partial_t u^p + u^p \partial_x u^p + v^p \partial_y u^p - \partial_y^2 u^p + \partial_x p^p = 0 \quad \text{in } \mathcal{S} \times (0, \infty), \\ \partial_y p^p = 0 \quad \text{in } \mathcal{S} \times (0, \infty), \\ \partial_x u^p + \partial_y v^p = 0 \quad \text{in } \mathcal{S} \times (0, \infty), \\ (u^p, v^p)|_{y=0,1} = 0, \\ u^p|_{t=0} = u_0 \quad \text{in } \mathcal{S}. \end{cases} \tag{1.2.47}$$

Consider the initial data of the form

$$u^\varepsilon(0, x, y) = u_0(x, y), \quad v^\varepsilon(0, x, y) = v_0(x, y),$$

which satisfy the compatibility conditions

$$\begin{cases} \partial_x u_0 + \partial_y v_0 = 0, \quad u_0(x, 0) = u_0(x, 1) = v_0(x, 0) = v_0(x, 1) = 0, \\ \int_0^1 \partial_x u_0 dy = 0, \quad \partial_y^2 u_0 \Big|_{y=0,1} = \int_0^1 \left(-\partial_x u_0^2 + \partial_y^2 u_0 \right) dy - \int_{\mathcal{S}} \partial_y^2 u_0, \end{cases}$$

and the convex condition

$$\inf_{\mathcal{S}} \partial_y^2 u_0 = 2\delta_0 > 0.$$

The authors of [138] further assumed that initial data fall into the Gevrey class with the bound

$$\left\| \partial_y u_0 \right\|_{X_{\sigma,\tau_0}^{N_0}} + \left\| \partial_y^3 u_0 \right\|_{X_{\sigma,\tau_0}^{N_0-4}} = M < +\infty.$$

Here the Gevrey class norm $\|\cdot\|_{X^r_{\sigma,\tau}}$ is defined by

$$\|f\|^2_{X^r_{\sigma,\tau}} = \left\|e^{\tau\langle D_x\rangle^\sigma} f\right\|^2_{H^{r,0}}, \text{ with } \|f\|_{H^{r,s}} = \|\|f\|_{H^r_x(\mathbb{T})}\|_{H^s_y(0,1)}.$$

Theorem 1.2.29 ([138]) *Let the initial data u_0 satisfy the assumptions with $\sigma \in \left[\frac{8}{9}, 1\right]$, $\tau_0 > 0$ and $N_0 \geq 10$. Then there exist $T > 0$ and a unique solution u^p of the former equation which satisfies*

$$\begin{aligned}
&\sup_{t\in[0,T]} \left(\left\|\partial_y u^p(t)\right\|_{X^{N_0-1}_{\sigma,\tau}} + \left\|\partial_y^3 u^p(t)\right\|_{X^{N_0-5}_{\sigma,\tau}} \right) < +\infty,\\
&\sup_{t\in[0,T]\times S} \partial_y^2 u^p > \delta_0.
\end{aligned}$$

Theorem 1.2.30 ([138]) *Let initial data u_0 satisfies the assumptions with $\sigma \in \left[\frac{8}{9}, 1\right]$, $\tau_0 > 0$ and $N_0 \geq 10$. Then there exists a unique solution of the Navier-Stokes equations in $[0, T]$, which satisfies*

$$\left\|\left(u^\varepsilon - u^p, \varepsilon v^\varepsilon - \varepsilon v^p\right)\right\|_{L^2_{x,y}\cap L^\infty_{x,y}} \leq C\varepsilon^2,$$

where (u^p, v^p) is given by the former theorem and $C > 0$ is a constant independent of ε.

At last, the following is the zero viscosity limit of the 3D incompressible Navier-Stokes equations.

Fei, Tao, and Zhang [30] in 2018 investigated the zero viscosity limit of the incompressible Navier-Stokes equations with non-slip boundary condition in $\mathbb{R}^3_+$ for the initial vorticity located away from the boundary by the energy method. In [30], they were concerned with the zero viscosity limit of the incompressible Navier-Stokes equations in the half-space $\mathbb{R}^3_+$:

$$\begin{cases}
u^\varepsilon_t - \varepsilon^2 \Delta u^\varepsilon + u^\varepsilon \cdot \nabla_x u^\varepsilon + v^\varepsilon \cdot \nabla_y u^\varepsilon + \nabla_x p = 0,\\
v^\varepsilon_t - \varepsilon^2 \Delta v^\varepsilon + u^\varepsilon \cdot \nabla_x v^\varepsilon + v^\varepsilon \cdot \partial_y v^\varepsilon + \nabla_x p = 0,\\
\nabla_x \cdot u^\varepsilon + \partial_y v^\varepsilon = 0,\\
(u^\varepsilon, v^\varepsilon)(t, x, 0) = (0, 0).
\end{cases} \tag{1.2.48}$$

Here $x = (x_1, x_2) \in \mathbb{T}$ and $y \in \mathbb{R}$ denote the tangential and the vertical components of the space variable, respectively. $\nabla_x = (\partial_{x_1}, \partial_{x_2})$ denotes the tangential gradient; $u = (u^1, u^2)(t, x, y)$ and $v = v(t, x, y)$ denote the tangential and the vertical velocities; ε^2 is the viscosity coefficient. For simplicity, we consider the initial data of the form

$$u^\varepsilon(0, x, y) = u_0(x, y), \;\; v^\varepsilon(0, x, y) = v_0(x, y),$$

which satisfies

$$\nabla_x \cdot u_0^\varepsilon + \partial_y v_0^\varepsilon = 0,\ u_0(x,0) = 0,\ v_0(x,0) = 0, \tag{1.2.49}$$

and the initial vorticity $\omega_0 = curl(u_0, v_0)$ satisfies

$$2d_0 = dist(supp\omega_0, y = 0) > 0. \tag{1.2.50}$$

Without loss of generality, we take $d_0 = 1$.

Theorem 1.2.31 ([30]) *There exist $T > 0$ and $C > 0$, independent of ε, such that for any $(u_0, v_0) \in H^{30}(\mathbb{R}^3_+)$ satisfying* (1.2.49)–(1.2.50), *there exists a unique solution $(u_\varepsilon, v_\varepsilon)$ of the Navier-Stokes equations* (1.2.48) *in $[0, T]$, which satisfies*

$$\begin{cases} \sup\limits_{0\le t\le T} \|u^\varepsilon(t,x,y) - u^e(t,x,y) - u^p(t,x,y/\varepsilon)\|_{L^2\cap L^2(\mathbb{R}^3_+)} \le C\varepsilon, \\ \sup\limits_{0\le t\le T} \|v^\varepsilon(t,x,y) - v^e(t,x,y) - v^p(t,x,y/\varepsilon)\|_{L^2\cap L^2(\mathbb{R}^3_+)} \le C\varepsilon. \end{cases} \tag{1.2.51}$$

1.3 Ill-Posedness of the 2D Prandtl Equations

E and Engquist [27] in 1997 considered the unsteady boundary layer equations (1.1.58) in the domain $\mathbb{R}^2 = \{(x,y)|x \in \mathbb{R}, y \in \mathbb{R}_+\}$. Let $U = p = 0$, they proved global blowup of solutions.

Theorem 1.3.1 ([27]) *Assume that u_0 takes the form $u_0(x,y) = -xb_0(x,y)$ where $a_0(\cdot) = -u_{0x}(0,\cdot)$ satisfies the conditions of nonnegative, compactly supported initial data such that $\int_0^\infty(\frac{1}{2}a_y^2 - \frac{1}{4}a^3)dy$; then smooth solutions of (1.1.58) do not exist globally in time.*

Gargano, Sammartino, and Sciacca [39] in 2009 investigated the numerical evidence of the ill-posedness of the Prandtl equations in H^1 and led to the solution blow-up whose time can be made arbitrarily short within the class.

The authors of [39] considered the Cauchy problem for the Prandtl equation. This problem is known to be well-posed for analytic data or for data with monotonicity properties. Gérard-Varet and Dormy [40] in 2010 proved the linearly ill-posed in Sobolev-type spaces. The Prandtl equations reduce to

$$\begin{cases} u_t + u\partial_x u + v\partial_y u - \partial_y^2 u = 0, \\ \partial_x u + \partial_y v = 0, \\ (u,v)|_{y=0} = 0,\ \lim\limits_{y\to+\infty} u = 0. \end{cases} \tag{1.3.1}$$

Let $u_s = u_s(t, y)$ be a smooth solution of the heat equation

$$\begin{cases} \partial_t u_s - \partial_y^2 u_s = 0, \\ u_s|_{y=0} = 0, \quad \lim\limits_{y\to+\infty} u = U_0, \\ u_s|_{t=0} = U_s(y) \end{cases} \tag{1.3.2}$$

with good decay as $y \to +\infty$. Clearly, the shear velocity profile $(u_s, v_s) = (u_s(t, y), 0)$ satisfies the system (1.3.1). Consider the linearization around (u_s, v_s), that is,

$$\begin{cases} u_t + u_s\partial_x u + v\partial_y u_s - \partial_y^2 u = 0, \\ \partial_x u + \partial_y v = 0, \\ (u, v)|_{y=0} = 0, \lim_{y\to+\infty} u = 0. \end{cases} \tag{1.3.3}$$

In the sequel, [39] studied well-posedness properties of problem (1.3.3), for a certain class of velocities u_s. In this view, [39] introduced the following functional spaces:

$$W_\alpha^{s,\infty}(\mathbb{R}_+) := \{f = f(y), e^{\alpha y} f \in W^{s,\infty}(\mathbb{R}_+), \ \forall\alpha, s \geq 0\},$$

with $\|f\|_{W_\alpha^{s,\infty}} := \|e^{\alpha y} f\|_{W_\alpha^{s,\infty}}$, and

$$E_{\alpha,\beta} := \Big\{u = u(x, y) = \sum_{k\in\mathbb{Z}} \bar{u}^k(y)e^{ikx}, \ \|\bar{u}^k\|_{W_\alpha^{s,\infty}} \leq C_{\alpha,\beta}e^{-\beta|k|}\Big\}, \forall\alpha, \beta > 0,$$

with $\|u\|_{E_{\alpha,\beta}} := \sup_k e^{\beta|k|}\|\bar{u}^k\|_{W_\alpha^{s,\infty}}$.

Theorem 1.3.2 ([40]) *Let $u_s \in C^0(\mathbb{R}_+; W_0^{4,\infty}) \cap C^1(\mathbb{R}_+; W_0^{2,\infty})$. Assume that the initial velocity has a non-degenerate critical point over $\mathbb{R}^+$. Then, there exists $\sigma > 0$, such that for all $\delta > 0$,*

$$\sup_{0\leq s\leq t\leq\delta} \|e^{-\sigma(t-s)\sqrt{|\partial_x|}}T(t, s)\|_{L(H^m,H^{m-\mu})} = +\infty, \forall\alpha, m \geq 0, \mu \in [0, \frac{1}{2}).$$

Moreover, one can find solutions u_s of (1.3.2) *and* $\sigma > 0$, *such that for all* $\delta > 0$.

$$\sup_{0\leq s\leq t\leq\delta} \|e^{-\sigma(t-s)\sqrt{|\partial_x|}}T(t, s)\|_{L(H^m,H^{m_2})} = +\infty, \forall m_1, m_2 \geq 0,$$

where $H^m = H^m(\mathbb{T}_x, W_\alpha^{0,\infty}(\mathbb{R}_y^+)), \ m \geq 0.$

Gérard-Varet and Dormy [40] in 2010 were concerned with the Cauchy problem for the Prandtl equation and proved that it is linearly ill-posed in Sobolev-type spaces.The key of the analysis is the construction, at high tangential frequencies, of unstable quasimodes for the linearization around solutions with non-degenerate critical points.

The authors of [40] restricted themselves to $(x, Y) \in \mathbb{T} \times \mathbb{R}_+$, and $\mathbf{u}^0 = 0$. To lighten notations, they wrote y instead of Y. The Prandtl equation comes down to

$$\begin{cases} \partial_t u + u\partial_x u + v\partial_y u - \partial_y^2 u = 0, \text{ in } \mathbb{T} \times \mathbb{R}_+ \\ \partial_x u + \partial_y v = 0, \text{ in } \mathbb{T} \times \mathbb{R}_+ \\ (u, v)|_{y=0} = (0, 0), \lim_{y \to +\infty} u = 0. \end{cases} \tag{1.3.4}$$

Let $u_s = u_s(t, y)$ be a smooth solution of the heat equation

$$e\partial_t u_s - \partial_y^2 u_s = 0, \quad u_s|_{y=0} = 0, \quad u_s|_{t=0} = U_s$$

with good decay as $y \to +\infty$. Clearly, the shear velocity profile $(u_s, v_s) = (u_s(t, y), 0)$ satisfies the system (1.3.4). The authors of [40] considered the linearization around (u_s, v_s), that is,

$$\begin{cases} \partial_t u + u_s\partial_x u + v\partial_y u_s - \partial_y^2 u = 0, \text{ in } \mathbb{T} \times \mathbb{R}_+, \\ \partial_x u + \partial_y v = 0, \text{ in } \mathbb{T} \times \mathbb{R}_+, \\ (u, v)|_{y=0} = (0, 0), \lim_{y \to +\infty} u = 0, \end{cases} \tag{1.3.5}$$

and introduced the following functional spaces:

$$W_\alpha^{s,\infty}(\mathbb{R}_+) := \left\{ f = f(y), \quad e^{\alpha y} f \in W^{s,\infty}(\mathbb{R}_+) \right\}, \quad \forall \alpha, s \geq 0$$

with $\|f\|_{W_\alpha^{s,\infty}} := \|e^{\alpha y} f\|_{W^{s,\infty}}$, and

$$E_{\alpha,\beta} := \left\{ u = u(x, y) = \sum_{k \in \mathbb{Z}} \hat{u}^k(y) e^{ikx}, \quad \left\| \hat{u}^k \right\|_{W_\alpha^{0,\infty}} \leq C_{\alpha,\beta} e^{-\beta|k|}, \forall k \right\},$$
$$\forall \alpha, \beta > 0$$

with $\|u\|_{E_{\alpha,\beta}} := \sup_k e^{\beta|k|} \left\| \hat{u}^k \right\|_{W_\alpha^{0,\infty}}$.

Theorem 1.3.3 ([40] Ill-posedness in the Sobolev Setting)

(i) Let $u_s \in C^0\left(\mathbb{R}_+; W_\alpha^{4,\infty}(\mathbb{R}_+)\right) \cap C^1\left(\mathbb{R}_+; W_\alpha^{2,\infty}(\mathbb{R}_+)\right)$. Assume that the initial velocity has a non-degenerate critical point over $\mathbb{R}_+$. Then, there exists $\sigma > 0$, such that for all $\delta > 0$,

$$\sup_{0 \leq s \leq t \leq \delta} \left\| e^{-\sigma(t-s)\sqrt{|\partial_x|}} T(t, s) \right\|_{\mathcal{L}(H^m, H^{m-\mu})} = +\infty, \quad \forall m \geq 0, \mu \in [0, 1/2).$$

(ii) Moreover, one can find solutions u_s of (1.3.5) and $\sigma > 0$ such that for all $\delta > 0$,

$$\sup_{0 \le s \le t \le \delta} \left\| e^{-\sigma(t-s)\sqrt{|\partial_x|}} T(t,s) \right\|_{\mathcal{L}(H^{m_1}, H^{m_2})} = +\infty, \quad \forall m_1, m_2 \ge 0.$$

Gérard-Varet and Nguyenin [44] in 2012 established various ill-posedness results for the Prandtl equation. By considering perturbations of stationary shear flows, they showed that for some linearizations of the Prandtl equation and some C^∞ initial data, local in time C^∞ solutions do not exist. At the nonlinear level, they proved that if a flow exists in the Sobolev setting, it cannot be Lipschitz continuous. Besides ill-posedness in time, they also established some ill-posedness in space that casts some light on the results obtained by Oleinik for monotonic data.

They were concerned in [44] with the famous Prandtl equations:

$$\begin{cases} \partial_t u + u\partial_x u + v\partial_y u - \partial_y^2 u + \partial_x P = f, & y > 0, \\ \partial_x u + \partial_y v = 0, & y > 0, \\ u = v = 0, & y = 0, \\ \lim_{y \to +\infty} u = U(t,x). \end{cases} \tag{1.3.6}$$

They had two settings:

The initial value problem (IVP): (t, x, y) in $[0, T) \times \mathbb{T} \times \mathbb{R}_+$, $u|_{t=0} = u_0(x, y)$.

The boundary value problem (BVP): (t, x, y) in $\mathbb{T} \times [0, X) \times \mathbb{R}_+$, $u|_{x=0} = u_1(t, y)$.

The system satisfied by the perturbation (u, v) of $(u_s, 0)$ reads

$$\begin{cases} \partial_t u + u_s\partial_x u + u_s' v + u\partial_x u + v\partial_y u - \partial_y^2 u = 0, & y > 0, \\ \partial_x u + \partial_y v = 0, & y > 0, \\ u = v = 0, & y = 0. \end{cases} \tag{1.3.7}$$

Theorem 1.3.4 ([44] Non-existence of Solutions for Linearized Prandtl Equations) *There exists a shear flow u_s with $u_s - U \in C_c^\infty(\mathbb{R}_+)$ such that for all $T > 0$, there exists an initial datum u_0 satisfying*

(i) $e^y u_0 \in H^\infty(\mathbb{T} \times \mathbb{R}_+)$;
(ii) The IVP has no distributional solution u with

$$u \in L^\infty((0,T); L^2(\mathbb{T} \times \mathbb{R}_+)), \quad \partial_y u \in L^2((0,T) \times \mathbb{T} \times \mathbb{R}_+).$$

Theorem 1.3.5 ([44] No Lipschitz Continuity of the Flow) *There exists a shear flow u_s with $u_s - U \in C_c^\infty(\mathbb{R}_+)$ such that for all $m \ge 0$, the Cauchy problem is not locally (H^m, H^1) Lipschitz well-posed.*

Ding [20] in 2012 concerned the linearized Prandtl equation around general stationary solutions with non-degenerate critical points depending on x which could be considered as the time-periodic solutions and showed some ill-posedness.

Consider the 2D Navier-Stokes equations on a half-space:

$$\begin{cases} \partial_t u^\nu + u^\nu \partial_x u^\nu + v^\nu \partial_y u^\nu + \partial_x p^\nu - \nu \Delta u^\nu = 0, \\ \partial_t v^\nu + u^\nu \partial_x v^\nu + v^\nu \partial_y v^\nu + \partial_y p^\nu - \nu \Delta v^\nu = 0, \\ \partial_x u^\nu + \partial_y v^\nu = 0, \\ (u^\nu, v^\nu)|_{y=0} = (0, 0). \end{cases} \tag{1.3.8}$$

When $\nu \to 0$, a natural question is that does the solution (u^ν, v^ν) convergence to the solution of Euler equation:

$$\begin{cases} \partial_t u^E + u^E \partial_x u^E + v^E \partial_y u^E + \partial_x p^E = 0, \\ \partial_t v^E + u^E \partial_x v^E + v^E \partial_y v^E + \partial_y p^E = 0, \\ \partial_x u^E + \partial_y v^E = 0, \\ v^\nu|_{y=0} = 0. \end{cases}$$

As there is the no-slip condition, $u^\nu|_{y=0} = 0$ in Navier-Stokes equations, the transition from zero velocity at the boundary to the full magnitude at some distance from it takes place in a very thin layer.

It is interesting to consider the linearized equation around stationary solution (u_0, v_0) where u_0 has non-degenerate critical points, that is

$$\begin{cases} \partial_t u + u_0 \partial_x u + v_0 \partial_y u + u \partial_x u_0 + v \partial_y u_0 - \partial_y^2 u = 0, \\ \partial_x u + \partial_y v = 0, \\ u|_{y=0} = v|_{y=0} = 0, \ u|_{x=0} = u_1, \ \lim\limits_{y \to +\infty} u = 0. \end{cases} \tag{1.3.9}$$

The author of [20] introduced the following function spaces:

$$W_\alpha^{s,\infty} := \left\{ f = f(y), e^{\alpha y} f \in W^{s,\infty}(\mathbb{R}_+) \right\}$$

with the norm: $\|f\|_{W_\alpha^{s,\infty}} = \|e^{\alpha y} f\|_{W^{s,\infty}}$,

$$H_\beta^m := H^m \left(\mathbb{T}_t, W_\beta^{0,\infty} \left(\mathbb{R}_y^+ \right) \right)$$

and

$$\bar{H}_\beta^m := H_\beta^m \cap C^1 \left(\mathbb{T}_t, W_\beta^{2,\infty} \left(\mathbb{R}_y^+ \right) \right),$$

where $\mathbb{T}_t$ denotes the time domain.

Theorem 1.3.6 ([20]) *Let* $u_0 - U \in C^0\big([0, X_0); W_\alpha^{4,\infty}(\mathbb{R}_+)\big) \cap C^1\big([0, X_0); W_\alpha^{2,\infty}(\mathbb{R}_+)\big)$, $u_0|_{x=0}$ *has a non-degenerate critical point. If there exists* $X > 0$, *such that for every* $u_1(t, y) \in \bar{H}_\beta^m$, *with* $\beta < \alpha$, *the equations* (1.3.9) *have a unique solution, let us denote* $u(x, \cdot) := \mathfrak{X}(x, \xi)u_1$, *where* $u(x, \cdot)$ *is the solution of the problem* (1.3.9) *with* $u|_{x=\xi} = u_1$, *then there exists a* $\delta > 0$, *such that for every* $\epsilon > 0$,

$$\sup_{0 \leqslant \xi \leqslant x \leqslant \epsilon} \left\| e^{-\delta(x-\xi)\sqrt{|\partial_t|}} \mathfrak{X}(x, \xi) \right\|_{\mathcal{L}\left(H_\beta^m, H_\beta^{m-\sigma}\right)} = \infty, \quad \forall m \geqslant 0, \sigma \in \left[0, \frac{1}{2}\right).$$

If $\lim_{y \to \infty} u_0(x, y) = C$, *where* $C \geqslant 0$ *is a constant, the result is valid for* $\alpha = \beta$.

Liu and Yang [91] in 2017 studied the ill-posedness of the following Prandtl equations (1.3.10) in Sobolev spaces around a shear flow with general decay:

$$\begin{cases} u_t + u\partial_x u + v\partial_y u - \partial_y^2 u = 0, \\ \partial_x u + \partial_y v = 0, \\ (u, v)|_{y=0} = 0, \quad \lim_{y \to +\infty} u = U_0. \end{cases} \tag{1.3.10}$$

Note that (1.3.10) has a special shear flow solution $(u_s(t, y), 0)$, where the function $u_s(t, y)$ is a smooth solution to the following heat equation:

$$\begin{cases} \partial_t u_s - \partial_y^2 u_s = 0, \\ u_s|_{y=0} = 0, \quad \lim_{y \to +\infty} u = U_0, \\ u_s|_{t=0} = U_s(y) \end{cases} \tag{1.3.11}$$

with an initial shear layer $U_s(y)$. Denote by $T(t, s)$ the linear solution operator

$$T(t, s)u_0 := u(t, \cdot).$$

Theorem 1.3.7 ([91]) *Let* $u_s(t, y)$ *be the solution of the problem* (1.3.11) *satisfying*

$$u_s - U_0 \in C^0(\mathbb{R}_+; W_0^{4,\infty}) \cap C^1(\mathbb{R}_+; W_0^{2,\infty})$$

and assume that the initial shear layer $U_s(y)$ *has a non-degenerate critical point in* $\mathbb{R}_+$. *Then, there exists* $\sigma > 0$ *such that for all* $\delta > 0$,

$$\sup_{0 \le s \le t \le \delta} \|e^{-\sigma(t-s)\sqrt{|\partial_x|}} T(t, s)\|_{\mathcal{L}(H_\alpha^m, H_0^{m-\mu})} = +\infty, \forall \alpha, m \ge 0, \mu \in [0, \frac{1}{2}).$$

Theorem 1.3.8 *Under the assumptions of Theorem 1.3.7, it holds that for any* $\delta > 0$ *and* $\alpha, m \geq 0$

$$\sup_{0\leq s\leq t\leq\delta} \|T(t,s)\|_{\mathcal{L}(H^m_\alpha, H^0_0)} = +\infty.$$

Kukavica, Vicol, and Wang [67] in 2017 proved a numerical conjecture by rigorously establishing the finite time blowup of the boundary layer thickness.They considered the 2D Prandtl boundary layer equations for the unknown velocity field $(u, v) = (u(t, x, y), v(t, x, y))$:

$$\begin{cases} u_t + u\partial_x u + v\partial_y u - \partial_y^2 u + \partial_x p^E = 0, \\ \partial_x u + \partial_y v = 0, \\ (u, v)|_{y=0} = 0, \quad \lim\limits_{y\to+\infty} u = U^E, \\ u|_{t=0} = u_0. \end{cases} \tag{1.3.12}$$

They [67] wanted to prove the formation of finite time singularities in the Prandtl boundary layer equations (1.3.12) when the underlying Euler flow is not trivial, i.e., when $U^E \neq 0$. For this purpose, consider the Euler trace

$$\begin{cases} U^E = ksinx, \\ -\partial_x P^E = \frac{k^2}{2} sin(2x). \end{cases} \tag{1.3.13}$$

Theorem 1.3.9 ([67]) *Consider the Cauchy problem for the Prandtl equations* (1.3.12)*, with boundary conditions at* $y = \infty$ *matching* (1.3.13)*, with* $k \neq 0$*. There exists a large class of initial conditions* (u_0, v_0) *which are real-analytic in x and y, such that the unique real-analytic solution* (u, v) *to the Prandtl equations* (1.3.12) *blows up in a finite time.*

Guo and Nguyen [52] in 2018 concerned nonlinear ill-posedness of the Prandtl equation and an invalidity of asymptotic boundary layer expansions of incompressible fluid flows near a solid boundary. Their analysis was built upon recent remarkable linear ill-posedness results established by Gérard-Varet and Dormy [40] and an analysis by Guo and Tice [54]. They showed that the asymptotic boundary layer expansion is not valid for nonmonotonic shear layer flows in Sobolev spaces. They also introduced a notion of weak well-posedness and proved that the nonlinear Prandtl equation is not well-posed in this sense near nonstationary and nonmonotonic shear flows. On the other hand, they were able to verify that Oleinik's monotonic solutions are well-posed.

The boundary layer or Prandtl equations for (u, v) then reads

$$\begin{cases} \partial_t u + u\partial_x u + v\partial_Y u - \partial_Y^2 u + \partial_x P = 0, & Y > 0, \\ \partial_x u + \partial_Y v = 0, & Y > 0, \\ u|_{t=0} = u_0(x, y), \\ u|_{Y=0} = v|_{Y=0} = 0, \\ \lim_{Y\to+\infty} u = U(t, x) \end{cases} \tag{1.3.14}$$

where $U = u^0(t, x, 0)$ and $P = P(t, x)$ are the normal velocity and pressure describing the Euler flow just outside the boundary layer and satisfy the Bernoulli's equation

$$\partial_t U + U\partial_x U + \partial_x P = 0.$$

Introduce the standard Sobolev spaces L^2 and H^m, $m \geq 0$, with the usual norms:

$$\|u\|_{L^2_{x,Y}} := \left(\int_{\mathbb{T}\times\mathbb{R}_+} |u|^2 dxdY\right)^{1/2}, \quad \|u\|_{H^m_{x,Y}} := \sum_{k=0}^{m} \sum_{i+j=k} \left\|\partial_x^i \partial_Y^j u\right\|_{L^2}.$$

Regarding the validity of the asymptotic boundary layer expansion, we can write

$$\begin{aligned} \begin{pmatrix} u^\nu \\ v^\nu \end{pmatrix}(t, x, y) &= \begin{pmatrix} u^0 - u^0|_{y=0} \\ 0 \end{pmatrix}(y) + \begin{pmatrix} u_s \\ 0 \end{pmatrix}\left(t, \frac{y}{\sqrt{v}}\right) \\ &\quad + (\sqrt{v})^\gamma \begin{pmatrix} \tilde{u}^\nu \\ \sqrt{v}\tilde{v}^\nu \end{pmatrix}\left(t, x, \frac{y}{\sqrt{v}}\right), \end{aligned}$$

and

$$p^\nu(t, x, y) = (\sqrt{v})^\gamma \tilde{p}^\nu\left(t, x, \frac{y}{\sqrt{v}}\right),$$

for shear flows u_s and for some $\gamma > 0$, where $\left(u^0(y), 0\right)^T$ is the Euler flow.

Theorem 1.3.10 ([52] Invalidity of Asymptotic Expansions) *The expansion is not valid in the sense of validity of asymptotic expansions for any $\gamma > 0$.*

Sun [128] in 2019 obtained a blow-up criterion for classical solutions to the two-dimensional Prandtl equations under a monotonicity assumption in weighted Sobolev spaces. The proof is based on nonlinear energy estimates inspired by Masmoudi and Wong [98] and maximum principles for parabolic equations.

Consider the following Prandtl system:

$$\begin{cases} \partial_t u + u\partial_x u + v\partial_y u - \partial_y^2 u = \partial_t U + U\partial_x U, & \text{in } [0,T]\times\mathbb{T}\times\mathbb{R}_+, \\ \partial_x u + \partial_y v = 0, & \text{in } [0,T]\times\mathbb{T}\times\mathbb{R}_+, \\ u|_{t=0} = u_0, & \text{on } \mathbb{T}\times\mathbb{R}_+, \\ u|_{y=0} = v|_{y=0} = 0, & \text{on } [0,T]\times\mathbb{T}, \\ \lim_{y\to\infty} u(t,x,y) = U(t,x), & \text{for all } (t,x)\in[0,T]\times\mathbb{T}. \end{cases} \tag{1.3.15}$$

Masmoudi and Wong [98] introduced the following space:

$$H^{s,\gamma}_{\sigma,\delta} := \left\{ \omega : \mathbb{T}\times\mathbb{R}_+ \to \mathbb{R} \,\middle|\, \|\omega\|_{H^{s,\nu}} < +\infty,\ (1+y)^\sigma \omega \geq \delta,\ \sum_{|\alpha|\leq 2} \right| \times (1+y)^{\sigma+\alpha_2} D^\alpha \omega\Big|^2 \leq \frac{1}{\delta^2} \Bigg\},$$

where $s \geq 4, \gamma \geq 1, \sigma > \gamma + \frac{1}{2}, \delta \in (0,1)$, $D^\alpha = \partial_x^{\alpha_1}\partial_y^{\alpha_2}$, and the weighted H^s norm $\|\cdot\|_{H^{s,\nu}}$ is defined by

$$\|\omega\|^2_{H^{sy}} := \sum_{|\alpha|\leq s} \left\|(1+y)^{\gamma+\alpha_2} D^\alpha \omega\right\|^2_{L^2(\mathbb{T}\times\mathbb{R}_+)}.$$

Moreover, another weighted norm $\|\cdot\|_{H_g^{sy}}$ introduced in [98], which is almost equivalent to $\|\cdot\|_{H^{s,\gamma}}$, reads as

$$\|\omega\|^2_{H_g^{s,y}} := \left\|(1+y)^\gamma g_s\right\|^2_{L^2(\mathbb{T}\times\mathbb{R}_+)} + \sum_{|\alpha|\leq s, \alpha_1\leq s-1} \left\|(1+y)^{\gamma+\alpha_2} D^\alpha \omega\right\|^2_{L^2(\mathbb{T}\times\mathbb{R}_+)},$$

where

$$g_s := \partial_x^s \omega - \frac{\partial_y \omega}{\omega}\partial_x^s (u - U), \quad u(t,x,y) := \int_0^y \omega(t,x,\tilde{y})d\tilde{y}.$$

Theorem 1.3.11 ([128]) *Let $s \geq 6$ be an even integer, $\gamma \geq 1, \sigma > \gamma + \frac{1}{2}$ and $0 < \delta < \frac{1}{2}$. Suppose that the outer flow U satisfies*

$$\sup_t \|U\|_{s+9,\infty} := \sup_t \sum_{l=0}^{\left[\frac{s+9}{2}\right]} \left\|\partial_t^l U\right\|_{W^{s-2l+,\infty}(\mathbb{T})} < +\infty.$$

Assume that $u_0 - U|_{t=0} \in H^{s,\gamma-1}$ and the initial vorticity $\omega_0 := \partial_y u_0 \in H^{s,\gamma}_{\sigma,2\delta}$. Then, there exist a time $T := T\left(s,\gamma,\sigma,\delta,\|\omega_0\|_{H_g^{s,}},U\right) > 0$ and a unique classical *solution* (u, v) *to the Prandtl system such that*

$$u - U \in L^\infty\left([0,T]; H^{s,\gamma-1}\right) \cap C\left([0,T]; H^s - \omega\right),$$

and the vorticity

$$\omega := \partial_y u \in L^\infty\left([0,T]; H^{s,\gamma}_{\sigma,\delta}\right) \cap C\left([0,T]; H^s - \omega\right),$$

where $H^s - \omega$ *is the space* H^s *endowed with its weak topology. Let* T^* *be a maximal existence time of the classical solution such that* $u - U \in H^{s,\gamma-1}$ *and* $\omega \in H^{s,\gamma}_{\sigma,\delta}$.

Theorem 1.3.12 ([128]) *Under the conditions of Theorem 1.3.11, and* $\left[\sigma - \frac{\gamma+1}{2}\right] \geq 5$. *If* $T^* < \infty$, *then*

$$\lim_{t\to T^*}\left(\left\|\frac{1}{(1+y)^\sigma \partial_y u}\right\|_{L^\infty(\mathbb{T}\times\mathbb{R}_+)} + \|\nabla u\|_{L^\infty(\mathbb{T}\times\mathbb{R}_+)} + \left\|\frac{v_x}{1+y}\right\|_{L^\infty(\mathbb{T}\times\mathbb{R}_+)}\right) = \infty.$$

Let us stress that $H^{s,\gamma}_{\sigma,\delta}$ *is not a linear space, but a subset of linear function space* $H^{s,\gamma}$. *That is, the function* $\omega \in H^{s,\gamma}_{\sigma,\delta}$ *satisfies not only the regularity condition*

$$\|\omega\|_{H^{s,\gamma}} < \infty,$$

but also the monotonicity and decay conditions

$$(1+y)^\sigma \omega \geq \delta, \quad \sum_{|\alpha|\leq 2} \left|(1+y)^{\sigma+\alpha_2} D^\alpha \omega\right|^2 \leq \frac{1}{\delta^2}.$$

Different from general blow up criterion, this blow-up criterion in Theorem 1.3.12 pays attention to both the regularity condition and the monotonicity and decay conditions. Combining this with the minimum principle, we arrive at a blow-up criterion in terms of the regularity condition of vorticity on the boundary.

Theorem 1.3.13 ([128]) *Under the conditions of Theorem 1.3.11, and* $\left[\sigma - \frac{\gamma+1}{2}\right] \geq 5$. *If* $T^* < \infty$, *then*

$$\lim_{t\to T^*}\left(\left\|\frac{1}{\omega|_{y=0}}\right\|_{L^\infty(\mathbb{T})} + \|\nabla u\|_{L^\infty(\mathbb{T}\times\mathbb{R}^+)} + \left\|\frac{v_x}{1+y}\right\|_{L^\infty(\mathbb{T}\times\mathbb{R}^+)}\right) = \infty.$$

1.4 3D Prandtl Equations

In this section, we survey results on the 3D Prandtl equations, taking the form as

$$\begin{cases} \partial_t u + (u\partial_x + v\partial_y + w\partial_z)u + \partial_x P = \partial_z^2 u, \\ \partial_t v + (u\partial_x + v\partial_y + w\partial_z)v + \partial_y P = \partial_z^2 v, \\ \partial_x u + \partial_y v + \partial_z w = 0, \\ (u, v)|_{t=0} = (u_0(x, y, z), v_0(x, y, z)), \\ (u, v, w)|_{z=0} = 0, \\ \lim_{z\to+\infty} (u, v) = (U(t, x, y), V(t, x, y)), \end{cases} \tag{1.4.1}$$

where $(U(t, x, y), V(t, x, y))$ satisfies

$$\begin{cases} \partial_t U + U\partial_x U + V\partial_y U + \partial_x p = 0, \\ \partial_t V + U\partial_x V + V\partial_y V + \partial_y p = 0. \end{cases} \tag{1.4.2}$$

1.4.1 3D Prandtl Equations-Local Existence

Oleinik and Samokhin [106] in 1999 considered the system of 3D nonstationary boundary layer

$$\begin{cases} \frac{\partial u}{\partial t} + u\frac{\partial u}{\partial x} + v\frac{\partial u}{\partial y} + w\frac{\partial u}{\partial z} - \nu\frac{\partial^2 u}{\partial z^2} = \frac{1}{\rho}\frac{\partial p}{\partial x}, \\ \frac{\partial v}{\partial t} + u\frac{\partial v}{\partial x} + v\frac{\partial v}{\partial y} + w\frac{\partial v}{\partial z} - \nu\frac{\partial^2 v}{\partial z^2} = \frac{1}{\rho}\frac{\partial p}{\partial y}, \\ \frac{\partial u}{\partial x} + \frac{\partial v}{\partial y} + \frac{\partial w}{\partial z} = 0, \end{cases} \tag{1.4.3}$$

in a domain $G = \{0 < t < T, 0 < x, X, 0 < y < Y, 0 < z < Z\}$ with the conditions

$$\begin{cases} u(t, x, y, 0)=0, \quad v(t, x, y, 0)=0, u(x, 0)=0, \quad w(t, x, y, 0) = w_0(t, x, y) \le 0, \\ u = u_0, \quad v = v_0, \quad \text{on } \Gamma, \end{cases} \tag{1.4.4}$$

where Γ is the part of the boundary of G belonging to the planes $t = 0, x = 0, y = 0, z = Z$.

Theorem 1.4.1 ([106]) *Assume that problems (1.4.3)–(1.4.4) have a solution $(u, v, w) \in C^2(\overline{G})$. Then $u \ge 0, w \ge 0$ everywhere in $\overline{G}$; and $u = 0, w = 0$ only for $z = 0$.*

Theorem 1.4.2 ([106]) *Let $(u, v, w) \in C^2(\overline{G})$ be a solution of problems (1.4.3)–(1.4.4). Then*

$$0 \le u \le \max u_0 + t \max(-\rho^{-1} p_x),$$
$$0 \le v \le \max v_0 + t \max(-\rho^{-1} p_y).$$

Theorem 1.4.3 ([106]) *Let $(u, v, w) \in C^2(\overline{G})$ be a solution of problems (1.4.3)–(1.4.4). Then the following inequalities hold in the domain G:*

$$\min\left\{\inf_{z>0}\frac{v_0}{u_0}, \min\frac{p_y}{p_x}\right\} \le \frac{v}{u} \le \max\left\{\sup_{z>0}\frac{v_0}{u_0}, \max\frac{p_y}{p_x}\right\}.$$

Under the assumption of Oleinik monotonicity, Liu, Wang, and Yang [86] in 2017 investigated the local well-posedness of solutions to the 3D Prandtl equations by Crocco transformation. In [86], they studied the well-posedness for the following problem of the Prandtl equations in the domain $Q_T = \{0 < t \le T, (x, y) \in D, z > 0\}$:

$$\begin{cases} \partial_t u + (u\partial_x u + v\partial_y u + w\partial_z u) - \partial_x^2 u = -\partial_x P, \\ \partial_t v + (u\partial_x v + v\partial_y v + w\partial_z v) - \partial_z^2 v = -\partial_y P, \\ \partial_x u + \partial_x v + \partial_x w = 0, \\ (u, w)|_{z=0} = 0, \lim_{z\to\infty} = (U(t, x, y), KU(t, x, y)), \\ (u, v)|_{\partial Q_T^-} = (u_1(t, x, y, z), k(x, y)u_1(t, x, y, z)), \\ (u, v)|_{t=0} = (u_0(t, x, y, z), k(x, y)u_0(t, x, y, z)) \end{cases} \tag{1.4.5}$$

where $\partial Q_T^- = (0, T]\times\gamma_-\times\mathbb{R}_+$ with $\gamma_- = \{(x, y) \in \partial D|(1, k(x, y))\cdot\vec{n}(x, y) < 0\}$. The main results on the well-posedness of the initial boundary value problem (1.4.5) are given in the following theorem.

Theorem 1.4.4 ([86]) *Under the above assumption (H) with $k \in C^{10}(D)$ and $(U, p) \in C^{10}((0, T] \times D)$ for a fixed $T > 0$, assume that*

$$u_0 \in C^{10}(\mathbb{R}_z^+ \times D), u_1 \in C^{15}(\partial Q_T^-)$$

have the following properties:

(1) $\partial_z u_0 > 0, u_1 > 0$ *for all* $z \ge 0$*, and there is a constant* $C_0 > 0$ *such that*

$$C_0^{-1}(U(0, x, y) - u_0(x, y, z)) \le \partial_z u_0(x, y, z) \le C_0(U(0, x, y) - u_0(x, y, z)),$$

and

$$
\begin{aligned}
&C_0^{-1}(U(t,x,y)-u_1(x,y,z)) \le \partial_z u_1(x,y,z)\\
&\le C_0(U(t,x,y)-u_1(x,y,z)), \ \text{on } \partial Q_T^-.
\end{aligned}
$$

(2) *the compatibility conditions of the problem* (1.4.5) *hold up to order 6 at* $\{t = 0\} \cap \partial Q_T^-$, *and up to order 3 (4 resp.) at* $\{t = 0\} \cap \{z = 0\}(\{t = 0\} \cap \{z = \infty\})$ *and* $\{z = 0\} \cap \partial Q_T^-(\{z = \infty\} \cap \partial Q_T^-$ *resp). Then, there exist* $0 < T_0 \le T$ *and a unique classical solution* (u, v, w) *to the problem* (1.4.5) *in the domain* Q_{T_0}. *Moreover, the solution is linearly stable with respect to any three-dimensional smooth perturbation.*

Li and Yang [75] in 2019 studied the 3D Prandtl equations without any monotonicity condition on the velocity field and proved that when one tangential component of the velocity field has a single curve of non-degenerate critical points with respect to the normal variable, the system is locally well-posed in the Gevrey function space with Gevrey index.

Denote by (u, v) the tangential component and by w the vertical component of the velocity field, then the 3D Prandtl system in Ω reads as

$$
\begin{cases}
\partial_t u + \left(u\partial_x + v\partial_y + w\partial_z\right) u - \partial_z^2 u + \partial_x p = 0, & t > 0, \quad (x,y,z) \in \Omega,\\
\partial_t v + \left(u\partial_x + v\partial_y + w\partial_z\right) v - \partial_z^2 v + \partial_y p = 0, & t > 0, (x,y,z) \in \Omega,\\
\partial_x u + \partial_y v + \partial_z w = 0, \quad t > 0, (x,y,z) \in \Omega,\\
u|_{z=0} = v|_{z=0} = w|_{z=0} = 0, \quad \lim_{z\to+\infty}(u,v) = (U(t,x,y), V(t,x,y)),\\
u|_{t=0} = u_0, \quad v|_{t=0} = v_0, \quad (x,y,z) \in \Omega,
\end{cases}
\tag{1.4.6}
$$

where $(U(t,x,y), V(t,x,y))$, and $p(t,x,y)$ are the boundary traces of the tangential velocity field and pressure of the outer flow, satisfying Bernoulli's law

$$
\begin{cases}
\partial_t U + U\partial_x U + V\partial_y U + \partial_x p = 0,\\
\partial_t V + U\partial_x V + V\partial_y V + \partial_y p = 0.
\end{cases}
\tag{1.4.7}
$$

Let U, V be the data given in the equation. With each pair (ρ, σ), $\rho > 0$ and $\sigma \ge 1$, a Banach space $X_{\rho,\sigma}$ consists of all smooth vector functions (u, v) such that $\|(u,v)\|_{\rho,\sigma} < +\infty$, where the Gevrey norm $\|\cdot\|_{\rho,\sigma}$ is defined below. For each multi-index $\alpha = (\alpha_1, \alpha_2) \in \mathbb{Z}_+^2$, denote $\partial^\alpha = \partial_x^{\alpha_1}\partial_y^{\alpha_2}$ and

$$
\psi = \partial_z(u - U) = \partial_z u, \quad \eta = \partial_z(v - V) = \partial_z v.
$$

Then the Gevrey norm is defined by

$$
\begin{aligned}
\|(u,v)\|_{\rho,\sigma} = & \sup_{|\alpha|\geq 7} \frac{\rho^{|\alpha|-6}}{[(|\alpha|-7)!]^{\sigma}} \left(\left\| \langle z\rangle^{\ell-1} \partial^{\alpha}(u-U) \right\|_{L^2} + \left\| \langle z\rangle^{\kappa} \partial^{\alpha}(v-V) \right\|_{L^2} \right) \\
& + \sup_{|\alpha|\leq 6} \left(\left\| \langle z\rangle^{\ell-1} \partial^{\alpha}(u-U) \right\|_{L^2} + \left\| \langle z\rangle^{\kappa} \partial^{\alpha}(v-V) \right\|_{L^2} \right) \\
& + \sup_{|\alpha|\geq 7} \frac{\rho^{|\alpha|-6}}{[(|\alpha|-7)!]^{\alpha}} \left\| \langle z\rangle^{\ell} \partial^{\alpha} \psi \right\|_{L^2} + \sup_{|\alpha|\leq 6} \left\| \langle z\rangle^{\ell} \partial^{\alpha} \psi \right\|_{L^2} \\
& + \sup_{|\alpha|\geq 7} \frac{\rho^{|\alpha|-5}}{[(|\alpha|-6)!]^{\sigma}} |\alpha| \left\| \langle z\rangle^{\kappa+2} \partial^{\alpha} \eta \right\|_{L^2} + \sup_{|\alpha|\leq 6} \left\| \langle z\rangle^{\kappa+2} \partial^{\alpha} \eta \right\|_{L^2} \\
& + \sup_{\substack{1\leq j\leq 4 \\ |\alpha|+j\geq 7}} \frac{\rho^{|\alpha|+j-6}}{[(|\alpha|+j-7)!]^{\sigma}} \left\| \langle z\rangle^{\ell+1} \partial^{\alpha} \partial_z^{j} \psi \right\|_{L^2} \\
& + \sup_{\substack{1<j\leq 4 \\ |\alpha|+j\leq 6}} \left\| \langle z\rangle^{\ell+1} \partial^{\alpha} \partial_z^{j} \psi \right\|_{L^2} \\
& + \sup_{\substack{1\leq j\leq 4 \\ |\alpha|+j\geq 7}} \frac{\rho^{|\alpha|+j-5}}{[(|\alpha|+j-6)!]^{\sigma}} |\alpha| \left\| \langle z\rangle^{\kappa+2} \partial^{\alpha} \partial_z^{j} \eta \right\|_{L^2} \\
& + \sup_{\substack{1\leq \leq \leq \\ |\alpha|+j\leq 6}} \left\| \langle z\rangle^{\kappa+2} \partial^{\alpha} \partial_z^{j} \eta \right\|_{L^2},
\end{aligned}
$$

where we use L^2 instead of $L^2(\Omega)$ without confusion. Moreover, define another Gevrey space $Y_{\rho,\sigma}$ consisting of smooth functions $F(x, y)$ such that $\|F\|_{\rho,\sigma} < +\infty$, where

$$
\|F\|_{\rho,\sigma} = \sup_{|\alpha|\geq 0} \frac{\rho^{|\alpha|}}{(|\alpha|!)^{\sigma}} \left\| \partial^{\alpha} F \right\|_{L^2(\mathbb{T}^2)} .
$$

The initial data (u_0, v_0) satisfy the following compatibility conditions:

$$
\begin{cases}
(u_0, v_0)|_{z=0} = (0,0), \quad \lim\limits_{z\to+\infty} (u_0, v_0) |= (U, V) \text{ and} \\
(\partial_z \psi_0, \partial_z \eta_0)|_{z=0} = \left(\partial_x p, \partial_y p\right), \\
\partial_z^3 \psi_0\big|_{z=0} = \psi_0 \left(\partial_x \psi_0 - \partial_y \eta_0\right)\big|_{z=0} + 2\eta_0 \partial_y \psi_0\big|_{z=0} + \partial_t \partial_x p, \\
\partial_z^3 \eta_0\big|_{z=0} = \eta_0 \left(\partial_y \eta_0 - \partial_x \psi_0\right)\big|_{z=0} + 2\psi_0 \partial_x \eta_0|_{z=0} + \partial_t \partial_y p,
\end{cases}
\tag{1.4.8}
$$

where $\psi_0 = \partial_z u_0$ and $\eta_0 = \partial_z v_0$.

Theorem 1.4.5 ([75]) *Let $1 < \sigma \leq 2$, under the compatibility conditions, suppose $U, V, p \in Y_{2\rho_0,\sigma}$ and $(u_0, v_0) \in X_{2\rho_0,\sigma}$ for some $\rho_0 > 0$. Moreover, suppose u_0*

satisfies the assumptions (1.4.8). *Then the Prandtl system admits a unique solution* $(u, v) \in L^\infty\left([0,T]; X_{\rho,\sigma}\right)$ *for some* $T > 0$ *and some* $0 < \rho < 2\rho_0$.

Li, Masmoudi, and Yang [70] in 2020 established the well-posedness in Gevrey function space with optimal class of regularity 2 for the 3D Prandtl system without any structural assumption.

Denote $\mathbb{R}^3_+ = \left\{(x, y, z) \in \mathbb{R}^3; z > 0\right\}$ and let (u, v) be the tangential component and w be the vertical component of the velocity field. Then the 3D Prandtl system in $\mathbb{R}^3_+$ reads as

$$\begin{cases} \left(\partial_t + u\partial_x + v\partial_y + w\partial_z - \partial_z^2\right) u + \partial_x p = 0, & t > 0, (x, y, z) \in \mathbb{R}^3_+, \\ \left(\partial_t + u\partial_x + v\partial_y + w\partial_z - \partial_z^2\right) v + \partial_y p = 0, & t > 0, (x, y, z) \in \mathbb{R}^3_+, \\ \partial_x u + \partial_y v + \partial_z w = 0, \quad t > 0, (x, y, z) \in \mathbb{R}^3_+, \\ u|_{z=0} = v|_{z=0} = w|_{z=0} = 0, \quad \lim\limits_{z\to+\infty} (u, v) = (U(t, x, y), V(t, x, y)), \\ u|_{t=0} = u_0, \quad v|_{t=0} = v_0, \quad (x, y, z) \in \mathbb{R}^3_+, \end{cases} \tag{1.4.9}$$

where $(U(t, x, y), V(t, x, y))$, and $p(t, x, y)$ are the boundary traces of the tangential velocity field and pressure of the outer flow, satisfying

$$\begin{cases} \partial_t U + U\partial_x U + V\partial_y U + \partial_x p = 0, \\ \partial_t V + U\partial_x V + V\partial_y V + \partial_y p = 0, \end{cases} \tag{1.4.10}$$

Here, p, U, V are given functions determined by the Euler flow. Then for the zero outer flow, the Prandtl system can be written as

$$\begin{cases} \left(\partial_t + u\partial_x + v\partial_y + w\partial_z - \partial_z^2\right) u = 0, & t > 0, (x, y, z) \in \mathbb{R}^3_+, \\ \left(\partial_t + u\partial_x + v\partial_y + w\partial_z - \partial_z^2\right) v = 0, & t > 0, (x, y, z) \in \mathbb{R}^3_+, \\ (u, v)|_{z=0} = (0, 0), \quad \lim\limits_{z\to+\infty} (u, v) = (0, 0), \\ (u, v)|_{t=0} = (u_0, v_0), \quad (x, y, z) \in \mathbb{R}^3_+, \end{cases}$$

with

$$w(t, x, y, z) = -\int_0^z \partial_x u(t, x, y, \tilde{z})d\tilde{z} - \int_0^z \partial_y v(t, x, y, \tilde{z})d\tilde{z}.$$

We will use without confusion $\|\cdot\|_{L^2}$ and $(\cdot,\cdot)_{L^2}$ to denote the norm and inner product of $L^2 = L^2\left(\mathbb{R}^3_+\right)$, and use the notations $\|\cdot\|_{L^2\left(\mathbb{R}^2_{x,y}\right)}$ and $(\cdot,\cdot)_{L^2\left(\mathbb{R}^2_{x,y}\right)}$ when the variables are specified. Similarly for L^∞. Moreover, we also use $L^\infty_{x,y}\left(L^2_z\right) = L^\infty\left(\mathbb{R}^2; L^2\left(\mathbb{R}_+\right)\right)$ to stand for the classical Sobolev space, so does the Sobolev space $L^2_{x,y}\left(L^\infty_z\right)$. In the following discussion by ∂^α, we always mean $\partial^\alpha = \partial_x^{\alpha_1}\partial_y^{\alpha_2}$ with each multi-index $\alpha = (\alpha_1, \alpha_2) \in \mathbb{Z}^2_+$.

Let $\ell > 1/2$ be a given number. With each pair (ρ, σ), $\rho > 0$, and $\sigma \geq 1$, a Banach space $X_{\rho,\sigma}$ consists of all smooth vector-valued functions (u, v) such that the Gevrey norm $\|(u,v)\|_{\rho,\sigma} < +\infty$, where $\|\cdot\|_{\rho,\sigma}$ is defined below. Recalling $\partial^\alpha = \partial_x^{\alpha_1}\partial_y^{\alpha_2}$, we define

$$\begin{aligned}\|(u,v)\|_{\rho,\sigma} = &\sup_{\substack{0\le j\le 5\\ |\alpha|+j\ge 7}} \frac{\rho^{|\alpha|+j-7}}{[(|\alpha|+j-7)!]^\sigma}\left(\left\|\langle z\rangle^{\ell+j}\partial^\alpha\partial_z^j u\right\|_{L^2} + \left\|\langle z\rangle^{\ell+j}\partial^\alpha\partial_z^j v\right\|_{L^2}\right)\\ &+ \sup_{\substack{0\le j\le 5\\ |\alpha|+j\le 6}}\left(\left\|\langle z\rangle^{\ell+j}\partial^\alpha\partial_z^j u\right\|_{L^2} + \left\|\langle z\rangle^{\ell+j}\partial^\alpha\partial_z^j v\right\|_{L^2}\right),\end{aligned}$$

where $\langle z\rangle = \left(1+|z|^2\right)^{1/2}$. We call σ the Gevrey index.

The initial data (u_0, v_0) satisfy the following compatibility conditions:

$$\begin{cases} (u_0,v_0)|_{z=0} = (0,0), \quad \lim_{z\to+\infty}(u_0,v_0)\,| = (0,0), \left(\partial_z^2 u_0, \partial_z^2 v_0\right)\big|_{z=0} = (0,0),\\ \partial_z^4 u_0\big|_{z=0} = (\partial_z u_0)\left(\partial_x\partial_z u_0 - \partial_y\partial_z v_0\right)\big|_{z=0} + 2(\partial_z v_0)\,\partial_y\partial_z u_0\big|_{z=0},\\ \partial_z^4 v_0\big|_{z=0} = (\partial_z v_0)\left(\partial_y\partial_z v_0 - \partial_x\partial_z u_0\right)\big|_{z=0} + 2(\partial_z u_0)\,\partial_x\partial_z u_0|_{z=0}.\end{cases}$$

Theorem 1.4.6 ([70]) *Let* $1 < \sigma \le 2$. *Suppose the initial datum* (u_0, v_0) *belongs to* $X_{2\rho_0,\sigma}$ *for some* $\rho_0 > 0$, *and satisfies the compatibility condition. Then the system* (1.4.10) *admits a unique solution* $(u,v) \in L^\infty\left([0,T]; X_{\rho,\sigma}\right)$ *for some* $T > 0$ *and some* $0 < \rho < 2\rho_0$.

1.4.2 3D Prandtl Equations-Global Existence

In this subsection, we survey the global existence of solutions to the 3D Prandtl equations. Liu, Wang, and Yang [85] in 2016 proved the global existence of weak solutions to the Prandtl equations (1.4.1) under special structure that takes the form

$$(u(t,x,y,z), k(x,y)u(t,x,y,z), w(t,x,y,z)),$$

in the domain $Q = \{t > 0, (x,y) \in D, z > 0\}$, $D \subseteq \mathbb{R}^2$, by the Crocco transformation,

$$\xi = x, \quad \eta = y, \quad \zeta = \frac{u(t,x,y,z)}{U(t,x,y)}, \; W(t,\xi,\eta,\zeta) = \frac{\partial_z u(t,x,y,z)}{U(t,x,y)},$$

the problem (1.4.1) becomes

$$\begin{cases} L(W) \triangleq \partial_t W + \zeta U(\partial_\xi + k\partial_\eta)W + A\partial_\zeta W + BW - W^2\partial_\zeta^2 W = 0,\\ W|_{\zeta=1} = 0, \quad W\partial_\zeta W|_{\zeta=0} = \frac{p_x}{U},\\ W|_{\Gamma_-} = W_1(t,\xi,\eta,\zeta) \triangleq \frac{\partial_z u_1}{U},\\ W|_{t=0} = W_0(\xi,\eta,\zeta) \triangleq \frac{\partial_z u_0}{U},\end{cases} \tag{1.4.11}$$

where

$$\begin{cases} \Omega_T = \{(t, \xi, \eta, \zeta) | 0 < t < T, (\xi, \eta) \in \mathbb{R}^2, 0 \le \zeta < 1\}, \\ \Gamma_- = \{(t, \xi, \eta, \zeta) | 0 < t < T, (\xi, \eta) \in \gamma_-, 0 \le \zeta < 1\}, \end{cases}$$

and

$$A = -\zeta(1-\zeta)\frac{U_t}{U} - (1-\zeta^2)\frac{p_x}{U}, \quad B = \frac{U_t}{U} + \zeta(U_x + kU_y) - \partial_y k \cdot \zeta U.$$

Assume that

$$\partial_z u > 0, \quad \partial_z u_0 > 0, \quad \text{for } z > 0,$$

and the favorable pressure condition holds

$$p_\xi(t, \xi, \eta) \le 0, \quad \forall t > 0, (\xi, \eta) \in \mathbb{R}^2. \tag{1.4.12}$$

Theorem 1.4.7 ([85]) *For the problem (1.4.11), and any $T > 0$, assume that $k \in C^2(D)$, $U \in C^2((0, T) \times D)$, $p_x \in C^1((0, T) \times D)$ satisfies (1.4.12), and the initial boundary data $W_0 \in C^1(\Omega)$, $W_1 \in C^3(\Gamma_-)$ satisfy*

$$C_0^{-1}(1-\zeta) \le W_0, W_1 \le C_0(1-\zeta),$$

for a positive constant C_0. Then, there exists a weak solution $W(t, \xi, \eta, \zeta) \in L^\infty(0, T; BV(\Omega))$ to the problem (1.4.11) in the sense of weak solution.

Zhang and Zhang [156] in 2016 investigated the long time existence and uniqueness of small solutions for the Prandtl system with small initial data by shear flow with $\mathbb{R}_+ \times \mathbb{R}_+^d$, $d = 2, 3$

$$\begin{cases} \partial_t u + u \cdot \nabla_{\mathrm{h}} u + v \partial_y u - \partial_{yy} u + \nabla_h p = 0, \quad (t, x, y) \in \mathbb{R}_+ \times \mathbb{R}^{d-1} \times \mathbb{R}_+, \\ \operatorname{div}_{\mathrm{h}} u + \partial_y v = 0, \\ u|_{y=0} = 0, \quad v|_{y=0} = 0, \quad \text{and} \quad \lim_{y \to \infty} u(t, x, y) = U(t, x), \\ u|_{t=0} = u_0, \end{cases} \tag{1.4.13}$$

where $u = u$, $\nabla_{\mathrm{h}} = \operatorname{div}_{\mathrm{h}} = \partial_x$ for $d = 2$, and $u = (u^1, u^2)$, $\nabla_{\mathrm{h}} = (\partial_{x_1}, \partial_{x_2})$, $\operatorname{div}_{\mathrm{h}} = \partial_{x_1} + \partial_{x_2}$ for $d = 3$. For simplicity, let $U = \varepsilon \mathbf{e}$ for some unit vector $\mathbf{e} \in \mathbb{R}^{d-1}$.

Let $u^s(t, y)$ be determined by

$$\begin{cases} \partial_t u^s - \partial_{yy} u^s = 0, (t, y) \in \mathbb{R}_+ \times \mathbb{R}_+, \\ u^s|_{y=0} = 0, \quad \lim_{y \to \infty} u^s(t, y) = \varepsilon \mathbf{e}, \\ u^s|_{t=0} = \varepsilon \chi(y) \mathbf{e}, \end{cases} \tag{1.4.14}$$

where $\chi(y) \in C^\infty(\mathbb{R})$, and $\chi(y) = 0$ for $y \leq 1$ and $\chi(y) = 1$ for $y \leq 2$, set $u = u^s + w$. Then w satisfies

$$\begin{cases} \partial_t w + (w + u^s) \cdot \nabla_{\mathrm{h}} w - \int_0^y \mathrm{div}_{\mathrm{h}}\, w dy' \partial_y w - \int_0^y \mathrm{div}_{\mathrm{h}}\, w dy' \partial_y u^s - \partial_{yy} w = 0, \\ w|_{y=0} = 0, \qquad \lim\limits_{y \to \infty} w = 0, \\ w|_{t=0} = u_0 - \varepsilon \chi \mathbf{e} \stackrel{\text{def}}{=} w_0. \end{cases} \tag{1.4.15}$$

Zhang and Zhang [156] introduced a key quantity $\theta(t)$ to describe the evolution of the analytic band of w:

$$\begin{cases} \dot{\theta}(t) = \langle t \rangle^{\frac{1}{4}} \left(\left\| e^\Psi \partial_y w_\Phi(t) \right\|_{\mathcal{B}^{\frac{d-1}{2},0}} + \left\| e^\Psi \partial_y u^s(t) \right\|_{L^2_v} \right), \\ \theta|_{t=0} = 0. \end{cases} \tag{1.4.16}$$

Here $\langle t \rangle = 1 + t$, the phase function Φ is defined by

$$\Phi(t, \xi) \stackrel{\text{def}}{=} (\delta - \lambda \theta(t))|\xi|, \tag{1.4.17}$$

and the weighted function $\Psi(t, y)$ is determined by

$$\Psi(t, y) \stackrel{\text{def}}{=} \frac{1 + y^2}{8\langle t \rangle}. \tag{1.4.18}$$

Theorem 1.4.8 ([156]) *Let $\delta > 0$ and ε be a sufficiently small positive constant. Assume that w_0 satisfies*

$$\left\| e^{\frac{1+y^2}{8\langle t \rangle}} e^{\delta |D_x|} w_0 \right\|_{\mathcal{B}^{\frac{d-1}{2},0}} \leq \varepsilon,$$

then there exists a positive time T_ε which is of size greater than $\varepsilon^{-\frac{4}{3}}$ so that the problem (1.4.15) has a unique solution w which satisfies

$$\|e^\Psi w_\Phi\|_{\widetilde{L}^\infty_t\left(\mathcal{B}^{\frac{d-1}{2},0}\right)} + c\sqrt{\lambda} \|e^\Psi w_\Phi\|_{\widetilde{L}^2_{t,\dot{\theta}(t)}\left(\mathcal{B}^{\frac{d}{2},0}\right)} + \|e^\Psi \partial_y w_\Phi\|_{\widetilde{L}^2_t\left(\mathcal{B}^{\frac{d-1}{2},0}\right)}$$
$$\leq \|e^{\frac{1+y^2}{8}} e^{\delta |D_x|} w_0\|_{\mathcal{B}^{\frac{d-1}{2},0}} \text{ for any } t \in [0, T_\varepsilon].$$

In particular, we have

$$e^{\Psi(t,y)} e^{\Phi(t,D)} w \in \widetilde{L}^\infty_T\left(\mathcal{B}^{\frac{d-1}{2},0}\right), \quad e^{\Psi(t,y)} e^{\Phi(t,D)}\Big|_{\partial y} w \in \widetilde{L}^2_T\left(\mathcal{B}^{\frac{d-1}{2},0}\right),$$

for any $T \le T_\varepsilon$*, and where the functions* $\Psi(t, y)$, $\Phi(t, \xi)$ *are determined by (1.4.16)–(1.4.18), respectively. Here define* $\widetilde{L}^p_T(\mathcal{B}^{s,0})$ *as the completion of* $C([0, T]; \mathcal{S}(\mathbb{R}^d_+))$ *by the norm*

$$\|a\|_{\widetilde{L}^p_T(\mathcal{B}^{s,0})} := \sum_{k\in\mathbb{Z}} 2^{ks} \left(\int_0^T \|\Delta^h_k a(t)\|_{L^2_+} dt\right)^{\frac{1}{p}},$$

$$\|a\|_{\widetilde{L}^p_{t,f}(\mathcal{B}^{s,0})} := \sum_{k\in\mathbb{Z}} 2^{ks} \left(\int_0^t f(t')\|\Delta^h_k a(t)\|_{L^2_+} dt\right)^{\frac{1}{p}},$$

with the usual change if $p = \infty$*. Here*

$$e^{\delta|D_x|} w_0 = \mathcal{F}^{-1}_{\xi\to x} e^{(\delta-\lambda\theta(t))|\xi|} \hat{w}_0(t, \xi, y).$$

Luo and Xin [96] in 2018 obtained Hölder continuous weak solutions to the 3D Prandtl equation with tangential velocities transverse to the outflow U. Consider the following 3D Prandtl system:

$$\begin{cases} u_t + u\partial_x u + v\partial_y u - \partial_y^2 u + \nabla_x p = 0, \\ \nabla_x u + \partial_y v = 0, \\ (u, v)|_{y=0} = 0, \qquad \lim\limits_{y\to+\infty} u = U, \\ u|_{t=0} = u_0. \end{cases} \tag{1.4.19}$$

Here $x = (x_1, x_2) \in \mathbb{T}$ and $y \in \mathbb{R}$ denote the tangential and the vertical components of the space variable, respectively. $\nabla_x = (\partial_{x_1}, \partial_{x_2})$ denotes the tangential gradient; $u = (u^1, u^2)(t, x, y)$ and $v = v(t, x, y)$ denote the tangential and the vertical velocities; $U(t, x)$ and $p(t, x)$ denote the tangential velocity and the pressure on the boundary $\{y = +\infty\}$ of the outer Euler flow, respectively, which satisfy

$$\partial_t U + U \cdot \nabla_x U + \nabla_x p = 0.$$

Theorem 1.4.9 ([96]) *Suppose that* (u_C, v_C) *is a classical solution to the system* (1.4.19) *and* (u, v) *is a smooth perturbation of* (u_C, v_C) *such that the difference* $(u - u_C, v - v_C) \in C^\infty_c(\mathbb{R}_+ \times \mathbb{T}^2 \times \mathbb{R}_+)$ *has compact support and satisfies*

$$\nabla_x u + \partial_y v = 0, \ in\ \mathbb{R}_+ \times \mathbb{T}^2 \times \mathbb{R}_+,$$

$$\int_{\mathbb{T}^2} (u - u_C)(t, x, y)dx = 0, \ for\ (t, y) \in \mathbb{R}_+ \times \mathbb{R}_+.$$

Let $\rho > 0$ *be a given positive number such that*

$$\overline{N(supp_{t,y}(u - u_C, v - v_C); \rho, \rho^{1/2})} \subset \mathbb{R}_+ \times \mathbb{R}_+,$$

where $supp_{t,y}$ *denotes the projection of the support to* $\mathbb{R}_+ \times \mathbb{R}_+$. *Then there exists a sequence of Hölder continuous weak solutions* $\{(u_k, v_k)\}_{=1}^{\infty}$ *to the system* (1.4.19) *and a sequence of positive numbers* $\{C_k\}$ *satisfying the estimates*

$$[u_k - u, v_k - v]_{\frac{1}{21}-\varepsilon, \frac{1}{10}-\varepsilon} \leq C_k,$$

and

$$supp_{t,y}(u_k - u, v_k - v) \subset \overline{N(supp_{t,y}(u - u_C, v - v_C); \rho, \rho^{1/2})}.$$

Furthermore, $(u_k, v_k) \rightharpoonup (u, v)$ *in the weak-* $*$ *topology on* $L^\infty(\mathbb{R}_+ \times \mathbb{T}^2 \times \mathbb{R}_+)$.

Here denote

$$N(E; \rho, \rho') = \{(t, y) : |t - t_0| < \rho, |y - y_0| < \rho', \text{for some } (t_0, y_0) \in E\},$$

$$[f]_{\alpha,\beta(\Omega)} := \sup_{(t,x,y),(t',x',y')\in\Omega} \frac{|f(t, x, y) - f(t', x'y')|}{|t - t'|^\alpha + |x - x'|^\alpha + |y - y'|^\beta}.$$

The constructions in the proof of Theorem 1.4.9 can also be adapted to some other models with vertical viscosities. In particular, consider the system

$$\begin{cases} u_t + u \cdot \nabla u - \partial_y^2 u + \nabla_x p = 0, \\ \nabla_x u + \partial_y v = 0, \\ u|_{t=0} = u_0. \end{cases} \tag{1.4.20}$$

Theorem 1.4.10 ([96]) *There exists a nontrivial Hölder continuous weak solutions* u *to the system* (1.4.20) *which is supported in a compact time interval, with*

$$[u]_{\frac{1}{21}-\varepsilon, \frac{1}{10}-\varepsilon} < +\infty.$$

Lin and Zhang [93] in 2020 proved the almost global existence of classical solutions to the 3D Prandtl system with the initial data which lie within ε of a stable shear flow. Using anisotropic Littlewood-Paley energy estimates in tangentially analytic norms and introducing new linearly good unknowns, they proved that the 3D Prandtl system has a unique solution with the lifespan of which is greater than $\exp\left(\varepsilon^{-1}/\log\left(\varepsilon^{-1}\right)\right)$. They considered the following Prandtl boundary layer equations in $\mathbb{R}_+ \times \mathbb{R}^3_+$:

$$\begin{cases} \partial_t u^p + \left(u^p\partial_x + v^p\partial_y + w^p\partial_z\right) u^p + \partial_x p^E = \partial_z^2 u^p, \\ \partial_t v^p + \left(u^p\partial_x + v^p\partial_y + w^p\partial_z\right) v^p + \partial_y p^E = \partial_z^2 v^p, \\ \partial_x u^p + \partial_y v^p + \partial_z w^p = 0, \\ (u^p, v^p, w^p)|_{z=0} = (0,0,0), \\ \lim\limits_{z\to+\infty} (u^p, v^p) = \left(U^E(t,x,y), V^E(t,x,y)\right), \\ (u^p, v^p)|_{t=0} = (u_0(x,y,z), v_0(x,y,z)). \end{cases} \tag{1.4.21}$$

Here and in what follows, $(t,x,y,z) \in \mathbb{R}_+ \times \mathbb{R} \times \mathbb{R} \times \mathbb{R}_+$, (u^p, v^p) and w^p denote the tangential and normal velocity of the boundary layer flow, the initial data $(u_0, v_0) := (u_0(x,y,z), v_0(x,y,z))$ and the far-field $\left(U^E(t,x,y), V^E(t,x,y)\right)$ are given. Furthermore, $\left(U^E(t,x,y), V^E(t,x,y)\right)$ and the given scalar pressure $p^E(t,x,y)$ are the tangential velocity field and pressure on the boundary $\{z=0\}$ of the Euler flow, satisfying

$$\begin{cases} \partial_t U^E + U^E\partial_x U^E + V^E\partial_y U^E + \partial_x p^E = 0, \\ \partial_t V^E + U^E\partial_x V^E + V^E\partial_y V^E + \partial_y p^E = 0. \quad t>0, (x,y) \in \mathbb{R}^2. \end{cases} \tag{1.4.22}$$

Set

$$\lim_{z\to+\infty} \left(u^p, v^p\right) = (\kappa_1, \kappa_2). \tag{1.4.23}$$

Write $u^p(t,x,y,z)$ and $v^p(t,x,y,z)$ as perturbations $u(t,x,y,z)$ and $v(t,x,y,z)$ of the lifts $\kappa_1\varphi(t,z)$ and $\kappa_2\varphi(t,z)$, respectively, via

$$u^p(t,x,y,z) = \kappa_1\varphi(t,z) + u(t,x,y,z), \quad v^p(t,x,y,z) = \kappa_2\varphi(t,z) + v(t,x,y,z)$$

where

$$\varphi(t,z) = \frac{1}{\sqrt{\pi}} \int_0^{z/\sqrt{\langle t\rangle}} \exp\left(-\frac{\tilde{z}^2}{4}\right) d\tilde{z}$$

and denote by $\langle t\rangle = 1+t$. Then introduce new linearly-good unknowns

$$\begin{cases} g_1(t,x,y,z) = \partial_z u(t,x,y,z) + \frac{z}{2\langle t\rangle} u(t,x,y,z), \\ g_2(t,x,y,z) = \partial_z v(t,x,y,z) + \frac{z}{2\langle t\rangle} v(t,x,y,z). \end{cases}$$

Through some simple calculations, the equations for the good unknowns $g = (g_1, g_2)^T$ in $\{t > 0, (x, y) \in \mathbb{R}^2, z \in \mathbb{R}_+\}$ are

$$\begin{cases} \partial_t g - \partial_z^2 g + \frac{1}{\langle t\rangle} g + \kappa_1 \partial_z \varphi \nabla_h^\perp v + \kappa_1 \varphi \partial_x g + g_1 \nabla_h^\perp v - \frac{z}{2\langle t\rangle} u \nabla_h^\perp v + u \partial_x g, \\ -\kappa_2 \partial_z \varphi \nabla_h^\perp u + \kappa_2 \varphi \partial_y g - g_2 \nabla_h^\perp u + \frac{z}{2\langle t\rangle} v \nabla_h^\perp u + v \partial_y g + w \partial_z g - \frac{1}{2\langle t\rangle} w\mathbf{u} = 0, \\ u = U(g_1), \quad v = U(g_2), \quad \text{and} \quad w = W(g_1, g_2), \\ \partial_z g|_{z=0} = \lim_{z\to+\infty} g = 0, \\ g|_{t=0} = (g_{10}, g_{20}), \end{cases} \tag{1.4.24}$$

where

$$\begin{cases} \mathbf{u} = (u, v)^T, \quad \nabla_h = \begin{pmatrix} \partial_x \\ \partial_y \end{pmatrix}, \quad \nabla_h^\perp = \begin{pmatrix} -\partial_y, \\ \partial_x \end{pmatrix}, \\ U(g_i) \doteq \exp\left(\frac{-z^2}{4\langle t\rangle}\right) \int_0^z g_i(t, x, y, \tilde{z}) \exp\left(\frac{\tilde{z}^2}{4\langle t\rangle}\right) d\tilde{z}, \quad i = 1, 2, \\ W(g_1, g_2) \doteq -\int_0^z \left[U(\partial_x g_1) + U(\partial_y g_2)\right] d\tilde{z}. \end{cases}$$

For any local bounded function Φ on $\mathbb{R}_+ \times \mathbb{R}^2$, define

$$\begin{aligned} u_\Phi(t, x, y, z) &= \mathscr{F}^{-1}_{\xi\to(x,y)}\left(e^{\Phi(t,\xi)}\widehat{u}(t, \xi, z)\right), \\ \widetilde{u_\Phi}(t, x, y, z) &= \mathscr{F}^{-1}_{\xi\to(x,y)}\left(\left|e^{\Phi(t,\xi)}\widehat{u}(t, \xi, z)\right|\right), \end{aligned}$$

where and in all that follows, $\mathscr{F}u$ and $\widehat{u}$ always denote the partial Fourier transform of the distribution u with respect to (x, y) variables, $\widehat{u}(\xi, z) = \mathscr{F}_{(x,y)\to\xi}(u)(\xi, z)$, and $\xi = (\xi_1, \xi_2)$. Define the phase function Φ as follows:

$$\Phi(t, \xi) \doteq (\tau_0 - \lambda\theta(t))\,|\xi|,$$

where λ is a large enough positive constant which will be chosen precisely later, and $\theta(t)$ is a key quantity to describe the analytic band of (g_1, g_2) :

$$\begin{cases} \frac{d}{dt}\theta(t) = \|(g_{1\Phi}, g_{2\Phi})\|_{B_{1,\alpha}} + (|\kappa_1| + |\kappa_2|)\,\langle t\rangle^{\frac{1}{4}} \left\|e^{\psi}\partial_z\varphi\right\|_{L_z^2}, \\ \theta|_{t=0} = 0. \end{cases}$$

Let s be in $\mathbb{R}$. For g in $\mathscr{S}_h'(\mathbb{R}_+^3)$, which means that g is in $\mathscr{S}'(\mathbb{R}_+^3)$ and satisfies $\lim_{k\to-\infty} \|S_k^h g\|_{L^\infty} = 0$, we set

$$\|g_\Phi\|_{X_{s,\alpha}} = \left\|\left(2^{ks}\left\|e^{\psi}\Delta_k^h g_\Phi\right\|_{L_+^2}\right)_k\right\|_{\ell^1(\mathbb{Z})}.$$

For compactness of notation, for a function g_Φ such that $g_\Phi, \zeta g_\Phi, \partial_z g_\Phi \in X_{s,\alpha}$, use the time weighted norm

$$\|g_\Phi\|_{B_{s,\alpha}} = \langle t\rangle^{\frac{1}{4}} \|g_\Phi\|_{X_{s,\alpha}} + \langle t\rangle^{\frac{1}{4}} \|\zeta g_\Phi\|_{X_{s,\alpha}} + \langle t\rangle^{\frac{3}{4}} \|\partial_z g_\Phi\|_{X_{s,\alpha}},$$

where $\zeta = \zeta(t,z) = \frac{z}{\langle t\rangle^{1/2}}$, the heat self-similar variable. In order to obtain a better description of the regularizing effect of the transport diffusion equation, use Chemin-Lerner-type spaces $\widetilde{L}^p_T\left(X_{s,\alpha}\left(\mathbb{R}^3_+\right)\right)$, with the norm and

$$\|g\|_{L^p_T(X_{s,\alpha})} \doteq \sum_{k\in\mathbb{Z}} 2^{ks}\left(\int_0^T \left\|e^{\psi}\Delta^h_k g_\Phi(t)\right\|^p_{L^2_+} dt\right)^{\frac{1}{p}}, \quad e^{\psi} = \exp\left(\frac{\alpha z^2}{4\langle t\rangle}\right),$$

with the usual change if $p = \infty$. Let $f(t) \in L^1_{loc}(\mathbb{R}_+)$ be a nonnegative function. Define

$$\|g\|_{\tilde{L}^p_{t,f}(X_{s,\alpha})} \doteq \sum_{k\in\mathbb{Z}} 2^{ks}\left(\int_0^t f\left(t'\right)\left\|e^{\psi}\Delta^h_k g_\Phi\left(t'\right)\right\|^p_{L^2_+} dt'\right)^{\frac{1}{p}},$$

$$\|g\|_{L^p_{t,f}(B_{s,\alpha})} \doteq \sum_{k\in\mathbb{Z}} 2^{ks}\left[\int_0^t f(t')\left(\left|t'\right\rangle^{\frac{p}{4}}\left(\left\|e^{\psi}\Delta^h_k g_\Phi\right\|^p_{L^2_+} + \left\|\zeta e^{\psi}\Delta^h_k g_\Phi\right\|^p_{L^2_+}\right)\right.\right.$$
$$\left.\left.+\left\langle t'\right\rangle^{\frac{3p}{4}}\left\|e^{\psi}\Delta^h_k \partial_z g_\Phi\right\|^p_{L^2_+}\right) dt'\right]^{\frac{1}{p}},$$

where $p < \infty$. When $p = \infty$, define

$$\|g\|\widetilde{L}^{\infty}_{T,f(t)}\left(X_{s,\alpha}\right) \doteq \sum_{k\in\mathbb{Z}} 2^{ks} \sup_{t\in[0,T]} f(t)\left\|e^{\psi}\Delta^h_k g_\Phi(t)\right\|_{L^2_+}.$$

Theorem 1.4.11 ([93]) *There exist $C_* > 0$ and $\varepsilon_* > 0$ such that for any $\varepsilon \in (0, \varepsilon_*]$, when $e^{\tau_0|D_h|}g_{i0} \in X_{1,\frac{1}{2}}$, $i = 1, 2$, and*

$$\left\|e^{\tau_0|D_h|}g_{10}\right\|_{X_{1,\frac{1}{2}}} + \left\|e^{\tau_0|D_h|}g_{20}\right\|_{X_{1,\frac{1}{2}}} \le \varepsilon,$$

with

$$\tau_0 \ge \frac{C_*}{\ln\frac{1}{\varepsilon}} + C_*\left(|\kappa_1| + |\kappa_2|\right)\exp\left(\frac{1}{\varepsilon\ln\frac{1}{\varepsilon}}\right),$$

the system has a unique solution (g_1, g_2) *on* $[0, T_\varepsilon]$, *where*

$$T_\varepsilon \geq \exp\left(\frac{\varepsilon^{-1}}{\ln\left(\varepsilon^{-1}\right)}\right).$$

Here $|D_h|$ *denotes the Fourier multiplier with the symbol* $|\xi|$. *Furthermore, the phase function* $\Phi(t, \xi)$ *of the solution* (g_1, g_2) *satisfies*

$$\Phi(t, \xi) \geq \frac{3\tau_0}{4}|\xi|,$$

for all $t \in [0, T_\varepsilon]$, *and the solution* (g_1, g_2) *obeys the bounds*

$$\begin{aligned}
\|(g_{1\Phi}, g_{2\Phi})\|_{\tilde{L}^\infty_{T_\varepsilon, f_3(t)}(X_{1,\alpha})} &\leq C\varepsilon, \\
\|(g_{1\Phi}, g_{2\Phi})\|_{\tilde{L}^2_{T_\varepsilon, f_2(t)}(B_{1,\alpha})} &\leq \frac{\varepsilon C\sqrt{C_0}}{\sqrt{\delta}},
\end{aligned}$$

where $\delta = \varepsilon \ln \frac{1}{\varepsilon}$, $\alpha = \frac{1-\delta}{2}$, $f_2(t) = \langle t\rangle^{1-2\delta}$, $f_3(t) = \langle t\rangle^{\frac{5}{4}-\delta}$, *and* C_0 *is a positive constant.*

1.4.3 Ill-Posedness of the 3D Prandtl Equations

Liu, Wang, and Yang [84] in 2016 gave an instability (ill-posedness) criterion for the 3D Prandtl equations, which shows that the monotonicity condition is not sufficient for the well-posedness of the 3D Prandtl equations, in contrast to the classical well-posedness theory of the 2D Prandtl equations under the Oleinik monotonicity assumption.

They considered the 3D Prandtl equations (1.4.1) in $\Omega \triangleq \{(t, x, y, z) : t > 0, (x, y) \in \mathbb{T}^2, z \in \mathbb{R}_+\}$. Let $u^s(t, z)$ and $v^s(t, z)$ be smooth solutions of the heat equations:

$$\begin{cases}
\partial_t u^s - \partial_z^2 u^s = 0, \ \ \partial_t v^s - \partial_z^2 v^s = 0, \\
(u^s, v^s)|_{z=0} = 0, \ \ \lim\limits_{z\to+\infty} (u^s, v^s) = (U_0, V_0), \\
(u^s, v^s)|_{t=0} = (U_s, V_s)(z).
\end{cases} \tag{1.4.25}$$

Theorem 1.4.12 ([84]) *Let* $(u^s, v^s)(t, z)$ *be the solution of the problem (1.4.25) satisfying*

$$\begin{aligned}
\left(u^s - U_0, v^s - V_0\right) \in & C^0\left(\mathbb{R}_+; W^{4,\infty}_\alpha(\mathbb{R}_+) \cap H^4_\alpha(\mathbb{R}_+)\right) \\
& \cap C^1\left(\mathbb{R}_+; W^{2,\infty}_\alpha(\mathbb{R}_+) \cap H^2_\alpha(\mathbb{R}_+)\right).
\end{aligned}$$

Assume that the initial data of the problem (1.4.25) satisfies that there exists a $z_0 > 0$*, such that*

$$V_s'(z_0)\,U_s''(z_0) \neq U_s'(z_0)\,V_s''(z_0)\,.$$

Then we have the following two instability statements:

(i) *There exists a* $\sigma > 0$ *such that for all* $\delta > 0$, $m > 0$ *and* $\mu \in \left[0, \frac{1}{4}\right)$,

$$\sup_{0 \leqq s \leqq t \leqq \delta} \left\| e^{-\sigma(t-s)\sqrt{|\partial_{\mathcal{T}}|}} T(t,s) \right\|_{\mathcal{L}\left(\mathcal{H}_\alpha, \mathcal{H}_\alpha^{m-\mu}\right)} = +\infty,$$

where the operator $\partial_{\mathcal{T}}$ *represents the tangential derivative* ∂_x *or* ∂_y,

(ii) *There exists an initial shear layer* (U_s, V_s) *to the problem (1.4.25) and* $\sigma > 0$, *such that for all* $\delta > 0$,

$$\sup_{0 \leqq s \leqq t \leqq \delta} \left\| e^{-\sigma(t-s)\sqrt{|\partial_{\mathcal{T}}|}} T(t,s) \right\|_{\mathcal{L}\left(\mathcal{H}_\alpha^{m_1}, \mathcal{H}_\alpha^{m_2}\right)} = +\infty, \quad \forall m_1, m_2 > 0.$$

Here denote by

$$\begin{cases} T(t,s)(u_0, v_0) := (u, v)(t, \cdot), \quad \mathcal{H}_\alpha^m := H^m\left(\mathbb{T}_{x,y}^2; L_\alpha^2(\mathbb{R}_+)\right), \\ \|T(t,s)\|_{\mathcal{L}\left(\mathcal{H}_\alpha^{m_1}, \mathcal{H}_\alpha^{m_2}\right)} := \sup\limits_{(u_0,v_0)\in E_{\alpha,\beta}} \dfrac{\|T(t,s)(u_0,v_0)\|_{\mathcal{H}_\alpha^{m_2}}}{\|(u_0,v_0)\|_{\mathcal{H}_\alpha^{m_1}}} \in \mathbb{R}_+ \cup \{\infty\}, \\ L_\alpha^2(\mathbb{R}_+) := \{f = f(z), z \in \mathbb{R}_+;\ \|f\|_{L_\alpha^2} = \|e^{\alpha z} f\|_{L^2} < \infty\}, \\ H_\alpha^m(\mathbb{R}_+) := \{f = f(z), z \in \mathbb{R}_+;\ \|f\|_{H_\alpha^m} = \|e^{\alpha z} f\|_{H^m} < \infty\}. \end{cases}$$

Theorem 1.4.13 ([84]) *The problem (1.4.1) with the initial data* $(u, v)|_{t=0} = (u_0, v_0)(x, y, z)$ *is locally well-posed, if there exist positive continuous functions* $T(\cdot,\cdot)$, $C(\cdot,\cdot)$, *some* $\alpha > 0$ *and integer* $m \geq 1$ *such that for any initial data* $\left(u_0^1, v_0^1\right)$ *and* $\left(u_0^2, v_0^2\right)$ *with*

$$\left(u_0^1 - U_0, v_0^1 - V_0\right) \in \mathcal{H}_\alpha^m, \quad \left(u_0^2 - U_0, v_0^2 - V_0\right) \in \mathcal{H}_\alpha^m,$$

there are unique distributional solutions $\left(u^1, v^1\right)$ *and* $\left(u^2, v^2\right)$ *satisfying that for* $(i = 1, 2)$, $\left(u^i, v^i\right)\big|_{t=0} = \left(u_0^i, v_0^i\right)$ *and*

$$\left(u^i - U_0, v^i - V_0\right) \in L^\infty\left(0, T; L^2\left(\mathbb{T}^2 \times \mathbb{R}_+\right)\right) \cap L^2\left(0, T; H^1\left(\mathbb{T}^2 \times \mathbb{R}_+\right)\right),$$

and the following estimate holds

$$\begin{aligned}&\left\|\left(u^1, v^1\right)-\left(u^2, v^2\right)\right\|_{L^\infty(0,T;L^2(\mathbb{T}^2\times\mathbb{R}_+))}+\left\|\left(u^1, v^1\right)-\left(u^2, v^2\right)\right\|_{L^2(0,T;H^1(\mathbb{T}^2\times\mathbb{R}_+))}\\&\quad\leqq C\left\|\left(u_0^1-u_0^2, v_0^1-V_0^2\right)\right\|_{\mathcal{H}_\alpha^m}.\end{aligned}$$

where both of C *and* T *depend on* $\left(u_0^1-U_0, v_0^1-V_0\right)_{\mathcal{H}_\alpha^m}$ *and* $\left(u_0^2-U_0, v_0^2-V_0\right)\in\mathcal{H}_\alpha^m$.

Theorem 1.4.14 ([84]) *Under the same assumption as given in Theorem 1.4.12, the problem (1.4.1) of the 3D nonlinear Prandtl equations is not locally well-posed in the sense of Definition 1.4.13.*

1.5 MHD Boundary Layer Equations

Liu, Xie, and Yang [88] in 2019 studied the high Reynolds numbers limit for the MHD system with Prandtl boundary layer expansion when no-slip boundary condition is imposed on velocity field and perfect conducting boundary condition on magnetic field. Under the assumption that the viscosity and resistivity coefficients are of the same order and the initial tangential magnetic field on the boundary is not degenerate, they justified the validity of the Prandtl boundary layer expansion and gave a L^∞ estimate on the error by multiscale analysis. Consider the following 2D incompressible viscous MHD equations in the domain $\{(t, x, y) \mid t>0, x\in\mathbb{T}, y\in\mathbb{R}_+\}$:

$$\begin{cases}\partial_t\mathbf{u}^\epsilon+(\mathbf{u}^\epsilon\cdot\nabla)\,\mathbf{u}^\epsilon+\nabla p^\epsilon-(\mathbf{H}^\epsilon\cdot\nabla)\,\mathbf{H}^\epsilon=\mu\epsilon\Delta\mathbf{u}^\epsilon,\\ \partial_t\mathbf{H}^\epsilon+(\mathbf{u}^\epsilon\cdot\nabla)\,\mathbf{H}^\epsilon-(\mathbf{H}^\epsilon\cdot\nabla)\,\mathbf{u}^\epsilon=\kappa\epsilon\Delta\mathbf{H}^\epsilon,\\ \nabla\cdot\mathbf{u}^\epsilon=0,\quad \nabla\cdot\mathbf{H}^\epsilon=0.\end{cases}\tag{1.5.1}$$

Here $\mathbf{u}^\epsilon=(u^\epsilon, v^\epsilon)$ and $\mathbf{H}^\epsilon=(h^\epsilon, g^\epsilon)$ stand for the velocity field and magnetic field, respectively; p^ϵ denotes the total pressure; the tangential variable is periodic: $x\in\mathbb{T}$, and the normal variable $y\in\mathbb{R}_+$. Also assume μ, κ are positive constants, and the viscosity and resistivity coefficients are of the same order in a small parameter ϵ. The initial data of the system (1.5.1) is given by

$$\left(\mathbf{u}^\epsilon, \mathbf{H}^\epsilon\right)\big|_{t=0}=(\mathbf{u}_0, \mathbf{H}_0)\,(x, y)=(u_0, v_0, h_0, g_0)\,(x, y),$$

independent of ϵ. The no-slip boundary condition is imposed on velocity field and the perfectly conducting boundary condition on magnetic field:

$$\mathbf{u}^\epsilon\big|_{y=0}=\mathbf{0},\quad \left(\partial_y h^\epsilon, g^\epsilon\right)\big|_{y=0}=\mathbf{0}.$$

Theorem 1.5.1 ([88]) *Let* $m \geqslant 36$ *be an integer. Let the initial data* $(\mathbf{u_0}, \mathbf{H_0})(x, y) \in H^m(\mathbb{T} \times \mathbb{R}_+)$ *satisfy:*

(i) $\mathbf{u_0} = (u_0, v_0)$ *is a divergence free vector field vanishing on the boundary, and* $\mathbf{H_0} = (h_0, g_0)$ *is a divergence free vector field tangent to the boundary;*
(ii) there exists a small $\epsilon_0 > 0$ *such that the following "strong" compatibility conditions hold for any* $\epsilon \in [0, \epsilon_0]$,

$$\partial_t^i \left(u^\epsilon, v^\epsilon, g^\epsilon\right)(0)\Big|_{y=0} = \mathbf{0},\ \partial_t^{i-1}\partial_y h^\epsilon(0)\Big|_{y=0} = 0, \quad 1 \leqslant i \leqslant \left[\frac{m}{2}\right] - 3,$$

where $[k], k \in \mathbb{R}$, *stands for the largest integer less than or equal to k, and* $\partial_t^i(u^\epsilon, v^\epsilon, h^\epsilon, g^\epsilon)(0)$ *is the i-th time derivative at* $\{t = 0\}$ *of any solution of the equation, as calculated from the equation to yield an expression in terms of derivatives of* $(\mathbf{u_0}, \mathbf{H_0})$;
(iii) the initial tangential magnetic field is non-degenerate: $h_0(x, 0) \geqslant \delta_0 > 0$ *for some constant* δ_0. *Then there exists* $T_* > 0$ *independent of* ϵ *such that the problem admits a solution* $(\mathbf{u}^\epsilon, \mathbf{H}^\epsilon)$ *in the time interval* $[0, T_*]$, *and there exists a smooth solution* $\left(\mathbf{u^0}, \mathbf{H^0}\right)(t, x, y) \in C([0, T_*], H^m)$ *to the problem with the initial data* $(\mathbf{u_0}, \mathbf{H_0})$, *and a boundary layer profile*

$$\left(u_b^0, v_b^0, h_b^0, g_b^0\right) \in C([0, T_*] \times \mathbb{T} \times \mathbb{R}_+),$$

such that for any arbitrarily small $\sigma > 0$,

$$\sup_{0 \leqslant t \leqslant T_*} \Bigg\| \left(\mathbf{u}^\epsilon, \mathbf{H}^\epsilon\right)(t, x, y) - \left(\mathbf{u^0}, \mathbf{H^0}\right)(t, x, y) - \left(u_b^0, \sqrt{\epsilon} v_b^0, h_b^0, \sqrt{\epsilon} g_b^0\right)\left(t, x, \frac{y}{\sqrt{\epsilon}}\right) \Bigg\|_{L_{xy}^\infty} \leqslant C\epsilon^{3/8-\sigma},$$

where the constant $C > 0$ *is independent of* ϵ.

Wang and Xin [141] in 2017 identified a nontrivial class of initial data for which the uniform stability of the Prandtl's type boundary layers can be established and proved rigorously that the solutions to the viscous and diffusive incompressible MHD systems converges strongly to the superposition of the solution to the ideal MHD systems with a Prandtl's type boundary layer corrector.

To this end, they considered zero viscosity-diffusion vanishing inviscid limit and zero magnetic diffusion vanishing limit for the 3D/2D incompressible viscous and diffusive magnetohydrodynamic (MHD) systems with Dirichlet boundary (no-slip characteristic) boundary conditions:

$$\begin{cases} u_t^\varepsilon - \varepsilon_1 \Delta u^\varepsilon + u^\varepsilon \cdot \nabla u^\varepsilon + \nabla p = b^\varepsilon \cdot \nabla b^\varepsilon, \\ b_t^\varepsilon - \varepsilon_2 \Delta b^\varepsilon + u^\varepsilon \cdot \nabla b^\varepsilon = b^\varepsilon \cdot \nabla u^\varepsilon, \\ divu^\varepsilon = divb^\varepsilon = 0, \\ u^\varepsilon = b^\varepsilon = 0, \ in \ \partial\Omega \times (0,T), \\ u^\varepsilon(t=0) = u_0^\varepsilon, \ b^\varepsilon(t=0) = b_0^\varepsilon. \end{cases} \tag{1.5.2}$$

Letting $\varepsilon \to 0$ in (1.5.2), we obtain formally the following inviscid MHD system:

$$\begin{cases} \partial_t u^0 + u^0 \cdot \nabla u^0 + \nabla p^0 = b^0 \cdot \nabla b^0, \ \Omega \times (0,T), \\ \partial_t b^0 + u^0 \cdot \nabla b^0 0 = b^0 \cdot \nabla u^0, \\ divu^0 = divb^0 = 0, \\ u^0 = 0, b^0 = 0, \partial\Omega \times (0,T), \\ (u^0, b^0)|_{t=0} = (u_0^0, b_0^0). \end{cases} \tag{1.5.3}$$

Theorem 1.5.2 ([141]) *Let (u^0, p^0, b^0) be the solution to the incompressible ideal MHD system (1.5.3). Assume that $(u_\varepsilon^0, b_\varepsilon^0)$ strongly converges in $L^2(\Omega)$ to (u_0^0, b_0^0), where $(u_0^0, b_0^0) \in H^s(\Omega)$ satisfies $divu_0^0 = divb_0^0 = 0$, and $u_0^0 \cdot n = b_0^0 \cdot n = 0$. Assume that $u_0^0(x,y,z) = b_0^0(x,y,z)$ or $u_0^0(x,y,z) = -b_0^0(x,y,z)$. Furthermore, assume that $\varepsilon, \varepsilon_1, \varepsilon_2$ satisfy the following convergence:*

$$\frac{\varepsilon_1 + \varepsilon_2}{\sqrt{\varepsilon}} \to 0, \quad \frac{(\varepsilon_1 - \varepsilon_2)^2}{\sqrt{\varepsilon}\varepsilon(\varepsilon_1 + \varepsilon_2)} \to 0, \quad \frac{(\varepsilon_1 - \varepsilon_2)^2}{\varepsilon(\varepsilon_1 + \varepsilon_2)} \le C \min\{\varepsilon_1, \varepsilon_2\}, \tag{1.5.4}$$

for some constant $C > 0$, independent of $\varepsilon, \varepsilon_1, \varepsilon_2$ as $\varepsilon \to 0$, $\varepsilon_1 \to 0$, $\varepsilon_1 \to 0$. Then there exists a global Leray-Hopf weak solutions $(u^\varepsilon, p^\varepsilon, b^\varepsilon)$ of the system (1.5.2) such that

$$(u^\varepsilon - u^0, b^\varepsilon - b^0) \to (0,0) \ in \ L^\infty(0,T; L^2(\Omega)), \tag{1.5.5}$$

for any $T : 0 < T < \infty$, as viscosity coefficient $\varepsilon_1 \to 0$ and diffusion coefficient $\varepsilon_2 \to 0$. Moreover, if

$$\|(u_0^\varepsilon - u_0^0, b_0^\varepsilon - b_0^0) \to (0,0)\|_{L^2(\Omega)} \le C\varepsilon^k, \ k \ge 1, \tag{1.5.6}$$

then there exists $C(T)$, independent of ε such that, for $0 < t \le T < +\infty$,

$$\|(u^\varepsilon - u^0, b^\varepsilon - b^0) \to (0,0)\|_{L^2(\Omega)} \le C(T)(\varepsilon^{k-1} + \varepsilon_1^2 + \varepsilon_2^2 + \tfrac{\varepsilon_1+\varepsilon_2}{\sqrt{\varepsilon}} + \tfrac{(\varepsilon_1-\varepsilon_2)^2}{\sqrt{\varepsilon}\varepsilon(\varepsilon_1+\varepsilon_2)}), \ k \ge 1. \tag{1.5.7}$$

Furthermore, we have the following stronger L^∞ convergence results for the viscous and diffusive incompressible 2D MHD systems: assume that $u_1^0(x,z=0)=b_1^0(x,z=0)=constant$ and $\varepsilon_1=\varepsilon_2$ or that $\varepsilon,\varepsilon_1,\varepsilon_2$ satisfy suitable relations (stated below). If, for some suitably large $k>2$,

$$\|(u_0^\varepsilon-u_0^0-u_B^\varepsilon(t=0),b_0^\varepsilon-b_0^0-b_B^\varepsilon(t=0))\|_{H^s(\Omega)}\le C\varepsilon^k,\ s\ge 3, \quad (1.5.8)$$

then there exists a $C(T)>0$, independent of ε, such that, for $0\le t\le T<\infty$,

$$\|(u^\varepsilon-u^0-u_B^\varepsilon,b^\varepsilon-b^0-b_B^\varepsilon)\|_{L^\infty(\Omega)}\le C(T)\varepsilon^{1/2},\ if\ \varepsilon_=\varepsilon_1=\varepsilon_2. \quad (1.5.9)$$

Wang, Wang, Liu, and Wang [136] in 2017 studied the boundary layer problem and zero viscosity-diffusion limit of the following initial boundary value problem for the incompressible viscous and diffusive magnetohydrodynamic (MHD) system with (no-slip characteristic) Dirichlet boundary conditions and proved that the corresponding Prandtl's type boundary layer is stable with respect to small viscosity-diffusion coefficients:

$$\begin{cases}\partial_t u^\epsilon+u^\epsilon\cdot\nabla u^\epsilon-\epsilon\Delta u^\epsilon=b^\epsilon\cdot\nabla b^\epsilon, in\ \Omega\times(0,T),\\ \partial_t b^\epsilon+u^\epsilon\cdot\nabla b^\epsilon-\epsilon\Delta b^\epsilon=b^\epsilon\cdot\nabla u^\epsilon,\\ divu^\epsilon=divb^\epsilon=0,\\ u^\epsilon=0, b^\epsilon=0, in\ \partial\Omega\times(0,T),\\ (u^\epsilon,b^\epsilon)|_{t=0}=(u_0^\epsilon,b_0^\epsilon)\end{cases} \quad (1.5.10)$$

and

$$\begin{cases}\partial_t u^{\eta,\nu}+u^{\eta,\nu}\cdot\nabla u^{\eta,\nu}-\nu_1\partial_z^2u^{\eta,\nu}-\eta_1\Delta_{x,y}u^{\eta,\nu}+\nabla p^{\eta,\nu}\\ \quad =b^{\eta,\nu}\cdot\nabla b^{\eta,\nu}, in\ \Omega\times(0,T),\\ \partial_t b^{\eta,\nu}+u^{\eta,\nu}\cdot\nabla b^{\eta,\nu}-\nu_2\partial_z^2b^{\eta,\nu}-\eta_2\Delta_{x,y}b^{\eta,\nu}=b^{\eta,\nu}\cdot\nabla u^{\eta,\nu},\\ divu^{\eta,\nu}=divb^{\eta,\nu}=0,\\ u^{\eta,\nu}=0, b^{\eta,\nu}=0, in\ \partial\Omega\times(0,T),\\ (u^{\eta,\nu},b^{\eta,\nu})|_{t=0}=(u_0^{\eta,\nu},b_0^{\eta,\nu})\end{cases} \quad (1.5.11)$$

First, letting $\epsilon\to 0$ in (1.5.10), we formally obtain the following 3D incompressible ideal MHD system

$$\begin{cases}\partial_t u^0+u^0\cdot\nabla u^0+\nabla p^0=b^0\cdot\nabla b^0, in\ \Omega\times(0,T),\\ \partial_t b^0+u^0\cdot\nabla b^0 0=b^0\cdot\nabla u^0,\\ divu^0=divb^0=0,\\ u^0=0, b^0=0, in\ \partial\Omega\times(0,T),\\ (u^0,b^0)|_{t=0}=(u_0^0,b_0^0).\end{cases} \quad (1.5.12)$$

Secondly, setting $\nu = (\eta, \nu) \to 0$ (which is called the anisotropic MHD limit in the following) in (1.5.10) one formally gets the following three-dimensional incompressible anisotropic MHD system

$$\begin{cases} \partial_t u^{\eta,0} + u^{\eta,0} \cdot \nabla u^{\eta,0} - \eta_1 \Delta_{x,y} u^{\eta,0} + \nabla p^{\eta,0} = b^{\eta,0} \cdot \nabla b^{\eta,0}, in\ \Omega \times (0,T), \\ \partial_t b^{\eta,0} + u^{\eta,0} \cdot \nabla b^{\eta,0} - \eta_2 \Delta_{x,y} b^{\eta,0} = b^{\eta,0} \cdot \nabla u^{\eta,0}, \\ div u^{\eta,0} = div b^{\eta,0} = 0, \\ u_3^{\eta,0} = 0, b_3^{\eta,0} = 0, in\ \partial\Omega \times (0,T), \\ (u^{\eta,0}, b^{\eta,0})|_{t=0} = (u_0^{\eta,0}, b_0^{\eta,0}). \end{cases} \tag{1.5.13}$$

For the MHD system (1.5.10), the authors of [136] had the following stability result of Prandtl's type boundary layer for a class of special initial data, which yields to the appearance of the Prandtl's type boundary layer.

Theorem 1.5.3 ([136]) *Let $(u^0, p^0, b^0)(x, y, z, t)$ be the solution of the incompressible ideal MHD equations* (1.5.12) *with the initial data $(u_0^0, b_0^0)(x, y, z)$ satisfying $u_0^0, b_0^0 \in H^s, s > \frac{3}{2}+1, div u^0 = div b^0 = 0$ and $u_0^0 \cdot n|_{\partial\Omega} = b_0^0 \cdot n|_{\partial\Omega} = 0$. Assume that $(u_0^\epsilon, b_0^\epsilon)$ strongly converges in (u_0^0, b_0^0). Also, assume that $u_0^0(x, y, z) = b_0^0(x, y, z)$ or $u_0^0(x, y, z) = -b_0^0(x, y, z)$. Then there exists a global Leray-Hopf weak solutions $(u^\epsilon, p^\epsilon, b^\epsilon)$ of the problem* (1.5.10) *such that*

$$(u^\epsilon - u^0, b^\epsilon - b^0) \to (0, 0), \ in \ \ L^\infty(0, T; L^2(\Omega))$$

for any $T : 0 < T < \infty$, as viscosity and diffusion coefficient $\epsilon \to 0$.

Here, in fact, $(u^0, p^0, b^0)(x, y, z, t) = (u_0^0, 0, b_0^0)(x, y, z)$ is a special steady state solution of the ideal MHD system (1.5.12)*. Moreover, if*

$$\|(u_0^\epsilon - u_0^0, b_0^\epsilon - b_0^0)\|_{L^2(\Omega)} \le C\epsilon^\kappa, \kappa > 1,$$

then there exists a $C(T) > 0$, independent of ϵ, such that

$$\|(u^\epsilon - u^0, b^\epsilon - b^0)\|_{L^2(\Omega)} \le C(T)\epsilon^{\min\{\kappa-1, \frac{1}{2}\}}.$$

For anisotropic viscous and diffusive MHD system (1.5.11) with the different horizontal and vertical viscosities and magnetic diffusions, the authors of [136] obtained the following result.

Theorem 1.5.4 ([136] The Ideal MHD Limit) *Assume that $(u_0^{\eta,\nu}, b_0^{\eta,\nu})$ strongly converges in $L^2(\Omega)$ to $(u_0^{0,0}, b_0^{0,0})$ where $u_0^{0,0}, b_0^{0,0} \in H^s, s > \frac{3}{2} + 1$. Assume that*

$$\eta_1 \to 0, \eta_2 \to 0, \nu_1 \to 0, \nu_2 \to 0,$$

$$\frac{\nu_1}{\sqrt{\min\{\eta_1, \eta_2\} \min\{\nu_1, \nu_2\}}} \to 0 \ and \ \frac{\nu_2}{\sqrt{\min\{\eta_1, \eta_2\} \min\{\nu_1, \nu_2\}}} \to 0.$$

Assume that $(u^{0,0}, p^{0,0}, b^{0,0})$ *is the smooth solution to the system* (1.5.12) *defined on* $[0, T_*)$ *with* $0 < T_* \le \infty$. *Then there exists a global Leray-Hopf weak solution* $(u^{\eta,\nu}, p^{\eta,\nu}, b^{\eta,\nu})$ *of the system* (1.5.11) *such that*

$$(u^{\eta,\nu}, b^{\eta,\nu}) - (u^{0,0}, b^{0,0}) \to (0,0), \ in \ L^\infty(0,T;L^2),$$

for any $T : 0 < T < T^*$.

Moreover, if

$$\|(u_0^{\eta,\nu} - u_0^{0,0}, b_0^{\eta,\nu} - b_0^{0,0})\|_{L^2(\Omega)} \le C\sqrt{\alpha},$$

where $\alpha = \max\left\{\nu_1+\nu_2+\eta_1+\eta_2, \sqrt{\min\{\eta_1,\eta_2\}\min\{\nu_1,\nu_2\}}, \frac{\nu_1}{\sqrt{\min\{\eta_1,\eta_2\}\min\{\nu_1,\nu_2\}}}, \frac{\nu_2}{\sqrt{\min\{\eta_1,\eta_2\}\min\{\nu_1,\nu_2\}}}\right\}$. *Then there exists a* $C = C(T) > 0$, *independent of* η, ν, *such that*

$$\|(u^{\eta,\nu} - u^{0,0}, b^{\eta,\nu} - b^{0,0})\|_{L^2(\Omega)} \le C\sqrt{\alpha}.$$

Theorem 1.5.5 ([136] The Anisotropic MHD Limit) *Let* $\eta_1 > 0$, $\eta_2 > 0$. *Assume that* $(u_0^{\eta,\nu}, b_0^{\eta,\nu})$ *strongly converges in* $L^2(\Omega)$ *to* $(u_0^{\eta,0}, b_0^{\eta,0})$ *where* $u_0^{\eta,0}, b_0^{\eta,0} \in H^s, s > \frac{3}{2} + 2$. *Assume that*

$$\nu_1 \to 0, \nu_2 \to 0,$$

$$\frac{\nu_1}{\sqrt{\min\{\nu_1,\nu_2\}}} \to 0 \text{ and } \frac{\nu_2}{\sqrt{\min\{\nu_1,\nu_2\}}} \to 0.$$

Assume that $(u^{\eta,0}, p^{\eta,0}, b^{\eta,0})$ *is the smooth solution to the system* (1.5.13) *defined on* $[0, T_*)$ *with* $0 < T_* \le \infty$. *Then there exists a global Leray-Hopf weak solution* $(u^{\eta,\nu}, p^{\eta,\nu}, b^{\eta,\nu})$ *of the system* (1.5.11) *such that*

$$(u^{\eta,\nu}, b^{\eta,\nu}) - (u^{\eta,0}, b^{\eta,0}) \to (0,0), \ in \ L^\infty(0,T;L^2),$$

for any $T : 0 < T < T^*$ *as* $\nu = (\nu_1, \nu_2) \to (0,0)$.

Moreover, if

$$\|(u_0^{\eta,\nu} - u_0^{\eta,0}, b_0^{\eta,\nu} - b_0^{\eta,0})\|_{L^2(\Omega)} \le C\sqrt{\beta},$$

where $\beta = \max\left\{\nu_1+\nu_2, \sqrt{\min\{\nu_1,\nu_2\}}, \frac{\nu_1}{\sqrt{\min\{\nu_1,\nu_2\}}}, \frac{\nu_2}{\sqrt{\min\{\nu_1,\nu_2\}}}\right\}$.

Then there exists a $C = C(T) > 0$, *independent of* η, ν, *such that*

$$\|(u^{\eta,\nu} - u^{\eta,0}, b^{\eta,\nu} - b^{\eta,0})\|_{L^2(\Omega)} \le C\sqrt{\beta}.$$

Gérand-Varet and Prestipino [45] in 2017 provided a systematic derivation of boundary layer models in magnetohydrodynamics (MHD), through an asymptotic analysis of the incompressible MHD system and recovered classical linear models, related to the famous Hartmann and Shercliff layers, as well as nonlinear ones, that called magnetic Prandtl models. An appropriate starting point to describe such dynamics is the classical incompressible MHD system

$$\begin{cases} \partial_t \mathbf{u} + \mathbf{u} \cdot \nabla \mathbf{u} + \nabla p - \frac{1}{Re} \Delta \mathbf{u} = S \mathbf{b} \cdot \nabla \mathbf{b}, \\ \partial_t \mathbf{b} - \text{curl}\,(\mathbf{u} \times \mathbf{b}) + \frac{1}{Rm} \text{curl curl} \mathbf{b} = 0, \\ \nabla \cdot \mathbf{u} = 0, \quad \nabla \cdot \mathbf{b} = 0, \quad t > 0, \quad x \in \Omega, \end{cases} \tag{1.5.14}$$

with

$$S = \frac{B_0^2}{\mu \rho U^2} \frac{Ha^2}{Re Rm}, \quad Ha = B_0 L (\frac{\sigma}{\eta})^{\frac{1}{2}}.$$

They considered the solutions of the system (1.5.14) behaving like

$$\begin{cases} \mathbf{u} \approx \left(u'_x(t, x, y, \lambda^{-1} z), u'_y(t, x, y, \lambda^{-1} z), \lambda u'_z(t, x, y, \lambda^{-1} z) \right), \\ \mathbf{b} \approx (e)_z + \delta \left(b'_x(t, x, y, \lambda^{-1} z), b'_y(t, x, y, \lambda^{-1} z), \lambda b'_z(t, x, y, \lambda^{-1} z) \right). \end{cases}$$

The 2D mixed Prandtl/Hartmann regime is as follows:

$$\begin{cases} \partial_t u + u \partial_x u + v \partial_z u - \frac{Ha^2}{Re} \partial_z^2 u + \frac{Ha^2}{Re} u = -\partial_x p, \\ \partial_x u + \partial_z v = 0, \\ u|_{z=0} = v|_{z=0} = 0, \quad \lim_{z \to +\infty} u = U(t, x), \end{cases} \tag{1.5.15}$$

and the 2D Mixed Prandtl/Shercliff regime is in the following:

$$\begin{cases} \partial_t u + u \partial_x u + v \partial_z u - \frac{Ha}{Re} \partial_z^2 u = \frac{Ha}{Re} \partial_x b - \partial_x p, \\ \partial_x u + \partial_z^2 b = 0, \\ \partial_x u + \partial_z v = 0, \\ u|_{z=0} = v|_{z=0} = b|_{z=0} = 0, \quad \lim_{z \to +\infty} u = U(t, x), \quad \lim_{b \to +\infty} u = B(t, x). \end{cases} \tag{1.5.16}$$

The linearized system of the problem (1.5.16) reads as

$$\begin{cases} \partial_t u + U \partial_x u + v U' - \frac{Ha}{Re} \partial_z^2 u = \frac{Ha}{Re} \partial_x b, \\ \partial_x u + \partial_z^2 b = 0, \\ \partial_x u + \partial_z v = 0, \\ u|_{z=0} = v|_{z=0} = b|_{z=0} = 0, \quad \lim_{z \to +\infty} u = \lim_{b \to +\infty} u = (0, 0). \end{cases} \tag{1.5.17}$$

Theorem 1.5.6 ([45]) *Assume that $U \in W^{2,\infty}(\mathbb{R}_+)$, $zU', zU'' \in L^\infty(\mathbb{R}_+)$. Let $u_0 \in L^2(\Omega)$ such that $w_0 = \partial_z u_0 \in L^2(\Omega)$, $u_0|_{z=0} = 0$. Let $b_0 \in L^2_{loc}(\Omega)$ such that $\partial_z b_0 \in L^2(\Omega)$, $b_0|_{z=0} = 0$ and with zero average in x. Then there exists a unique solution (u, v, b) of the problem* (1.5.17) *satisfying $(u, b)|_{t=0} = (u_0, b_0)$.*

The 2D nonlinear MHD layer ($Re \sim Rm \sim Ha$) is the following:

$$\begin{cases} \partial_t u + u\partial_x u + v\partial_z u - \partial_z^2 u = S\mathbf{b}\cdot\nabla\mathbf{b} - \partial_x p, \\ \partial_t \mathbf{b} - \nabla^T(\mathbf{u}\times\mathbf{b}) - \frac{Re}{Rm}\partial_z^2\mathbf{b} = 0, \\ \partial_x u + \partial_z v = div\mathbf{b} = 0, \\ u|_{z=0} = v|_{z=0} = b|_{z=0} = 0, \ \lim\limits_{z\to+\infty} u = U(t,x), \ \lim\limits_{b\to+\infty} b = B(t,x), \end{cases} \tag{1.5.18}$$

whose linearized system reads as

$$\begin{cases} \partial_t u + U\partial_x u + U'v - \partial_z^2 u = S\partial_x b, \\ \partial_t \mathbf{b} - \nabla^T(v - Uc) - \frac{Re}{Rm}\partial_z^2\mathbf{b} = 0, \\ \partial_x u + \partial_z v = div\mathbf{b} = 0, \\ u|_{z=0} = v|_{z=0} = b|_{z=0} = 0, \ \lim\limits_{z\to+\infty} u = \lim\limits_{b\to+\infty} b = 0. \end{cases} \tag{1.5.19}$$

Theorem 1.5.7 ([45]) *Assume that $U \in W^{3,\infty}(\mathbb{R}_+)$, $zU', zU'', zU''' \in L^\infty(\mathbb{R}_+)$. Let $u_0 \in L^2(\Omega)$, $\phi_0 \in L^2_{loc}(\Omega)$, such that $b_0 = \partial_y\phi_0 \in L^2(\Omega)$, $\phi_0|_{z=0} = 0$ and with zero average in x. Then there exists a unique solution of the problem* (1.5.19) *satisfying $(u, b)|_{t=0} = (0, -\nabla^T\phi_0)$.*

1.5.1 MHD Boundary Layer Equations-Local Existence

Oleinik and Samokhin [106] in 1999 considered the boundary layer theory which, in fact, can also be applied to some problems connected with the boundary layer in magnetohydrodynamics (MHD). They considered the stationary boundary layer system for a plane-parallel conducting flow in the presence of magnetic and electric fields in the domain $D = \{0 < x < X, 0 < y < \infty\}$ by the von Mises variables that has the form

$$\begin{cases} u\frac{\partial u}{\partial x} + v\frac{\partial u}{\partial y} = \nu\frac{\partial^2 u}{\partial y^2} - \frac{\partial p}{\partial x} - jB, \\ \frac{\partial u}{\partial x} + \frac{\partial v}{\partial y} = 0 \end{cases} \tag{1.5.20}$$

where $j = \sigma(E + B)$ is the z-component of the electric current vector; $\mathbf{E} = (0, 0, E)$ is the electric field; $\mathbf{B} = (0, B, 0)$ is the magnetic field; and p, B, E are given functions that depend only on x. System (1.5.20) is supplemented with the conditions

$$\begin{aligned} &u(0, y) = u_0(y), \ \ u(x, 0) = 0, \ \ v(x, 0) = v_0(x), \\ &u(x, y) \to U(x), \ \ \text{as } \ y \to \infty. \end{aligned} \tag{1.5.21}$$

The velocity $U(x)$ of the outer flow is related to the pressure $p(x)$ and the electromagnetic field components by

$$U\frac{dU}{dx} = -\frac{dp}{dx} - \sigma EB - \sigma B^2 U.$$

The above equation is a generalization of the Bernoulli's law. Taking into account this relation, we can rewrite (1.5.20) in the form

$$\begin{cases} u\frac{\partial u}{\partial x} + v\frac{\partial u}{\partial y} = \nu\frac{\partial^2 u}{\partial y^2} + U\frac{dU}{dx} + \sigma B^2(U-u), \\ \frac{\partial u}{\partial x} + \frac{\partial v}{\partial y} = 0. \end{cases} \tag{1.5.22}$$

Theorem 1.5.8 ([106]) *For some $X > 0$, problems (1.5.21)–(1.5.22) in D admit a solution u, v with the following properties: u continuous and bounded in $\overline{D}$; $u > 0$ for $y > 0$; $\partial u/\partial y > m > 0$ for $0 \le y \le y_0$; m, y_0 =const.; $\partial u/\partial y$, $\partial^2 u/\partial y^2$ are bounded and continuous in D; $\partial u/\partial x$, v, $\partial v/\partial y$ are continuous and bounded in any finite subdomain of D. If*

$$\frac{dU}{dx} \ge 0 \ \text{and} \ v_0(x) \le 0 \ \text{or} \ U\left(\frac{dU}{dx} + \sigma B^2\right) \ge \alpha = \text{const.} > 0,$$

then the solution of problems (1.5.21)–(1.5.22) exists for any $X > 0$.

Theorem 1.5.9 ([106]) *Let $u(x, y)$, $v(x, y)$ be a solution of problems (1.5.21)–(1.5.22) with the following properties: u, v are continuous in $\overline{D}$; $0 < u < C_0$ for $y > 0$;*

$$k_1 y \le u \le k_2 y \ \text{for} \ 0 < y < y_0, \tag{1.5.23}$$

$$\partial^2 u/\partial y^2 \le k_3 \ \text{in} \ D, \tag{1.5.24}$$

where $C_0, k_i, (i = 1, 2, 3)$, y_0 positive constants. Such a solution of problems (1.5.21)–(1.5.22) is unique. If $U(dU/dx + \sigma B^2) \ge 0$ and $v_0(x) \le 0$, then conditions (1.5.23)–(1.5.24) can be dropped. If, however, $v_0(x)$ is allowed to take positive values, then condition (1.5.23) should be retained.

Theorem 1.5.10 ([106]) *Assume the following conditions holds*

$$v_0(x) \le 0, \ \ 0 \le \frac{dU}{dx} \le \frac{M_0}{(x+1)^{1+\gamma_0}}, \ \ \gamma_0 > 0, \ \ \lim_{x\to\infty} U(x) = U_0.$$

Let $u(x, y)$, $v(x, y)$ be the solution of problems (1.5.21)–(1.5.22) in $D = \{0 < x < \infty, 0 < y < \infty\}$. Assume also that

$$\sigma B^2(x) = \frac{U_0 b(x)}{2(x+1)^{1+\delta_1}}, \quad v_0(x) = \frac{\sqrt{\nu U_0} q(x)}{\sqrt{2}(x+1)^{\frac{1}{2}+\delta_2}},$$

where $-1 \le \delta_1 \le \infty, -1/2 \le \delta_2 \le \infty$; $b(x), q(x)$ are bounded functions. If $\delta_1 > 0$, $b(x) \ge 0$. In the case $\delta_1 \le 0$, if it is assumed that $b(x) > 0$ and, for some N =const.> 0, we have

$$|b(x) - N| \le \frac{M_1}{(x+1)^{1+\delta_3}}, \quad \delta_3 > 0.$$

If $\delta_2 > 0$, we assume that $q(x) \le 0$; if $\delta_2 \le 0$, we assume that $q(x) < 0$ and, for some α =const.≤ 0, the following inequality holds:

$$|q(x) - \alpha| \le \frac{M_2}{(x+1)^{1+\delta_4}}, \quad \delta_4 > 0.$$

For definiteness, it is also assumed that $B(x) \ge 0$ and that for $\delta_1 < 0, \delta_2 < 0$, we have

$$\frac{B(x)}{v_0(x)} = O((x+1)^{1+\delta_5}) \text{ as } x \to \infty.$$

Under these assumptions, it is claimed that

$$|u(x, y) - \phi(x, y)| \to 0 \text{ as } x \to \infty$$

uniformly with respect to y, where the function $\phi(x, y)$ is defined as follows:

(1) $\phi(x, y) = u_{0,0}(x, y)$ if $\delta_1 > 0, \delta_2 > 0$;
(2) $\phi(x, y) = u_{0,N}(x, y)$ if $\delta_1 = 0, \delta_2 > 0$;
(3) $\phi(x, y) = u_{\alpha,0}(x, y)$ if $\delta_1 > 0, \delta_2 = 0$;
(4) $\phi(x, y) = u_{\alpha,N}(x, y)$ if $\delta_1 = \delta_2 = 0$;
(5) if $-1 \le \delta_1 < 0$ and either $\delta_2 \ge 0$ or $\delta_2 < 0$ and $\delta_5 > 0$, then

$$\phi(x, y) = h_1(x, y) = U_0 \left(1 - \exp\left\{-\frac{y\sqrt{U_0 N}}{\sqrt{2\nu(x+1)^{1+\delta_2}}}\right\}\right);$$

(6) if $-\frac{1}{2} \le \delta_2 < 0$ and either $\delta_1 \ge 0$ or $\delta_1 < 0$ and $\delta_5 < 0$, then

$$\phi(x, y) = h_2(x, y) = U_0 \left(1 - \exp\left\{\frac{\alpha y\sqrt{U_0}}{\sqrt{2\nu(x+1)^{1+2\delta_2}}}\right\}\right);$$

(7) if $-1 \le \delta_1 < 0$, $-\frac{1}{2} \le \delta_2 < 0$, $\delta_5 = 0$, *then*

$$\phi(x, y) = h_3(x, y) = U_0\left(1 - \exp\{C(x)y\}\right),$$

where

$$C(x) = \frac{\sqrt{U_0}(\alpha - \sqrt{\alpha^2 + 4N})}{2\sqrt{2\nu(x+1)^{1+2\delta_2}}}.$$

Theorem 1.5.11 ([106]) *Suppose the assumptions of Theorem 1.5.10 hold for* $U(x), v_0(x), b(x)$, *and* $(u(x, y), v(x, y))$ *is a solution of problems (1.5.21)–(1.5.22). If* $v_0(x)$ *and* $B(x)$ *ensure the convergence* $|u(x, y) - \phi(x, y)| \to 0$, *uniformly in* y, *as* $x \to \infty$ *(where* $\phi(x, y)$ *is the same as in Theorem 1.5.10), then*

$$\left|\tau - \nu\frac{\partial \phi}{\partial y}\right| \to 0 \quad as \quad x \to \infty.$$

Oleinik and Samokhin [106] in 1999 also considered the MHD boundary in pseudo-plastic fluids as follows:

$$\begin{cases} u\frac{\partial u}{\partial x} + v\frac{\partial u}{\partial y} = \nu\frac{\partial}{\partial y}\left(|\frac{\partial u}{\partial y}|^{n-1}\frac{\partial u}{\partial y}\right) + U(x)\frac{dU}{dx} + \sigma B^2(U - u), \quad n > 0, \\ \frac{\partial u}{\partial x} + \frac{\partial v}{\partial y} = 0, \end{cases} \tag{1.5.25}$$

in the domain $D = \{0 < x < X, 0 < y < \infty\}$, with the boundary conditions

$$\begin{array}{l} u(0, y) = u_0(y), \quad u(x, 0) = 0, \quad v(x, 0) = v_0(x), \\ u(x, y) \to U(x), \quad \text{as} \quad y \to +\infty, \end{array} \tag{1.5.26}$$

where $U(x), B(x), u_0(x), v_0(x)$ are given functions. Assume that

$$U(x) = xV(x), \quad v_0(x) = x^{\frac{n-1}{n+1}}v_1(x),$$

where $V(x) > 0$, $v_1(x)$ are bounded functions. In (1.5.25), introduce new independent variables

$$\xi = x, \quad \eta = \frac{u}{U}, \quad w(\xi, \eta) = \frac{|u_y|^{n-1}u_y}{x^{\frac{n-1}{n+1}}U},$$

then problems (1.5.25)–(1.5.26) reduce to the equation

$$\nu n V^{\frac{1-n}{n}}|w|^{\frac{1+n}{n}}w_{\eta\eta} - \eta\xi V w_\xi + A w_\eta + B w = 0, \tag{1.5.27}$$

in the domain $\Omega = \{0 < \xi < X, 0 < \eta < 1\}$, with the boundary conditions

$$w|_{\eta=1} = 0, \quad (\nu w|w|_{\eta}^{\frac{1-n}{n}} - v_1 w|w|^{\frac{1-n}{n}} + C)|_{\eta=0} = 0, \tag{1.5.28}$$

where

$$A = (\eta^2 - 1)(V + \xi V_x) + (\eta - 1)\sigma B^2, \quad B = -\eta\left(\frac{2nV}{1+n} + \xi V_x\right) - \sigma B^2,$$

$$C = V^{\frac{n-1}{n}}(V + \xi V_x + \sigma B^2).$$

Theorem 1.5.12 ([106]) *Assume that*

$$U(x) = x(a + xa_1(x)), \quad v_0(x) = x^{\frac{n-1}{n+1}}(b + xb_1(x)),$$

a =const.> 0, b, d_0 =const.; $U > 0$ for $x > 0$; $a_1, a_{1x}, a_{1xx}, b_1, b_{1x}$ are bounded for $0 \le x \le X$. Then problems (1.5.25)–(1.5.26) in the domain D, with X depending on U, v_0, B, n, ν, admit a solution $(u(x, y), v(x, y))$ with the following properties: u/U is bounded and continuous in $\overline{D}$; v is continuous in D; v is continuous with respect to y in $\overline{D}$ and bounded for bounded y; $u(x, y) > 0$ for $y > 0$ and $x > 0$; the derivatives u_x, u_y, u_{yy}, u_y are continuous and bounded in D and continuous with respect to y in $\overline{D}$; $u_y > 0$ for $y \ge 0$ and $x > 0$; u and v satisfy equations (1.5.25) at every point of D; $u(x, y) \to U(x)$ as $y \to \infty$, uniformly on the segment [0,X]; the boundary conditions (1.5.26) are satisfied. Moreover, $u_y^n / \left(x^{\frac{n-1}{n+1}} U\right)$ is bounded and continuous in $\overline{D}$, $u_y \to 0$ as $y \to \infty$; $u_{yy}^n / \left(x^{\frac{n-1}{n+1}} u_y^{2-n}\right)$ is continuous with respect to y in $\overline{D}$; u_{xy}, u_{yyy} are continuous in D; the following inequalities hold:

$$x^{\frac{n-1}{n+1}} U(x) Y(\frac{u}{U}) e^{-K_1 x} \le u_y^n \le x^{\frac{n-1}{n+1}} U(x) Y(\frac{u}{U}) e^{K_3 x},$$

$$-K_6 \le \frac{n(n-2)(u_{yy})^2 + n u_y u_{yyy}}{x^{\frac{n-1}{n}} U^{\frac{n-1}{n}} u_y^{3-n}} \le -K_7,$$

$$\left(M_2^{\frac{1}{n}} \frac{1-n}{n+1} e^{\frac{K_2}{n}} x^{\frac{n-1}{n(n+1)}} U^{\frac{1-n}{n}} y + 1\right)^{\frac{n+1}{n-1}} \le 1 - \frac{u}{U}$$

$$\le \left(M_1^{\frac{1}{n}} \frac{1-n}{n+1} e^{-\frac{K_1}{n}} x^{\frac{n-1}{n(n+1)}} U^{\frac{1-n}{n}} y + 1\right)^{\frac{n+1}{n-1}}.$$

The solution of problems (1.5.25)–(1.5.26) with the above properties is unique.

Now consider a more general problem, with $U(x) = O(x^m)$ as $x \to 0$, $m > 0$, that is, system (1.5.25) with the boundary conditions

$$\begin{cases} u(0, y) = 0, \quad u(x, 0) = 0, \quad v(x, 0) = v_0(x), \\ u(x, y) \to U(x), \quad \text{as} \quad y \to +\infty. \end{cases} \tag{1.5.29}$$

Assume that

$$U(x) = x^m V(x), \quad v_0(x) = x^{\frac{2mn-n-m}{n+1}} v_1(x), \quad d^2(x) = \sigma B(x)^2, \quad d(x) = x^{\frac{m-1}{2}} W(x),$$

where $V(x) > 0$ for $0 \le x \le X$, $m > 0$; $W(x) > 0$ for $0 \le x \le X$; moreover, the functions $V(x)$, $W(x)$, $v_1(x)$ are smooth, $V(0) = a > 0$, $v_1(0) = b$, $W(0) = d_0$. In (1.5.25), introduce new independent variables

$$\xi = x, \quad \eta = \frac{u}{U}, \quad w(\xi, \eta) = \frac{|u_y|^{n-1} u_y}{x^{\frac{2mn-n-m}{n+1}} U}.$$

Theorem 1.5.13 ([106]) *Under the assumptions*

$$x(U_x + \sigma B(x)^2) + \frac{2mn - n - m}{n+1} U > 0, \quad \frac{(3m-1)na}{n+1} + \sigma B^2(0) > 0,$$
$$V(x) = x(a + x a_1(x)), W(x) = d_0 + x d_1(x), \quad v_1(x) = b + x b_1(x),$$

where $a > 0$ and the functions $a_1, a_{1x}, a_{1xx}, b_1, b_{1x}, b_1, b_{1x}$ are bounded for $0 \le x \le X$. Then the problem (1.5.25), (1.5.29) in the domain D, with X depending on U, v_0, d, n, ν has a solution $u(x, y), v(x, y)$ with the following properties: u/U is bounded and continuous in $\overline{D}$; v is continuous in D; v is continuous with respect to y in $\overline{D}$ and bounded for bounded y; $u(x, y) > 0$ for $y > 0$ and $x > 0$; the derivatives u_x, u_y, u_{yy}, u_y are continuous and bounded in D and continuous with respect to y in $\overline{D}$; $u_y > 0$ for $y \ge 0$ and $x > 0$; equations (1.5.25) hold for u, v at each point of D; $u(x, y) \to U(x)$ as $y \to \infty$, uniformly on the segment $[0, X]$, the boundary conditions (1.5.29) hold for u, v. Moreover, $u_y^n / \left(x^{\frac{2mn-n-m}{n+1}} U(x)\right)$ is bounded and continuous in $\overline{D}$, $u_y \to 0$ as $y \to \infty$; $u_{yy}^n / \left(x^{\frac{n-1}{n+1}} u_y^{2-n}\right)$ is continuous with respect to y in $\overline{D}$; u_{xy}, u_{yyy} are continuous in D; the following inequalities hold:

$$x^{\frac{2n-m-n}{n+1}} U(x) Y(\frac{u}{U}) e^{-K_8 x} \le u_y^n \le x^{\frac{2mn-m-n}{n+1}} U(x) Y(\frac{u}{U}) e^{K_9 x},$$

$$-K_{13} \le \frac{n(n-2)(u_{yy})^2 + n u_y u_{yyy}}{x^{\frac{2mn-m-n}{n}} U^{\frac{n-1}{n}} u_y^{3-n}} \le -K_{14},$$

$$\left(M_8^{\frac{1}{n}}\frac{1-n}{n+1}e^{\frac{K_9}{n}}x^{\frac{2mn-m-n}{n(n+1)}}U^{\frac{1-n}{n}}y+1\right)^{\frac{n+1}{n-1}} \le 1-\frac{u}{U}$$

$$\le \left(M_7^{\frac{1}{n}}\frac{1-n}{n+1}e^{-\frac{K_8}{n}}x^{\frac{2mn-m-n}{n(n+1)}}U^{\frac{1-n}{n}}y+1\right)^{\frac{n+1}{n-1}}.$$

The solution of problem (1.5.25), (1.5.29) with the above properties is unique.

Oleinik and Samokhin [106] in 1999 continued to consider the system of equations of the MHD boundary layer for a dilatable fluid:

$$\begin{cases} u\frac{\partial u}{\partial x}+v\frac{\partial u}{\partial y}=\frac{k}{\rho}\frac{\partial}{\partial y}\left(|\frac{\partial u}{\partial y}|^{n-1}\frac{\partial u}{\partial y}\right)+U(x)\frac{dU}{dx}+\frac{\sigma B^2(x)}{\rho}(U-u), \quad n>1, \\ \frac{\partial u}{\partial x}+\frac{\partial v}{\partial y}=0 \end{cases} \tag{1.5.30}$$

in the domain $D=\{0<x<X, 0<y<\infty\}$, with the boundary conditions

$$\begin{cases} u(0,y)=u_0(y), \quad u(x,0)=0, \quad v(x,0)=0, \\ u(x,y)\to U(x), \quad \text{as } y\to+\infty. \end{cases} \tag{1.5.31}$$

Assume that

$$U(x)=C_0x^m, \quad B(x)=bx^{\frac{m-1}{2}},$$

where C_0 =const.$>$ 0, m =const., b =const.. In (1.6.28), introduce new independent variables

$$\eta=yx^{\frac{m-n}{n+1}}\left(\frac{r\rho C_0^{2-n}}{n(n+1)k}\right)^{\frac{1}{n+1}}, \quad \psi(x,y)=C_0x^{\frac{r}{n+1}}\left(\frac{n(n+1)k}{r\rho C_0^{2-n}}\right)^{\frac{1}{n+1}}f(\eta),$$

and take the flow function of the form

$$u=\partial\psi/\partial y, \quad v=-\partial\psi/\partial x,$$

then problems (1.5.30)–(1.5.31) reduce to the following problem:

$$|f''|^{n-1}f'''+ff''+\beta(1-f'^2)+N(1-f')=0, \tag{1.5.32}$$

with the boundary conditions

$$f(0)=0, \quad f'(0)=0, \quad f'(\eta)\to 1 \text{ as } \eta\to\infty, \tag{1.5.33}$$

where $\beta=m(n+1)/r$, $N=(n+1)\sigma b^2/(C_0\rho r)$.

Theorem 1.5.14 ([106]) *Let $f(\eta)$ be a solution of problem (1.5.32)–(1.5.33) such that $f'(\eta) \to C$ as $\eta \to \infty$. Assume that $\eta^{-1}(f'')^n \to 0$ as $\eta \to \infty$. Then $C = 1$.*

Oleinik and Samokhin [106] in 1999 considered the dilatable fluid in a transversed magnetic field system of equations

$$\begin{cases} \frac{\partial u}{\partial t} + u\frac{\partial u}{\partial x} + v\frac{\partial u}{\partial y} = \nu\frac{\partial}{\partial y}\left(|\frac{\partial u}{\partial y}|^{n-1}\frac{\partial u}{\partial y}\right) + U(x)\frac{dU}{dx} + d^2(U-u) + \frac{\partial U}{\partial t}, \quad n > 1, \\ \frac{\partial u}{\partial x} + \frac{\partial v}{\partial y} = 0 \end{cases} \tag{1.5.34}$$

in the domain $D = \{0 < x < X, 0 < y < \infty\}$, with the boundary conditions

$$\begin{cases} u(0, y) = 0, \quad u(x, 0) = 0, \quad v(x, 0) = v_0(x), \\ u(x, y) \to U(x), \quad \text{as } y \to \infty. \end{cases} \tag{1.5.35}$$

Theorem 1.5.15 ([106]) *Suppose that $U'(x), U''(x), v_0'(x), d_x^2$ are continuous functions and $U(0) = 0$, $U'(0) > 0$,*

$$U'(x) + d^2(x)/2 \geq a_0 > 0, \quad -M_0 x^{2n-1} \leq v_0(x) \leq 0,$$

for $0 \leq x \leq X$, where M_0 =const.> 0. Then there exist $u(x, y), v(x, y)$ with the following properties: u, v have locally square summable derivatives involved in (1.5.34); equation (1.5.34) holds almost everywhere; u is continuous in $\overline{D}$; v is square summable on any compact set inside D; $u(x, 0) = 0, u(0, y) = 0, u(x, y) \to U(x)$ as $y \to \infty$; $u(x, y) > 0$ for $x > 0$ and $y > 0$; $0 \leq u_y \leq M_1$, M_1 =const.> 0; $u_y > 0$ for $y = 0$ and $x > 0$; u_x, v are continuous in y for $y = 0$; $v(x, 0) = v_0(x)$ for $0 < x < X$; $u(x, y) \equiv U(x)$ for $y \geq C_0$, where $C_0 > 0$ is a constant depending on X and the other data of the problem. If $d^2(x)/U^{2n+1}(x) \geq a_1 > 0$, then C_0 can be chosen independent of X. The solution of problems (1.5.34)–(1.5.35) with these properties is unique.

Oleinik and Samokhin [106] in 1999 considered the stationary MHD boundary layer in a rapidly oscillating magnetic field, the system reads as

$$\begin{cases} u\frac{\partial u}{\partial x} + v\frac{\partial u}{\partial y} = \nu\frac{\partial^2 u}{\partial y^2} + d^2(x)(U(x) - u) + U(x)\frac{dU}{dx}, \\ \frac{\partial u}{\partial x} + \frac{\partial v}{\partial y} = 0, \end{cases} \tag{1.5.36}$$

in the domain $D = \{0 < x < X, 0 < y < \infty\}$, with the boundary conditions

$$\begin{cases} u(0, y) = u_0(y), \quad u(x, 0) = 0, \quad v(x, 0) = v_0(x), \\ u(x, y) \to U(x), \quad \text{as } y \to \infty. \end{cases} \tag{1.5.37}$$

Theorem 1.5.16 ([106]) *Suppose that $U'(x) \geq 0$ for $0 \leq x \leq X$; $U(x), u_0(y), v_0(x), d^2(x, \varepsilon^{-1}x)$ satisfy the compatibility conditions for any $\varepsilon \in (0, \varepsilon_0]$. Then for $d(x) = d(x, \varepsilon^{-1}x)$ and $\varepsilon \to 0$, the solution of* *problems (1.5.36)–(1.5.37) in the sense that*

$$u_\varepsilon \to \overline{u} \text{ uniformly for } 0 \le x \le X,\ \ 0 \le y \le y_1 < \infty,$$
$$v_\varepsilon \to \overline{v} \text{ uniformly for } x_0 \le x \le X,\ \ 0 \le y \le y_1,$$
$$\text{grad}\, u_\varepsilon \to \text{grad}\, \overline{u} \text{ weakly in } L_2(\{0 \le x \le X, 0 \le y \le y_1\}),$$

where $y_1 > 0$ and $x_0 \in (0, X)$ are arbitrary.

Gao, Guo, and Huang [34] in 2017 studied the local existence and uniqueness of solutions to the 2D MHD boundary layer where the initial tangential magnetic field has a lower bound $\delta > 0$ that plays an important role. They studied the following initial boundary value problem for the nonlinear MHD boundary layer equations:

$$\begin{cases} \partial_t u_1 + u_1 \partial_x u_1 + u_2 \partial_y u_1 - h_1 \partial_x h_1 - h_2 \partial_y h_1 = \mu \partial_y((\theta + 1)\partial_y u_1) - P_x, \\ \partial_t \theta + u_1 \partial_x \theta + u_2 \partial_y \theta_1 - \kappa \partial_y^2 \theta = \mu(\theta + 1)(\partial_y u_1)^2 + \nu(\partial_y h_1)^2, \\ \partial_x u_1 + \partial_y u_2 = 0, \quad \partial_x h_1 + \partial_y h_2 = 0, \\ u_1|_{t=0} = u_{10}(x, y), \quad h_1|_{t=0} = h_{10}(x, y), \\ (u_1, u_2, \partial_y h_1, h_2)|_{y=0} = \mathbf{0}, \quad \lim_{y \to +\infty} (u_1, \theta, h_1) = (U, \Theta, H)(t, x). \end{cases} \tag{1.5.38}$$

Theorem 1.5.17 ([34]) *Let $m \ge 5$ be an integer, and $l \ge 0$ a real number. Assume that the outer flow $(U, \theta, H, P_x)(t, x)$ satisfies that for some $T > 0$,*

$$M_0 := \sum_{i=0}^{2m+2} \Big(\sup_{0 \le t \le T} \|\partial_t^i (U, \theta, H, P)(t, \cdot)\|_{H^{2m+2-i}(\mathbb{T}_x)} + \|\partial_t^i (U, H, P)\|_{L^2(0,T; H^{2m+2-i}(\mathbb{T}_x))} \Big) < +\infty. \tag{1.5.39}$$

Also, we suppose the initial data $(u_{10}, \theta_{10}, h_{10})(x, y)$ satisfies

$$(u_{10}(x, y) - U(0, x), \theta_{10}(x, y) - \Theta(0, x), h_{10}(x, y) - H(0, x)) \in H_l^{3m+2}(\Omega), \tag{1.5.40}$$

and the compatibility conditions up to m-th order. Moreover, there exists a sufficiently small constant $\delta_0 > 0$ such that

$$\left| < y >^{l+1} \partial_y^i (u_{10}, h_{10})(x, y) \right| \le (2\delta_0)^{-1}, \quad h_{10}(x, y) \ge 2\delta_0, \quad for\ i = 1, 2,\ \ (x, y) \in \Omega. \tag{1.5.41}$$

Then, there exist a positive time $0 < T_* \le T$ *and a unique solution* (u_1, u_2, h_1, h_2) *to the initial boundary value problem* (1.5.38), *such that*

$$(u_1 - U, \theta - \Theta, h_1 - H) \in \bigcap_{i=0}^{m} W^{i,\infty}(0, T_*; H_l^{m-i}(\Omega)), \tag{1.5.42}$$

and

$$(u_2 + U_x y, h_2 + H_x y) \in \bigcap_{i=0}^{m-1} W^{i,\infty}(0, T_*; H_l^{m-1-i}(\Omega)), \tag{1.5.43}$$

$$(\partial_y u_2 + U_x, \partial_y h_2 + H_x) \in \bigcap_{i=0}^{m-1} W^{i,\infty}(0, T_*; H_l^{m-i}(\Omega)). \tag{1.5.44}$$

Moreover, if $l > \frac{1}{2}$,

$$(u_2 + U_x y, h_2 + H_x y) \in \bigcap_{i=0}^{m-1} W^{i,\infty}(0, T_*; L^\infty(\mathbb{R}_{y,+}; H^{m-1-i}(\mathbb{T}_x))), \tag{1.5.45}$$

$$(\partial_y u_2 + U_x, \partial_y h_2 + H_x) \in \bigcap_{i=0}^{m-1} W^{i,\infty}(0, T_*; H_l^{m-i}(\Omega)), \tag{1.5.46}$$

where $< y >= 1 + y$, *and*

$$H_l^m(\Omega) := \Big\{ f(x, y) : \Omega \to \mathbb{R}, \|f\|^2_{H_l^m(\Omega)} = \sum_{m_1+m_2 \le m} \| < y >^{l+m_2} \partial_x^{m_1} \partial_y^{m_2} f\|^2_{L^2(\Omega)} < +\infty \Big\}.$$

Xie and Yang [148] in 2019 considered the lifespan of solution to the MHD boundary layer system as an analytic perturbation of general shear flow. Since there is no restriction on the strength of the shear flow and the lifespan estimate is larger than the one obtained for the classical Prandtl system in this setting, it reveals the stabilizing effect of the magnetic field on the electrically conducting fluid near the boundary.

Consider the high Reynolds number limit to the MHD system near a no-slip boundary, when both of the Reynolds number and the magnetic Reynolds number have the same order in two space dimensions. Precisely, they considered the MHD system in the domain $\{(x, Y) \mid x \in \mathbb{R}, Y \in \mathbb{R}_+\}$ with $Y = 0$ being the boundary

$$\begin{cases} \partial_t u^\varepsilon + (u^\varepsilon \partial_x + v^\varepsilon \partial_Y) u^\varepsilon + \partial_x p^\varepsilon - (h^\varepsilon \partial_x + g^\varepsilon \partial_Y) h^\varepsilon = \varepsilon \left(\partial_x^2 u^\varepsilon + \partial_Y^2 u^\varepsilon\right), \\ \partial_t v^\varepsilon + (u^\varepsilon \partial_x + v^\varepsilon \partial_Y) v^\varepsilon + \partial_Y p^\varepsilon - (h^\varepsilon \partial_x + g^\varepsilon \partial_Y) g^\varepsilon = \varepsilon \left(\partial_x^2 v^\varepsilon + \partial_Y^2 v^\varepsilon\right), \\ \partial_t h^\varepsilon + (u^\varepsilon \partial_x + v^\varepsilon \partial_Y) h^\varepsilon - (h^\varepsilon \partial_x + g^\varepsilon \partial_Y) u^\varepsilon = \kappa\varepsilon \left(\partial_x^2 h^\varepsilon + \partial_Y^2 h^\varepsilon\right), \\ \partial_t g^\varepsilon + (u^\varepsilon \partial_x + v^\varepsilon \partial_Y) g^\varepsilon - (h^\varepsilon \partial_x + g^\varepsilon \partial_Y) v^\varepsilon = \kappa\varepsilon \left(\partial_x^2 g^\varepsilon + \partial_Y^2 g^\varepsilon\right), \\ \partial_x u^\varepsilon + \partial_Y v^\varepsilon = 0, \quad \partial_x h^\varepsilon + \partial_Y g^\varepsilon = 0, \end{cases} \tag{1.5.47}$$

where both the viscosity and resistivity coefficients are denoted by a small positive parameter ε, $(u^\varepsilon, v^\varepsilon)$ and $(h^\varepsilon, g^\varepsilon)$ represent the velocity and the magnetic field, respectively. The no-slip boundary condition is imposed on the velocity field:

$$\left(u^\varepsilon, v^\varepsilon\right)\big|_{Y=0} = \mathbf{0}. \tag{1.5.48}$$

Note that systems (1.5.47)–(1.5.48) govern the fluid behavior in the leading order of approximation near the boundary derived in

$$\begin{cases} \partial_t u_1 + u_1 \partial_x u_1 + u_2 \partial_y u_1 = b_1 \partial_x b_1 + b_2 \partial_y b_1 + \partial_y^2 u_1, \\ \partial_t b_1 + \partial_y (u_2 b_1 - u_1 b_2) = k \partial_y^2 b_1, \\ \partial_x u_1 + \partial_y u_2 = 0, \quad \partial_x b_1 + \partial_y b_2 = 0, \\ u_1(t, x, y)|_{t=0} = u_0(x, y), \quad b_1(t, x, y)|_{t=0} = b_0(x, y), \\ \lim\limits_{y \to \infty} u_1 = \bar{u}, \quad \lim\limits_{y \to \infty} b_1 = \bar{b}. \end{cases} \tag{1.5.49}$$

Assume the shear flow $u^s(t, y)$ has the following properties:

$$(H) \qquad \|\partial_y^i u_s(t, \cdot)\|_{L_y^\infty} \le \frac{C}{\langle t \rangle^{\frac{i}{2}}}, \quad \int_0^\infty |\partial_t u_s(t, y)| dy \le C,$$

$$\|\theta_\alpha \partial_y^2 u_s(t, \cdot)\|_{L_y^2} \le \frac{C}{\langle t \rangle^{\frac{3}{4}}}, \quad \langle t \rangle = 1 + t.$$

To define the function space of the solution considered in [148], the following Gaussian weighted function θ_α will be used:

$$\theta_\alpha(t, y) = \exp\left(\frac{\alpha z(t, y)^2}{4}\right), \quad \text{with } z(t, y) = \frac{y}{\sqrt{\langle t \rangle}} \text{ and } \alpha \in [1/4, 1/2].$$

With this and

$$M_m = \frac{\sqrt{m+1}}{m!},$$

define the Sobolev weighted semi-norms by

$$X_m = X_m(f,\tau) = \left\|\theta_\alpha \partial_x^m f\right\|_{L^2} \tau^m M_m, \quad D_m = D_m(f,\tau) = \left\|\theta_\alpha \partial_y \partial_x^m f\right\|_{L^2} \tau^m M_m,$$
$$Y_m = Y_m(f,\tau) = \left\|\theta_\alpha \partial_x^m f\right\|_{L^2} \tau^{m-1} m M_m.$$

Then the following space of analytic functions in the tangential variable x and weighted Sobolev in the normal variable y is defined by

$$X_{\tau,\alpha} = \left\{ f(t,x,y) \in L^2(\mathbb{H}; \theta_\alpha dxdy) : \|f\|_{X_{\tau,\alpha}} < \infty \right\},$$

with $\tau > 0$ and the norm

$$\|f\|_{X_{\tau,\alpha}} = \sum_{m\geq 0} X_m(f,\tau).$$

In addition, the following two semi-norms will also be used:

$$\|f\|_{D_{\tau,\alpha}} = \sum_{m\geq 0} D_m(f,\tau) = \left\|\partial_y f\right\|_{X_{\tau,\alpha}}, \quad \|f\|_{Y_{\tau,\alpha}} = \sum_{m\geq 1} Y_m(f,\tau).$$

Theorem 1.5.18 ([148]) *For any $\lambda \in [3/2, 2)$, there exists a small positive constant ε_* depending on $2-\lambda$. Under the assumption (H) on the background shear flow $\left(u_s(t,y), 0, \bar{b}, 0\right)$ with $\bar{b} \neq 0$, assume the initial data u_0 and b_0 satisfy*

$$\|u_0 - u_s(0,y)\|_{X_{2\tau_0,1/2}} \leq \varepsilon, \quad \left\|b_0 - \bar{b}\right\|_{X_{2\tau_0,1/2}} \leq \varepsilon,$$

for some given $\varepsilon \in (0, \varepsilon_*]$. Then there exists a unique solution (u_1, u_2, b_1, b_2) to the MHD boundary layer equations (1.5.49) *such that*

$$\left(u_1 - u_s(t,y), b_1 - \bar{b}\right) \in X_{\tau,\alpha}, \quad \alpha \in [1/4, 1/2],$$

with analyticity radius τ larger than $\tau_0/4$ in the time interval $[0, T_\varepsilon]$. And the lifespan T_ε has the following low bound estimate

$$T_\varepsilon \geq C\varepsilon^{-\lambda},$$

where the constant $C > 0$ is independent of ε.

Huang, Liu, and Yang [58] in 2019 were concerned with the motion of electrically conducting fluid governed by the 2D nonisentropic viscous compressible MHD system on the half plane with no-slip condition on the velocity field, perfectly conducting wall condition on the magnetic field and Dirichlet boundary condition on the temperature on the boundary, and obtained the local-in-time well-posedness of the following boundary layer system (1.5.50) in weighted Sobolev spaces:

$$\begin{cases} \partial_t \rho^e + \nabla \cdot (\rho^e \mathbf{u}^e) = 0, \\ \rho^e (\partial_t \mathbf{u}^e + (\mathbf{u}^e \cdot \nabla) \mathbf{u}^e) + \nabla p (\rho^e, \theta^e) - (\nabla \times \mathbf{H}^e) \times \mathbf{H}^e = 0, \\ c_V \rho^e (\partial_t \theta^e + (\mathbf{u}^e \cdot \nabla) \theta^e) + p (\rho^e, \theta^e) \nabla \cdot \mathbf{u}^e = 0, \\ \partial_t \mathbf{H}^e - \nabla \times (u^e \times \mathbf{H}^e) = 0, \quad \nabla \cdot \mathbf{H}^e = 0, \end{cases} \tag{1.5.50}$$

with the following natural and sufficient boundary conditions:

$$\left(u_2^e, h_2^e\right)\big|_{y=0} = \mathbf{0}.$$

Theorem 1.5.19 ([58]) *For the initial-boundary value problem with a smooth outflow $(U, \Theta, H, P)(t, x)$, assume that the initial data $\left(u_{1,0}, \theta_0, h_{1,0}\right)(x, y)$ and the boundary value $\theta^*(t, x)$ are smooth, compatible and satisfy*

$$\theta^*(t, x), \theta_0(x, y), h_{1,0}(x, y) \geq 2\delta, \quad \frac{1}{2}\left(h_{1,0}(x, y)\right)^2 \leq P(0, x) - 2\delta,$$

for $t \in [0, T]$, $(x, y) \in \mathbb{T} \times \mathbb{R}_+$ with some constant $\delta > 0$. Then there exists a $T_* \in (0, T]$ and a unique classical solution $(u_1, u_2, \theta, h_1, h_2)$ to the problem (1.5.50) in the domain

$D_{T_*} := \{(t, x, y) \mid t \in [0, T_*], x \in \mathbb{T}, y \in \mathbb{R}_+\}$ with the following properties:

(1) $\theta(t, x, y), h_1(t, x, y) \geq \delta$ and $(h_1(x, y))^2/2 \leq P(t, x) - \delta$ in D_{T_*};
(2) (u_1, θ, h_1), $\partial_y(u_1, \theta, h_1)$ and $\partial_y^2(u_1, \theta, h_1)$ are continuous and bounded in D_{T_*}, $\partial_t(u_1, \theta, h_1)$ and $\partial_x(u_1, \theta, h_1)$ are continuous and bounded in any compact set of D_{T_*};
(3) (u_2, h_2) and $\partial_y(u_2, h_2)$ are continuous and bounded in any compact set of D_{T_*}.

Liu, Xie, and Yang [89] in 2019 studied the well-posedness theory for the MHD boundary layer. The boundary layer equations are governed by the Prandtl-type equations that are derived from the incompressible MHD system with non-slip boundary condition on the velocity and perfectly conducting condition on the magnetic field. Under the assumption that the initial tangential magnetic field is not zero, they established the local-in time existence, uniqueness of solutions for the nonlinear MHD boundary layer equations.

One important problem about magnetohydrodynamics (MHD) is to understand the high Reynolds number limits in a domain with boundary. To this end, they considered the following initial boundary value problem for the 2D viscous MHD equations in a periodic domain $\{(t, x, y) : t \in [0, T], x \in \mathbb{T}, y \in \mathbb{R}_+\}$:

$$\begin{cases} \partial_t \mathbf{u}^\epsilon + (\mathbf{u}^\epsilon \cdot \nabla) \mathbf{u}^\epsilon - (\mathbf{H}^\epsilon \cdot \nabla) \mathbf{H}^\epsilon + \nabla p^\epsilon = \mu\epsilon \Delta \mathbf{u}^\epsilon, \\ \partial_t \mathbf{H}^\epsilon - \nabla \times (\mathbf{u}^\epsilon \times \mathbf{H}^\epsilon) = \kappa \, \epsilon \, \Delta \mathbf{H}^\epsilon, \\ \nabla \cdot \mathbf{u}^\epsilon = 0, \quad \nabla \cdot \mathbf{H}^\epsilon = 0. \end{cases} \tag{1.5.51}$$

Here, assume the viscosity and resistivity coefficients have the same order of a small parameter ϵ. $\mathbf{u}^\epsilon = (u_1^\epsilon, u_2^\epsilon)$ denotes the velocity vector, $\mathbf{H}^\epsilon = (h_1^\epsilon, h_2^\epsilon)$ denotes the magnetic field, and $p^\epsilon = \tilde{p}^\epsilon + |\mathbf{H}^\epsilon|^2/2$ denotes the total pressure with $\tilde{p}^\epsilon$ the pressure of the fluid. On the boundary, the non-slip boundary condition is imposed on velocity field

$$\mathbf{u}^\epsilon\big|_{y=0} = \mathbf{0},$$

and the perfectly conducting boundary condition on magnetic field

$$h_2^\epsilon\big|_{y=0} = \partial_y h_1^\epsilon\big|_{y=0} = 0.$$

Denote

$$\Omega := \{(x, y) : x \in \mathbb{T}, y \in \mathbb{R}+\}.$$

For any $l \in \mathbb{R}$, denote by $L_l^2(\Omega)$ the weighted Lebesgue space with respect to the spatial variables

$$L_l^2(\Omega) := \left\{ f(x, y) : \Omega \to \mathbb{R}, \|f\|_{L_l^2(\Omega)} := \left(\int_\Omega \langle y\rangle^{2l} |f(x, y)|^2 dxdy \right)^{\frac{1}{2}} < +\infty \right\},$$

with $\langle y\rangle = 1 + y$. Then, for any given $m \in \mathbb{N}$, denote by $H_l^m(\Omega)$, the weighted Sobolev spaces,

$$H_l^m(\Omega) := \left\{ f(x, y) : \Omega \to \mathbb{R}, \|f\|_{H_l^m(\Omega)} < +\infty \right\},$$

with the norm

$$\|f\|_{H_l^m(\Omega)} := \left(\sum_{m_1+m_2 \le m} \left\| \langle y\rangle^{l+m_2} \partial_x^{m_1} \partial_y^{m_2} f \right\|_{L^2(\Omega)}^2 \right)^{\frac{1}{2}}.$$

Theorem 1.5.20 ([89]) *Let $m \ge 5$ be an integer and $l \ge 0$ a real number. Assume that the outer flow $(U, H, P_x)(t, x)$ satisfies that for some $T > 0$,*

$$M_0 := \sum_{i=0}^{2m+2} \left(\sup_{0\le t\le T} \left\| \partial_t^i (U, H, P)(t, \cdot) \right\|_{H^{2m+2-i}(\mathbb{T}_x)} \right) < +\infty.$$

Also, assume the initial data $(u_{10}, h_{10})(x, y)$ satisfies

$$(u_{10}(x, y) - U(0, x), h_{10}(x, y) - H(0, x)) \in H_l^{3m+2}(\Omega),$$

and the compatibility conditions up to m-th order. Moreover, there exists a sufficiently small constant $\delta_0 > 0$ *such that* $\left|\langle y\rangle^{l+1}\partial_y^i (u_{10}, h_{10})(x, y)\right| \le (2\delta_0)^{-1}$ *for* $i = 1, 2$, $(x, y) \in \Omega$, $h_{10}(x, y) \ge 2\delta_0$. *Then, there exist a positive time* $0 < T_* \le T$ *and a unique solution* (u_1, u_2, h_1, h_2) *to the initial boundary value problem (1.5.51) such that*

$$(u_1 - U, h_1 - H) \in \bigcap_{i=0}^{m} W^{i,\infty}\left(0, T_*; H_l^{m-i}(\Omega)\right),$$

and

$$(u_2 + U_x y, h_2 + H_x y) \in \bigcap_{i=0}^{m-1} W^{i,\infty}\left(0, T_*; H_{-1}^{m-1-i}(\Omega)\right),$$

$$\left(\partial_y u_2 + U_x, \partial_y h_2 + H_x\right) \in \bigcap_{i=0}^{m-1} W^{i,\infty}\left(0, T_*; H_l^{m-1-i}(\Omega)\right).$$

Moreover, if $l > \frac{1}{2}$,

$$(u_2 + U_x y, h_2 + H_x y) \in \bigcap_{i=0}^{m-1} W^{i,\infty}\left(0, T_*; L^\infty\left(\mathbb{R}_{y,+}; H^{m-1-i}(\mathbb{T}_x)\right)\right).$$

Chen, Ren, Wang, and Zhang [10] in 2020 studied the long time well-posedness of 2D MHD boundary layer equation and proved that if the initial data satisfies

$$\|(u_0, h_0 - 1)\|_{H_\mu^{3,0} \cap H_\mu^{1,2}} \le \varepsilon,$$

then the life span of the solution is at least of order $\varepsilon^{2-\eta}$ for $\eta > 0$.

Let $h(t, x, y) = 1 + \widetilde{h}(t, x, y)$. Then $(u, \widetilde{h})$ satisfies the following system:

$$\begin{cases} \partial_t u + u\partial_x u + v\partial_y u - h\partial_x \tilde{h} - g\partial_y \tilde{h} - \partial_y^2 u = 0, \\ \partial_t \widetilde{h} + u\partial_x \widetilde{h} + v\partial_y \widetilde{h} - h\partial_x u - g\partial_y u - \partial_y^2 \widetilde{h} = 0, \\ \partial_x u + \partial_y v = 0, \quad \partial_x \widetilde{h} + \partial_y g = 0, \\ \left(u, v, \partial_y \widetilde{h}, g\right)\big|_{y=0} = 0 \quad \text{and} \quad \lim_{y\to+\infty} (u, \widetilde{h}) = (0, 0), \\ (u, \widetilde{h})\big|_{t=0} = \left(u_0, \widetilde{h}_0\right). \end{cases} \tag{1.5.52}$$

Now introduce the following weighted Sobolev space. For $k, \ell \in \mathbb{N}$, the space $H_\omega^{k,\ell}(\mathbb{R}_+^2)$ consists of all functions $f \in L_\omega^2$ satisfying

$$\|f\|_{H_\omega^{k,\ell}}^2 \stackrel{\text{def}}{=} \sum_{\alpha=0}^{k} \sum_{\beta=0}^{\ell} \left\| \partial_x^\alpha \partial_y^\beta f \right\|_{L_\omega^2}^2 < +\infty,$$

where $\|f\|_{L_\omega^p} = \|\omega(y) f(x, y)\|_{L^p}$ with $\omega(y)$ a positive weight function.

Theorem 1.5.21 ([10]) *Let $\mu = \exp\left(\frac{1+y^2}{8\langle t\rangle}\right)$ with $\langle t\rangle = 1 + t$. For any $\eta \in (0, 1)$, there exists an $\varepsilon > 0$ such that if the initial data $(u_0, \widetilde{h}_0)$ satisfies*

$$\left\|(u_0, \widetilde{h}_0)\right\|_{H_\mu^{3,0}} + \left\|(u_0, \widetilde{h}_0)\right\|_{H_\mu^{1,2}} \le \varepsilon,$$

then there exists a time $T_\varepsilon \ge \varepsilon^{-(2-\eta)}$ such that the system has a solution $(u, \widetilde{h})$ on $[0, T_\varepsilon]$, which satisfies

$$\begin{aligned}(u, \widetilde{h}) &\in L^\infty\left([0, T_\varepsilon]; H_\mu^{3,0}\left(\mathbb{R}_+^2\right) \cap H_\mu^{1,2}\left(\mathbb{R}_+^2\right)\right)\\ &\cap L^2\left([0, T_\varepsilon]; H_\mu^{3,1}\left(\mathbb{R}_+^2\right) \cap H_\mu^{1,3}\left(\mathbb{R}_+^2\right)\right).\end{aligned}$$

Liu, Wang, Xie, and Yang [83] in 2020 were concerned with the magnetic effect on the Sobolev solvability of boundary layer equations for the 2D incompressible MHD system without resistivity. The MHD boundary layer is described by the Prandtl type equations derived from the incompressible viscous MHD system without resistivity under the no-slip boundary condition on the velocity. Assuming that the initial tangential magnetic field does not degenerate, a local-in-time well-posedness in Sobolev spaces is proved without the monotonicity condition on the velocity field. Moreover, the authors of [83] showed that if the tangential magnetic field of shear layer is degenerate at one point, then the linearized MHD boundary layer system around the shear layer profile is ill-posed in the Gevrey function space provided.

Consider the initial-boundary value problem for the following 2D MHD boundary layer equations in the domain $\{(t, x, y) : t \in [0, T], x \in \mathbb{T}, y \in \mathbb{R}_+\}$:

$$\begin{cases} \partial_t u + u\partial_x u + v\partial_y u + \partial_x p - \partial_y^2 u - b_1\partial_x b_1 - b_2\partial_y b_1 = 0,\\ \partial_t b_1 + u\partial_x b_1 + v\partial_y b_1 - b_1\partial_x u - b_2\partial_y u = 0,\\ \partial_x u + \partial_y v = 0, \quad \partial_x b_1 + \partial_y b_2 = 0,\\ (u, v, b_2)|_{y=0} = 0, \quad \lim\limits_{y\to+\infty} (u, b_1)(t, x, y) = (U, B)(t, x),\\ (u, b_1)|_{t=0} = (u_0, b_0)(x, y), \end{cases} \tag{1.5.53}$$

where $\mathbb{T}$ stands for a torus or a periodic domain, $\mathbb{R}_+ = [0, +\infty)$, (u, v) and (b_1, b_2) are the velocity and magnetic boundary layer functions, respectively, and the known functions U, B, and p satisfy the Bernoulli law:

$$\begin{cases} \partial_t U + U\partial_x U + \partial_x p = B\partial_x B, \\ \partial_t B + U\partial_x B = B\partial_x U. \end{cases}$$

Theorem 1.5.22 ([83]) *Suppose that the outflow $(U, p, B)(t, x)$ in the equation is smooth, and the initial data and boundary conditions are smooth and compatible and satisfy*

$$b_0(x, y) \geq \delta_0,$$

for some positive constant δ_0. Then there exists a time T, such that the initial-boundary value problem (1.5.53) admits a unique smooth solution $(u, v, b_1, b_2)(t, x, y)$ satisfying

$$b_1(t, x, y) \geq \delta_0/2,$$

for all $t \in [0, T]$, $(x, y) \in \mathbb{T} \times \mathbb{R}_+$.

Li and Yang [77] in 2020 established the well-posedness of the MHD boundary layer system in Gevrey function space without any structural assumption. Compared to the classical Prandtl equation, the loss of tangential derivative comes from both the velocity and magnetic fields that are coupled with each other. By observing a new type of cancellation mechanism in the system for overcoming the loss derivative degeneracy, they showed that the MHD boundary layer system is well-posed with Gevrey index up to 3/2 in both two- and three-dimensional spaces. Consider the system

$$\begin{cases} \left(\partial_t + \vec{u} \cdot \nabla - \nu\partial_z^2\right) u_h - (\vec{f} \cdot \nabla) f_h = 0, \\ \partial_t \vec{f} - \nabla \times (\vec{u} \times \vec{f}) - \mu\partial_z^2 \vec{f} = 0, \\ \operatorname{div} \vec{u} = \operatorname{div} \vec{f} = 0, \\ \vec{u}|_{z=0} = (\partial_z f_h, f_z)|_{z=0} = \mathbf{0}, \quad (u_h, f_h)|_{z\to+\infty} = \mathbf{0}, \\ u_h|_{t=0} = u_{h,0}, \quad f_h|_{t=0} = f_{h,0}. \end{cases} \tag{1.5.54}$$

Let $\ell \geq 1$ be a given number. With a given integer $N \geq 0$ and a pair (ρ, σ), $\rho > 0$ and $\sigma \geq 1$, a Banach space $X_{\rho,\sigma,N}$ consists of all smooth vector-valued functions $\boldsymbol{A} = \boldsymbol{A}(x_h, z)$ with $(x_h, z) \in \mathbb{R}_+^d$ such that the Gevrey norm $\|\boldsymbol{A}\|_{\rho,\sigma,N} < +\infty$, where $\|\cdot\|_{\rho,\sigma,N}$ is defined below. Denote $\partial_{x_h}^{\alpha} = \partial_{x_1}^{\alpha_1} \cdots \partial_{x_{d-1}}^{\alpha_{d-1}}$ and define

$$\begin{aligned} \|\boldsymbol{A}\|_{\rho,\sigma,N} = & \sup_{\substack{0\leq j\leq N \\ |\alpha|+j\geq 7}} \frac{\rho^{|\alpha|+j-7}}{[(|\alpha|+j-7)!]^{\sigma}} \left\| \langle z \rangle^{\ell+j} \partial_{x_h}^{\alpha} \partial_z^j \boldsymbol{A} \right\|_{L^2(\mathbb{R}_+^d)} \\ & + \sup_{\substack{0\leq j\leq N \\ |\alpha|+j\leq 6}} \left\| \langle z \rangle^{\ell+j} \partial_{x_h}^{\alpha} \partial_z^j \boldsymbol{A} \right\|_{L^2(\mathbb{R}_+^d)}, \end{aligned}$$

where $\langle z\rangle = \left(1+|z|^2\right)^{1/2}$ and

$$\|\boldsymbol{A}\|_{L^2(\mathbb{R}^d_+)} \overset{\text{def}}{=} \left(\sum_{1\le j\le k} \|A_j\|^2_{L^2(\mathbb{R}^d_+)}\right)^{1/2},$$

for $\boldsymbol{A} = (A_1, \cdots, A_k)$. Here, σ is the Gevrey index.

Theorem 1.5.23 ([77]) *Let the dimension $d = 2$ or 3. Suppose the initial data $\left(u_{h,0}, f_{h,0}\right)$ in the system (1.5.54) belong to $X_{2\rho_0,\sigma,8}$ for some $1 < \sigma \le 3/2$ and some $0 < \rho_0 \le 1$, compatible with the boundary condition. Then the system (1.5.54) admits a unique solution $(u_h, f_h) \in L^\infty\left([0,T]; X_{\rho,\sigma,4}\right)$ for some $T > 0$ and some $0 < \rho < 2\rho_0$.*

1.5.2 MHD Boundary Layer Equations-Global Existence

In this subsection, we survey the results on the 2D MHD boundary layer equations.

Xie and Yang [147] in 2018 proved global existence of solutions with analytic regularity to the 2D MHD boundary layer equations in the mixed Prandtl and Hartmann regime. The following mixed Prandtl and Hartmann boundary layer equations from the classical incompressible MHD system:

$$\begin{cases} \partial_t u_1 + u_1\partial_x u_1 + u_2\partial_y u_1 = \partial_y b_1 + \partial_y^2 u_1, \\ \partial_y u_1 + \partial_y^2 b_1 = 0, \\ \partial_x u_1 + \partial_y u_2 = 0, \quad x \in \mathbb{R},\ y \in \mathbb{R}_+, \\ u_1(t=0,x,y) = u_{10}(x,y), \\ u_1|_{y=0} = 0, \quad u_2|_{y=0} = 0, \\ \lim\limits_{y\to+\infty} u_1 = \bar{u}, \qquad \lim\limits_{y\to+\infty} b_1 = \bar{b}. \end{cases} \tag{1.5.55}$$

Here, (u_1, u_2) denotes the velocity field of the boundary layer and b_1 is the corresponding tangential magnetic component. Integrating the equation of $(1.5.55)_2$ over y yields

$$-u_1(t,x,y) + \bar{u} = \partial_y b_1. \tag{1.5.56}$$

Thus, the equation (1.5.55) can be rewritten as

$$\begin{cases} \partial_t u_1 + u_1\partial_x u_1 + u_2\partial_y u_1 = -u_1(t,x,y) + \bar{u} + \partial_y^2 u_1, \\ \partial_x u_1 + \partial_y u_1 = 0. \end{cases} \tag{1.5.57}$$

Recall that the classical Hartmann boundary layer is given by

$$u_1 = (1 - e^{-y})\bar{u}, \quad u_2 = 0, \tag{1.5.58}$$

which is a steady solution to the system (1.5.57). Without loss of generality, set $\bar{u} = 1$ and denote the perturbation by (u, v):

$$u_1 = (1 - e^{-y}) + u, \quad u_2 = v. \tag{1.5.59}$$

Obviously, (u, v) satisfies

$$\begin{cases} \partial_t u + \left(1 - e^{-y} + u\right) \partial_x u + v \partial_y \left(-e^{-y} + u\right) = -u + \partial_y^2 u, \\ \partial_x u + \partial_y v = 0, \\ u_0(x, y) = u_{10}(x, y) - (1 - e^{-y}), \\ u_1|_{y=0} = 0, \quad u_2|_{y=0} = 0. \end{cases} \tag{1.5.60}$$

For some $r > 1$, denote an analytic weight M_m by

$$M_m = \frac{(m+1)r}{m!}.$$

Define

$$\begin{cases} X_m = \left\| e^{\alpha y} \partial_x^m g \right\|_{L^2(\mathbb{R}_+^2)} \tau^m M_m, \quad Z_m = \left\| e^{\alpha y} \partial_y \partial_x^m g \right\|_{L^2(\mathbb{R}_+^2)} \tau^m M_m, \\ Y_m = \left\| e^{\alpha y} \partial_x^m g \right\|_{L^2(\mathbb{R}_+^2)} \tau^{m-1/2} m^{1/2} M_m, \quad D_m = \left\| e^{\alpha y} \partial_x^m g \right\|_{L_y^\infty L_x^2} \tau^m M_m, \end{cases} \tag{1.5.61}$$

and

$$\|g\|_{X'_{\tau,\alpha}}^2 = \sum_{m \geq 0} X_m^2, \quad \|g\|_{Z^2_{\tau,\alpha}}^2 = \sum_{m \geq 0} Z_m^2, \tag{1.5.62}$$

$$\|g\|_{Y'_{\tau,\alpha}}^2 = \sum_{m \geq 0} Y_m^2, \quad \|g\|_{D^2_{\tau,\alpha}}^2 = \sum_{m \geq 0} D_m^2. \tag{1.5.63}$$

Here, τ denotes the analytic radius.

Theorem 1.5.24 ([147]) *Let the initial data $u_{10}(x, y)$ be a small perturbation of the Hartmann profile $(\bar{u}(1 - e^{-y}), 0)$ satisfying the compatibility conditions and*

$$\|\partial_y u_{10} + u_{10} - \bar{u}\|_{X^r_{\tau_0,\alpha}} \leq \delta_0, \tag{1.5.64}$$

with $r > 1, 0 < \alpha < \frac{\sqrt{2}}{2}$ *for some small constant* $\delta_0 > 0$. *Then there exists a unique global-in- time solution* (u_1, u_2) *to the problem* (1.5.55) *satisfying*

$$\|g\|_{X^r_{\tau(t),\alpha}} \le e^{-2(1-2\alpha^2)t}\delta_0, \quad \text{with} \quad \tau(t) > \tau_0/2, \tag{1.5.65}$$

for all time $t \ge 0$, *where* $g = \partial_y u_1 + u_1 - \bar{u}$.

Chen, Ren, Wang, and Zhang [9] in 2020 proved the global well-posedness of the 2D magnetic Prandtl model in the mixed Prandtl/Hartmann regime when the initial data is a small perturbation of the Hartmann layer in Sobolev spaces. They considered a 2D magnetic Prandtl model in $\mathbb{R}^2_+$:

$$\begin{cases} \partial_t u + u\partial_x u + v\partial_y u - \partial_y^2 u = \partial_y b, \\ \partial_y u + \partial_y^2 b = 0, \\ \partial_x u + \partial_y v = 0, \\ u|_{y=0} = v|_{y=0} = b|_{y=0} = 0, \\ \lim\limits_{y\to+\infty} u(t,x,y) = u^\infty, \qquad \lim\limits_{y\to+\infty} b(t,x,y) = b^\infty, \end{cases} \tag{1.5.66}$$

where (u, v) denotes the velocity field and b denotes the tangential magnetic field.

Thanks to the boundary conditions and $\partial_y u + \partial_y^2 b = 0$, we get

$$\partial_y b = -\left(u - u^\infty\right)$$

which, substituted into the first equation, gives a Prandtl type equation with an extra damping term $(u - u^\infty)$:

$$\begin{cases} \partial_t u + u\partial_x u + v\partial_y u - \partial_y^2 u + (u - u^\infty) = 0, \\ \partial_x u + \partial_y v = 0, \\ u|_{y=0} = v|_{y=0} = 0 \quad \text{and} \quad \lim\limits_{y\to+\infty} u(t,x,y) = u^\infty. \end{cases} \tag{1.5.67}$$

Now introduce the following weighted Sobolev space. For $k, \ell \in \mathbb{N}$, the space $H^{k,\ell}_\omega$ consists of all functions $f \in L^2_\omega$ satisfying

$$\|f\|^2_{H^{k,\ell}_\omega} := \sum_{\alpha=0}^{k}\sum_{\beta=0}^{\ell} \left\|\partial_x^\alpha \partial_y^\beta f\right\|^2_{L^2_\omega} < +\infty,$$

where $\|f\|_{L^p_\omega} = \|\omega(y) f(x,y)\|_{L^p}$ with $\omega(y)$ being a positive weight function.

The authors of [9] took the global stability of the Hartmann layer $u^s = u^\infty(1 - e^{-y})$, without loss of generality, let $u^\infty = 1$ and introduce the perturbation $\widetilde{u} = u - u^s$, which satisfies

$$\begin{cases} \partial_t \widetilde{u} + u\partial_x \widetilde{u} + v\partial_y u - \partial_y^2 \widetilde{u} + \widetilde{u} = 0, \\ \partial_x \widetilde{u} + \partial_y v = 0, \\ \widetilde{u}|_{y=0} = v|_{y=0} = 0 \quad \text{and} \quad \lim\limits_{y\to+\infty} \widetilde{u}(t,x,y) = 0, \\ \widetilde{u}|_{t=0} = \widetilde{u}_0(x,y). \end{cases} \tag{1.5.68}$$

Theorem 1.5.25 ([9]) *Assume that* $\widetilde{u}_0 \in H_\mu^{3,1} \cap H_\mu^{1,2} \left(\mu(y) = e^{\frac{y}{2}}\right)$ *with* $\tilde{u}_0|_{y=0} = 0$ *satisfies*

$$\|\tilde{u}_0\|_{H_\mu^{3,1}} + \|\tilde{u}_0\|_{H_\mu^{1,2}} \leqslant \varepsilon, \quad |\partial_x \tilde{u}_0(x,y)| + \left|\partial_x\partial_y \tilde{u}_0(x,y)\right| + \left|\partial_y^2 \tilde{u}_0(x,y)\right| \leqslant \varepsilon e^{-y}$$

for $(x, y) \in \mathbb{R}_+^2$.

Then there exists $\varepsilon_0 > 0$*, such that for any* $\varepsilon \in (0, \varepsilon_0)$, *the magnetic Prandtl model (1.5.68) has a unique global in time solution. Moreover, there holds that for any* $(t, x, y) \in [0, \infty) \times \mathbb{R}_+^2$,

$$\begin{aligned} &ce^{-y} \leqslant \partial_y u(t,x,y) \leqslant Ce^{-y}, \\ &|\partial_x \widetilde{u}(t,x,y)| + \left|\partial_x\partial_y \tilde{u}(t,x,y)\right| + \left|\partial_y^2 \tilde{u}(t,x,y)\right| \leqslant C\varepsilon e^{-y}, \\ &e^{\frac{t}{4}} \left(\|\widetilde{u}(t)\|_{H_\mu^{3,1}}^2 + \|\tilde{u}(t)\|_{H_\mu^{1,2}}^2 \right) + \int_0^t e^{\frac{s}{4}} \|\tilde{u}(s)\|_{H_\mu^{1,3}}^2 ds \leqslant C\varepsilon^2. \end{aligned}$$

Gong and Wang [47] in 2021 obtained the global existence of a weak solution to the following mixed Prandtl-Hartmann boundary layer problem (1.5.69) under the Crocco transformation, and the existence of a back-flow point of the Prandtl-Hartmann boundary layer when the initial velocity satisfies certain growth condition in the domain $D_T = \{0 \le t < T, 0 \le x \le X, y \ge 0\}$,

$$\begin{cases} \partial_t u + u\partial_x u + v\partial_y u = \partial_y^2 u + \sigma B_2 \partial_y b_1 - p_x, \\ \sigma B_2 \partial_y u + \partial_y^2 b_1 = 0, \\ \partial_x u + \partial_y v = 0, \\ u|_{t=0} = u_0(x,y), \quad u|_{x=0} = u_1(t,y), \\ (u,v)|_{y=0} = (0,0), \quad \lim\limits_{y\to+\infty} (u, b_1) = (U(t,,x), B_1(t,x)), \end{cases} \tag{1.5.69}$$

where (u, v) is the velocity field and b_1 is the tangential magnetic component. Here $\sigma > 0$ is the magnetic conductivity parameter, the given functions U, p, $\overrightarrow{B}(t, x) = (B_1, B_2)$ are the tangential velocity, hydrodynamic pressure, and magnetic field of the outer flow, respectively, which obey Bernoulli's law (1.0.3). The pressure is favorable for the classical Prandtl equation, i.e.,

$$\partial_x p \le 0. \tag{1.5.70}$$

Assume that

$$U(t,x) \ge 0, \;\; u_0(x,y) > 0, \;\; u_1(x,y) > 0, \;\; \forall 0 \le x \le X, \;\; y \ge 0, \;\; t \ge 0, \tag{1.5.71}$$

and the initial tangential velocity $u_0(x, y)$ and the upstream velocity $u_1(t, y)$ satisfy the following monotonicity condition

$$\partial_y u_0(x,y) > 0, \;\; \partial_y u_1(x,y) > 0, \;\; \forall 0 \le x \le X, \;\; y \ge 0, \;\; t \ge 0, \tag{1.5.72}$$

and

$$\sigma B_2^2 U - \partial_x p \ge 0. \tag{1.5.73}$$

The Crocco transformation is

$$\tau = t, \;\; \xi = x, \;\; \eta = \frac{u(t,x,y)}{U(t,x)}, \;\; w(\tau,\xi,\eta) = \frac{u_y(t,x,y)}{U(t,x)},$$

and denoting by $\alpha = \sigma B_2^2(t,x)$ and $\Omega = \{0 \le \xi \le X, 0 \le \eta < 1\}$, the original problem (1.5.69) can be transformed into the following initial-boundary value problem in $Q_T = [0,T) \times \Omega$:

$$\begin{cases} w_\tau + \eta U w_\xi + A w_\eta + B w = w^2 w_{\eta\eta}, \\ w|_{\tau=0} = w_0, \;\; w|_{\xi=0} = w_1, \\ w|_{\eta=1} = 0, \;\; (w_\eta + \frac{C}{w})|_{\eta=0} = 0, \end{cases} \tag{1.5.74}$$

where

$$A = (1-\eta)\left(\frac{U_t}{U} + \alpha\right) + (1-\eta^2)U_x, \;\; w_0 = \frac{u_{0y}}{U}, \;\; w_1 = \frac{u_{1y}}{U},$$

$$B = \frac{U_t}{U} + \eta U_x + \alpha, \;\; C = \alpha + U_x + \frac{U_t}{U} \ge 0.$$

Definition 1.5.26 ([47]) A function $w \in L^\infty(0,T; BV(\Omega)) \cap L^\infty(Q_T)$ is a weak solution to the problem (1.5.74), if the following conditions hold:

(1) There exist some positive constants c_1 and c_2 such that

$$c_1(1-\eta) \le w(\tau,\xi,\eta) \le c_2(1-\eta), \;\; \forall (\tau,\xi,\eta) \in Q_T; \tag{1.5.75}$$

(2) $w(\tau, \xi, \eta)$ satisfies the problem (1.5.74) in the weak sense:

$$\int_{Q_T} \left(\frac{1}{w}(\phi_\tau + \eta(U\phi)_\xi + B\phi) - w\phi_{\eta\eta} \right) d\xi d\eta d\tau$$
$$+ \int_0^T \int_0^X (A\frac{\phi}{w})(\tau, \xi, 0) d\xi d\tau$$
$$= - \int_0^T \int_0^1 \eta \frac{U(\tau, 0)}{w_1} \phi(\tau, 0, \eta) d\eta d\tau - \int_\Omega \frac{1}{w_0} \phi(0, \xi, \eta) d\xi d\eta$$
$$+ \int_0^T \int_0^X (\frac{C\phi}{w})(\tau, \xi, 0) d\xi d\tau$$

for any test function $\phi \in C^\infty(\overline{Q_T})$ satisfying

$$\phi|_{\tau=T} = \phi|_{\xi=T} = \phi|_{\eta=1} = \partial_\eta \phi|_{\eta=0} = 0.$$

Theorem 1.5.27 ([47] Well-posedness) *For any $T > 0$, assume that*

$$U \in C^2([0, T] \times [0, X]), \alpha \in C^1([0, T] \times [0, X]),$$

and the initial and boundary data $w_0 \in C^1(\Omega)$, $w_1 \in C^3([0, T] \times [0, 1])$ satisfy the compatibility condition. If the condition (1.5.73) holds for all $\tau \geq 0$ and $0 \leq \xi \leq L$, then problem (1.5.74) admits a global weak solution $w \in L^\infty((0, T); BV(\Omega))$ in the sense of Definition 1.5.26.

The condition (1.5.73) is fulfilled if the pressure $p(t, x)$ is favorable in the sense of (1.5.70). When the pressure $p(t, x)$ is not favorable, the condition (1.5.73) may still hold as long as $UB_2^2(t, x)$ is properly large. Thus this result shows that the magnetic field has certain stabilizing effect on hydrodynamic flow. When investigating the occurrence of a back-flow point in the Prandtl-Hartmann boundary layer, we have the following result.

Theorem 1.5.28 ([47])

(1) For any fixed $T > 0$ and $X > 0$, assume that the tangential velocity of the outer flow $U(t, x)$, and the initial and boundary data $u_0(x, y)$, $u_1(t, y)$ satisfy the conditions (1.5.71) and (1.5.72), and

$$U \in C^1([0, T] \times [0, X]), \quad B_2 \in C^1([0, T] \times [0, X]).$$

Then, for the solution $(u, v, b_1) \in C^2(D_T)$ to the problem (1.5.69), a first zero point of $\partial_y u(t, x, y)$, when the time evolves, must appear at the boundary $y = 0$ if it exists for some time $0 < t < T$.

(2) Furthermore, assume that

$$C(t,x)=\sigma B_2^2(t,x)+U_x+\frac{U_t}{U}\le 0,\quad \forall (t,x)\in[0,T]\times[0,X],$$

and for some constant C_* depending on X, T, U and B_2, if the initial velocity $u_0(x,y)$ satisfies

$$\int_0^X\int_0^{+\infty}\frac{u_{0y}(X-x)^{\frac{3}{2}}}{\sqrt{u_{oy^2+u_0^2}}}dxdy\ge C_*,$$

then there exists a back-flow point $(t^*,x^*)\in(0,T)\times[0,X]$, such that

$$\begin{cases}\partial_y u(t^*,x^*)=0,\\ \partial_y u(t,x,y)>0,\quad \forall 0<t<t^*,\quad x\in[0,X],\quad y\ge 0.\end{cases}$$

Ding, Ji, and Lin [22] in 2021 established the boundary layer theory for 2D steady viscous incompressible MHD equations under the assumption of a moving boundary at $\{y=0\}$ and convergence rates in Sobolev sense.

The following steady viscous incompressible magnetohydrodynamics system in $\Omega=[0,L]\times\mathbb{R}_+$ with moving boundary conditions on velocity field and the perfect conducting boundary conditions on magnetic field at $\{y=0\}$:

$$\begin{cases}(u\partial_x+v\partial_y)u-(h\partial_x+g\partial_y)h+p_x=\nu\varepsilon(\partial_{xx}+\partial_{yy})u,\\ (u\partial_x+v\partial_y)v-(h\partial_x+g\partial_y)g+p_y=\nu\varepsilon(\partial_{xx}+\partial_{yy})v,\\ (u\partial_x+v\partial_y)h-(h\partial_x+g\partial_y)u=k\varepsilon(\partial_{xx}+\partial_{yy})h,\\ (u\partial_x+v\partial_y)g-(h\partial_x+g\partial_y)v=k\varepsilon(\partial_{xx}+\partial_{yy})g,\\ \partial_x u+\partial_y v=0,\quad \partial_x h+\partial_y g=0,\\ (u,v,\partial_y h,g)|_{y=0}=(u_b,0,0,0),\end{cases}\tag{1.5.76}$$

where (u,v) and (h,g) are velocity and magnetic field, respectively, and the given constant u_b stands for the moving speed of the plate. Assume that the viscosity and resistivity coefficients have the same order of a small parameter $\varepsilon>0$. Suppose that the outer ideal MHD flows are prescribed by

$$(u_e^0,v_e^0,h_e^0,g_e^0,p_e^0)(x,y)$$

as $\varepsilon\to 0$, $(u_e^0,h_e^0)(x,0)=(\overline{u}_e^0,\overline{h}_e^0)$.

Following the idea of Prandtl with the scaling boundary layer variable (x',y')

$$x'=x,\quad y'=\frac{y}{\sqrt{\varepsilon}},$$

and letting

$$(u^\varepsilon, v^\varepsilon, g^\varepsilon, h^\varepsilon, p^\varepsilon)(x', y') = (u, \frac{v}{\sqrt{\varepsilon}}, h, \frac{g}{\sqrt{\varepsilon}}, p)(x, y),$$

we can reduce problem (1.5.76) to the form

$$\begin{cases} (u^\varepsilon\partial_x + v^\varepsilon\partial_y)u^\varepsilon - (h^\varepsilon\partial_x + g^\varepsilon\partial_y)h^\varepsilon + p_x^\varepsilon = \nu\varepsilon\partial_{xx}u^\varepsilon + \nu\partial_{yy}u^\varepsilon, \\ (u^\varepsilon\partial_x + v^\varepsilon\partial_y)v^\varepsilon - (h^\varepsilon\partial_x + g^\varepsilon\partial_y)g^\varepsilon + \frac{p_y^\varepsilon}{\varepsilon} = \nu\varepsilon\partial_{xx}v^\varepsilon + \nu\partial_{yy}v^\varepsilon, \\ (u^\varepsilon\partial_x + v^\varepsilon\partial_y)h^\varepsilon - (h^\varepsilon\partial_x + g^\varepsilon\partial_y)u^\varepsilon = k\varepsilon\partial_{xx}h^\varepsilon + k\partial_{yy}h^\varepsilon, \\ (u^\varepsilon\partial_x + v^\varepsilon\partial_y)g^\varepsilon - (h^\varepsilon\partial_x + g^\varepsilon\partial_y)v^\varepsilon = k\varepsilon\partial_{xx}g^\varepsilon + k\partial_{yy}g^\varepsilon, \\ \partial_x u^\varepsilon + \partial_y v^\varepsilon = 0, \quad \partial_x h^\varepsilon + \partial_y g^\varepsilon = 0, \\ (u^\varepsilon, v^\varepsilon, \partial_y h^\varepsilon, g^\varepsilon)|_{y=0} = (u_b, 0, 0, 0). \end{cases} \tag{1.5.77}$$

Define the following norm of $\mathcal{X}$:

$$\begin{aligned} \|u^\varepsilon, v^\varepsilon, h^\varepsilon, g^\varepsilon\|_{\mathcal{X}} :=& \|\{u_y^\varepsilon, v_y^\varepsilon, \sqrt{\varepsilon}(h_x^\varepsilon, g_x^\varepsilon)\} \cdot y\|_{L^2} + \|u_y^\varepsilon, v_y^\varepsilon, \sqrt{\varepsilon}(h_x^\varepsilon, g_x^\varepsilon) \cdot y\|_{L^2} \\ &+ \|\{u_{yy}^\varepsilon, v_{yy}^\varepsilon, \sqrt{\varepsilon}(h_{xy}^\varepsilon, g_{xy}^\varepsilon), \varepsilon(h_{xx}^\varepsilon, g_{xx}^\varepsilon)\} \cdot y\|_{L^2} \\ &+ \|u^\varepsilon, v^\varepsilon, h^\varepsilon, g^\varepsilon\|_B + \varepsilon^{\frac{\gamma}{2}}\|u_y^\varepsilon, v_y^\varepsilon, \sqrt{\varepsilon}(h^\varepsilon, g^\varepsilon)\|_{L^\infty}, \end{aligned}$$

where the boundary term is defined by

$$\|u^\varepsilon, v^\varepsilon, h^\varepsilon, g^\varepsilon\|_B := \|\{u_y^\varepsilon, \sqrt{\varepsilon}(u_x^\varepsilon, h_x^\varepsilon)\} \cdot y\|_{L^2(x=L)} + \|\sqrt{\varepsilon}(u_x^\varepsilon, h_x^\varepsilon)\|_{L^2(x=L)}.$$

Theorem 1.5.29 ([22]) *Let $u_b > 0$ be a constant tangential velocity of the viscous MHD flows on the boundary $\{y = 0\}$, and the given smooth non-shear ideal MHD flows $(u_e^0, v_e^0, h_e^0, g_e^0, p_e^0)(x, y)$ satisfy the following hypotheses:*

$$\begin{aligned} &0 < c_0 \le h_e^0 \ll u_e^0 \le C_0 \le \infty, \\ &\|\tfrac{v_e^0}{y}\|_{L^\infty} \ll 1, \\ &\|y^k\nabla^m(v_e^0, g_e^0)\|_{L^\infty} \le \infty \ \textit{for sufficiently large}\ k, m \ge 0, \\ &\|y^k\nabla^m(u_e^0, h_e^0)\|_{L^\infty} \le \infty \ \textit{for sufficiently large}\ k \ge 0, m \ge 1, \\ &\|\langle y\rangle\partial_y(u_e^0, h_e^0)\|_{L^\infty} \le \delta_0 \quad \textit{suitable small}\ \delta_0 \ge 0. \end{aligned}$$

In addition, let $m \ge 5$ be an integer, the outer ideal MHD flows are assumed to enjoy

$$M_0 := \sum_{i=0}^{m+2} \|(\overline{u}_e^0, \overline{h}_e^0, \overline{p}_e^0)(x)\|_{H^{m+2-i}(0,L)} < +\infty.$$

Moreover, for some positive constants ϑ_0, η_0 *and small* σ_0, *suppose that*

$$
\begin{cases}
\overline{u}_e^0 + u_p^0(0, y) > \overline{h}_e^0 > \overline{h}_e^0 + h_p^0(0, y) \geq \vartheta_0, \\
|u_e^0(0, y) + u_p^0(0, y)| \gg |h_e^0(0, y) + h_p^0(0, y)| \geq \eta_0, \\
|\langle y \rangle^{l+1} \partial_y (u_p^0, h_p^0)(0, y)| \leq \frac{1}{2}\sigma_0, \\
|\langle y \rangle^{l+1} \partial_y^2 (u_p^0, h_p^0)(0, y)| \leq \frac{1}{2}\vartheta_0^{-1}.
\end{cases}
$$

Then there exist the remainder solutions $(u^\varepsilon, v^\varepsilon, h^\varepsilon, g^\varepsilon)$ *in the space* $\mathcal{X}$ *satisfying*

$$
\|(u^\varepsilon, v^\varepsilon, h^\varepsilon, g^\varepsilon)\|_{\mathcal{X}} \lesssim 1
$$

in $[0, L] \times [0, +\infty)$, *where the positive number* L *is sufficiently small relative to universal constant.*

Theorem 1.5.30 ([22]) *Under the hypotheses stated in Theorem 1.5.29 with the profile* (u_p^0, h_p^0), *it holds that*

$$
\|(u - u_e^0 - u_p^0, h - h_e^0 - h_p^0)\|_{L^\infty} + \|(v - v_e^0 v, g - g_e^0)\|_{L^\infty} \lesssim \sqrt{\varepsilon}
$$

where (u_p^0, h_p^0) *is the solution of Prandtl equations (1.5.77).*

Liu and Zhang [94] in 2021 proved the global existence and the large time decay estimate of solutions to the 2D MHD boundary layer equations (1.5.78) with small initial data in the upper space $\mathbb{R}_+^2 = \{(x, y) : x \in \mathbb{R}, y \in \mathbb{R}_+\}$:

$$
\begin{cases}
\partial_t u_1 - \partial_y^2 u_1 + u_1 \partial_x u_1 + u_2 \partial_y u_1 + \partial_x p = b_1 \partial_x b_1 + b_2 \partial_y b_1, \\
\partial_t b_1 - k \partial_y^2 b_1 + u_1 \partial_x b_1 + u_2 \partial_y b_1 = b_1 \partial_x u_1 + b_2 \partial_y u_1, \\
u_1|_{y=0} = u_2|_{y=0} = 0, \quad \partial_y b_1|_{y=0} = b_2|_{y=0} = 0, \\
\lim\limits_{y \to +\infty} u_1 = U_1, \quad \lim\limits_{y \to +\infty} b_1 = B_1, \\
u_1|_{t=0} = u_{1,0}, \quad b_1|_{t=0} = b_{1,0},
\end{cases}
\tag{1.5.78}
$$

where (u_1, u_2) and (b_1, b_2) represent the velocity of fluid and the magnetic field, respectively, $k > 0$ is a constant which represents the difference between the Reynolds number and the magnetic Reynolds number, and $(U_1, B_1, p)(t, x)$ are the traces of the tangential fields and pressure of the outflow on the boundary, which satisfy Bernoulli's law

$$
\begin{cases}
\partial_t U_1 + U_1 \partial_x U_1 + \partial_x p = B_1 \partial_x B_1, \\
\partial_t B_1 + U_1 \partial_x B_1 = B_1 \partial_x U_1.
\end{cases}
\tag{1.5.79}
$$

Take a cutoff function $\mathcal{X} \in C^\infty[0, \infty]$ with

$$
\mathcal{X}(y) = \begin{cases} y & \text{if } y \geq 2, \\ 0 & \text{if } y \leq 2, \end{cases}
$$

and make the following function transformations:

$$
\begin{cases} u = u_1 - \mathcal{X}'(y)U \quad \text{and} \quad v = u_2 + \mathcal{X}(y)\partial_x U, \\ b = b_1 - \mathcal{X}'(y)B - \overline{B}_x \quad \text{and} \quad h = b_2 + \mathcal{X}(y)\partial_x B, \end{cases} \tag{1.5.80}
$$

where $U = U_1$ and $B = B_1 - \overline{B}_x$. Then in view of (1.5.78) and (1.5.79), (u, v, b, h) solves

$$
\begin{cases} \partial_t u - \partial_y^2 u - \overline{B}_k \partial_x b + u\partial_x u - b\partial_x b + v\partial_y u - h\partial_y b \\ \qquad + \mathcal{X}'(U\partial_x u - B\partial_x b) + \mathcal{X}'(\partial_x U u - \partial_x B b) \\ \qquad + \mathcal{X}(-\partial_x U \partial_y u + \partial_x B \partial_y b) + \mathcal{X}''(Uv - Bh) = m_U, \\ \partial_t b - k\partial_y^2 b - \overline{B}_k \partial_x u + u\partial_x b - b\partial_x u + v\partial_y b - h\partial_y u \\ \qquad + \mathcal{X}'(U\partial_x b - B\partial_x u) + \mathcal{X}'(\partial_x B u - \partial_x U b) \\ \qquad + \mathcal{X}(-\partial_x U \partial_y b + \partial_x B \partial_y u) + \mathcal{X}''(Bv - Uh) = m_B, \\ \partial_x u + \partial_y v = 0, \quad \partial_x b + \partial_y h = 0, \\ u|_{y=0} = v|_{y=0} = 0, \quad \partial_y b|_{y=0} = h|_{y=0} = 0, \\ \lim\limits_{y\to+\infty} u = 0, \quad \lim\limits_{y\to+\infty} b = 0, \\ \lim\limits_{y\to+\infty} v = 0, \quad \lim\limits_{y\to+\infty} h = 0, \\ u|_{t=0} = u_0 = u_{1,0} - \mathcal{X}'U_0, \quad b|_{t=0} = b_0 = b_{1,0} - \mathcal{X}'B_0 - \overline{B}_k, \end{cases} \tag{1.5.81}
$$

where $(U_0(x), B_0(x)) = (U(0, x), B(0, x))$ and the terms (m_U, m_B) are given by

$$
\begin{cases} m_U = (1 - \mathcal{X}')(\partial_t U - \overline{B}_k \partial_x B) + \mathcal{X}'''U + (1 - \mathcal{X}'^2 + \mathcal{X}\mathcal{X}'')(U\partial_x U - B\partial_x B), \\ m_B = (1 - \mathcal{X}')(\partial_t B - \overline{B}_k \partial_x U) + \mathcal{X}'''B + (1 - \mathcal{X}'^2 - \mathcal{X}\mathcal{X}'')(U\partial_x B - B\partial_x U). \end{cases} \tag{1.5.82}
$$

Note that there exist two potential functions (ϕ, φ) so that $(u, b) = \partial_y(\phi, \varphi)$ and $(v, h) = \partial_x(\phi, \varphi)$, then integrating (u, b) equations of (1.5.81) with respect to y variable over $[y, \infty]$, we get

$$
\begin{cases} \partial_t \varphi - \partial_y^2 \varphi - \overline{B}_k \partial_x \psi + u\partial_x \varphi - b\partial_x \psi + 2\int_y^\infty (\partial_x \varphi \partial_y u - \partial_x \psi \partial_y b) dy' \\ \qquad + \mathcal{X}'(U\partial_x \varphi - B\partial_x \psi) \\ \qquad + 2\int_y^\infty \mathcal{X}''(U\partial_x \varphi - B\partial_x \psi) dy' + \mathcal{X}'(\partial_x U \varphi - \partial_x B \psi) \\ \qquad + \mathcal{X}(-\partial_x U \partial_y \varphi + \partial_x B \partial_y \psi) = m_U, \\ \partial_t \psi - k\partial_y^2 \psi - \overline{B}_k \partial_x \varphi + u\partial_x \psi - b\partial_x \varphi + \mathcal{X}'(U\partial_x \varphi - B\partial_x \varphi) \\ \qquad + \mathcal{X}(-\partial_x U b + \partial_x B u) = m_B, \\ \varphi|_{y=0} = \psi|_{y=0} = 0, \quad \partial_y b|_{y=0} = h|_{y=0} = 0, \\ \lim\limits_{y\to+\infty} u = 0, \quad \lim\limits_{y\to+\infty} \varphi = \lim\limits_{y\to+\infty} \psi = 0, \\ \varphi|_{t=0} = \varphi_0 = -\int_y^\infty u_0 dy', \quad \psi|_{t=0} = \psi_0 = -\int_y^\infty b_0 dy'. \end{cases} \tag{1.5.83}
$$

Define

$$\Delta_k^h a = \mathcal{F}^{-1}(\varphi 2^{-k}|\xi|\hat{a}), \quad S_k^h a = \mathcal{F}^{-1}(\mathcal{X}2^{-k}|\xi|\hat{a}),$$

where $\mathcal{F}^{-1}$ and $\hat{a}$ denote the partial Fourier transform of the distribution a with respect to x variable, and

$$\|u\|_{\mathcal{B}^{s,0}} := \|(2^{ks}\|\Delta_k^h a(t)\|_{L_+^2})_{k\in\mathbb{Z}}\|_{L^1(\mathbb{Z})}, \tag{1.5.84}$$

$$\|\|_{\tilde{L}^p(T_0,T;\mathcal{B}^{s,0})} := \sum_{k\in\mathbb{Z}} 2^{ks}\left(\int_{T_0}^T \|\Delta_k^h a(t)\|_{L_+^2}^p\,dt\right)^{\frac{1}{p}}. \tag{1.5.85}$$

Here the phase function Φ is defined by

$$\Phi(t,\xi) := (\delta - \lambda\theta(t))|\xi|,$$

and the weighted function $\Psi(t, y)$ is determined by

$$\Psi(t,y) := \frac{y^2}{8\langle t\rangle}, \quad \langle t\rangle = 1+t,$$

which satisfies

$$\partial_t\Psi + 2(\partial_y\Psi)^2 = 0.$$

Theorem 1.5.31 ([94]) *Let* $k \in [0, 12]$, $\overline{B}_k = \begin{cases} 1, \ if\ k=1, \\ 0, \ otherwise, \end{cases}$ *and* $\varepsilon, \delta > 0$. *Assume that the far field states* (U, B) *satisfy*

$$\|\langle t\rangle^{\frac{9}{4}} e^{\delta|D_x|}(U,B)\|_{\widetilde{L}^\infty(\mathbb{R}^+;\mathcal{B}_h^{\frac{3}{2}})} + \|\langle t\rangle^{\frac{7}{4}} e^{\delta|D_x|}(\partial_t U, \partial_t B, U, B)\|_{\widetilde{L}^\infty(\mathbb{R}^+;\mathcal{B}_h^{\frac{1}{2}})} \le \varepsilon.$$

Let the initial data $(u_0, b_0, \varphi_0, \psi_0)$ *satisfy the compatibility condition:* $u_0|_{y=0} = \partial_y b_0|_{y=0} = 0$, $\int_0^\infty u_0 dy = \int_0^\infty b_0 dy = 0$ *and*

$$\|e^{\frac{y^2}{8}} e^{\delta|D_x|}(u_0, b_0, \varphi_0, \psi_0)\|_{\mathcal{B}^{\frac{1}{2},0}} < \infty, \quad \|e^{\frac{y^2}{8}} e^{\delta|D_x|}(G_0, H_0)\|_{\mathcal{B}^{\frac{1}{2},0}} \le \sqrt{\varepsilon},$$

where $G_0 = u_0 + \frac{y}{2\langle t\rangle}\varphi_0$ *and* $H_0 = b_0 + \frac{y}{2k\langle t\rangle}\psi_0$. *Then there exist positive constants* $\lambda\varsigma$ *and* $\varepsilon_0(\lambda, \varsigma, k\delta)$ *so that for* $\varepsilon \le \varepsilon_0$ *and* $l_k = \frac{k(2-k)}{4} \in [0, \frac{1}{4}]$, *the system (1.5.83) has a unique global solution* (u, b) *which satisfies* $\sup\limits_{t\in[0,\infty]} \theta(t) \le \frac{\delta}{2\lambda}$, *and*

$$\|e^{\Psi}(u,b)\Phi\|_{\widetilde{L}^\infty_t(\mathcal{B}^{\frac{1}{2},0})} + \sqrt{l_k}\|e^{\Psi}\partial_y(u,b)\Phi\|_{\widetilde{L}^2_t(\mathcal{B}^{\frac{1}{2},0})} \le \|e^{\frac{y^2}{8}}e^{\delta|D_x|}(u_0,b_0)\|_{\mathcal{B}^{\frac{1}{2},0}} + C\sqrt{\varepsilon}.$$

Furthermore, for any $t > 0$ *and* $\gamma \in [0, 1]$, *there holds that*

$$\|\langle\tau\rangle^{\frac{1}{2}+l_k-\varsigma\varepsilon}e^{\Psi}(u,b)\Psi\|_{\widetilde{L}^\infty_t(\mathcal{B}^{\frac{1}{2},0})} + \|\langle\tau\rangle^{\frac{1}{2}+l_k-\varsigma\varepsilon}e^{\Psi}\partial_y(u,b)\Psi\|_{\widetilde{L}^2_{[\frac{t}{2},t]}(\mathcal{B}^{\frac{1}{2},0})}$$
$$\le C(\|e^{\frac{y^2}{8}}e^{\delta|D_x|}(u_0,b_0,\varphi_0,\psi_0)\|_{\mathcal{B}^{\frac{1}{2},0}} + \sqrt{\varepsilon}),$$
$$\|\langle\tau\rangle^{\frac{1}{2}+l_k-\varsigma\varepsilon}e^{\Psi}(G,H)\Psi\|_{\widetilde{L}^\infty_t(\mathcal{B}^{\frac{1}{2},0})} + \|\langle\tau\rangle^{\frac{1}{2}+l_k-\varsigma\varepsilon}e^{\Psi}\partial_y(G,H)\Psi\|_{\widetilde{L}^2_{[\frac{t}{2},t]}(\mathcal{B}^{\frac{1}{2},0})}$$
$$\le C\sqrt{\varepsilon}(1 + \|e^{\frac{y^2}{8}}e^{\delta|D_x|}(u_0,b_0,\varphi_0,\psi_0)\|_{\mathcal{B}^{\frac{1}{2},0}}),$$
$$\|\langle\tau\rangle^{\frac{1}{2}+l_k-\varsigma\varepsilon}e^{\gamma\Psi}(u,b)\Psi\|_{\widetilde{L}^\infty_t(\mathcal{B}^{\frac{1}{2},0})} + \|\langle\tau\rangle^{\frac{1}{2}+l_k-\varsigma\varepsilon}e^{\gamma\Psi}\partial_y(u,b)\Psi\|_{\widetilde{L}^2_{[\frac{t}{2},t]}(\mathcal{B}^{\frac{1}{2},0})}$$
$$\le C\sqrt{\varepsilon}(1 + \|e^{\frac{y^2}{8}}e^{\delta|D_x|}(u_0,b_0,\varphi_0,\psi_0)\|_{\mathcal{B}^{\frac{1}{2},0}}).$$

1.5.3 Compressible MHD Boundary Layer Equations

Jiang and Zhang [61] in 2017 studied the non-resistive limit of the global solutions with large data and the global well-posedness of the following compressible non-resistive MHD equations:

$$\begin{cases} \rho_t + (\rho u)_x = 0, \\ (\rho u)_t + (\rho u^2 + P(\rho) + \frac{1}{2}b^2)_x = \lambda u_{xx}, \\ b_t + (ub)_x = \nu b_{xx}, \end{cases} \tag{1.5.86}$$

with the following initial and boundary conditions:

$$\begin{cases} (\rho, u, b)(x, 0) = (\rho_0, u_0, b_0)(x), x \in [0, 1], \\ u(0,t) = u(1,t) = 0, b(0,t) = b_1(t), b(1,t) = b_2(t), t \ge 0, \end{cases} \tag{1.5.87}$$

where $P(\rho) = A\rho^\gamma$, $A > 0$, $\gamma > 1$.

The one-dimensional compressible, viscous, non-resistive MHD equations are as follows:

$$\begin{cases} \bar{\rho}_t + (\bar{\rho}\bar{u})_x = 0, \\ (\bar{\rho}\bar{u})_t + (\bar{\rho}\bar{u}^2 + P(\bar{\rho}) + \frac{1}{2}\bar{b}^2)_x = \lambda \bar{u}_{xx}, \\ \bar{b}_t + (\bar{u}\bar{b})_x = \nu \bar{b}_{xx}, \end{cases} \tag{1.5.88}$$

which subjects to the following initial and boundary conditions:

$$\begin{cases} (\bar{\rho}, \bar{u}, \bar{b})(x, 0) = (\rho_0, u_0, b_0)(x), x \in [0, 1], \\ \bar{u}(0, t) = \bar{u}(1, t) = 0, t \geq 0. \end{cases} \tag{1.5.89}$$

Theorem 1.5.32 ([61]) *Assume that* $P(\bar{\rho}) = A\bar{\rho}^\gamma$, *with* $A > 0$, $\gamma > 1$, *and that the initial data* (ρ_0, u_0, b_0) *satisfy*

$$\inf_{x\in[0,1]} \rho_0(x) \geq 0, \ (\rho_0, b_0) \in H^1, \ u_0 \in H_0^1, \ u_0(0) = u_0(1) = 0. \tag{1.5.90}$$

Then for any $0 < T < \infty$, *there exist a positive constant* C *and a unique global strong solution* $(\bar{\rho}, \bar{u}, \bar{b})$ *to the initial-boundary value problems* (1.5.88)–(1.5.89) *on* $(0, 1) \times [0, T]$, *such that*

$$0 < C^{-1} \leq \bar{\rho} \leq C, \ \forall (x, t) \in (0, 1) \times [0, T], \tag{1.5.91}$$

and

$$(\bar{\rho}, \bar{b}) \in H^1, \ \bar{u} \in H_0^1, \ (\bar{\rho}_t, \bar{b}_t) \in L^2, \ (\bar{u}_t, \bar{u}_{xx}) \in L^2(0, T; L^2(\Omega)). \tag{1.5.92}$$

Assume that $P(\bar{\rho}) = A\bar{\rho}^\gamma$, *with* $A > 0$, $\gamma > 1$. *Moreover, in addition to* (1.5.90), *suppose that for any given* $T \in (0, \infty)$,

$$(b_1, b_2) \in C^1([0, T]), \ b_0(0) = b_1(0), \ b_0(1) = b_2(0). \tag{1.5.93}$$

Then for each fixed $\nu > 0$, *there exist a positive constant* C *and a unique global strong solution to the initial-boundary value problem* (1.5.86)–(1.5.87) *on* $(0, 1) \times [0, T]$, *such that*

$$0 < C^{-1} \leq \rho \leq C, \ \forall (x, t) \in (0, 1) \times [0, T],$$

and

$$\sup_{0\leq t\leq T} (\|u_x\|_{L^2}^2 + \|b\|_{L^\infty}^2 + \nu^{1/2}\|\rho_x\|_{L^2}^2 + \nu^{1/2}\|b_x\|_{L^2}^2)$$
$$+ \int_0^T \left(\|u_t\|_{L^2}^2 + \nu^{1/2}\|u_{xx} + \nu^{3/2}\|b_{xx}\|_{L^2}^2\right) ds \leq C.$$

Moreover, as $\nu \to 0$,

$$\begin{cases} (\rho, u, b) \to (\bar{\rho}, \bar{u}, \bar{b}), \ \ strongly\ in\ L^\infty(0,T;L^2), \\ \nu b_x \to 0,\ u_x \to \bar{u}_x\ \ strongly\ in\ L^\infty(0,T;L^2) \end{cases}$$

and

$$\sup_{0\le t\le T} (\|u-\bar{u}\|_{L^2}^2 + \|b-\bar{b}\|_{L^2}^2 + \|\rho-\bar{\rho}\|_{L^2}^2) + \int_0^T \|u_x-\bar{u}_x\|_{L^2}^2 ds \le C\nu^{1/2}.$$

Here, $C > 0$ *is a positive constant independent of* ν.

Theorem 1.5.33 ([61]) *In addition to* (1.5.90) *and* (1.5.93)*, assume further that* $(\rho_0, u_0, b_0) \in H^2$. *Then any function* $\delta(\nu) \in (0, 1/2)$*, satisfying*

$$\delta(\nu) \to 0\ and\ \frac{\delta(\nu)}{\nu^{1/2}} \to \infty,\ as\ \nu \to 0,$$

such that

$$\lim_{\nu\to 0}\left(\|\rho-\bar{\rho}\|^2_{L^\infty(0,T;C(\bar{\Omega}_{\delta(\nu)}))} + \|b-\bar{b}\|^2_{L^\infty(0,T;C(\bar{\Omega}_{\delta(\nu)}))}\right) = 0,$$

and

$$\lim_{\nu\to 0}\left(\|\rho-\bar{\rho}\|^2_{L^\infty(0,T;C(\bar{\Omega}))} + \|b-\bar{b}\|^2_{L^\infty(0,T;C(\bar{\Omega}))}\right) = 0,$$

provided $b_i(t) \ne \bar{b}_i(t)(i = 1, 2)$*, where* $\bar{b}_1(t)$ *and* $\bar{b}_2(t)$ *denote the boundary values of* $\bar{b}(x,t)$ *on the boundaries* $x = 0, 1$ *, respectively. Here* $C(\bar{\Omega}_{\delta(\nu)}) := \{x \in \Omega | \delta < x < 1-\delta\}$.

Under the assumption that initial tangential magnetic field is not zero and density is a small perturbation of the outer constant flow in supernorm, Gao, Huang, and Yao [35] in 2018 established the local-in-time existence and uniqueness of inhomogeneous incompressible MHD boundary layer equation in weighted Sobolev spaces by the energy method:

$$\begin{cases} \partial_t\rho + u_1\partial_x\rho + u_2\partial_y\rho = 0, \\ \rho\partial_t u_1 + \rho u_1\partial_x u_1 + \rho u_2\partial_y u_1 - \kappa\partial_y^2 u_1 = h_1\partial_x h_1 + h_2\partial_y h_1 = 0, \\ \partial_t h_1 - \partial_y(u_2h_1 - u_1h_2) - \kappa\partial_y^2 h_1 = 0, \\ \partial_x u_1 + \partial_y u_2 = 0,\ \partial_x h_1 + \partial_y h_2 = 0, \end{cases} \tag{1.5.94}$$

where the density $\rho := \rho(t, x, y)$, velocity field (u_1, u_2), the magnetic field (h_1, h_2) are unknown functions. The boundary conditions for equation (1.5.94) are given by

$$\begin{cases} (u_1, u_2, \partial_y h_1, h_2)|_{y=0} = 0, \\ \lim\limits_{y\to+\infty} \rho = \lim_{y\to+\infty} u_1 = \lim_{y\to+\infty} h_1 = 1. \end{cases} \tag{1.5.95}$$

We supplement the MHD boundary layer equations (1.5.94) with the initial data

$$(\rho, u_1, h_1)(0, x, y) = (\rho_0, u_{10}, h_{10})(x, y), \tag{1.5.96}$$

satisfying

$$0 < m_0 \le \rho_0 \le M_0 < +\infty, \tag{1.5.97}$$

and

$$\|(\rho_0, u_{10}, h_{10})\|_{B_l^m} \le C_0 < +\infty, \tag{1.5.98}$$

where $m_0, M_0, C_0 > 0$ are positive constants. Define the functional space B_l^m for a pair of function $(\rho, u_1, h_1)(t, x, y)$ as follows:

$$\begin{aligned} B_l^m &= \Big\{(\rho - 1, u_1 - 1, h_1 - 1) \in L^\infty([0, T]; \\ &\qquad L_l^2(\Omega)) : esssup_{0\le t\le T} \|(\rho, u_1, h_1)\|_{B_l^m} < +\infty\Big\}, \\ B^{m,l}_{BL,app} &= \Big\{(\rho - 1, u_1 - 1, h_1 - 1) \in H_l^{4k} | \partial_t(\rho, u_1, h_1), k = 1, 2, \ldots, m\Big\} \end{aligned}$$

and

$$\begin{cases} B^{m,l}_{BL} = \text{the closure of } B^{m,l}_{BL,app} \text{in the norm } \|\cdot\|_{B_l^m}, \\ L_l^2(\Omega) \\ \quad = \Big\{f(x, y) : \Omega \to \mathbb{R}, \|f\|^2_{L_l^2(\Omega)} := \int_\Omega \langle y\rangle^{2l} |f|^2 dxdy < +\infty, \langle y\rangle = 1 + y\Big\}, \\ L_l^\infty(\Omega) \\ \quad = \Big\{f(x, y) : \Omega \to \mathbb{R}, \|f\|_{L_l^\infty(\Omega)} := \sup\limits_{(x,y)\in\Omega} |\langle y\rangle^l f| dxdy < +\infty, \langle y\rangle = 1 + y\Big\}. \end{cases} \tag{1.5.99}$$

Theorem 1.5.34 ([35]) *Let $m \geq 5$ be an integer and $l \geq 2$ be a real number. Assume the initial data $(\rho_0, u_{10}, h_{10}) \in B_{BL}^{m,l}$ given in (1.5.99) and satisfying (1.5.97)–(1.5.98). Moreover, there exists a small constant $\delta_0 > 0$ such that*

$$h_{10} \geq 2\delta_0, \quad for\ all\ (x, y) \in \Omega$$

and

$$\|\rho_0 - 1\|_{L_0^\infty(\Omega)} \leq \frac{2l-1}{16}\delta_0^2, \quad \|\partial_y u_{10}\|_{L_1^\infty(\Omega)} \leq (2\delta_0)^{-1}.$$

Then, there exist a positive time $0 < T_0$ and a unique solution $(\rho, u_1, u_2, h_1, h_2)$ to the initial boundary value problems (1.5.94)–(1.5.95), such that

$$\begin{aligned}
&\sup_{0\leq t\leq T_0} \|(\rho-1, u_1-1, h_1-1)(t)\|_{H^m}^2 + \|\partial_y(\rho, u_1, h_1)(t)\|_{H_1^{m-1}}^2 + \|\partial_y \rho(t)\|_{H_1^{1,\infty}}^2 \\
&+ \int_0^{T_0} \|\partial_y(\sqrt{\mu}u_1, \sqrt{\kappa}h_1)(t)\|_{H_1^m}^2 dt + \int_0^{T_0} \|\partial_y(\sqrt{\mu}u_1, \sqrt{\kappa}h_1)(t)\|_{H_1^{m-1}}^2 dt \\
&\leq C_0 < +\infty,
\end{aligned}$$

here C_0 depends only on l, δ_0.

Ye and Zhang [155] in 2020 considered that the 3D MHD flow with spatial variables (x, y, z) being moving only in the longitudinal direction x and uniformly in the transverse direction (y, z). Then, based on the specific choice of dependent variables $\rho = \rho(x, t)$, $\boldsymbol{u} = (u, \boldsymbol{w})(x, t)$, $\boldsymbol{B} = (b, \boldsymbol{b})(x, t)$ and $\theta = \theta(x, t)$, the compressible viscous heat-conducting MHD equations can be written in the following form:

$$\begin{cases}
\rho_t + (\rho u)_x = 0, \\
(\rho u)_t + \left(\rho u^2 + R\rho\theta + \frac{|\boldsymbol{b}|^2}{2}\right)_x = (\lambda u_x)_x, \\
(\rho \boldsymbol{w})_t + (\rho u \boldsymbol{w} - \boldsymbol{b})_x = (\mu \boldsymbol{w}_x)_x, \\
\boldsymbol{b}_t + (u\boldsymbol{b} - \boldsymbol{w})_x = (v\boldsymbol{b}_x)_x, \\
c_V(\rho\theta)_t + c_V(\rho u\theta)_x + R\rho\theta u_x = (\kappa(\theta)\theta_x)_x + \lambda u_x^2 + \mu|\boldsymbol{w}_x|^2 + v|\boldsymbol{b}_x|^2,
\end{cases} \tag{1.5.100}$$

where the unknown functions $\rho, u \in \mathbb{R}$, $\boldsymbol{w} = (w^1, w^2) \in \mathbb{R}^2$, $\boldsymbol{b} = (b^1, b^2) \in \mathbb{R}^2$ and θ are the density of the fluid, the longitudinal velocity, the transverse velocity, the transverse magnetic field, and the temperature, respectively. The constants $\lambda > 0$, $\mu \geq 0$ are the bulk and shear viscosity coefficients, respectively, $v > 0$ is the resistivity coefficient, $\kappa = \kappa(\theta) > 0$ is the heat-conductivity coefficient, and $R, c_V > 0$ are the perfect gas constant and the specific heat at constant volume, respectively.

Theorem 1.5.35 ([155]) *For any given $0 < T < \infty$, assume that*

$$\begin{cases} \rho_0 > 0, \quad \theta_0 > 0, \quad (\rho_0, u_0, \boldsymbol{w}_0, \boldsymbol{b}_0, \theta_0) \in H^2, \quad (\boldsymbol{w}_1, \boldsymbol{w}_2)(t) \in C^2([0, T]), \\ (u_0, \boldsymbol{b}_0)|_{x=0,1} = 0, \quad \theta_{0x}|_{x=0,1} = 0, \quad \boldsymbol{w}_0|_{x=0} = \boldsymbol{w}_1(0), \quad \boldsymbol{w}_0|_{x=1} = \boldsymbol{w}_2(0). \end{cases}$$

Assume also that there exist two positive numbers $\underline{\kappa}$ and $\bar{\kappa}$ such that for some $q > 0$,

$$\kappa(\cdot) \in C^2(0, \infty) \text{ and } \underline{\kappa}\xi^q \le \kappa(\xi) \le \bar{\kappa}\xi^q, \quad \forall \xi \ge 0.$$

Then there exists a global unique strong solution $(\rho, u, \boldsymbol{w}, \boldsymbol{b}, \theta)$ (resp.$(\bar{\rho}, \bar{u}, \overline{\boldsymbol{w}}, \overline{\boldsymbol{b}}, \bar{\theta})$) to the problem (1.5.100) with fixed $\mu > 0$ on $(0, 1) \times [0, T]$, such that

$$\begin{aligned} &\sup_{0\le t\le T} \left(\|(\rho - \bar{\rho}, u - \bar{u}, \boldsymbol{b} - \overline{\boldsymbol{b}}, \theta - \bar{\theta})(t)\|_{H^1}^2 + \|(\boldsymbol{w} - \overline{\boldsymbol{w}})(t)\|_{L^2}^2\right) \\ &+ \int_0^T \left(\left\|(u - \bar{u}, \boldsymbol{b} - \overline{\boldsymbol{b}}, \theta - \bar{\theta})_t\right\|_{L^2}^2 + \|(u - \bar{u}, \theta - \bar{\theta})\|_{H^2}^2\right) dt \le C\mu^{1/2}, \end{aligned}$$

where $C > 0$ is a positive constant independent of μ.

1.5.4 Ill-Posedness of the Boundary Layer of MHD Equations

Liu, Xie, and Yang [87] in 2018 investigated the sufficient degeneracy in the tangential magnetic field at a non-degenerate critical point of the tangential velocity field of shear flow, which indeed yields instability as for the classical Prandtl equations without magnetic field. They also partially showed the necessity of the non-degeneracy in the tangential magnetic field for the stability of the boundary layer of MHD in 2D at least in Sobolev spaces:

$$\begin{cases} u_t + u_s \partial_x u + v \partial_y u_s - b_s \partial_x b - g \partial_y b_s = \kappa \partial_y^2 u, \\ b_t - b_s \partial_x u + v \partial_y b_s + u_s \partial_x b - g \partial_y u_s = \nu \partial_y^2 b, \\ \partial_x u + \partial_y v = 0, \ \partial_x b + \partial_y g = 0, \\ (u, v, b, g)|_{y=0} = 0, \quad \lim\limits_{y\to+\infty} u = \lim\limits_{y\to+\infty} b = 0, \\ u(0, x, y) = u_0(x, y), \ b(0, x, y) = b_0(x, y). \end{cases} \tag{1.5.101}$$

Theorem 1.5.36 ([87] Ill-Posedness in the Sobolev-Type Spaces)

(1) Let

$$u_s - \bar{u} \in \bigcap_0^1 C^0(\mathbb{R}_+; W_\alpha^{4-2i,\infty}), \ b_s - \bar{b} \in \bigcap_0^2 C^0(\mathbb{R}_+; W_\alpha^{6-2i,\infty}).$$

Suppose the initial velocity u_0^s *has a non-degenerate critical point* $a > 0$, *i.e.,*

$$u'_{s0}(a) = 0, \; u''_{s0}(a) \neq 0.$$

Moreover

$$\partial_y^j b_{s0}(a) = 0, \quad j = 0, \ldots 6.$$

Then, there exists $\sigma > 0$, *such that for any* $\gamma > 0$,

$$\sup_{0 \le s \le t \le \gamma} \| e^{-\sigma(t-s)\sqrt{|\partial_x|}} T(t,s) \|_{L(H^m, H^{m-\mu})} = +\infty, \forall m \ge 0, \mu \in [0, \frac{1}{2}).$$

(2) *Moreover, there exist some background shear flow* $(u_s(t, y), b_s(t, y))$ *and* $\sigma > 0$ *such that for any* $\gamma > 0$, *it holds that*

$$\sup_{0 \le s \le t \le \gamma} \| e^{-\sigma(t-s)\sqrt{|\partial_x|}} T(t,s) \|_{L(H_1^m, H_2^m)} = +\infty, \forall m_1, m_2 \ge 0.$$

Theorem 1.5.37 ([87] Well-Posedness in the Sobolev-Type Spaces) *Let* $u_s - \bar{u} \in C^1(\mathbb{R}_+; W_\alpha^{1,\infty})$, $b_s = u_s$ *(or* $-u_s$*) and* $\kappa = \nu$. *Then the linearized operator* $T(t,s)$ *satisfies*

$$\|T(t,s)\|_{L(H^{m+1}, H^m)} = \sup_{(u_0, b_0) \in K_{\alpha,\beta}} \frac{\|T(t,s)(u_0, b_0)\|_{H^m}}{\|(u_0, b_0)\|_{H^{m+1}}} \le C < \infty, \forall m \ge 0.$$

Here, the constant $C > 0$ *depends only on time* t *and the background shear flows* u_s *and* b_s.

1.6 Other Boundary Layer Equations

Korpela [64] in 1985 used the Dorodnitsyn-Howarth transformation across the boundary layer, which reduces to the Blasius equation

$$f''' + ff'' = 0, \quad f(0) = f'(0) = f'(\infty) - 1 = 0, \tag{1.6.1}$$

the solution of which for small values of η is

$$f = \frac{1}{2}\alpha\eta^2 - \frac{1}{5!}\alpha^2\eta^5 + \frac{11}{8!}\alpha^3\eta^8 - \ldots.$$

From the coefficients, it is obvious that the large Prandtl number expansion breaks down in the range occupied by gases. Thus the result has little utility in aeronautical applications.

Joseph and Lefloch [62] in 1999 considered the initial-boundary-value problem of conservation laws:

$$\partial_t u + \partial_x f(u) = 0, \quad x > 0, \quad t > 0. \tag{1.6.2}$$

They derived conditions satisfied by the boundary layer, which take the form of a family of boundary entropy inequalities and a boundary-layer equation.

Theorem 1.6.1 ([62]) *The following statements hold for all convex entropy pairs U, F associated with the system (1.6.2), all functions $\eta \in BV(\mathbb{R}_+)$, and any bounded interval (T_1, T_2).*

(1) When $\eta(t) \geq 0$, the distribution

$$y \mapsto \int_{T_1}^{T_2} \langle \mu_{y,t}, F \rangle \eta(t) dt - \frac{\partial}{\partial y} \int_{T_1}^{T_2} \langle \mu_{y,t}, U \rangle \eta(t) dt$$

is in fact a function of locally bounded variation and thus is defined pointwise as a right-continuous function. There exists a Young measure $\mu_{0,t}$, such that the following limit exists and is given by $\mu_{0,t}$:

$$\lim_{y \to 0^+} \int_{T_1}^{T_2} \langle \mu_{y,t}, U \rangle \eta(t) dt = \int_{T_1}^{T_2} \langle \mu_{0,t}, U \rangle \eta(t) dt.$$

When $\eta(t) \geq 0$, the function

$$x \mapsto \int_{T_1}^{T_2} \langle \nu_{x,t}, F \rangle \eta(t) dt$$

has locally bounded variation. There exists a Young measure $\nu_{0,t}$, the "trace" of $\nu_{x,t}$ at $x = 0$, such that the following limit exists and is given by $\nu_{0,t}$:

$$\lim_{x \to 0^+} \int_{T_1}^{T_2} \langle \nu_{x,t}, U \rangle \eta(t) dt = \int_{T_1}^{T_2} \langle \nu_{0,t}, F \rangle \eta(t) dt.$$

When $(U, F) = (u_j, f_j)$, $1 \leq J \leq N$, all of the results above still hold when the function η has no specific sign.

(2) *For all* $0 < y_1 < y_2$ *and for* $t \in \mathbb{R}_+$, *the following inequalities hold in the sense of distributions:*

$$\begin{aligned} F(u_B) + \nabla U(u_B)\big(\langle \nu_{0,t}, f\rangle - f(u_B)\big) &\geq \langle \mu_{y_1,t}, F\rangle - \partial_y \mu_{y_1,t}, U\rangle \\ &\geq \langle \mu_{y_2,t}, F\rangle - \partial_y \mu_{y_2,t}, U\rangle \\ &\geq \langle \nu_{0,t}, F\rangle. \end{aligned}$$

(3) *Moreover,*

$$\mu_{0,t} = \delta_{uB(t)} \quad a.e.\ t \in \mathbb{R}_+$$

and, when $\eta \geq 0$,

$$\lim_{y\to\infty}\left(\int_{T_1}^{T_2} \langle \mu_{y,t}, F\rangle \eta(t)dt - \frac{d}{\partial y}\int_{T_1}^{T_2} \langle \mu_{y,t}, U\rangle \eta(t)dt\right) \geq \int_{T_1}^{T_2} \langle \nu_{0,t}, F\rangle \eta(t)dt.$$

Here $\langle \cdot, \cdot\rangle$ *denotes inner product of* $BV(\mathbb{R}_+)$.

Oleinik and Samokhin [106] in 1999 considered the boundary layer in non-Newtonian flows. They considered the stationary plane-parallel symmetric flow of a pseudo-plastic fluid in the boundary layer is described by the system

$$\begin{cases} u\frac{\partial u}{\partial x} + v\frac{\partial u}{\partial y} = \nu\frac{\partial}{\partial y}\left(|\frac{\partial u}{\partial y}|^{n-1}\frac{\partial u}{\partial y}\right) + U(x)\frac{dU}{dx}, \quad 0 < n < 1, \\ \frac{\partial u}{\partial x} + \frac{\partial v}{\partial y} = 0, \end{cases} \tag{1.6.3}$$

in the domain $D = \{0 < x < X, 0 < y < \infty\}$, with the boundary conditions

$$\begin{aligned} &u(0, y) = 0, \quad u(x, 0) = 0, \quad v(x, 0) = 0, \\ &u(x, y) \to U(x), \quad \text{as } y \to +\infty. \end{aligned} \tag{1.6.4}$$

Assume that $U(0) = 0$, $U(x) > 0$ for $x > 0$; $\nu > 0$, and

$$U(x) = xV(x), \quad v_0(x) = x^{\frac{n-1}{n+1}} v_1(x),$$

where $V(x) > 0$, $v_1(x)$ are bounded functions. In (1.6.3), introduce new independent variables

$$\xi = x, \quad \eta = \frac{u}{U}, \quad w(\xi, \eta) = \frac{|u_y|^{n-1}u_y}{x^{\frac{n-1}{n+1}} U},$$

then problems (1.6.3)–(1.6.4) reduce to the equation

$$\nu n V^{\frac{1-n}{n}} |w|^{\frac{1-n}{n}} w_{\eta\eta} - \eta\xi V w_\xi + A w_\eta + B w = 0, \tag{1.6.5}$$

in the domain $\Omega = \{0 < \xi < X, 0 < \eta < 1\}$, with the boundary conditions

$$w|_{\eta=1} = 0, \quad (\nu w|w|_\eta^{\frac{1+n}{n}} - v_1 w|w|^{\frac{1-n}{n}} + C)|_{\eta=0} = 0, \tag{1.6.6}$$

where

$$A = (\eta^2 - 1)(V + \xi V_x), \quad B = -\eta\left(\frac{2n}{1+n}V + \xi V_x\right), \quad C = V^{\frac{n-1}{n}}(V + \xi V_x).$$

Theorem 1.6.2 ([106]) *Assume that*

$$U(x) = x(a + xa_1(x)), \quad v_0(x) = x^{\frac{n-1}{n+1}}(b + xb_1(x)),$$

where a *=const.*> 0, b *=const.; $U > 0$ for $x > 0$; $a_1, a_{1x}, a_{1xx}, b_1, b_{1x}$ are bounded. Then problems (1.6.3)–(1.6.4) in the domain D, with X depending on U, v_0, n, admit one and only one solution u, v with the following properties: $u_y > 0$ for $y \geq 0$ and $x > 0$;*

$$\frac{u}{U}, \quad \frac{u_y^n}{x^{\frac{n-1}{n+1}}U}$$

are bounded and continuous in $\overline{D}$; $u > 0$ for $y > 0$ and $x > 0$;

$$u \to U, \quad \frac{u_y^n}{x^{\frac{n-1}{n+1}}U} \to 0 \ \ as \ \ y \to \infty;$$

u_x, u_y, u_{yy} are bounded and continuous in D; v is continuous in $\overline{D}$ with respect to y for $x > 0$; v is continuous in x and y inside D;

$$\frac{u_{yy}}{x^{\frac{n-1}{n+1}}u_y^{2-n}}$$

is continuous in D with respect to y; the following estimates hold:

$$x^{\frac{n-1}{n+1}}UY\left(\frac{u}{U}\right)e^{-C_3x} \leq u_y \leq x^{\frac{n-1}{n+1}}UY\left(\frac{u}{U}\right)e^{C_4x},$$

$$x^{\frac{n-1}{n+1}}Y_n\left(\frac{u}{U}\right)e^{C_5x} \leq \frac{u_{yy}}{u_y^{2-n}} \leq x^{\frac{n-1}{n+1}}Y_n\left(\frac{u}{U}\right)e^{-C_7x},$$

$$\left(M_2^{\frac{1}{n}}\frac{n-1}{n+1}e^{\frac{C_4}{n}}x^{\frac{n-1}{n(n+1)}}U^{\frac{1-n}{n}}y + 1\right)^{\frac{n+1}{n-1}} \leq 1 - \frac{u}{U}$$

$$\leq \left(M_1^{\frac{1}{n}}\frac{n-1}{n+1}e^{-\frac{C_3}{n}}x^{\frac{n-1}{n(n+1)}}U^{\frac{1-n}{n}}y + 1\right)^{\frac{n+1}{n-1}}.$$

Oleinik and Samokhin [106] in 1999 considered system (1.6.3) in the domain $D = \{0 < x < X, 0 < y < \infty\}$, whose the boundary conditions are more complicated than those in (1.6.4),

$$\begin{aligned} &u(0,y) = u_0(y), \quad u(x,0) = 0, \quad v(x,0) = v_0(x), \\ &u(x,y) \to U(x), \quad \text{as } y \to +\infty. \end{aligned} \tag{1.6.7}$$

Theorem 1.6.3 ([106]) *Let $U(x), U_x(x), v_0(x)$ be continuous functions on the interval $0 \le x \le X$, and let $U_x(x) \ge$const., and assume also that $u_0(y)$ is continuous for $0 \le y < \infty$; $u_0(0) = 0$; $u_0(y) > 0$ for $y > 0$; $u_0(y) \to U(0)$ as $y \to \infty$; the derivative $u_{0y}(y)$ is continuous; $u_{0y}(y) > 0$; there exists a weak derivative $u_{0yy}(y)$ and*

$$K_1^{\frac{1}{n}} U^{\frac{1}{n}}(0) \left(1 - \frac{u_0(y)}{U(0)}\right)^{\frac{2}{n+1}} \le u_{0y}(y) \le K_2^{\frac{1}{n}} U^{\frac{1}{n}}(0) \left(1 - \frac{u_0(y)}{U(0)}\right)^{\frac{2}{n+1}},$$

$$\int_0^1 u_{0y}^{2n-3}(y) u_{0yy}^2(y) \left(1 - \frac{u_0(y)}{U(0)}\right)^{\frac{2(1-n)}{n+1}} dy < \infty.$$

Then problems (1.6.3) and (1.6.7) have a weak solution $u(x,y), v(x,y)$ in the following sense: $u(x,y)$ is bounded, measurable, and continuous in $\overline{D}$ with respect to y; $u > 0$ for $y > 0$; $u = 0$ for $y = 0$; $u_y(x,y)$ is bounded in D,

$$K_{11}^{\frac{1}{n}} U^{\frac{1}{n}}(x) \left(1 - \frac{u(x,y)}{U(x)}\right)^{\frac{2}{n+1}} \le u_y(x,y) \le K_{12}^{\frac{1}{n}} U^{\frac{1}{n}}(x) \left(1 - \frac{u(x,y)}{U(x)}\right)^{\frac{2}{n+1}},$$

$$U(x) - U(x) \left(1 + K_{12}^{\frac{1}{n}} \frac{1-n}{n+1} U^{\frac{1-n}{n}}(x) y\right)^{\frac{n+1}{n-1}} \le u(x,y)$$

$$\le U(x) - U(x) \left(1 + K_{12}^{\frac{1}{n}} \frac{1-n}{n+1} U^{\frac{1-n}{n}}(x) y\right)^{\frac{n+1}{n-1}};$$

then the weak derivatives u_{yy}, u_{yyy} exist and satisfy the inequalities

$$\int_D u_y^{2n-3}(y) u_{yy}^2(y) \left(1 - \frac{u}{U}\right)^{\frac{2(1-n)}{n+1}} dxdy < \infty,$$

$$\int_D u^2 \left(1 - \frac{u}{U}\right)^{\frac{6-2n}{n+1}} \left(\frac{n u_y u_{yyy} + n(n-1) u_{yy}}{u_y^{4-n}}\right)^2 u_y dxdy < \infty;$$

$$\int_{D \cap \{(x,y): 0 \le x \le X - \varepsilon\}} v(x,y)(1+y)^{\frac{n+3}{n-1}} dxdy < \infty, \quad \varepsilon > 0,$$

here $v(x, y)$ is measurable in D and equations (1.6.3) and the boundary conditions (1.6.7) for $x = 0$, $y = 0$ hold in terms of the following integral identities

$$\int_D \left[\nu U^{\frac{1-n}{n}} \frac{u_y \phi_{yy} - \phi_y u_{yy}}{u_y^{2-n}} - u\phi_x U^{\frac{1-n}{n}} + \left(\frac{u^2}{U^2} - 1 \right) \frac{\phi_y}{u_y} U^{\frac{1}{n}} U_x \right.$$
$$\left. + \frac{n-1}{n} u\phi\, U^{\frac{1-2n}{n}} U_x \right] dxdy$$
$$- \int_0^X U^{\frac{1-n}{n}}(x)\phi(x,0)v_0(x)dx - \int_0^\infty u_0(y)\phi(0,y)dy = 0,$$
$$\int_D \left(\nu n u_y^{n-1} u_{yy}\phi + \frac{1}{2}u^2\phi_x - v u_y \phi + U U_x \phi \right) dxdy$$
$$+ \int_0^\infty \frac{1}{2} u_0^2(y)\phi(0,y)U^{\frac{1-n}{n}}(0)dy = 0,$$

for any $\phi(x, y)$ such that the derivatives $\phi_x, \phi_y, \phi_{yy}$ are continuous and bounded in $\overline{D}$, and $\phi(X, y) = 0$, $\phi_y(x, 0) = 0$,

$$|\phi(x, y)| \le K_{13}(1 + K_{14}y)^{\frac{n+1}{n-1}} \quad as \quad y \to \infty.$$

The solution (u, v) of problems (1.6.3) and (1.6.7) with all these properties is unique.

Oleinik and Samokhin [106] in 1999 considered the nonstationary motion of a pseudo-plastic fluid in a symmetric boundary layer is described by a system of the form $(0 < n < 1)$:

$$\begin{cases} \frac{\partial u}{\partial t} + u\frac{\partial u}{\partial x} + v\frac{\partial u}{\partial y} = \nu \frac{\partial}{\partial y}\left(|\frac{\partial u}{\partial y}|^{n-1} \frac{\partial u}{\partial y} \right) + \frac{\partial U}{\partial t} + U(x)\frac{dU}{dx}, \\ \frac{\partial u}{\partial x} + \frac{\partial v}{\partial y} = 0, \end{cases} \tag{1.6.8}$$

in the domain $D = \{0 < t < \infty, 0 < x < X, 0 < y < \infty\}$, with the boundary conditions

$$\begin{aligned} &u(0,x,y) = u_0(x,y), \;\; u(t,0,y) = 0, \;\; u(t,x,0) = 0, \;\; v(t,x,0) = v_0(t,x), \\ &u(t,x,y) \to U(x,y), \;\; \text{as} \;\; y \to \infty. \end{aligned} \tag{1.6.9}$$

Assume that

$$U(x) = xV(t,x), \;\; V(t,x) > 0, \;\; v_0(t,x) = x^{\frac{n-1}{n+1}} v_1(x),$$

where $V, v_1(x), V_x, V_t$ are bounded functions. In (1.6.8), introduce new independent variables

$$\tau = t, \ \ \xi = x, \ \ \eta = \frac{u}{U}, \ \ w(\tau, \xi, \eta) = \frac{|u_y|^{n-1}u_y}{x^{\frac{n-1}{n+1}}U},$$

then problems (1.6.8)–(1.6.9) reduce to the equation

$$\nu n V^{\frac{1-n}{n}}|w|^{\frac{1+n}{n}}w_{\eta\eta} - w_\tau - \eta\xi V w_\xi + A w_\eta + B w = 0, \tag{1.6.10}$$

in the domain $\Omega = \{0 < \tau < \infty, 0 < \xi < X, 0 < \eta < 1\}$, with the boundary conditions

$$\begin{aligned} &w|_{\tau=0} = w_0 = \frac{|u_{0y}|^{n-1}u_{0y}}{x^{\frac{n-1}{n+1}}U}, \\ &w|_{\eta=1} = 0, \ \ (\nu w|w|_\eta^{\frac{1-n}{n}} - v_1 w|w|^{\frac{1-n}{n}} + C)|_{\eta=0} = 0, \end{aligned} \tag{1.6.11}$$

where

$$A = (\eta^2 - 1)(V + \xi V_x) + (\eta - 1)\frac{V_t}{V}, \ \ B = -\eta\left(\frac{2n}{1+n}V + \xi V_x\right) - \frac{V_t}{V},$$

$$C = V^{\frac{n-1}{n}}(V + \xi V_x + \frac{V_t}{V}).$$

Theorem 1.6.4 ([106]) *Assume that*

$$U(t, x) = x(a + x a_1(t, x)), \ \ v_0(t, x) = x^{\frac{n-1}{n+1}}(b + x b_1(t, x)),$$

where a =const.> 0, b =const., $a_1(t, x)$ has bounded second-order derivatives, $b_1(t, x)$ has bounded first-order derivatives for $0 \le t < \infty, 0 \le x \le X$. Let the function $u_0(x, y)$ be such that

$$\frac{|u_{0y}|^{n-1}u_{0y}}{x^{\frac{n-1}{n+1}}U}$$

satisfies the conditions

$$Ye^{-M_4\xi} \le w_0 \le Ye^{M_5\xi}, \ \ |w_{0\xi}| \le M_{11}(1-\eta)^{\frac{2n}{1+n}}, \ \ Y_\eta e^{M_{12}\xi} \le w_{0\eta} \le Y_\eta e^{-M_{15}\xi},$$

$$-M_{14}(1-\eta)^{\frac{2n}{1+n}} \le \nu n V^{\frac{1-n}{n}}(0, x) w_0^{\frac{1+n}{n}} w_{0\eta\eta} + A(0, \xi, \eta) w_{0\eta}$$

$$+B(0, \xi, \eta) w_0 \le M_{15} x (1-\eta)^{\frac{2n}{1+n}},$$

$$(\nu w_{0\eta} w_0^{\frac{1-n}{n}} - v_1 w_0^{\frac{1-n}{n}} + C)|_{\tau=0, \eta=0} = 0.$$

Then for some X depending on U, u_0, v_0, problems (1.6.8)–(1.6.9) in D have one and only one solution u, v with the following properties: $u > 0$ for $y > 0$ and $x > 0$; $u \to U$ as $y \to \infty$; $u_y > 0$ for $y \geq 0$;

$$\frac{u}{U}, \quad \frac{u_y^n}{x^{\frac{n-1}{n+1}}U}$$

are bounded and continuous in $\overline{D}$;

$$\frac{u_y^n}{x^{\frac{n-1}{n+1}}U} \to 0 \ \ as \ \ y \to \infty;$$

$u_y, u_x, u_t, u_{yy}, v_y$ are bounded and continuous in D with respect to y; v is continuous in $\overline{D}$ with respect to y and bounded for bounded y; equations (1.6.8) hold almost everywhere in D; the following inequalities are satisfied:

$$x^{\frac{n-1}{n+1}}UY(\frac{u}{U})e^{-C_3x} \leq u_y^n \leq x^{\frac{n-1}{n+1}}UY(\frac{u}{U})e^{C_4x},$$

$$U - U\left(1 + M_1^{\frac{1}{n}}\frac{1-n}{n+1}\exp\left(-\frac{C_3x}{n}\right)x^{\frac{n-1}{n(n+1)}}U^{\frac{1-n}{n}}(x)y\right)^{\frac{n+1}{n-1}} \leq u(x, y)$$

$$\leq U - U\left(1 + M_2^{\frac{1}{n}}\frac{1-n}{n+1}\exp\left(\frac{C_4x}{n}\right)x^{\frac{n-1}{n(n+1)}}U^{\frac{1-n}{n}}(x)y\right)^{\frac{n+1}{n-1}},$$

where M_1 and M_2 are positive constants.

Assume that

$$U(x) = x\widetilde{V}(t, x), \quad \widetilde{V}(t, x) > 0,$$

where $\widetilde{V} = V$ is bounded functions, and $\widetilde{V}(t, x)$ is the corresponding solution $\widetilde{u}(t, x)$.

Theorem 1.6.5 ([106]) *Assume that the functions*

$$V - \widetilde{V}, \ V_x - \widetilde{V}_x, \ \frac{V_t}{V} - \frac{\widetilde{V}_t}{\widetilde{V}}, \ v_1 - \widetilde{v}_1 \tag{1.6.12}$$

vanish for $t \geq t_0$ =const.> 0. Then

$$\left|\frac{u}{V} - \frac{\widetilde{u}}{\widetilde{V}}\right| \leq xK_3e^{-\alpha t} \ \ in \ \ D$$

for $0 \leq t \leq y_0 < \infty$, where α, K are positive constants.

Theorem 1.6.6 ([106]) *Assume that the functions (1.6.12) have their absolute values less than* ε, *and* $V, \widetilde{V}, u_0, \widetilde{u}_0$ *are such that* $|w - \widetilde{w}| \leq \varepsilon$, *where* u *and* $\widetilde{u}$ *are the solutions of problems (1.6.8)–(1.6.9) with* V, v_1, u_0 *and* $\widetilde{V}, \widetilde{v}_1, \widetilde{u}_{0,s}$ *respectively. Then*

$$\left|\frac{u}{V} - \frac{\widetilde{u}}{\widetilde{V}}\right| \leq x K_6 \varepsilon \ \ in \ \ D$$

for $y \leq y_0 < \infty$ *and all* $t \geq 0$, *where* K_6 *is a positive constant.*

Oleinik and Samokhin [106] in 1999 considered the continuation problem for the following stationary boundary layer equations in a dilatable medium by the von Mises variables:

$$\begin{cases} u\frac{\partial u}{\partial x} + v\frac{\partial u}{\partial y} = \nu\frac{\partial}{\partial y}\left(|\frac{\partial u}{\partial y}|^{n-1}\frac{\partial u}{\partial y}\right) + U(x)\frac{dU}{dx}, \quad 1 < n < \infty, \\ \frac{\partial u}{\partial x} + \frac{\partial v}{\partial y} = 0, \end{cases} \tag{1.6.13}$$

in the domain $D = \{0 < x < X, 0 < y < \infty\}$, with the boundary conditions

$$\begin{cases} u(0, y) = u_0(y), \quad u(x, 0) = 0, \quad v(x, 0) = v_0(x), \\ u(x, y) \to U(x), \quad \text{as} \quad y \to +\infty. \end{cases} \tag{1.6.14}$$

Theorem 1.6.7 ([106]) *Assume that* $u_0(y) > 0$ *for* $y > 0$, $u_0 = 0$, $u\prime_0(0) > 0$, $u_0(y) \to U(0) \neq 0$ *as* $y \to \infty$, $U(x) > 0$, $U_x(x)$ *and* $v_0(x)$ *are continuously differentiable on* $[0, X]$, $u_0(y), u\prime_0(y), u\prime\prime_0(y)$ *are bounded for* $0 \leq y \leq \infty$ *and satisfy the Hölder condition; the compatibility condition*

$$\nu n\left|\frac{du_0}{dy}\right|^{n-1}\frac{d^2u_0}{dy^2} - v_0(0)\frac{du_0}{dy} + U(0)U_x(0) = O(y^2)$$

holds as $y \to 0$. *Then, for some* $X > 0$, *there exists a weak solution* $(u(x, y), v(x, y))$ *of problems (1.6.13)–(1.6.14). If* $U_x \geq 0$ *and* $v_0 \leq 0$, *or* $U_x(x) > 0$, *then the solution exists for any* $X > 0$. *If we assume, in addition, that* $U_x > 0$, $v_0(x) \leq 0$, $u_0(y) \equiv U(0)$ *for* $y \geq y_1 > 0$, *then there is a* $y_2 \geq y_1$ *such that* $u(x, y) \equiv U(x)$ *for* $y \geq y_2$ *and this solution is unique.*

Oleinik and Samokhin [106] in 1999 considered system (1.6.13) in the domain $D = \{0 < x < X, 0 < y < \infty\}$ by the von Mises variables, with the boundary conditions

$$\begin{cases} u(0, y) = 0, \quad u(x, 0) = 0, \quad v(x, 0) = v_0(x), \\ u(x, y) \to U(x), \quad \text{as} \quad y \to +\infty. \end{cases} \tag{1.6.15}$$

Theorem 1.6.8 ([106]) *Suppose that* $U(x), U_x(x), U_{xx}(x), v_0(x), v_{0x}(x)$ *are continuous for* $0 \leq x \leq X$, *and let* $U_x > 0$. *Then problems (1.6.13) and (1.6.15)* *have a weak solution $u(x, y), v(x, y)$ with the following properties: $0 \leq u_y \leq M_1$, M_1 =const.> 0; $u_y > 0$ for $y = 0$ and $x > 0$; the function $u_y^{n-1}(u_{yy})^2$ is summable in any finite portion of the domain Ω; for any $x_1 > 0$ and $0 < x_1 \leq x \leq X$, the functions $(u_x(x, y), v(x, y))$ are continuous at $y = 0$ and $v(x, 0) = v_0(x)$. The solution $(u(x, y), v(x, y))$ with these properties is unique and $u(x, y) \equiv U(x)$ for*

$$y \geq \int_0^{Cx} \frac{ds}{u(x, s)},$$

where C is a positive constant depending on $U(x)$ and $v_0(x)$.

Hou, Liu, Wang, and Wang [57] in 2003 studied the stability of boundary layer solutions for a viscous hyperbolic system transformed via a Cole-Hopf transformation from a singular chemotactic system modeling the initiation of tumor angiogenesis. They proved the stability of boundary layer solutions and identify the precise structure of boundary layer solutions.

The prototypical chemotaxis model is known as the Keller-Segel (KS) mode as follows:

$$\begin{cases} u_t = [Du_x - \chi u(\ln c)_x]_x, \\ c_t = \varepsilon c_{xx} + \mu u c, \end{cases} \tag{1.6.16}$$

where $u(x, t)$ and $c(x, t)$ denote the cell density and chemical (signal) concentration at position x and time t, respectively. The function χ is called the chemotactic sensitivity function accounting for the signal response mechanism. $D > 0$ and $\varepsilon > 0$ are cell and chemical diffusion coefficients, respectively.

The common approach currently used to overcome this singularity is the Cole-Hopf type transformation

$$v = -\frac{\sqrt{\chi\mu}}{\mu}(\ln c)_x = -\frac{\sqrt{\chi\mu}}{\mu}\frac{c_x}{c}, \tag{1.6.17}$$

which transforms the model (1.6.16) into a nonsingular system of conservation laws

$$\begin{cases} u_t - (uv)_x = u_{xx}, \\ v_t - \left(u - \frac{\varepsilon}{\chi}v^2\right)_x = \frac{\varepsilon}{D}v_{xx}, \qquad (x, t) \in (0, 1) \times (0, +\infty), \\ (u, v)(x, 0) = (u_0, v_0)(x). \end{cases} \tag{1.6.18}$$

For illustration, let's first consider the following initial-boundary value problem of system (1.6.19) in an interval (0, 1):

$$\begin{cases} u_t - (uv)_x = u_{xx}, \\ v_t - \left(u - \varepsilon|v|^2\right)_x = \varepsilon v_{xx}, & (x,t) \in (0,1) \times (0,+\infty), \\ (u,v)(x,0) = (u_0, v_0)(x), \\ u_x|_{x=0,1} = v|_{x=0,1} = 0. \end{cases} \tag{1.6.19}$$

Precisely, we postulate that the initial data $(u_0, v_0) \in H^3(0,1) \times H^3(0,1)$ satisfying

$$\begin{cases} (u_0, v_0)|_{x=0,1} = (\bar{u}, \bar{v}), \\ u_{0x}|_{x=0,1} = 0, \\ |(u_0 v_0)_x + u_{0xx}|_{x=0,1} = 0. \end{cases} \tag{1.6.20}$$

Theorem 1.6.9 ([57]) *Assume that* $(u_0, v_0) \in H^3(0,1) \times H^3(0,1)$ *with* $u_0 \geq 0$ *satisfies the compatibility conditions* (1.6.20). *Denote by* $(u^\varepsilon, v^\varepsilon)$ *the unique global solution of problem* (1.6.19) *with* $\varepsilon \geq 0$. *Then as* $\varepsilon \to 0$, *the following asymptotic expansions hold in space* $L^\infty([0,1] \times [0,T])$ *for any fixed* $0 < T < \infty$,

$$u^\varepsilon(x,t) = u^0(x,t) + O(\varepsilon^{1/2}),$$

$$v^\varepsilon(x,t) = v^0(x,t) + v^{B,0}\left(\frac{x}{\sqrt{\varepsilon}}, t\right) + v^{b,0}\left(\frac{x-1}{\sqrt{\varepsilon}}, t\right) + O(\varepsilon^{1/2})$$

where $(u^0, v^0) = (u^{I,0}, v^{I,0})$ *denotes the outer layer profile and the inner layer profile* $(v^{B,0}, v^{b,0})$ *is given by*

$$\begin{cases} v^{B,0} = \int_0^t \int_{-\infty}^0 \frac{1}{\sqrt{\pi(t-s)}} e^{-(\frac{(z-y)^2}{4(t-s)}) + \bar{u}(t-s)} [\bar{u}(\bar{v} - v^0(0,s)) - v^0(0,s)] dy ds, \\ v^{b,0} = \int_0^t \int_{-\infty}^0 \frac{1}{\sqrt{\pi(t-s)}} e^{-(\frac{(z-y)^2}{4(t-s)}) + \bar{u}(t-s)} [\bar{u}(\bar{v} - v^0(1,s)) - v^0(1,s)] dy ds. \end{cases}$$

The counterpart of the original system (1.6.16) in [0, 1] corresponding to the initial-boundary value problem of the transformed system (1.6.19) reads as follows:

$$\begin{cases} u_t = (u_x - u(lnc)_x)_x, \\ c_t = (uc - \varepsilon c_{xx}), & (x,t) \in (0,1) \times (0,+\infty), \\ (u,v)(x,0) = (u_0, v_0)(x), \\ u_x|_{x=0,1} = \bar{u}, \ \frac{c_x}{c}|_{x=0,1} = -\bar{c}, \ if \ \varepsilon > 0, \\ u|_{x=0,1} = \bar{u}. \end{cases} \tag{1.6.21}$$

Theorem 1.6.10 ([57]) *Suppose that the initial data* $(u_0, lnc_0) \in H^3 \times H^4$ *with* $u_0 \geq 0$, $c_0 \geq 0$ *and the compatibility conditions* (1.6.20) *with* $v_0 = -(lnc_0)_x$ *and*

$\bar{v} = \bar{c}$. *Let* $(u^\varepsilon, v^\varepsilon)$ *be the unique global solution of problem* (1.6.21) *with* $\varepsilon \geq 0$. *Then for any fixed* $0 < T < \infty$, *we have in space* $L^\infty([0, 1] \times (0, T))$ *that*

$$u^\varepsilon(x, t) = u^0(x, t) + O(\varepsilon^{1/2}),$$

$$c^\varepsilon(x, t) = c^0(x, t) + O(\varepsilon^{1/2}),$$

and

$$c^\varepsilon(x, t) = c^0(x, t) - c^0(x, t)\Big(c^{B,0}\left(\frac{x}{\sqrt{\varepsilon}}, t\right) + v^{b,0}\left(\frac{x-1}{\sqrt{\varepsilon}}, t\right)\Big) + O(\varepsilon^{1/2}).$$

Renardy and Wang [120] in 2014 considered the flow of the Maxwell fluid in the boundary layer and proved the well-posedness of these equations. They started with the upper-convected Maxwell model in dimensionless form:

$$\begin{cases} \partial_t \mathbf{v} + (\mathbf{v} \cdot \nabla)\mathbf{v} = \frac{1}{R}\nabla \cdot \mathbf{T} - \nabla p, \\ \mathbf{T}_t + (\mathbf{v} \cdot \nabla)\mathbf{T} - (\nabla \mathbf{v})\mathbf{T} - \mathbf{T}(\nabla \mathbf{v})^T + \frac{1}{W}\mathbf{T} = \frac{1}{W}(\nabla \mathbf{v} + (\nabla \mathbf{v})^T). \end{cases} \tag{1.6.22}$$

Here v denotes the velocity, T the extra stress and p the pressure, where R is the Reynolds number and W is the Weissenberg number measuring the elasticity. Set $\mathbf{S} = \mathbf{T} + \frac{1}{W^2}$, they assumed a flat boundary given by $y = 0$, set

$$t' = t,\ \ x' = x,\ \ y' = Wy,\ \ u' = u,\ \ v' = Wv,\ S\prime_{11} = S_{11},$$

$$S\prime_{12} = WS_{12},\ \ S\prime_{22} = W^2 S_{22},\ \ p' = p.$$

Then the momentum equations reduce to the problem

$$\begin{cases} u_t + uu_x + vu_y = E(S_{11}x + S_{11}y) - p_x, \\ p_y = 0, \end{cases} \tag{1.6.23}$$

here $E = \frac{R}{W}$. Let

$$x = x(\xi_1, \xi_2, t),\ \ y = y(\xi_1, \xi_2, t).$$

Theorem 1.6.11 ([120]) *Consider the boundary layer system (1.6.23) on the domain* $-\infty < \xi_1 < \infty$, $0 \leq \xi_2 < \infty$, *for* $t \geq 0$. *Let* $m \geq 1$ *be an integer. We assume that their derivatives up to order* m *are bounded for* $(\xi_1, \xi_2, t) \in (-\infty, \infty) \times [0, \infty) \times [0, T]$. *Moreover they are all periodic in* ξ_1 *with period* 1. *Let* $\Omega = [0, 1] \times [0, \infty)$, $Q = \Omega \times [0, T]$, *and denote by* $H_p^m(Q)$, $m \in \mathbb{N}$, *the space of all periodic (in* ξ_1*) functions which have* H^m *regularity. If* $\Phi \in H_p^m(Q)$, *then there exists some* $T' \in (0, T]$ *such that a unique solution of the problem (1.6.23) exists and satisfies* $\in H_p^m(Q)$.

Renardy [119] in 2015 proved the local well-posedness of the Prandtl boundary layer equations for the upper-convected Maxwell fluid (1.6.22). He considered boundary layers in the case where the Reynolds number tends to infinity, but the Weissenberg number remains finite. As before, Renardy and Wang [120] in 2014 considered boundary layers for the upper-convected Maxwell model in a limit where both the Reynolds and Weissenberg number become infinite.

Let $X(\xi, t)$ denote the outer flow solution, i.e., the solution of the equation

$$\frac{\partial^2 X}{\partial t^2} = -P(X, t),$$

with initial conditions

$$X(\xi, 0) = \xi, \quad \frac{\partial X}{\partial t}(\xi, 0) = \lim_{\eta \to \infty} u_0(\xi, \eta).$$

Let $\alpha > 0$ be a constant. We shall look for solutions of the form

$$x(\xi, \eta, t) = X(\xi, t) + e^{-\alpha\eta}(Z(\xi, t) + w(\xi, \eta, t)),$$

$$C(\xi, \eta, t) = \begin{pmatrix} e^{-2\alpha\eta}(G_{11}(\xi, t) + H_{11}(\xi, \eta, t) & e^{-\alpha\eta}(G_{12}(\xi, t) + H_{12}(\xi, \eta, t)) \\ (e^{-\alpha\eta}(G_{12}(\xi, t) + H_{12}(\xi, \eta, t) & (G_{22}(\xi, t) + H_{22}(\xi, \eta, t))) \end{pmatrix},$$

where it is assumed that w and H tend to zero as $\eta \to \infty$. With this ansatz, we find the equations

$$\begin{cases} \dfrac{\partial^2 Z}{\partial t^2} = \alpha^2 G_{22} Z - \alpha G_{12} \dfrac{\partial X}{\partial \xi} - \dfrac{\partial P}{\partial X}(X, t) Z, \\ \dfrac{\partial G}{\partial t} = -\dfrac{1}{W} G + \dfrac{1}{W^2} \begin{pmatrix} \alpha^2 Z^2 & \alpha Z \frac{\partial X}{\partial \xi} \\ \alpha Z \frac{\partial X}{\partial \xi} & \left(\frac{\partial X}{\partial \xi}\right)^2 \end{pmatrix}. \end{cases}$$

After a series of transformations, then the momentum equations becomes

$$\begin{aligned} \frac{\partial^2 w}{\partial t^2} =& e^{-\alpha\eta} \frac{\partial}{\partial \xi}\left((G_{11} + H_{11}) \frac{\partial}{\partial \xi}\left(X + e^{-\alpha\eta}(Z + w)\right)\right) \\ &+ e^{-\alpha\eta} \frac{\partial}{\partial \xi}\left((G_{12} + H_{12})\left(\frac{\partial}{\partial \eta} - \alpha\right)(Z + w)\right) + \left(\frac{\partial}{\partial \eta} - \alpha\right) H_{12} \frac{\partial X}{\partial \xi} \\ &+ \left(\frac{\partial}{\partial \eta} - \alpha\right)\left((G_{12} + H_{12}) e^{-\alpha\eta} \frac{\partial}{\partial \xi}(Z + w)\right) \\ &- \alpha\left(\left(\frac{\partial}{\partial \eta} - \alpha\right) H_{22}\right) Z + e^{\alpha\eta} \frac{\partial}{\partial \eta}\left((G_{22} + H_{22}) e^{-\alpha\eta}\left(\frac{\partial w}{\partial \eta} - \alpha w\right)\right) \\ &+ e^{\alpha\eta}\left(-P\left(X + e^{-\alpha\eta}(Z + w), t\right) + P(X, t)\right) + \frac{\partial P}{\partial X}(X, t) Z, \end{aligned} \tag{1.6.24}$$

$$
\begin{aligned}
\frac{\partial H_{11}}{\partial t} &= -\frac{1}{W}H_{11} - \frac{2}{W^2}\alpha Z\left(\frac{\partial w}{\partial \eta} - \alpha w\right) + \frac{1}{W^2}\left(\frac{\partial w}{\partial \eta} - \alpha w\right)^2, \\
\frac{\partial H_{12}}{\partial t} &= -\frac{1}{W}H_{12} - \frac{1}{W^2}\frac{\partial X}{\partial \xi}\left(\frac{\partial w}{\partial \eta} - \alpha w\right) \\
&\quad - \frac{1}{W^2}e^{-\alpha\eta}\left(\frac{\partial Z}{\partial \xi} + \frac{\partial w}{\partial \xi}\right)\left(-\alpha Z + \frac{\partial w}{\partial \eta} - \alpha w\right), \\
\frac{\partial H_{22}}{\partial t} &= -\frac{1}{W}H_{22} + \frac{2}{W^2}e^{-\alpha\eta}\frac{\partial X}{\partial \xi}\left(\frac{\partial Z}{\partial \xi} + \frac{\partial w}{\partial \xi}\right) + \frac{1}{W^2}e^{-2\alpha\eta}\left(\frac{\partial Z}{\partial \xi} + \frac{\partial w}{\partial \xi}\right)^2.
\end{aligned} \tag{1.6.25}
$$

Theorem 1.6.12 ([119]) *Consider problems (1.6.24)–(1.6.25) with initial conditions $\omega(\xi,\eta,0) = 0$, $\omega_t(\xi,\eta,0) = w_1(\xi,\eta)$, $H(\xi,\eta,0) = H_0(\xi,\eta)$. Assume that the functions P, X, G and G are smooth and periodic in X or, respectively, ξ with period L. Assume, moreover, that ω_1, $H_0 \in \overline{\omega}_{n-1}$, where $n \geq 5$, and that appropriate compatibility conditions between the initial and boundary data for w hold. Finally, assume that $G(\xi,0) + H(\xi,\eta,0)$ is strictly and uniformly positive definite. Then, for some $T > 0$, there exists a unique solution with the regularity*

$$
\begin{cases}
w \in L^\infty([0,T], V_n) \cap W^{1,\infty}([0,T], W_{n-1}) \cap \bigcap_{k=2}^{n} W^{k,\infty}([0,T], H_{n-k}), \\
H \in L^\infty([0,T], W_{n-1}) \cap \bigcap_{k=1}^{n-1} W^{k,\infty}([0,T], H_{n-1-k}).
\end{cases}
$$

Li and Wang [71] in 2018 investigated the global solutions to initial-boundary value problem for the following nonlinear evolution system (1.6.26) on the domain $[0,1] \times [0,\infty)$ with damping and diffusion and gave the convergence rates and boundary layer thickness

$$
\begin{cases}
\phi_t^\alpha = -(k-1)\alpha\phi^\alpha - k\alpha\theta_x^\alpha + \alpha\phi_{xx}^\alpha, 0 \leq x \leq 1,\ t > 0, \\
\theta_t^\alpha = -(1-\beta)\theta^\alpha + \mu\alpha\phi_x^\alpha + 2\phi^\alpha\theta^\alpha + \beta\theta_{xx}^\alpha, \\
(\phi^\alpha, \theta^\alpha)(x,0) = (\phi_0^\alpha, \theta_0^\alpha)(x),\ 0 \leq x \leq 1, \\
(\phi_x^\alpha, \theta^\alpha)(0,t) = (\phi_x^\alpha, \theta^\alpha)(1,t) = (0,0),\ t \geq 0,
\end{cases} \tag{1.6.26}
$$

where k, α, β and μ are positive constants with $k > 1$ and $0 < \beta < 1$. The limit problem of the vanishing parameter $\alpha \to 0^+$ in (1.6.26) is the following:

$$
\begin{cases}
\phi_t^0 = 0, \\
\theta_t^0 = -(1-\beta)\theta^0 + 2\phi^0\theta_{xx}^0,\ 0 < x < 1,\ t > 0, \\
(\phi^0, \theta^0)(x,0) = (\phi_1(x), \phi_2(x)),\ 0 \leq x \leq 1, \\
\theta^0(0,t) = \theta^0(1,t) = 0,\ t \geq 0.
\end{cases} \tag{1.6.27}
$$

We first show that the initial-boundary value problem (1.6.26) admits a unique global smooth solution $(\phi^\alpha, \theta^\alpha)$, which is stated in the next theorem.

Theorem 1.6.13 ([71]) *Suppose that the initial data satisfy the conditions:* $(\phi^0, \theta^0) \in H^2$, $(\phi^0, \theta^0)(0, t) = (\phi^0, \theta^0)(1, t) = (0, 0)$ *and* $\|(\phi^0, \theta^0)\|_{H^1}$ *is sufficiently small, then there exists a unique solution* $(\phi^\alpha, \theta^\alpha)$ *to the initial boundary value problem* (1.6.26) *satisfying*

$$\theta^\alpha \in L^\infty(0, T; H^2) \cap L^2(0, T; H^2),\ \phi^\alpha \in L^\infty(0, T; H^2) \cap L^2(0, T; H^1),$$

where the norms are all uniform in α.

The second result in the following shows that the initial-boundary value problem (1.6.27) admits a unique global smooth solution (ϕ^0, θ^0).

Theorem 1.6.14 ([71]) *Suppose that the initial data satisfy the conditions:* $(\phi_1(x), \phi_2(x)) \in H^2$, $\phi_1(0, t) = 0$, $\phi_2(1, t) \le \varepsilon$ *and* ε *is sufficiently small, then there exists a unique solution* (ϕ^0, θ^0) *to the initial boundary value problem* (1.6.27) *satisfying*

$$\begin{aligned} &\phi^0 \in L^\infty(0, T; H^2),\ \theta^0 \in L^\infty(0, T; H^2) \cap L^2(0, T; H^2),\\ &\theta_t^0 \in L^\infty(0, T; H^2) \cap L^2(0, T; H^1). \end{aligned}$$

Furthermore, Li and Wang [71] gave the convergence rates and boundary layer thickness.

Theorem 1.6.15 ([71]) *Under the assumptions of the above two theorems, suppose that*

$$\|\phi_0(x) - \phi_1(x)\|_{H^1}^2 \le C\alpha, \quad \|\theta_0(x) - \phi_2(x)\|_{H^1}^2 \le C\alpha,$$

then any function $\delta(\alpha)$ *satisfying the condition* $\delta(\alpha) \to 0$ *and* $\frac{\alpha}{\delta(\alpha)} \to 0$ *as* $\alpha \to 0^+$ *is a BL-thickness such that*

$$\|\theta^\alpha - \theta^0\|_{L^\infty(0,T;L^\infty[0,1])} \le C\alpha, \quad |\phi^\alpha - \phi^0\|_{L^\infty(0,T;L^\infty[\delta,1-\delta])} \le C\frac{\alpha}{\delta(\alpha)},$$

$$\liminf_{\alpha \to 0} \|\phi^\alpha - \phi^0\|_{L^\infty(0,T;L^\infty[0,1])} > 0.$$

Consequently,

$$\lim_{\alpha \to 0} \|\theta^\alpha - \theta^0\|_{L^\infty(0,T;L^\infty[0,1])} = 0, \quad \lim_{\alpha \to 0} \|\phi^\alpha - \phi^0\|_{L^\infty(0,T;L^\infty[\delta,1-\delta])} = 0.$$

Oleinik and Samokhin [106] in 1999 considered the nonstationary symmetric flow of an electrically conducting pseudo-plastic fluid within the boundary layer

$$\begin{cases} \frac{\partial u}{\partial t} + u\frac{\partial u}{\partial x} + v\frac{\partial u}{\partial y} = \nu\frac{\partial}{\partial y}\left(|\frac{\partial u}{\partial y}|^{n-1}\frac{\partial u}{\partial y}\right) + U(x)\frac{dU}{dx} + d^2(U-u) + \frac{\partial U}{\partial t}, \\ \quad 0 < n < 1, \\ \frac{\partial u}{\partial x} + \frac{\partial v}{\partial y} = 0. \end{cases} \tag{1.6.28}$$

Here $d^2 = \sigma B^2(x)$ in the domain $D = \{0 < t < \infty, 0 < x < X, 0 < y < \infty\}$, with the boundary conditions

$$\begin{aligned} &u(0,x,y) = u_0(x,y), \;\; u(t,0,y) = 0, \;\; u(x,0) = 0, \;\; v(x,0) = v_0(t,x), \\ &u(t,x,y) \to U(t,x), \;\; \text{as} \;\; y \to \infty. \end{aligned} \tag{1.6.29}$$

Assume that

$$U(t,x) = x^m V(t,x), \;\; v_0(t,x) = x^{\frac{2mn-n-m}{n+1}} v_1(t,x), \;\; d(t,x) = x^{\frac{m-1}{2}} W(t,x),$$

where the functions V, W, v_1 are bounded in D, $V(t,0) = a$ =const.> 0, $v_1(t,0) = b$ =const., $W(t,0) = d_0$ =const.. The assumptions about the smoothness of V, W, v_1, and the properties of their derivatives will be stated later. In (1.6.28), introduce new independent variables

$$\tau = t, \;\; \xi = x, \;\; \eta = \frac{u}{U}, \;\; w(\tau,\xi,\eta) = \frac{|u_y|^{n-1}u_y}{x^{\frac{2mn-n-m}{n+1}}U}.$$

Theorem 1.6.16 ([106]) *Suppose that* $U(t,x) = x^m(a + xa_1(t,x))$, $d(t,x) = x^{\frac{m-1}{2}}(d_0 + xd_1(t,x))$, $v_0(t,x) = x^{\frac{2mn-n-m}{n+1}}(b + xb_1(t,x))$, $m > 0$, $a_1(t,x), d_1(t,x)$ *have bounded second-order derivatives;* $b_1(t,x)$ *has bounded first-order derivatives;* $d_0^2 + (3m-1)na/(n+1) > 0$*; the functions* $u_0(x,y)$ *and* $v_0(t,x)$ *satisfy*

$$w_0(\xi,\eta) = |u_{0y}|^{n-1}u_{0y}\left(\xi^{\frac{2mn-n-m}{n+1}}U(0,\xi)\right)^{-1}.$$

Then problems (1.6.28)–(1.6.29) in D, with X depending on U, v_0, d, ν, m, n*, admit a solution* $u(t,x,y)$, $v(t,x,y)$ *with the following properties:* u/U *is continuous in* $\overline{D}$*;* v *is continuous in* $\overline{D}$ *with respect to y and bounded for bounded y, the derivatives* $u_y, u_{yy}, u_x, u_t, v_y$ *are continuous in D with respect to y; equations (1.6.28) hold for* u, v *almost everywhere in D;* u, v *satisfy the boundary conditions (1.6.29).*

Moreover, $u > 0$ *for* $y > 0$ *and* $x > 0$*;* $u_y > 0$ *for* $y \le 0$*;* $u_y \to 0$ *as* $y \to \infty$*;* u_{yyy} *is bounded in D; the following inequalities hold:*

$$x^{\frac{2n-m-n}{n+1}} U(t,x) Y\left(\frac{u}{U}\right) e^{-K_{15}x} \le u_y^n(t,x,y) \le x^{\frac{2mn-m-n}{n+1}} U(t,x) Y\left(\frac{u}{U}\right) e^{K_{16}x},$$

$$-K_{22} \le \frac{n(n-2)(u_{yy})^2 + nu_y u_{yyy}}{x^{\frac{2mn-m-n}{n}} U^{\frac{1-n}{n}} u_y^{3-n}} \le -K_{23},$$

$$\left(M_8^{\frac{1}{n}} \frac{1-n}{n+1} e^{\frac{K_{14}}{n}} x^{\frac{2mn-m-n}{n(n+1)}} U^{\frac{1-n}{n}}(t,x) y + 1\right)^{\frac{n+1}{n-1}} \le 1 - \frac{u}{U}$$

$$\le \left(M_7^{\frac{1}{n}} \frac{1-n}{n+1} e^{-\frac{K_{15}}{n}} x^{\frac{2mn-m-n}{n(n+1)}} U^{\frac{1-n}{n}}(t,x) y + 1\right)^{\frac{n+1}{n-1}}.$$

The solution of problems (1.6.28)–(1.6.29) with the above properties is unique.

Chapter 2
Global Well-Posedness of Solutions to the 2D Prandtl-Hartmann Equations in an Analytic Framework

In this chapter, we shall consider the 2D Prandtl-Hartmann equations on the half plane and prove the global existence and uniqueness of solutions to the 2D Prandtl-Hartmann equations by using the classical energy methods in an analytic framework. We prove that the lifespan of the solutions to the 2D Prandtl-Hartmann equations can be extended up to T_ε (see Theorem 2.2.1) when the strength of the perturbation is of the order of ε. The difficulty of solving the Prandtl-Hartmann equations in the analytic framework is the loss of x-derivative in the term $v\partial_y u$. To overcome this difficulty, we show the Gaussian-weighted Poincaré inequality (see Lemma 2.2.3). Compared to the existence and uniqueness of solutions to the classical Prandtl equations where the monotonicity condition of the tangential velocity, playing a key role, is required there, in this chapter, it is not needed for the 2D Prandtl-Hartmann equations in an analytic framework. Besides, compared with the existence and uniqueness of solutions to the 2D MHD boundary layer where the initial tangential magnetic field has a lower bound, also playing an important role there, which is also not needed in this chapter for 2D Prandtl-Hartmann equations in an analytic framework. The content of this chapter is chosen from [24].

2.1 Introduction

The following Prandtl-Hartmann equations were derived in [45] by the classical 2D incompressible MHD system under a transverse magnetic field on the boundary for the parameters of the system in the mixed Prandtl-Hartmann regime. In this chapter, we shall investigate the global existence and uniqueness of solutions to the

Y. Qin et al., *Prandtl Equations and Related Boundary Layer Equations*,
https://doi.org/10.1007/978-981-97-4565-4_2

following initial boundary value problem for the 2D Prandtl-Hartmann system in the upper half plane $\mathbb{R}^2_+ = \{(x, y) : x \in \mathbb{R},\ y \in \mathbb{R}_+\}$, which reads as

$$\begin{cases} \partial_t u_1 + u_1 \partial_x u_1 + u_2 \partial_y u_1 - \partial_y^2 u_1 + u_1 = 0, \\ \partial_x u_1 + \partial_y u_2 = 0, \end{cases} \tag{2.1.1}$$

where the velocity field $(u_1(t, x, y), u_2(t, x, y))$ of the boundary layer is a unknown vector function.

The system (2.1.1) subjects to the initial data and boundary conditions

$$\begin{cases} u_1(t, x, y)|_{t=0} = u_{10}(x, y), \\ u_1|_{y=0} = u_2|_{y=0} = 0. \end{cases} \tag{2.1.2}$$

The far field is represented by a positive constant $\bar{u}$

$$\lim_{y \to +\infty} u_1(t, x, y) = \bar{u}. \tag{2.1.3}$$

2.2 Global Well-Posedness of Solutions

As in [59, 148], we use the following Gaussian weighted function: θ_α to introduce the function space of the solutions,

$$\theta_\alpha = \exp\left(\frac{\alpha z^2}{4}\right) \text{ with } z = \frac{y}{\sqrt{\langle t \rangle}},\ \langle t \rangle = (1 + t),\ \alpha \in [1/4, 1/2],$$

which, combined with

$$M_m = \frac{\sqrt{(m + 1)}}{m!},$$

defines the Sobolev weighted semi-norms by

$$\begin{cases} X_m = X_m(f, \tau) = \|\theta_\alpha \partial_x^m f\|_{L^2(\mathbb{R}^2_+)} \tau^m M_m, \\ D_m = D_m(f, \tau) = \|\theta_\alpha \partial_y \partial_x^m f\|_{L^2(\mathbb{R}^2_+)} \tau^m M_m, \\ Y_m = Y_m(f, \tau) = \|\theta_\alpha \partial_x^m f\|_{L^2(\mathbb{R}^2_+)} \tau^{m-1} m M_m. \end{cases} \tag{2.2.1}$$

To define the following analytic function space in the tangential variable x and weighted Sobolev space in the normal variable y by

$$X_{\tau,\alpha} = \Big\{ f(t, x, y) \in L^2(\mathbb{R}^2_+; \theta_\alpha dxdy) : \|f\|_{X_{\tau,\alpha}} < \infty \Big\},$$

with $\tau > 0$ and the norms

$$\begin{cases} \|f\|_{X_{\tau,\alpha}} = \sum\limits_{m\geq 0} X_m(f,\tau), \\ \|f\|_{D_{\tau,\alpha}} = \sum\limits_{m\geq 0} D_m(f,\tau), \\ \|f\|_{Y_{\tau,\alpha}} = \sum\limits_{m\geq 1} Y_m(f,\tau). \end{cases}$$

Theorem 2.2.1 *Let $\frac{1}{4} \leq \alpha \leq \frac{1}{2}$ and assume that the initial datum $u_1(0)$ satisfies*

$$\|u_1(0) - \bar{u}\|_{X_{\tau_0,\alpha}} \leq \varepsilon, \tag{2.2.2}$$

for some small constant $0 < \varepsilon \ll 1$ and assume there exists an analyticity radius τ_0 such that

$$\max\left\{\frac{7\varepsilon}{8C_1}, 128\sqrt{2}C\varepsilon, \frac{256\sqrt{2}CC_2^{\frac{3}{4}}}{\alpha}\right\} \leq \tau_0^{\frac{3}{2}}. \tag{2.2.3}$$

Then there exists a unique solution (u_1, u_2) to the damping Prandtl equations (2.1.1)–(2.1.3) *such that*

$$\|u\|_{X_{\tau,\alpha}} \leq e^{-t}\varepsilon, \tag{2.2.4}$$

with analyticity radius τ larger than $\frac{\tau_0}{4}$ in the time interval $[0, T_\varepsilon]$, where $u = u_1 - \bar{u}$, $T_\varepsilon = C_2\varepsilon^{-\frac{4}{3}} - 1$ and $\bar{u}$ is given by (2.1.3), *$C_1 = \frac{3C(\alpha+2)}{2\alpha}$ and $C_2 = \left(\frac{7}{8C_1}\tau_0^{\frac{3}{2}}\right)^{\frac{4}{3}}$.*

The above theorem is the main part of this chapter, and its proof follows from the discussion in Sects. 2.2.1 and 2.2.2.

Before proving the theorem, we first list some important inequalities as follows. The following main estimates on the functions in the norms defined in the previous section will be used frequently, whose proofs can be found in [56, 59, 148].

Lemma 2.2.2 (Agmon Inequality [59]) *Let $f \in H^1(\mathbb{R}_+^2)$, then*

$$\|f\|_{L_x^\infty L_y^2} \leq C\|f\|_{L^2(\mathbb{R}_+^2)}^{\frac{1}{2}}\|f\|_{H^1(\mathbb{R}_+^2)}^{\frac{1}{2}}.$$

Lemma 2.2.3 ([56, 59]) *Let f be a function such that $f|_{y=0}$(or $\partial_y f|_{y=0}$) and $f|_{y=\infty} = 0$. Then, for $m \geq 0$ and $t \geq 0$, there holds that*

$$\begin{cases} \frac{\alpha}{\langle t\rangle}\|\theta_\alpha\partial_x^m f\|_{L_y^2}^2 \leq \|\theta_\alpha\partial_y\partial_x^m f\|_{L_y^2}^2, \\ \sum\limits_{m\geq 0}\frac{\|\theta_\alpha\partial_y\partial_x^m f\|_{L^2(\mathbb{R}_+^2)}^2}{\|\theta_\alpha\partial_x^m f\|_{L^2(\mathbb{R}_+^2)}}\tau^m M_m \geq \frac{\alpha}{2\langle t\rangle^{\frac{1}{2}}}\|f\|_{D_{\tau,\alpha}} + \frac{\alpha(2-\sqrt{\alpha})}{\langle t\rangle}\|f\|_{X_{\tau,\alpha}}, \end{cases} \tag{2.2.5}$$

with $\alpha \in [1/4, 1/2]$, $\langle t\rangle = 1+t$.

Proof For the convenience of the readers, we will prove the second inequality as follows. Actually, by the first inequality, we have

$$\begin{aligned}
\frac{\|\theta_\alpha \partial_y \partial_x^m f\|_{L^2(\mathbb{R}^2_+)}^2}{\|\theta_\alpha \partial_x^m f\|_{L^2(\mathbb{R}^2_+)}} &\geq \frac{\sqrt{\alpha}}{4}\frac{\|\theta_\alpha \partial_y \partial_x^m f\|_{L^2(\mathbb{R}^2_+)}^2}{\|\theta_\alpha \partial_x^m f\|_{L^2(\mathbb{R}^2_+)}} + \frac{2-\frac{\sqrt{\alpha}}{2}}{2}\frac{\sqrt{\alpha}}{\sqrt{\langle t\rangle}}\|\theta_\alpha \partial_y \partial_x^m f\|_{L^2(\mathbb{R}^2_+)} \\
&\geq \frac{\sqrt{\alpha}}{4}\frac{\|\theta_\alpha \partial_y \partial_x^m f\|_{L^2(\mathbb{R}^2_+)}^2}{\|\theta_\alpha \partial_x^m f\|_{L^2(\mathbb{R}^2_+)}} + \frac{\alpha}{4\sqrt{\langle t\rangle}}\|\theta_\alpha \partial_y \partial_x^m f\|_{L^2(\mathbb{R}^2_+)} \\
&\quad + \frac{\alpha(2-\sqrt{\alpha})}{2\langle t\rangle}\|\theta_\alpha \partial_x^m f\|_{L^2(\mathbb{R}^2_+)} \\
&\geq \frac{\alpha}{2\sqrt{\langle t\rangle}}\|\theta_\alpha \partial_y \partial_x^m f\|_{L^2(\mathbb{R}^2_+)} + \frac{\alpha(2-\sqrt{\alpha})}{2\langle t\rangle}\|\theta_\alpha \partial_x^m f\|_{L^2(\mathbb{R}^2_+)}.
\end{aligned}$$

□

Lemma 2.2.4 ([148]) *Let f be a function such that $f|_{y=0}$. Then, it holds that*

$$\begin{cases}
\|f\|_{L^2_x L^\infty_y} \leq C\|\theta_\alpha f\|_{L^2(\mathbb{R}^2_+)}^{\frac{1}{2}}\|\theta_\alpha \partial_y f\|_{L^2(\mathbb{R}^2_+)}^{\frac{1}{2}}, \\
\|f\|_{L^\infty_{xy}} \leq C\|\theta_\alpha f\|_{L^2(\mathbb{R}^2_+)}^{\frac{1}{4}}\|\theta_\alpha \partial_x f\|_{L^2(\mathbb{R}^2_+)}^{\frac{1}{4}}\|\theta_\alpha \partial_y f\|_{L^2(\mathbb{R}^2_+)}^{\frac{1}{4}}\|\theta_\alpha \partial_x\partial_y f\|_{L^2(\mathbb{R}^2_+)}^{\frac{1}{4}}.
\end{cases} \tag{2.2.6}$$

2.2.1 *Uniform Estimates*

In this subsection, we will prove the existence of solutions for the 2D Prandtl-Hartmann regime in an analytic framework defined in Sect. 2.2 by the standard energy method.

Now, we are ready to establish uniform estimates for the solutions to systems (2.1.1)–(2.1.3). For our purpose, we first rewrite the solutions (u_1, u_2) to problems (2.1.1)–(2.1.3) as a perturbation (u, v) around the point $(\bar{u}, 0)$ by standing for

$$\begin{cases}
u_1 = u + \bar{u}, \\
u_2 = v.
\end{cases} \tag{2.2.7}$$

Then system (2.1.1) becomes

$$\begin{cases}
\partial_t u + (\bar{u}+u)\partial_x u + v\partial_y(u+\bar{u}) - \partial_y^2 u + u + \bar{u} = 0, \\
\partial_x u + \partial_y v = 0,
\end{cases} \tag{2.2.8}$$

which subjects to the initial datum and boundary data

$$\begin{cases} u(t,x,y)|_{t=0} = u_0(x,y) - \bar{u}, \\ u|_{y=0} = v|_{y=0} = 0, \end{cases} \tag{2.2.9}$$

and the corresponding far field condition

$$\lim_{y\to+\infty} u = 0. \tag{2.2.10}$$

Next, we will establish the estimates of solutions to the system (2.2.8)–(2.2.10). For $m \geq 0$, applying the operator ∂_x^m on $(2.2.8)_1$, multiplying the resulting equation by $\theta_\alpha^2 \partial_x^m u$ and integrating it by parts over $\mathbb{R}_+^2$, we derive that

$$\int_{\mathbb{R}_+^2} \partial_x^m (\partial_t u + (u+\bar{u})\partial_x u + v\partial_y u - \partial_y^2 u + u + \bar{u})\theta_\alpha^2 \partial_x^m u dxdy = 0. \tag{2.2.11}$$

We now deal with each term in (2.2.11) as follows. For the first term, the second term, and the fourth term, we have, respectively,

$$\begin{aligned} \int_{\mathbb{R}_+^2} \partial_x^m \partial_t u \theta_\alpha^2 \partial_x^m u dxdy &= \frac{1}{2}\int_{\mathbb{R}_+^2} \partial_t (\partial_x^m u)^2 \theta_\alpha^2 dxdy \\ &= \frac{1}{2}\frac{d}{dt}\int_{\mathbb{R}_+^2} (\partial_x^m u)^2 \theta_\alpha^2 dxdy - \int_{\mathbb{R}_+^2} (\partial_x^m u)^2 \theta_\alpha \frac{d}{dt}\theta_\alpha dxdy \\ &= \frac{1}{2}\frac{d}{dt}\|\theta_\alpha \partial_x^m u\|_{L^2(\mathbb{R}_+^2)}^2 + \frac{\alpha}{4\langle t\rangle}\|\theta_\alpha z \partial_x^m u\|_{L^2(\mathbb{R}_+^2)}^2, \end{aligned} \tag{2.2.12}$$

$$\int_{\mathbb{R}_+^2} \partial_x^m (u+\bar{u})\theta_\alpha^2 \partial_x^m u dxdy = \|\theta_\alpha \partial_x^m u\|_{L^2(\mathbb{R}_+^2)}^2, \tag{2.2.13}$$

and

$$\begin{aligned} &-\int_{\mathbb{R}_+^2} \partial_y^2 \partial_x^m u \theta_\alpha^2 \partial_x^m u dxdy \\ &= \|\theta_\alpha \partial_x^m \partial_y u\|_{L^2(\mathbb{R}_+^2)}^2 + \int_{\mathbb{R}_+^2} \partial_y \partial_x^m u \partial_y(\theta_\alpha^2)\partial_x^m u dxdy \\ &= \|\theta_\alpha \partial_x^m \partial_y u\|_{L^2(\mathbb{R}_+^2)}^2 - \frac{1}{2}\int_{\mathbb{R}_+^2} \partial_y (\partial_x^m u)^2 \partial_y(\theta_\alpha^2) dxdy \\ &= \|\theta_\alpha \partial_x^m \partial_y u\|_{L^2(\mathbb{R}_+^2)}^2 - \frac{\alpha}{2\langle t\rangle}\|\theta_\alpha \partial_x^m u\|_{L^2(\mathbb{R}_+^2)}^2 - \frac{\alpha^2}{2\langle t\rangle}\|\theta_\alpha z \partial_x^m u\|_{L^2(\mathbb{R}_+^2)}^2, \end{aligned} \tag{2.2.14}$$

where we have used the fact $\partial_y^2(\theta_\alpha^2) = \frac{\alpha}{\langle t\rangle}\theta_\alpha^2 + \frac{\alpha^2}{\langle t\rangle}z^2\theta_\alpha^2$ and boundary conditions $\partial_x^m u|_{y=0} = 0$. Next, we build the estimate of the nonlinear terms in (2.2.11) as follows

$$
\begin{aligned}
R_1 &= \int_{\mathbb{R}_+^2} \partial_x^m((u+\bar{u})\partial_x u)\theta_\alpha^2\partial_x^m u dxdy = \sum_{i=0}^{m}\binom{m}{i}\int_{\mathbb{R}_+^2}\partial_x^{m-i}u\partial_x^{i+1}u\theta_\alpha^2\partial_x^m u dxdy \\
&\le \sum_{i=0}^{[m/2]}\binom{m}{i}\|\partial_x^{m-i}u\|_{L_x^2L_y^\infty}\|\theta_\alpha\partial_x^{i+1}u\|_{L_x^\infty L_y^2}\|\theta_\alpha\partial_x^m u\|_{L^2(\mathbb{R}_+^2)} \\
&\quad + \sum_{i=[m/2]+1}^{m}\binom{m}{i}\|\partial_x^{m-i}u\|_{L_{xy}^\infty}\|\theta_\alpha\partial_x^{i+1}u\|_{L^2(\mathbb{R}_+^2)}\|\theta_\alpha\partial_x^m u\|_{L^2(\mathbb{R}_+^2)}.
\end{aligned} \tag{2.2.15}
$$

For $0 \le i \le [m/2]$, using Lemmas 2.2.4 and 2.2.2 (the Agmon inequality) in x, we deduce

$$
\|\partial_x^{m-i}u\|_{L_x^2L_y^\infty} \le C\|\theta_\alpha\partial_x^{m-i}u\|_{L^2(\mathbb{R}_+^2)}^{\frac{1}{2}}\|\theta_\alpha\partial_y\partial_x^{m-i}u\|_{L^2(\mathbb{R}_+^2)}^{\frac{1}{2}}, \tag{2.2.16}
$$

and

$$
\|\theta_\alpha\partial_x^{i+1}u\|_{L_x^\infty L_y^2} \le C\|\theta_\alpha\partial_x^{i+1}u\|_{L^2(\mathbb{R}_+^2)}^{\frac{1}{2}}\|\theta_\alpha\partial_x^{i+2}u\|_{L^2(\mathbb{R}_+^2)}^{\frac{1}{2}}. \tag{2.2.17}
$$

For $[m/2] \le i \le m$, by using Lemma 2.2.4, we arrive at

$$
\begin{aligned}
\|\partial_x^{m-i}u\|_{L_{xy}^\infty} &\le C\|\theta_\alpha\partial_x^{m-i}u\|_{L^2(\mathbb{R}_+^2)}^{\frac{1}{4}}\|\theta_\alpha\partial_x^{m-i+1}u\|_{L^2(\mathbb{R}_+^2)}^{\frac{1}{4}} \\
&\quad \times\|\theta_\alpha\partial_y\partial_x^{m-i}u\|_{L^2(\mathbb{R}_+^2)}^{\frac{1}{4}}\|\theta_\alpha\partial_x^{m-i+1}\partial_y u\|_{L^2(\mathbb{R}_+^2)}^{\frac{1}{4}}.
\end{aligned} \tag{2.2.18}
$$

Accordingly,

$$
\begin{aligned}
\frac{|R_1|\tau^m M_m}{\|\theta_\alpha\partial_x^m u\|_{L^2(\mathbb{R}_+^2)}} &\le \frac{C}{\tau(t)^{\frac{1}{2}}}\Big(\sum_{i=0}^{[m/2]}\binom{m}{i}X_{m-i}^{\frac{1}{2}}D_{m-i}^{\frac{1}{2}}Y_{i+1}^{\frac{1}{2}}Y_{i+2}^{\frac{1}{2}} \\
&\quad + \sum_{i=[m/2]+1}^{m}\binom{m}{i}X_{m-i}^{\frac{1}{4}}X_{m-i+1}^{\frac{1}{4}}D_{m-i}^{\frac{1}{4}}D_{m-i+1}^{\frac{1}{4}}Y_{i+1}\Big).
\end{aligned} \tag{2.2.19}
$$

Similarly, we also have

$$
\begin{aligned}
R_2 &= \int_{\mathbb{R}_+^2} \partial_x^m (v\partial_y u)\theta_\alpha^2 \partial_x^m u dx dy = \sum_{i=0}^{m} \binom{m}{i} \int_{\mathbb{R}_+^2} \partial_x^{m-i} v \partial_x^i \partial_y u \theta_\alpha^2 \partial_x^m u dx dy \\
&\leq \sum_{i=0}^{[m/2]} \binom{m}{i} \|\partial_x^{m-i} v\|_{L_x^2 L_y^\infty} \|\theta_\alpha \partial_x^i \partial_y u\|_{L_x^\infty L_y^2} \|\theta_\alpha \partial_x^m u\|_{L^2(\mathbb{R}_+^2)} \\
&\quad + \sum_{i=[m/2]+1}^{m} \binom{m}{i} \|\partial_x^{m-i} v\|_{L_{xy}^\infty} \|\theta_\alpha \partial_x^i \partial_y u\|_{L^2(\mathbb{R}_+^2)} \|\theta_\alpha \partial_x^m u\|_{L^2(\mathbb{R}_+^2)}. \qquad (2.2.20)
\end{aligned}
$$

For $0 \leq i \leq [m/2]$, using Lemmas 2.2.3–2.2.4 and 2.2.2 (the Agmon inequality) in x, we attain

$$
\begin{aligned}
\|\partial_x^{m-i} v\|_{L_x^2 L_y^\infty} &\leq C \|\theta_\alpha \partial_x^{m-i} v\|_{L^2(\mathbb{R}_+^2)}^{\frac{1}{2}} \|\theta_\alpha \partial_y \partial_x^{m-i} v\|_{L^2(\mathbb{R}_+^2)}^{\frac{1}{2}} \\
&\leq C\langle t\rangle^{\frac{1}{4}} \|\theta_\alpha \partial_x^{m-i+1} u\|_{L^2(\mathbb{R}_+^2)}^{\frac{1}{2}} \|\theta_\alpha \partial_x^{m-i+1} u\|_{L^2(\mathbb{R}_+^2)}^{\frac{1}{2}} \\
&= C\langle t\rangle^{\frac{1}{4}} \|\theta_\alpha \partial_x^{m-i+1} u\|_{L^2(\mathbb{R}_+^2)}, \qquad (2.2.21)
\end{aligned}
$$

and

$$
\|\theta_\alpha \partial_x^i \partial_y u\|_{L_x^\infty L_y^2} \leq C \|\theta_\alpha \partial_x^i \partial_y u\|_{L^2(\mathbb{R}_+^2)}^{\frac{1}{2}} \|\theta_\alpha \partial_x^{i+1} \partial_y u\|_{L^2(\mathbb{R}_+^2)}^{\frac{1}{2}}. \qquad (2.2.22)
$$

For $[m/2] \leq i \leq m$, using Lemmas 2.2.3–2.2.4, we find

$$
\begin{aligned}
\|\partial_x^{m-i} v\|_{L_{xy}^\infty} &\leq C \|\theta_\alpha \partial_x^{m-i} v\|_{L^2(\mathbb{R}_+^2)}^{\frac{1}{4}} \|\theta_\alpha \partial_x^{m-i+1} v\|_{L^2(\mathbb{R}_+^2)}^{\frac{1}{4}} \|\theta_\alpha \partial_y \partial_x^{m-i} v\|_{L^2(\mathbb{R}_+^2)}^{\frac{1}{4}} \\
&\quad \times \|\theta_\alpha \partial_x^{m-i+1} \partial_y v\|_{L^2(\mathbb{R}_+^2)}^{\frac{1}{4}} \\
&\leq C\langle t\rangle^{\frac{1}{4}} \|\theta_\alpha \partial_x^{m-i+1} u\|_{L^2(\mathbb{R}_+^2)}^{\frac{1}{2}} \|\theta_\alpha \partial_x^{m-i+2} u\|_{L^2(\mathbb{R}_+^2)}^{\frac{1}{2}}. \qquad (2.2.23)
\end{aligned}
$$

Hence,

$$
\begin{aligned}
\frac{|R_2|\tau^m M_m}{\|\theta_\alpha \partial_x^m u\|_{L^2(\mathbb{R}_+^2)}} &\leq \frac{C}{\tau(t)^{\frac{1}{2}}} \Big(\sum_{i=0}^{[m/2]} \binom{m}{i} \langle t\rangle^{\frac{1}{4}} Y_{m-i+1} D_i^{\frac{1}{2}} D_{i+1}^{\frac{1}{2}} \\
&\quad + \sum_{i=[m/2]+1}^{m} \binom{m}{i} \langle t\rangle^{\frac{1}{4}} Y_{m-i+1}^{\frac{1}{2}} Y_{m-i+2}^{\frac{1}{2}} D_i \Big). \qquad (2.2.24)
\end{aligned}
$$

Inserting (2.2.12)–(2.2.14), (2.2.19), and (2.2.24) into (2.2.11) and summing over $m \geq 0$ and choosing $\alpha \leq \frac{1}{2}$, we obtain

$$
\begin{aligned}
&\frac{d}{dt}\sum_{m\geq 0}\|\theta_\alpha\partial_x^m u\|_{L^2(\mathbb{R}^2_+)}\tau^m M_m+\sum_{m\geq 0}\frac{\alpha(1-2\alpha)}{4\langle t\rangle}\tau^m M_m\frac{\|\theta_\alpha z\partial_x^m u\|^2_{L^2(\mathbb{R}^2_+)}}{\|\theta_\alpha\partial_x^m u\|_{L^2(\mathbb{R}^2_+)}}\\
&\quad+(1-\frac{\alpha}{2\langle t\rangle})\sum_{m\geq 0}\|\theta_\alpha\partial_x^m u\|_{L^2(\mathbb{R}^2_+)}\tau^m M_m+\sum_{m\geq 0}\tau^m M_m\frac{\|\theta_\alpha\partial_x^m\partial_y u\|^2_{L^2(\mathbb{R}^2_+)}}{\|\theta_\alpha\partial_x^m u\|_{L^2(\mathbb{R}^2_+)}}\\
&\leq\frac{C}{\tau(t)^{\frac{1}{2}}}\Big((\sum_{m\geq 0}\|\theta_\alpha\partial_x^m u\|_{L^2(\mathbb{R}^2_+)}\tau^m M_m)^{\frac{1}{2}}(\sum_{m\geq 0}\|\theta_\alpha\partial_y\partial_x^m u\|_{L^2(\mathbb{R}^2_+)}\tau^m M_m)^{\frac{1}{2}}\\
&\quad\times\sum_{m\geq 0}\|\theta_\alpha\partial_x^m u\|_{L^2(\mathbb{R}^2_+)}\tau^{m-1}mM_m\\
&\quad+\langle t\rangle^{\frac{1}{4}}\sum_{m\geq 0}\|\theta_\alpha\partial_y\partial_x^m u\|_{L^2(\mathbb{R}^2_+)}\tau^m M_m\sum_{m\geq 0}\|\theta_\alpha\partial_x^m u\|_{L^2(\mathbb{R}^2_+)}\tau^{m-1}mM_m\Big)\\
&\quad+\dot{\tau}(t)\sum_{m\geq 0}\|\theta_\alpha\partial_x^m u\|_{L^2(\mathbb{R}^2_+)}\tau^{m-1}mM_m\\
&\leq\frac{C}{\tau(t)^{\frac{1}{2}}}\Big((\sum_{m\geq 0}\|\theta_\alpha\partial_x^m u\|_{L^2(\mathbb{R}^2_+)}\tau^m M_m+\sum_{m\geq 0}\|\theta_\alpha\partial_y\partial_x^m u\|_{L^2(\mathbb{R}^2_+)}\tau^m M_m)\\
&\quad\times\sum_{m\geq 0}\|\theta_\alpha\partial_x^m u\|_{L^2(\mathbb{R}^2_+)}\tau^{m-1}mM_m\\
&\quad+\langle t\rangle^{\frac{1}{4}}\sum_{m\geq 0}\|\theta_\alpha\partial_y\partial_x^m u\|_{L^2(\mathbb{R}^2_+)}\tau^m M_m\sum_{m\geq 0}\|\theta_\alpha\partial_x^m u\|_{L^2(\mathbb{R}^2_+)}\tau^{m-1}mM_m\Big)\\
&\quad+\dot{\tau}(t)\sum_{m\geq 0}\|\theta_\alpha\partial_x^m u\|_{L^2(\mathbb{R}^2_+)}\tau^{m-1}mM_m,
\end{aligned}
\tag{2.2.25}
$$

where we have used the fact that for any sequences $\{a_j\}_{j\geq 0}$ and $\{b_j\}_{j\geq 0}$

$$
\sum_{m\geq 0}\sum_{j\geq 0}^{m}a_j b_{m-j}\leq\sum_{j\geq 0}a_j\sum_{j\geq 0}b_j,
$$

and the discrete Hölder's inequality for any sequences $\{a_{ij}\}_{i,j\geq 0}$ and $\{b_{ij}\}_{i,j\geq 0}$

$$
\sum_{j=1}^{m}a_{1j}a_{2j}\cdots a_{mj}\leq\Big(\sum_{j=1}^{m}(a_{1j})^{p_1}\Big)^{\frac{1}{p_1}}\cdots\Big(\sum_{j=1}^{m}(a_{mj})^{p_m}\Big)^{\frac{1}{p_m}},
$$

where $\{p_i\}_{i\le m}$ are positive integers and satisfy $\sum\limits_{i=1}^{m}\frac{1}{p_i}=1$.

It follows from (2.2.25) that

$$\frac{d}{dt}\|u\|_{X_{\tau,\alpha}}+\left(1-\frac{\alpha}{2\langle t\rangle}\right)\|u\|_{X_{\tau,\alpha}}+\sum_{m\ge 0}\tau^m M_m\frac{\|\theta_\alpha\partial_x^m\partial_y u\|^2_{L^2(\mathbb{R}^2_+)}}{\|\theta_\alpha\partial_x^m u\|_{L^2(\mathbb{R}^2_+)}}$$
$$\le\left(\dot{\tau}(t)+\frac{C}{\tau(t)^{\frac{1}{2}}}\left(\|u\|_{X_{\tau,\alpha}}+\langle t\rangle^{\frac{1}{4}}\|u\|_{D_{\tau,\alpha}}\right)\right)\|u\|_{Y_{\tau,\alpha}}. \tag{2.2.26}$$

Then, using Lemma 2.2.3, we discover

$$\sum_{m\ge 0}\frac{\|\theta_\alpha\partial_y\partial_x^m u\|^2_{L^2(\mathbb{R}^2_+)}}{\|\theta_\alpha\partial_x^m u\|_{L^2(\mathbb{R}^2_+)}}\tau^m M_m\ge\frac{\alpha}{2\langle t\rangle^{\frac{1}{2}}}\|u\|_{D_{\tau,\alpha}}+\frac{\alpha(2-\sqrt{\alpha})}{2\langle t\rangle}\|u\|_{X_{\tau,\alpha}}, \tag{2.2.27}$$

which, combined with (2.2.26), gives

$$\frac{d}{dt}\|u\|_{X_{\tau,\alpha}}+\Big(1+\frac{\alpha(1-\sqrt{\alpha})}{2\langle t\rangle}\Big)\|u\|_{X_{\tau,\alpha}}+\frac{\alpha}{2\langle t\rangle^{\frac{1}{2}}}\|u\|_{D_{\tau,\alpha}}$$
$$\le\left(\dot{\tau}(t)+\frac{C}{\tau(t)^{\frac{1}{2}}}\left(\|u\|_{X_{\tau,\alpha}}+\langle t\rangle^{\frac{1}{4}}\|u\|_{D_{\tau,\alpha}}\right)\right)\|u\|_{Y_{\tau,\alpha}}. \tag{2.2.28}$$

We will choose suitable function $\tau(t)$ such that the following ordinary differential equation holds:

$$\frac{d}{dt}(\tau(t))^{\frac{3}{2}}+\frac{3C}{2}\left(\|u\|_{X_{\tau,\alpha}}+\langle t\rangle^{\frac{1}{4}}\|u\|_{D_{\tau,\alpha}}\right)=0. \tag{2.2.29}$$

Inserting (2.2.29) into (2.2.28) and applying $\alpha\in[1/4,1/2]$, we arrive at

$$\begin{aligned}&\frac{d}{dt}\|u\|_{X_{\tau,\alpha}}+\|u\|_{X_{\tau,\alpha}}+\frac{\alpha}{2\langle t\rangle^{\frac{1}{2}}}\|u\|_{D_{\tau,\alpha}}\\&\le\frac{2}{3\tau(t)^{\frac{1}{2}}}\Big(\frac{d}{dt}(\tau(t))^{\frac{3}{2}}+\frac{3C}{2}\left(\|u\|_{X_{\tau,\alpha}}+\langle t\rangle^{\frac{1}{4}}\|u\|_{D_{\tau,\alpha}}\right)\Big)\|u\|_{Y_{\tau,\alpha}}\\&=0,\end{aligned}$$

which gives

$$\frac{d}{dt}\left(e^t\|u\|_{X_{\tau,\alpha}}\right)+\frac{\alpha e^t}{2\langle t\rangle^{\frac{1}{2}}}\|u\|_{D_{\tau,\alpha}}\le 0.$$

Integrating above inequality in time t yields

$$\|u\|_{X_{\tau,\alpha}} + \int_0^t \frac{\alpha e^{-(t-s)}}{2\langle s\rangle^{\frac{1}{2}}}\|u\|_{D_{\tau,\alpha}} ds \leq e^{-t}\|u(0)\|_{X_{\tau_0,\alpha}},$$

implying

$$\|u\|_{X_{\tau,\alpha}} + \frac{\alpha e^{-t}}{2}\int_0^t \frac{1}{\langle s\rangle^{\frac{1}{2}}}\|u\|_{D_{\tau,\alpha}} ds \leq e^{-t}\|u(0)\|_{X_{\tau_0,\alpha}}. \tag{2.2.30}$$

From (2.2.29), we easily derive for all $t \geq 0$ and $\alpha \in [1/4, 1/2]$,

$$\begin{aligned}
\tau(t)^{\frac{3}{2}} - \tau(0)^{\frac{3}{2}} &= -\frac{3C}{2}\int_0^t \left(\|u\|_{X_{\tau,\alpha}} + \langle s\rangle^{\frac{1}{4}}\|u\|_{D_{\tau,\alpha}}\right) ds \\
&\geq -\frac{3C\|u(0)\|_{X_{\tau_0,\alpha}}}{2}\left(1 + \frac{2}{\alpha}\langle t\rangle^{\frac{3}{4}}\right) \\
&\geq -\frac{3C(\alpha+2)\|u(0)\|_{X_{\tau_0,\alpha}}}{2\alpha}\langle t\rangle^{\frac{3}{4}} \\
&\geq -C_1\varepsilon\langle t\rangle^{\frac{3}{4}},
\end{aligned} \tag{2.2.31}$$

where the positive constant $C_1 = \frac{3C(\alpha+2)}{2\alpha}$.

Therefore, we choose the initial perturbation datum is so small enough that

$$\tau(t) \geq \frac{\tau_0}{4}, \tag{2.2.32}$$

in the time interval $[0, T_\varepsilon]$, where T_ε satisfies

$$T_\varepsilon = C_2\varepsilon^{-\frac{4}{3}} - 1,$$

with $C_2 = \left(\frac{7}{8C_1}\tau_0^{\frac{3}{2}}\right)^{\frac{4}{3}}$. The proof is thus complete. □

We have completed the proof of the existence of solutions to the 2D Prandtl-Hartmann equations (2.2.8)–(2.2.10) with damping term and derived the lifespan T_ε of solutions.

2.2.2 Uniqueness of Solutions

In this subsection, we shall prove the uniqueness of solutions to Eqs. (2.2.8)–(2.2.10). Let $u_1(t, x, y)$ and $u_2(t, x, y)$ are two solutions to the Eqs. (2.2.8)–(2.2.10) with same initial data $u_1(0, x, y) = u_2(0, x, y) \in X_{\tau_0,\alpha}$. Represent the tangential radii of analytic regularity of u_1 and u_2 by $\tau_1(t)$ and $\tau_2(t)$, respectively, which satisfy the bounds in Theorem 2.2.1. Define $\tau(t)$ to solve

$$\dot{\tau}(t) + \frac{C}{\tau(t)^{\frac{1}{2}}}\big(\|u_1\|_{X_{\tau_1,\alpha}} + \langle t\rangle^{\frac{1}{4}}\|u_1\|_{D_{\tau_1,\alpha}}\big) = 0, \ \tau(0) = \frac{\tau_0}{8}. \tag{2.2.33}$$

From the above equation, we can deduce

$$\begin{aligned}
&\dot{\tau}(t) + \frac{C}{\tau(t)^{\frac{1}{2}}}\big(\|u_1\|_{X_{\tau,\alpha}} + \langle t\rangle^{\frac{1}{4}}\|u_1\|_{D_{\tau,\alpha}}\big)\\
&\leq \dot{\tau}(t) + \frac{C}{\tau(t)^{\frac{1}{2}}}\big(\|u_1\|_{X_{\tau_1,\alpha}} + \langle t\rangle^{\frac{1}{4}}\|u_1\|_{D_{\tau_1,\alpha}}\big) = 0,
\end{aligned} \tag{2.2.34}$$

where we have used the fact that $\tau(t) \leq \tau_1(t)$ and the norms $X_{\tau,\alpha}$ and $D_{\tau,\alpha}$ are increasing with respect to τ.

In view of the estimate (2.2.30) for u_1 and the lower bounds (2.2.32) for $\tau_1(t)$ and $\tau_2(t)$, we conclude

$$\frac{\tau_0}{32} \leq \tau(t) \leq \frac{\tau_0}{8} \leq \frac{\min\{\tau_1(t), \tau_2(t)\}}{2}, \tag{2.2.35}$$

for all $t \in [0, T_\varepsilon]$.

On the one hand, from [59], the following inequality holds:

$$\|u\|_{Y_{\tau,\alpha}} \leq \tau^{-1}\|u\|_{X_{2\tau,\alpha}}. \tag{2.2.36}$$

On the other hand, using (2.2.30) and (2.2.35)–(2.2.36), we also have

$$\|u_2\|_{Y_{\tau,\alpha}} \leq \frac{1}{\tau(t)}\|u_2\|_{X_{2\tau,\alpha}} \leq \frac{1}{\tau(t)}\|u_2\|_{X_{\tau_2,\alpha}} \leq \frac{\varepsilon}{\tau(t)}. \tag{2.2.37}$$

Setting $u = u_1 - u_2$, then the system (2.2.8) reads as

$$\begin{cases}
\partial_t u + (\bar{u} + u_1)\partial_x u + u\partial_x u_2 + v\partial_y(u_1 + \bar{u}) + v_2\partial_y u - \partial_y^2 u + u = 0,\\
\partial_x u + \partial_y v = 0.
\end{cases} \tag{2.2.38}$$

For $m \geq 0$, applying the operator ∂_x^m on $(2.2.38)_1$, multiplying the resulting equation by $\theta_\alpha^2\partial_x^m u$ and integrating it by parts over $\mathbb{R}_+^2$, we derive that

$$\int_{\mathbb{R}_+^2} \partial_x^m\big(\partial_t u + u_1\partial_x u + u\partial_x u_2 + v\partial_y u_1 + v_2\partial_y u - \partial_y^2 u + u\big)\theta_\alpha^2\partial_x^m u dxdy = 0. \tag{2.2.39}$$

Now, we deal with this integral equality. Similar to the estimates given in Sect. 2.2.1, we can deduce

$$\int_{\mathbb{R}^2_+} \partial_t \partial_x^m u \theta_\alpha^2 \partial_x^m u dxdy = \frac{1}{2}\frac{d}{dt}\|\theta_\alpha \partial_x^m u\|^2_{L^2(\mathbb{R}^2_+)} + \frac{\alpha}{4\langle t\rangle}\|\theta_\alpha z \partial_x^m u\|^2_{L^2(\mathbb{R}^2_+)}, \tag{2.2.40}$$

$$\int_{\mathbb{R}^2_+} \partial_x^m u \theta_\alpha^2 \partial_x^m u dxdy = \|\theta_\alpha \partial_x^m u\|^2_{L^2(\mathbb{R}^2_+)}, \tag{2.2.41}$$

and

$$\begin{aligned}
&-\int_{\mathbb{R}^2_+} \partial_y^2 \partial_x^m u \theta_\alpha^2 \partial_x^m u dxdy \\
=&\|\theta_\alpha \partial_x^m \partial_y u\|^2_{L^2(\mathbb{R}^2_+)} - \frac{\alpha}{2\langle t\rangle}\|\theta_\alpha \partial_x^m u\|^2_{L^2(\mathbb{R}^2_+)} - \frac{\alpha^2}{2\langle t\rangle}\|\theta_\alpha z \partial_x^m u\|^2_{L^2(\mathbb{R}^2_+)}.
\end{aligned} \tag{2.2.42}$$

Next, we estimate the nonlinear terms in (2.2.39) in which the first term can be estimated as follows:

$$\begin{aligned}
\tilde{R}_1 =& \int_{\mathbb{R}^2_+} \partial_x^m (u_1 \partial_x u) \theta_\alpha^2 \partial_x^m u dxdy = \sum_{i=0}^{m} \binom{m}{i} \int_{\mathbb{R}^2_+} \partial_x^{m-i} u_1 \partial_x^{i+1} u \theta_\alpha^2 \partial_x^m u dxdy \\
\le& \sum_{i=0}^{[m/2]} \binom{m}{i} \|\partial_x^{m-i} u_1\|_{L^2_x L^\infty_y} \|\theta_\alpha \partial_x^{i+1} u\|_{L^\infty_x L^2_y} \|\theta_\alpha \partial_x^m u\|_{L^2(\mathbb{R}^2_+)} \\
&+ \sum_{i=[m/2]+1}^{m} \binom{m}{i} \|\partial_x^{m-i} u_1\|_{L^\infty_{xy}} \|\theta_\alpha \partial_x^{i+1} u\|_{L^2(\mathbb{R}^2_+)} \|\theta_\alpha \partial_x^m u\|_{L^2(\mathbb{R}^2_+)}.
\end{aligned} \tag{2.2.43}$$

For $0 \le i \le [m/2]$, using Lemmas 2.2.4 and 2.2.2 (the Agmon inequality) in x, we deduce

$$\|\partial_x^{m-i} u_1\|_{L^2_x L^\infty_y} \le C\|\theta_\alpha \partial_x^{m-i} u_1\|^{\frac{1}{2}}_{L^2(\mathbb{R}^2_+)} \|\theta_\alpha \partial_y \partial_x^{m-i} u_1\|^{\frac{1}{2}}_{L^2(\mathbb{R}^2_+)}, \tag{2.2.44}$$

and

$$\|\theta_\alpha \partial_x^{i+1} u\|_{L^\infty_x L^2_y} \le C\|\theta_\alpha \partial_x^{i+1} u\|^{\frac{1}{2}}_{L^2(\mathbb{R}^2_+)} \|\theta_\alpha \partial_x^{i+2} u\|^{\frac{1}{2}}_{L^2(\mathbb{R}^2_+)}. \tag{2.2.45}$$

For $[m/2] \le i \le m$, using Lemma 2.2.4, we arrive at

$$\begin{aligned}
\|\partial_x^{m-i} u_1\|_{L^\infty_{xy}} \le& C\|\theta_\alpha \partial_x^{m-i} u_1\|^{\frac{1}{4}}_{L^2(\mathbb{R}^2_+)} \|\theta_\alpha \partial_x^{m-i+1} u_1\|^{\frac{1}{4}}_{L^2(\mathbb{R}^2_+)} \\
&\times \|\theta_\alpha \partial_y \partial_x^{m-i} u_1\|^{\frac{1}{4}}_{L^2(\mathbb{R}^2_+)} \|\theta_\alpha \partial_x^{m-i+1} \partial_y u_1\|^{\frac{1}{4}}_{L^2(\mathbb{R}^2_+)}.
\end{aligned} \tag{2.2.46}$$

Therefore,

$$
\begin{aligned}
\frac{|\tilde{R}_1|\tau^m M_m}{\|\theta_\alpha \partial_x^m u\|_{L^2(\mathbb{R}_+^2)}} \le \frac{C}{\tau(t)^{\frac{1}{2}}}\Big(&\sum_{i=0}^{[m/2]} \binom{m}{i} \tilde{X}_{m-i}^{\frac{1}{2}} \tilde{D}_{m-i}^{\frac{1}{2}} Y_{i+1}^{\frac{1}{2}} Y_{i+2}^{\frac{1}{2}} \\
&+ \sum_{i=[m/2]+1}^{m} \binom{m}{i} \tilde{X}_{m-i}^{\frac{1}{4}} \tilde{X}_{m-i+1}^{\frac{1}{4}} \tilde{D}_{m-i}^{\frac{1}{4}} \tilde{D}_{m-i+1}^{\frac{1}{4}} Y_{i+1}\Big). \quad (2.2.47)
\end{aligned}
$$

From now on, we use $\tilde{X}_i$, $\tilde{Y}_i$, $\tilde{D}_i$ to stand for the semi-norms of function u_1 and $\bar{X}_i$, $\bar{Y}_i$, $\bar{D}_i$ to stand for the semi-norms of function u_2. Similarly,

$$
\begin{aligned}
\frac{|\tilde{R}_2|\tau^m M_m}{\|\theta_\alpha \partial_x^m u\|_{L^2(\mathbb{R}_+^2)}} \le \frac{C}{\tau(t)^{\frac{1}{2}}}\Big(&\sum_{i=0}^{[m/2]} \binom{m}{i} X_{m-i}^{\frac{1}{2}} D_{m-i}^{\frac{1}{2}} \bar{Y}_{i+1}^{\frac{1}{2}} \bar{Y}_{i+2}^{\frac{1}{2}} \\
&+ \sum_{i=[m/2]+1}^{m} \binom{m}{i} X_{m-i}^{\frac{1}{4}} X_{m-i+1}^{\frac{1}{4}} D_{m-i}^{\frac{1}{4}} D_{m-i+1}^{\frac{1}{4}} \bar{Y}_{i+1}\Big), \quad (2.2.48)
\end{aligned}
$$

$$
\begin{aligned}
\frac{|\tilde{R}_3|\tau^m M_m}{\|\theta_\alpha \partial_x^m u\|_{L^2(\mathbb{R}_+^2)}} \le \frac{C}{\tau(t)^{\frac{1}{2}}}\Big(&\sum_{i=0}^{[m/2]} \binom{m}{i} \langle t\rangle^{\frac{1}{4}} Y_{m-i+1} \tilde{D}_i^{\frac{1}{2}} \tilde{D}_{i+1}^{\frac{1}{2}} \\
&+ \sum_{i=[m/2]+1}^{m} \binom{m}{i} \langle t\rangle^{\frac{1}{4}} Y_{m-i+1}^{\frac{1}{2}} Y_{m-i+2}^{\frac{1}{2}} \tilde{D}_i\Big), \quad (2.2.49)
\end{aligned}
$$

and

$$
\begin{aligned}
\frac{|\tilde{R}_4|\tau^m M_m}{\|\theta_\alpha \partial_x^m u\|_{L^2(\mathbb{R}_+^2)}} \le \frac{C}{\tau(t)^{\frac{1}{2}}}\Big(&\sum_{i=0}^{[m/2]} \binom{m}{i} \langle t\rangle^{\frac{1}{4}} \bar{Y}_{m-i+1} D_i^{\frac{1}{2}} D_{i+1}^{\frac{1}{2}} \\
&+ \sum_{i=[m/2]+1}^{m} \binom{m}{i} \langle t\rangle^{\frac{1}{4}} \bar{Y}_{m-i+1}^{\frac{1}{2}} Y_{m-i+2}^{\frac{1}{2}} D_i\Big). \quad (2.2.50)
\end{aligned}
$$

Similarly to (2.2.25), inserting (2.2.40)–(2.2.50) into (2.2.39) and summing over $m \ge 0$, we attain

$$
\begin{aligned}
&\frac{d}{dt}\sum_{m\ge 0} \|\theta_\alpha \partial_x^m u\|_{L^2(\mathbb{R}_+^2)} \tau^m M_m + \sum_{m\ge 0} \frac{\alpha(1-2\alpha)}{4\langle t\rangle} \tau^m M_m \frac{\|\theta_\alpha z \partial_x^m u\|_{L^2(\mathbb{R}_+^2)}^2}{\|\theta_\alpha \partial_x^m u\|_{L^2(\mathbb{R}_+^2)}} \\
&+(1-\frac{\alpha}{2\langle t\rangle}) \sum_{m\ge 0} \|\theta_\alpha \partial_x^m u\|_{L^2(\mathbb{R}_+^2)} \tau^m M_m + \sum_{m\ge 0} \tau^m M_m \frac{\|\theta_\alpha \partial_x^m \partial_y u\|_{L^2(\mathbb{R}_+^2)}^2}{\|\theta_\alpha \partial_x^m u\|_{L^2(\mathbb{R}_+^2)}}
\end{aligned}
$$

$$
\begin{aligned}
&\leq \frac{C}{\tau(t)^{\frac{1}{2}}}\Big(\big(\sum_{m\geq 0}\|\theta_\alpha\partial_x^m u_1\|_{L^2(\mathbb{R}^2_+)}\tau^m M_m+\sum_{m\geq 0}\|\theta_\alpha\partial_y\partial_x^m u_1\|_{L^2(\mathbb{R}^2_+)}\tau^m M_m\big)\\
&\times\sum_{m\geq 0}\|\theta_\alpha\partial_x^m u\|_{L^2(\mathbb{R}^2_+)}\tau^{m-1}mM_m\\
&+\langle t\rangle^{\frac{1}{4}}\sum_{m\geq 0}\|\theta_\alpha\partial_y\partial_x^m u_1\|_{L^2(\mathbb{R}^2_+)}\tau^m M_m\sum_{m\geq 0}\|\theta_\alpha\partial_x^m u\|_{L^2(\mathbb{R}^2_+)}\tau^{m-1}mM_m\Big)\\
&+\frac{C}{\tau(t)^{\frac{1}{2}}}\Big(\big(\sum_{m\geq 0}\|\theta_\alpha\partial_x^m u\|_{L^2(\mathbb{R}^2_+)}\tau^m M_m+\sum_{m\geq 0}\|\theta_\alpha\partial_y\partial_x^m u\|_{L^2(\mathbb{R}^2_+)}\tau^m M_m\big)\\
&\times\sum_{m\geq 0}\|\theta_\alpha\partial_x^m u_2\|_{L^2(\mathbb{R}^2_+)}\tau^{m-1}mM_m\\
&+\langle t\rangle^{\frac{1}{4}}\sum_{m\geq 0}\|\theta_\alpha\partial_y\partial_x^m u\|_{L^2(\mathbb{R}^2_+)}\tau^m M_m\sum_{m\geq 0}\|\theta_\alpha\partial_x^m u_2\|_{L^2(\mathbb{R}^2_+)}\tau^{m-1}mM_m\Big)\\
&+\dot{\tau}(t)\sum_{m\geq 0}\|\theta_\alpha\partial_x^m u\|_{L^2(\mathbb{R}^2_+)}\tau^{m-1}mM_m. \qquad (2.2.51)
\end{aligned}
$$

From (2.2.51) and using Lemma 2.2.3, for $\alpha\in[1/4,1/2]$, we can deduce that

$$
\begin{aligned}
&\frac{d}{dt}\|u\|_{X_{\tau,\alpha}}+\|u\|_{X_{\tau,\alpha}}+\frac{\alpha}{2\langle t\rangle^{\frac{1}{2}}}\|u\|_{D_{\tau,\alpha}}\\
&\leq\Big(\dot{\tau}(t)+\frac{C}{\tau(t)^{\frac{1}{2}}}\big(\|u_1\|_{X_{\tau,\alpha}}+\langle t\rangle^{\frac{1}{4}}\|u_1\|_{D_{\tau,\alpha}}\big)\Big)\|u\|_{Y_{\tau,\alpha}}\\
&\quad+\frac{C}{\tau(t)^{\frac{1}{2}}}\big(\|u\|_{X_{\tau,\alpha}}+\langle t\rangle^{\frac{1}{4}}\|u\|_{D_{\tau,\alpha}}\big)\|u_2\|_{Y_{\tau,\alpha}}\\
&\leq\frac{C\varepsilon}{\tau^{\frac{3}{2}}(t)}\big(\|u\|_{X_{\tau,\alpha}}+\langle t\rangle^{\frac{1}{4}}\|u\|_{D_{\tau,\alpha}}\big). \qquad (2.2.52)
\end{aligned}
$$

According to the inequality (2.2.3), we can derive the following inequalities:

$$
\frac{C\varepsilon}{\tau^{\frac{3}{2}}(t)}\leq\frac{128\sqrt{2}C\varepsilon}{\tau_0^{\frac{3}{2}}}\leq 1,\quad \frac{C\varepsilon\langle t\rangle^{\frac{3}{4}}}{\tau^{\frac{3}{2}}(t)}\leq\frac{128\sqrt{2}CC_2^{\frac{3}{4}}}{\tau_0^{\frac{3}{2}}}\leq\frac{\alpha}{2}. \qquad (2.2.53)
$$

Inserting (2.2.53) into (2.2.52) that

$$
\frac{d}{dt}\|u\|_{X_{\tau,\alpha}}+\delta\|u\|_{X_{\tau,\alpha}}\leq 0, \qquad (2.2.54)
$$

where $\delta \in (0, 1)$ is a small constant. This implies the uniqueness of solutions since $u(0) = 0$. The proof of uniqueness is thus complete. □

Combining (2.2.30) with (2.2.54), we complete the proof of Theorem 2.2.1. □

2.3 Bibliographic Comments

Let us briefly review some known results of problems (2.1.1)–(2.1.3). Especially, when the damping term u_1 does not exist, systems (2.1.1)–(2.1.3) reduce to the classical Prandtl equations which was firstly introduced formally by Prandtl [111] in 1904. This system is the foundation of the boundary layer equations, which describes that the away from the boundary part can be considered as general ideal fluid, but the near a rigid wall part is deeply affected by the viscous force. Formally, the asymptotic limit of the Navier-Stokes equations can be denoted by the Prandtl equations within the boundary layer and by the Euler equations away from boundary. About 60 years later, Oleinik [104] established the first systematic work in strictly mathematical analysis, in which she pointed out that the local-in-time well-posedness of the Prandtl system can be proved in 2D by using the Crocco transformation under the monotonicity condition on the tangential velocity field in the normal variable to the boundary. This result together with an expanded introduction to the boundary layer theory was showed in Oleinik-Samokhin's classical book [106]. In addition to Oleinik′s monotonicity assumption on the tangential velocity field, Xin and Zhang [149] obtained the existence of global weak solutions to the Prandtl equation (2.3.1) when the pressure is favorable ($\partial_x p \leq 0$) where the 2D Prandtl system (2.3.1) takes the form

$$\begin{cases} \partial_t u + u\partial_x u + v\partial_y u + \partial_x p = \partial_y^2 u, \, (x, y, t) \in \mathbb{R}_+^2 \times \mathbb{R}_+, \\ \partial_x u + \partial_y v = 0, \ \ (x, y, t) \in \mathbb{R}_+^2 \times \mathbb{R}_+, \\ (u, v)|_{y=0} = 0, \ \ \lim\limits_{y \to +\infty} u = U, \ \ \text{on } \mathbb{R}^2 \times \mathbb{R}_+, \\ u|_{t=0} = u_0(x, y), \, \text{on } \mathbb{R}_+^2, \end{cases} \tag{2.3.1}$$

There have been some results in the analytic framework with analytic radii $\tau(t)$ for 2D boundary layer equations. First, authors of [65] investigated the local well-posedness of solutions to the 2D Prandtl and hydrostatic Euler equations by energy methods, which was the original result for boundary layer problem in the analytic framework with analytic radii $\tau(t)$. Kukavica and Vicol [66] considered the local well-posedness of solutions to the 2D Prandtl boundary layer equations with general initial data by using analytic energy estimates in the tangential variables. Later on, Ignatova and Vicol [59] investigated the almost global existence for the two-dimensional Prandtl equations when the small initial datum lies in an exponential weighted space by applying analytic energy estimates in the tangential variables and got the unique solution can be extended at least up to $T_\varepsilon \geq \exp(\varepsilon^{-1}/\log(\varepsilon^{-1}))$.

There have also been a lot of results on the 2D Prandtl equation in different frameworks, for example, weighted Sobolev space framework and Gevrey framework, etc. Under the Oleinik′s monotonicity assumption, some authors in [3, 98, 151] proved the well-posedness of solution for 2D Prandtl equations by using energy method and constructing a new unknown function to eliminate the troublesome term from the convection term. Qin and Wang [117] investigated the local existence of solutions to 2D magnetic Prandtl model in the Prandtl-Hartmann regime by the energy method; see also Chap. 5 of this book. So far, besides the well-posedness of solutions to Prandtl equation in weighted Sobolev space, there are so many results on the existence theory of solutions, for example, authors in [11, 19, 43, 72, 107, 154] investigated the well-posedness of the Prandtl equations for the initial data in Gevrey framework. There have also been some results for the Prandtl equations by using the methods of different from the energy method. For interested readers, we refer to [6–8, 40, 95, 107, 156] and the references therein for the recent progress.

Xu and Zhao [152] obtained the global existence and uniqueness of weak solutions to the 2D Prandtl equation (3.1.1) (see, Chap. 3) with non-stationary boundary layer by the Crocco transformation and assumed the outer flow has a structure $U(t,x,y) = x^m U_1(t,x,y)$ where the $x \in [0,L]$, $m \geq 1$ and $U_1(t,x,y) > 0$ and obtained the existence and uniqueness of weak solutions to the Prandtl equations for any time T and any interval L.

Motivated by Xie and Yang [148] and Ignatova and Vicol [59], we investigate the global well-posedness of solutions to the 2D Prandtl-Hartmann equations (2.1.1)–(2.1.3) in an analytic framework by the standard energy method, the proof of which can be divided into two sections. First, we prove the existence of solutions for Eqs. (2.1.1)–(2.1.3) by using the classical energy method. The difficulty of solving these problems (2.1.1)–(2.1.3) in the analytic framework is the loss of x-derivative in the term $v\partial_y u$. To overcome this difficulty, we show the Gaussian-weighted Poincaré inequality (see Lemma 2.2.3). In other words, the Poincaré inequality does not holds in unbounded domains. Then, similar to the estimates of the existence of solutions, we verify the uniqueness of solutions for Eqs. (2.1.1)–(2.1.3) as usual. Compared to the existence and uniqueness of solutions to the classical Prandtl equations, the monotonicity condition of the tangential velocity, which is not needed for the 2D Prandtl-Hartmann equations (2.1.1)–(2.1.3) in analytic framework, plays a key role. Besides, compared with the existence and uniqueness of solutions to the 2D MHD boundary layer where the initial tangential magnetic field has a lower bound, which is also not needed for the 2D Prandtl-Hartmann equations (2.1.1)–(2.1.3) in analytic framework, this condition plays an important role.

In addition to the above results for 2D Prandtl problem (2.3.1), there are some results on the vanishing viscosity limit for 2D Prandtl problem (2.3.1); we refer to [5, 41, 42, 51, 53, 97, 121] and the references therein for the recent progress. So far, besides the existence of solutions to Prandtl equation (2.3.1) in weighted Sobolev spaces, there are so many results on the existence theory of solutions, which have been investigated for the Prandtl boundary layer system (2.3.1) when the initial data belong to the following different suitable function spaces:

(1) The well-posedness of solutions were investigated for the Prandtl equations (2.3.1) when the initial data with respect to the tangential variable are analytic; see [6–8, 59, 65, 66, 95, 156].
(2) The authors in [11, 19, 43, 72, 107, 154] investigated the well-posedness of the Prandtl equations (2.3.1) for the initial data with Gevrey class in the horizontal variable x and Sobolev regularity in normal variable y. Therefore, it is not necessary to use the monotonicity condition of tangential velocity to overcome the loss of tangential derivative.

It is noteworthy that the above results are mainly devoted to the 2D Prandtl equations and there also have been some results [30, 79, 84–86, 96] for the 3D Prandtl equations as follows:

$$
\begin{cases}
\partial_t u + \left(u\partial_x + v\partial_y + w\partial_z\right) u - \partial_z^2 u + \partial_x p = 0, \text{ in } Q,\\
\partial_t v + \left(u\partial_x + v\partial_y + w\partial_z\right) v - \partial_z^2 v + \partial_y p = 0, \text{ in } Q,\\
\partial_x u + \partial_y v + \partial_z w = 0, \text{ in } Q,\\
(u, v, w)|_{z=0} = 0, \quad \lim_{z\to+\infty} (u, v) = (U(t, x, y), V(t, x, y)),\\
(u, v, w)(t) = (u_0, v_0, w_0), \text{ on } D \times \mathbb{R}_+,
\end{cases}
\tag{2.3.2}
$$

where $Q = \{t > 0, (x, y) \in D \subset \mathbb{R}^2, z > 0\}$. Functions $(U(t, x, y), V(t, x, y))$ and $p(t, x, y)$ are the values on the boundary of the Euler's tangential velocity and Euler's pressure of the outflow, which satisfy Bernoulli's law

$$
\begin{cases}
\partial_t U(t, x, y) + U(t, x, y)\partial_x U(t, x, y) + V(t, x, y)\partial_y U(t, x, y) + \partial_x p(t, x, y) = 0,\\
\partial_t V(t, x, y) + U(t, x, y)\partial_x V(t, x, y) + V(t, x, y)\partial_y V(t, x, y) + \partial_y p(t, x, y) = 0.
\end{cases}
$$

Under a special structure on the velocity of the outer Euler flow, Liu, Wang, and Yang [86] studied the local well-posedness of classical solutions of the 3D Prandtl equations (2.3.2) by using the Crocco transformation. The local classical solutions structure are the same as the structure of the velocity of the outer Euler flow. Moreover, they proved that this structured flow is linearly stable for any smooth 3D perturbation. A continuous result was obtained with the same structure conditions [86] for 3D Prandtl equation (2.3.2) in [85]; they proved the global existence of weak solutions by the Crocco transformation to the 3D Prandtl equation (2.3.2). Li and Yang [75] investigated the locally well-posedness in the Gevrey function space with Gevrey index in $[1, 2]$ to 3D Prandtl equation (2.3.2) without any monotonicity condition on the velocity field and improved some new observation of cancellation mechanism in the three space dimensional system.

For general MHD equations, there have been some results in the analytic framework. Xie and Yang [147] studied the global existence of solutions to the 2D MHD boundary layer equations in the mixed Prandtl and Hartmann regime when initial datum is a small perturbation of the Hartmann profile and proved the solution in an analytic norm is an exponential decay in time. Recently, Liu and Zhang [94] established the global existence and the asymptotic decay estimate of solutions to the 2D MHD boundary layer equations with small initial data. Inspired by Ignatova and Vicol [59], Xie and Yang [148] investigated the lifespan of solution to the 2D MHD boundary layer system by using the cancellation mechanism and obtained that the lifespan of solution has a lower bound. Motivated by Kukavica and Vicol [66], Lin and Zhang [78] studied the local existence of solutions for 2D incompressible magneto-micropolar boundary layer equations when the initial data are analytic in the x variable by using the energy methods. There have also been some results for the MHD equations by using the methods of the different from energy method. For interested readers, we refer to [35, 58, 83, 87, 89, 90] and the references therein for the recent progress. Liu, Xie, and Yang [87] studied the well-posedness of solutions to the 2D MHD in an analytic framework, and inspired by Gérard-Varet and Dormy [40] on the classical Prandtl equations, they proved that if the tangential magnetic field is degenerate sufficiently, then the non-degenerate critical point in the tangential velocity does not prevent the formation of singularity. Liu, Xie, and Yang [89] investigated the local existence and uniqueness of solutions in weighed Sobolev space for the two-dimensional nonlinear MHD boundary layer equations by using energy method under the assumption that the initial tangential magnetic field has a lower bound $\delta_0 > 0$. As a continuation of [89], the same authors in [90] proved the validity of the Prandtl boundary layer expansion and gave a L^∞ estimate on the error by multi-scale analysis under the assumptions that both the viscosity and resistivity coefficients with same order and the initial tangential magnetic field on the boundary is not degenerate. Liu, Wang, Xie, and Yang [83] proved the local well-posedness of the 2D MHD boundary layer equations in Sobolev spaces and used the initial tangential magnetic field has a lower bound $\delta_0 > 0$ instead of the monotonicity condition of the velocity field. Finally, they got the linear instability of the 2D MHD boundary layer when the tangential magnetic field is degenerate at one point. So far, besides the well-posedness of solutions in the frameworks of Sobolev space and analytic class, there have been some results on the vanishing limits for the incompressible MHD systems; we refer to [134, 136, 141] and the references therein for the recent progress.

For the following 2D compressible MHD boundary layer, Huang, Liu, and Yang [58] attained the local well-posedness of solutions to the 2D MHD system in weighted Sobolev spaces by applying the classical iteration scheme under the tangential magnetic field that is the non-degeneracy instead of monotonicity of velocity field and assuming the viscosity, heat conductivity, and magnetic diffusivity coefficients tend to zero in the same rate

$$\begin{cases} \partial_t \rho + \nabla \cdot (\rho \mathbf{u}) = 0, \\ \rho\,(\partial_t \mathbf{u} + (\mathbf{u} \cdot \nabla)\mathbf{u}) + \nabla p(\rho, \theta) - (\nabla \times \mathbf{H}) \times \mathbf{H} \\ = \mu \Delta \mathbf{u} + (\lambda + \mu)\nabla(\nabla \cdot \mathbf{u}), \\ c_V \rho\,(\partial_t \theta + (\mathbf{u} \cdot \nabla)\theta) + p(\rho, \theta)\nabla \cdot \mathbf{u} \\ = \kappa \Delta \theta + \lambda(\nabla \cdot \mathbf{u})^2 + 2\mu|\mathfrak{D}(\mathbf{u})|^2 + \nu|\nabla \times \mathbf{H}|^2, \\ \partial_t \mathbf{H} - \nabla \times (u \times \mathbf{H}) = \nu \Delta \mathbf{H}, \quad \nabla \cdot \mathbf{H} = 0, \end{cases}$$

where unknown functions ρ, $\mathbf{u}$, and $\mathbf{H}$ stand for the density, velocity, absolute temperature, and magnetic field, respectively, and $p(\rho, \theta)$ represents the pressure of fluid. The constant c_V is the specific heat capacity, λ and μ are the viscosity coefficients satisfying $\mu > 0$, $\lambda + \mu > 0$, κ is the heat conductivity coefficient, and ν is the magnetic diffusivity coefficient. The deformation tensor $\mathfrak{D}(\mathbf{u})$ is given by

$$\mathfrak{D}(\mathbf{u}) = \frac{\nabla \mathbf{u} + (\nabla \mathbf{u})'}{2}.$$

For the following 2D incompressible MHD boundary layer equations, Gao, Huang, and Yao [35] investigated the local well-posedness of solutions in weighted conormal Sobolev spaces to the 2D MHD boundary layer equations with any large initial data by the energy method under the assumption that initial tangential magnetic field has a lower bound $\delta > 0$

$$\begin{cases} \partial_t \rho + \nabla \cdot (\rho \mathbf{u}) = 0, \\ \rho\,(\partial_t \mathbf{u} + (\mathbf{u} \cdot \nabla)\mathbf{u}) + \nabla p - \mu \Delta \mathbf{u} = (\mathbf{H} \cdot \nabla)\mathbf{H}, \\ \partial_t \mathbf{H} - \nabla \times (u \times \mathbf{H}) = \kappa \Delta \mathbf{H}, \\ \nabla \cdot \mathbf{u} = 0, \quad \nabla \cdot \mathbf{H} = 0. \end{cases}$$

Chapter 3
Local Existence of Solutions to the 2D Prandtl Equations in a Weighted Sobolev Space

In this chapter, we shall investigate two-dimensional nonlinear Prandtl equations on the half plane and prove the local existence of solutions by energy methods in an exponential weighted Sobolev space. We use the skill of cancellation mechanism and construct a new unknown function to overcome some difficulties, respectively. The content of this chapter is chosen from [115].

3.1 Introduction

In this chapter, we investigate the local existence of the initial boundary value problem for the 2D Prandtl system which reads as follows:

$$\begin{cases} \partial_t u + u\partial_x u + v\partial_y u + \partial_x p = \partial_y^2 u, \ (x, y, t) \in \mathbb{R}_+^2 \times \mathbb{R}_+, \\ \partial_x u + \partial_y v = 0, \ \ (x, y, t) \in \mathbb{R}_+^2 \times \mathbb{R}_+, \\ (u, v)|_{y=0} = 0, \ \ \lim\limits_{y\to+\infty} u = U, \ \ \text{on}\ \mathbb{R}^2 \times \mathbb{R}_+, \\ u|_{t=0} = u_0(x, y), \ \text{on}\ \mathbb{R}_+^2, \end{cases} \tag{3.1.1}$$

where $\mathbb{R}_+^2 = \{(x, y) \in \mathbb{R}^2; y > 0\}$ and $\mathbb{R}_+ = [0, +\infty)$, $u(x, y, t)$ and $v(x, y, t)$ stand for the tangential velocity and normal velocity, respectively. Functions $U(x, t)$ and $p(x, t)$ are the values on the boundary of the Euler's tangential velocity and Euler's pressure of the outflow, respectively, which satisfy Bernoulli's law

$$\partial_t U(x, t) + U(x, t)\partial_x U(x, t) + \partial_x p(x, t) = 0.$$

Y. Qin et al., *Prandtl Equations and Related Boundary Layer Equations*,
https://doi.org/10.1007/978-981-97-4565-4_3

3.2 Local Well-Posedness of Solutions

As a preparation, we define some weighted Sobolev spaces and Sobolev norms as follows:

$$\|f\|_{L^2_\mu}^2 = \|e^{\mu} f\|_{L^2}^2, \quad \|f\|_{H_\mu^{m,m-1}}^2 = \sum_{0\le\alpha+\beta\le m,\ \alpha\le m-1} \|\partial_x^\alpha \partial_y^\beta f\|_{L^2_\mu}^2,$$

$$\|f\|_{H_\mu^m}^2 = \|f\|_{H_\mu^{m,m-1}}^2 + \|\partial_x^m f\|_{L^2_\mu}^2, \quad D_\mu^m(\mathbb{R}_+^2) = \left\{u : \partial_y u \in H_\mu^m(\mathbb{R}_+^2)\right\},$$

where $\mu = \frac{y}{4}$. If $f|_{y=0} = 0$, then we get the following inequality:

$$\|f\|_{L^\infty(\mathbb{R}_+^2)} \le C\|\partial_y f\|_{L^2_\mu}^{\frac{1}{2}} \|\partial_x \partial_y f\|_{L^2_\mu}^{\frac{1}{2}}. \tag{3.2.1}$$

Suppose that u_0^s satisfies the following conditions:

$$\begin{cases} \partial_y^{2j} u_0^s(0) = \tilde{u}_0|_{y=0} = 0, \ 0 \le j \le 2, \\ C^{-1}e^{-\frac{y}{4}} \le \partial_y u_0^s(y), \ \text{for } y \in [0, +\infty), \\ |\partial_y^p u_0^s(y)| \le Ce^{-\frac{y}{4}}, \ \text{for } y \in [0, +\infty), \ 0 \le p \le 5. \end{cases} \tag{3.2.2}$$

We can now state our main result of this chapter as follows.

Theorem 3.2.1 *Assume that $u_0^s(y)$ satisfies* (3.2.2)*, the initial datum $\tilde{u}_0(y) = u_0(y) - u_0^s(y) \in D_\mu^4(\mathbb{R}_+^2)$ satisfies the compatibility condition up to order 6. Then there exists* $0 \le T \le T_{\max} = \sup\{t \in [0, T], 1 - C\|\tilde{u}_0\|_{D_\mu^4(\mathbb{R}_+^2)} t - \frac{t^2}{2} > 0\}$*, if*

$$\|\tilde{u}_0(y)\|_{D_\mu^4(\mathbb{R}_+^2)} \le \eta,$$

for some $\eta \in [0, \sqrt{2}(e^{-\frac{1}{2}} - \frac{1}{2})^{\frac{1}{2}}]$, then the system (3.1.1) *admits a solution $(\tilde{u}, \tilde{v})$ with*

$$\tilde{u}(t, x, y) = u(t, x, y) - u^s(t, y) \in L^\infty([0, T]; D_\mu^4(\mathbb{R}_+^2)),$$

$$\tilde{v}(t, x, y) \in L^\infty([0, T]; L^\infty(\mathbb{R}_{+,y}; H^3(\mathbb{R}_x))).$$

3.2.1 The Decay of Shear Flow $u^s(t, y)$ in y

For the simplicity, we consider the case of a uniform outflow $U(x,t) = 1$, which implies pressure $\partial_x p(x,t) = 0$. Then Prandtl equation (3.1.1) can be rewritten as

$$\begin{cases} u_t + u\partial_x u + v\partial_y u = \partial_{yy}u, t > 0, x \in \mathbb{R}, y > 0, \\ \partial_x u + \partial_y v = 0, \\ (u, v)|_{y=0} = 0, \ \lim\limits_{y\to+\infty} u = 1, \\ u|_{t=0} = u_0(x, y). \end{cases} \tag{3.2.1}$$

We consider the datum around a shear flow, i.e.,

$$u_0(x, y) = u_0^s(y) + \tilde{u}_0(x, y).$$

First of all, a shear flow $(u^s(t, y), 0)$ is a trivial solution to Eq. (3.2.1) with $u^s(t, y)$ being the solution of heat equation

$$\begin{cases} \partial_t u^s - \partial_y^2 u^s = 0, \\ u^s|_{y=0} = 0, \text{ and } \lim\limits_{y\to+\infty} u^s = 1, \\ u^s|_{t=0} = u_0^s(y). \end{cases} \tag{3.2.2}$$

We rewrite the solution to (3.2.1) as a perturbation u of the $u^s(t, y)$ by denoting $u(t, x, y) = u^s(t, y) + \tilde{u}(t, x, y)$, then $\tilde{u}(t, x, y)$ satisfies the following equations:

$$\begin{cases} \tilde{u}_t + (u^s + \tilde{u})\partial_x \tilde{u} + \tilde{v}(\partial_y u^s + \partial_y \tilde{u}) - \partial_y^2 \tilde{u} = 0, \\ \partial_x \tilde{u} + \partial_y \tilde{v} = 0, \\ (\tilde{u}, \tilde{v})|_{y=0} = 0 \text{ and } \lim\limits_{y\to+\infty} \tilde{u} = 0, \\ \tilde{u}|_{t=0} = \tilde{u}_0. \end{cases} \tag{3.2.3}$$

The properties of a monotonic shear flow $u^s(t, y)$ will be first given, and we also investigate the decay of shear flow in y as follows.

Lemma 3.2.2 *Assume that initial datum $u_0^s(y)$ satisfies (3.2.2), then for any $T_1 > 0$, there exists a constant $C > 0$ such that the solution $u^s(t, y)$ of the initial boundary value problem (3.2.2) satisfies*

$$\begin{cases} C^{-1}e^{-\frac{y}{4}} \le \partial_y u^s(t, x, y), \text{ for } (t, x, y) \in [0, T_1] \times \mathbb{R}_+^2, \\ |\partial_y^p u^s(t, x, y)| \le Ce^{-\frac{y}{4}}, \text{ for } (t, x, y) \in [0, T_1] \times \mathbb{R}_+^2, \end{cases} \tag{3.2.4}$$

for some $0 \le p \le 5$, where a constant $C > 0$ depends on T_1.

Proof To deduce the pointwise estimates of the solutions, we can use the representation of the solution of (3.2.2),

$$
\begin{aligned}
u^s(t,y) &= \frac{1}{2\sqrt{\pi t}}\int_0^{\infty}\Big(e^{-\frac{(y-\bar y)^2}{4t}} - e^{-\frac{(y+\bar y)^2}{4t}}\Big)u_0^s(\bar y)d\bar y \\
&= \frac{1}{\sqrt{\pi}}\Big(\int_{-\frac{y}{2\sqrt{t}}}^{+\infty} e^{-\xi^2}u_0^s(2\sqrt{t}\xi + y)d\xi - \int_{\frac{y}{2\sqrt{t}}}^{+\infty} e^{-\xi^2}u_0^s(2\sqrt{t}\xi - y)d\xi\Big).
\end{aligned}
\tag{3.2.5}
$$

By using $\partial_y^{2j}u_0^s(0) = 0$ for $0 \le j \le 2$, it follows

$$
\begin{aligned}
\partial_y^p u^s(t,y) &= \frac{1}{\sqrt{\pi}}\Big(\int_{-\frac{y}{2\sqrt{t}}}^{+\infty} e^{-\xi^2}(\partial_y^p u_0^s)(2\sqrt{t}\xi + y)d\xi + (-1)^{p+1} \\
&\quad\times \int_{\frac{y}{2\sqrt{t}}}^{+\infty} e^{-\xi^2}(\partial_y^p u_0^s)(2\sqrt{t}\xi - y)d\xi\Big) \\
&= \frac{1}{2\sqrt{\pi t}}\int_0^{\infty}\Big(e^{-\frac{(y-\bar y)^2}{4t}} + (-1)^{p+1}e^{-\frac{(y+\bar y)^2}{4t}}\Big)\partial_y^p u_0^s(\bar y)d\bar y := A + B.
\end{aligned}
\tag{3.2.6}
$$

We now estimate A and B by using the monotonicity of $\partial_y^p u_0^s(y)$, respectively. We can derive that

$$
\begin{aligned}
A &= \frac{1}{2\sqrt{\pi t}}\int_0^{\infty} e^{-\frac{(y-\bar y)^2}{4t}}\partial_y^p u_0^s(\bar y)d\bar y \\
&\le Ce^{-\frac{y}{4}}\frac{1}{2\sqrt{\pi t}}\int_0^{+\infty} e^{-\frac{(y-\bar y)^2}{4t}}e^{\frac{(y-\bar y)^2}{8t}}e^{\frac{t}{8}}d\bar y \\
&\le Ce^{-\frac{y}{4}}\frac{1}{2\sqrt{\pi t}}\int_0^{+\infty} e^{-\frac{(y-\bar y)^2}{8t}}d\bar y \\
&\le Ce^{-\frac{y}{4}}.
\end{aligned}
\tag{3.2.7}
$$

Next, we establish the estimate of B. The result is obvious for $|y| \le T_1$. Thus, we assume $y \ge T_1 \ge t$. Thanks to $\partial_y^p u_0^s(y) \le Ce^{-\frac{y}{4}}$, it follows that

$$
\begin{aligned}
B &= (-1)^{p+1}\frac{1}{2\sqrt{\pi t}}\int_0^{+\infty} e^{-\frac{(y+\bar y)^2}{4t}}\partial_{\bar y}^p u_0^s(\bar y)d\bar y \\
&\le C\frac{1}{2\sqrt{\pi t}}\int_0^{+\infty} e^{-\frac{y^2}{4t}}e^{-\frac{2y\bar y+\bar y^2}{4t}}e^{-\bar y}d\bar y
\end{aligned}
$$

$$
\begin{aligned}
&\le Ce^{-\frac{y}{4}}\frac{1}{2\sqrt{\pi t}}\int_0^{+\infty} e^{-\frac{(\bar{y})^2}{4t}}e^{-\bar{y}}d\bar{y}\\
&\le Ce^{-\frac{y}{4}+t}\frac{1}{\sqrt{\pi}}\int_{\sqrt{t}}^{+\infty} e^{-x^2}dx\\
&\le Ce^{-\frac{y}{4}}.
\end{aligned}
\tag{3.2.8}
$$

For $p = 1$, we consider the estimate of another side, since the monotonic assumptions of $\partial_y u_0^s(y)$, we can conclude that

$$
\begin{aligned}
\partial_y u^s(t,y) &= \frac{1}{2\sqrt{\pi t}}\int_0^{\infty}\left(e^{-\frac{(y-\bar{y})^2}{4t}} + e^{-\frac{(y+\bar{y})^2}{4t}}\right)\partial_{\bar{y}}^p u_0^s(\bar{y})d\bar{\bar{y}}\\
&\ge \frac{1}{2\sqrt{\pi t}}\int_0^{\infty} e^{-\frac{(y-\bar{y})^2}{4t}}\partial_{\bar{y}}^p u_0^s(\bar{y})d\bar{\bar{y}}\\
&= \frac{1}{\sqrt{\pi}}\int_{\frac{y}{2\sqrt{t}}}^{+\infty} e^{-\xi^2}(\partial_\xi^p u_0^s)(2\sqrt{t}\xi + y)d\xi\\
&\ge \frac{C^{-1}}{\sqrt{\pi}}\int_0^1 e^{-\xi^2}e^{-\frac{1}{4}(2\sqrt{t}\xi+y)}d\xi\\
&\ge C^{-1}e^{-\frac{1}{4}(2\sqrt{t}+y)}\\
&\ge C^{-1}(T_1)e^{-\frac{y}{4}}, \text{ for } y\in[0,+\infty).
\end{aligned}
\tag{3.2.9}
$$

The proof is thus complete. □

Remark 3.2.1 From this lemma, we know that the decay of the gradient of shear flow is the same as that of the gradient of initial data of shear flow. The gradient of shear flow is the polynomial decay when the gradient of initial data of shear flow is the polynomial decay.

In this position, we show the precise version of the compatibility condition for the nonlinear system (3.2.3). Since authors in [151] studied the well-posedness of Prandtl equations when the weighted function is a polynomial function, we investigate the local existence of Prandtl equations when the weighted function is an exponential function in this chapter. Thus, the compatibility conditions and Propositions 3.2.1–3.2.3 below are the same as those in [151].

Proposition 3.2.1 ([151]) *Assume that $\tilde{u}(t,x,y)$ is a smooth solutions of the system* (3.2.3), *then the initial datum $\tilde{u}_0(x,y)$ have to satisfy the following compatibility conditions up to order 6:*

$$\begin{cases} \tilde{u}_0(x,0)=0, \ (\partial_y^2\tilde{u}_0)(x,0)=0, \ \forall x\in\mathbb{R}, \\ (\partial_y^4\tilde{u}_0)(x,0)=\big(\partial_y u_0^s(0)+\partial_y\tilde{u}_0(x,0)\big)\partial_x\partial_y\tilde{u}_0(x,0), \ \forall x\in\mathbb{R}, \\ (\partial_y^6\tilde{u}_0)(x,0)=\Big(\partial_y^3 u_0^s(0)+\partial_y^3\tilde{u}_0(x,0)\Big)\partial_x\partial_y\tilde{u}_0(x,0) \\ +\sum\limits_{1\le j\le 3} C_4^j\Big(\partial_y^j(u_0^s+\tilde{u}_0)\partial_y^{4-j}\partial_x\tilde{u}_0-(\partial_y^{j-1}\partial_x\tilde{u}_0)\partial_y^{4-j}(\partial_y u_0^s+\partial_y\tilde{u}_0)\Big)(x,0). \end{cases} \tag{3.2.10}$$

3.2.2 *The Approximate Solutions*

In this subsection, we investigate the existence of a parabolic regular equation by using classical energy method to prove the existence of solutions of the Prandtl equation.

Specifically, for a small parameter $\varepsilon\in(0,1]$, we study the existence of solution $\tilde{u}_\varepsilon$ to the regularized Prandtl equation

$$\begin{cases} \partial_t\tilde{u}_\epsilon+(u^s+\tilde{u}_\epsilon)\partial_x\tilde{u}_\epsilon+\tilde{v}_\epsilon(u_y^s+\partial_y\tilde{u}_\epsilon)-\partial_y^2\tilde{u}_\epsilon-\epsilon\partial_x^2\tilde{u}_\epsilon=0, \text{ in } \mathbb{R}_+^2\times\mathbb{R}_+, \\ \partial_x\tilde{u}_\varepsilon+\partial_y\tilde{v}_\varepsilon=0, \text{ in } \mathbb{R}_+^2\times\mathbb{R}_+, \\ (\tilde{u}_\epsilon,\tilde{v}_\varepsilon)|_{y=0}=0, \text{ and } \lim\limits_{y\to+\infty}\tilde{u}_\epsilon=0, \text{ on } \mathbb{R}\times\mathbb{R}_+, \\ \tilde{u}_\epsilon|_{t=0}=\tilde{u}_0+\varepsilon\mu_\varepsilon, \text{ on } \mathbb{R}_+^2, \end{cases} \tag{3.2.11}$$

where $\varepsilon\mu_\varepsilon$ is corrector and $\tilde{u}_0+\varepsilon\mu_\varepsilon$ satisfies the compatibility condition for the regularized system (3.2.11), $\tilde{u}_0$ is the initial datum of the original problem (3.2.3).

We show the boundary data of the solutions for the regularized system (3.2.11) which give also the accurate edition of the compatibility conditions for the system (3.2.11).

Proposition 3.2.2 *Assume that $\tilde{u}_0(x,y)$ satisfies the compatibility conditions* (3.2.10) *for the system* (3.2.11), *and $\mu_\varepsilon\in H_\mu^5(\mathbb{R}_+^2)$ such that $\tilde{u}_0+\varepsilon\mu_\varepsilon$ satisfies the compatibility conditions up to order 6 for the regularized system* (3.2.11). *If*

$\tilde{u}_\varepsilon(t,x,y) \in L^\infty([0,T]; D^4_\mu(\mathbb{R}^2_+)) \cap Lip([0,T]; D^2_\mu(\mathbb{R}^2_+))$ *is a solution of the system* (3.2.11)*, then we have*

$$\begin{cases} \tilde{u}_\varepsilon(t,x,0)=0, \ \ (\partial_y^2\tilde{u}_\varepsilon)(t,x,0)=0, \ \forall (t,x)\in[0,T]\times\mathbb{R}, \\ (\partial_y^4\tilde{u}_\varepsilon)(t,x,0)=\big(\partial_y u^s_\varepsilon(t,0)+\partial_y\tilde{u}_\varepsilon(t,x,0)\big)\partial_x\partial_y\tilde{u}_\varepsilon(t,x,0), \\ \forall (t,x)\in[0,T]\times\mathbb{R}, \\ (\partial_y^6\tilde{u}_\varepsilon)(t,x,0)=\Big(\partial_y^3 u^s_\varepsilon(t,0)+\partial_y^3\tilde{u}_\varepsilon(t,x,0)\Big)\partial_x\partial_y\tilde{u}_\varepsilon(t,x,0) \\ +\sum\limits_{1\le j\le 3} C_4^j\Big(\partial_y^j(u^s_\varepsilon(t,0)+\tilde{u}_\varepsilon)\partial_y^{4-j}\partial_x\tilde{u}_\varepsilon-(\partial_y^{j-1}\partial_x\tilde{u}_\varepsilon(t,x,0))\partial_y^{4-j}(\partial_y u^s_\varepsilon(t,0) \\ +\partial_y\tilde{u}_\varepsilon(t,x,0))\Big) \\ -2\varepsilon\partial_x\partial_y\tilde{u}_\varepsilon(t,x,0)\partial_x^2\partial_y\tilde{u}_\varepsilon(t,x,0). \end{cases} \tag{3.2.12}$$

Proof The proof is similar to that of Proposition 3.1 in [151], so we omit it here. □

Proposition 3.2.3 *Assume that* $\tilde{u}_0(x,y)$ *satisfies the compatibility conditions* (3.2.10) *for the system* (3.2.3)*, and* $\partial_y\tilde{u}_0(x,y)\in H^4_\mu(\mathbb{R}^2_+)$*, then there exists a constant* $\varepsilon_0>0$ *such that for any* $\varepsilon\in(0,\varepsilon_0]$*, there exists a* $\mu_\varepsilon\in H^5_\mu(\mathbb{R}^2_+)$ *such that* $\tilde{u}_0+\varepsilon\mu_\varepsilon$ *satisfies the compatibility conditions up to order 6 for the regularized system* (3.2.11)*. Moreover,*

$$\|\partial_y\tilde{u}_{0,\varepsilon}\|^2_{H^4_\mu(\mathbb{R}^2_+)} \le 2\|\partial_y\tilde{u}_0\|^2_{H^4_\mu(\mathbb{R}^2_+)},$$

and

$$\lim_{\varepsilon\to0^+}\|\partial_y\tilde{u}_{0,\varepsilon}-\partial_y\tilde{u}_0\|^2_{H^4_\mu(\mathbb{R}^2_+)}=0.$$

Proof The proof is similar to that of Corollary 3.3 in [151], thus we omit it here. □

We shall now prove the existence of approximate solutions of the system (3.2.11) by using vorticity equation $w_\varepsilon=\partial_y\tilde{u}_\varepsilon$, which reads

$$\begin{cases} \partial_t\tilde{w}_\epsilon+(u^s+\tilde{u}_\epsilon)\partial_x\tilde{w}_\epsilon+\tilde{v}_\epsilon(u^s_{yy}+\partial_y\tilde{w}_\epsilon)-\partial_y^2\tilde{w}_\epsilon-\epsilon\partial_x^2\tilde{w}_\epsilon=0 \text{ in } \mathbb{R}^2_+\times\mathbb{R}_+, \\ \partial_y\tilde{w}_\varepsilon=0 \text{ on } \mathbb{R}\times\mathbb{R}_+, \\ \tilde{w}_\epsilon|_{t=0}=\tilde{w}_0+\varepsilon\partial_y\mu_\varepsilon \text{ on } \mathbb{R}^2_+. \end{cases} \tag{3.2.13}$$

Using $(3.2.13)_1$ and boundary conditions $(3.2.13)_2$, we can get following the higher derivative of boundary data:

$$\partial_y^3\tilde{w}_\varepsilon|_{y=0}=((u^s_y+\tilde{w}_\varepsilon)\partial_x\tilde{w}_\varepsilon)|_{y=0},$$

and

$$\begin{aligned}\partial_y^5\tilde{w}_\varepsilon|_{y=0} &= (\partial_y^3 u^s + \partial_y^2\tilde{w}_\varepsilon + \varepsilon\partial_x^2\tilde{w}_\varepsilon)\partial_x\tilde{w}_\varepsilon|_{y=0}\\ &-(u_y^s + \tilde{w}_\varepsilon)(\partial_x\partial_y^2\tilde{w}_\varepsilon + \varepsilon\partial_x^3\tilde{w}_\varepsilon)|_{y=0} - \partial_x\partial_y\tilde{w}_\varepsilon(u_y^s + \tilde{w}_\varepsilon)|_{y=0}\\ &+\sum_{1\le j\le 4} C_4^j\Big(\partial_y^j(u^s + \tilde{u}_\varepsilon)\partial_y^{4-j}\partial_x\tilde{u}_\varepsilon - \partial_y^{j-1}\partial_x\tilde{u}_\varepsilon\partial_y^{4-j}(u^s + \tilde{w}_\varepsilon)\Big)|_{y=0}\\ &-\varepsilon\partial_x^2\Big((u_y^s + \tilde{w}_\varepsilon)\partial_x\tilde{w}_\varepsilon\Big)|_{y=0},\end{aligned}$$

where $\tilde{u}_\varepsilon = -\int_y^{+\infty}\tilde{w}_\varepsilon(t, x, \tilde{y})d\tilde{y}$ and $\tilde{v}_\varepsilon = -\int_0^y \partial_x\tilde{u}_\varepsilon(t, x, \tilde{y})d\tilde{y}$.

We can state the existence of approximate solutions as follows.

Theorem 3.2.3 *Let $\partial_y\tilde{u}_0 \in H_\mu^4(\mathbb{R}_+^2)$ satisfies the compatibility conditions up to order 6 for system* (3.2.3) *and assume $u_0^s(y)$ satisfies* (3.2.2)*; then for any $\varepsilon \in (0, 1]$ and $\eta \in [0, \sqrt{2}(e^{-\frac{1}{2}} - \frac{1}{2})^{\frac{1}{2}}]$, there exists a $T_\varepsilon > 0$ such that if*

$$\|\tilde{w}_0\|_{H_\mu^4(\mathbb{R}_+^2)} \le \eta,$$

then the system (3.2.13) *admits a solution*

$$\tilde{w}_\varepsilon \in L^\infty([0, T_\varepsilon]; H_\mu^4(\mathbb{R}_+^2)),$$

which satisfies

$$\|\tilde{w}_\varepsilon\|^2_{L^\infty([0,T_\varepsilon];H_\mu^4(\mathbb{R}_+^2))} \le 2\|\tilde{w}_{0,\varepsilon}\|^2_{L^\infty([0,T_\varepsilon];H_\mu^4(\mathbb{R}_+^2))} \le 4\|\tilde{w}_0\|^2_{L^\infty([0,T_\varepsilon];H_\mu^4(\mathbb{R}_+^2))}. \tag{3.2.14}$$

First, we will use the following lemma, since it is helpful to deal with boundary value.

Lemma 3.2.4 ([2]) *Let $1 < p < +\infty$. If $U \in W^{m,p}(\mathbb{R}^{n+1})$; then its trace u belongs to the space $B = B^{m-\frac{1}{p};p,p}(\mathbb{R}^n)$ and*

$$\|u\|_B \le K\|U\|_{W^{m,p}(\mathbb{R}^{n+1})},$$

with the constant $K > 0$ independent of U.

Corollary 3.2.1 *Let $1 < p < +\infty$. If $U \in W^{m,p}(\mathbb{R}^{n+1})$, then its trace u belongs to the space $W^{m-1,p}(\mathbb{R}^n)$ and*

$$\|u\|_{W^{m-1,p}(\mathbb{R}^n)} \le K\|U\|_{W^{m,p}(\mathbb{R}^{n+1})},$$

with the constant $K > 0$ independent of U.

Proof Since $1 < p < +\infty$, it follows from the fact $W^{m-1,p}(\mathbb{R}^n) = F^{m-1}_{p,2}(\mathbb{R}^n)$ and the embedding theorem in [130], $B^{m-1,p,1}(\mathbb{R}^n) \hookrightarrow F^{m-1}_{p,2}(\mathbb{R}^n)$ that $B^{m-\frac{1}{p};p,p}(\mathbb{R}^n) \hookrightarrow B^{m-1,p,1}(\mathbb{R}^n) \hookrightarrow W^{m-1,p}(\mathbb{R}^n)$, which gives

$$\|u\|_{W^{m-1,p}(\mathbb{R}^n)} \leq C\|U\|_{B^{m-\frac{1}{p};p,p}(\mathbb{R}^n)},$$

which, together with Lemma 3.2.4, gives the desired inequality. □

Next, we will use the following lemmas to prove Theorem 3.2.1.

Lemma 3.2.5 *Let* $m = 1, 2, 3, 4$ *and under the assumptions of Theorem 3.2.1, for a smooth solution* $\tilde{w}_\varepsilon$ *to Eqs.* (3.2.13), *then*

$$\begin{aligned}&\frac{d}{dt}\|\tilde{w}_\varepsilon\|^2_{H^{m,m-1}_\mu(\mathbb{R}^2_+)} + \varepsilon\|\partial_x\tilde{w}_\varepsilon\|^2_{H^{m,m-1}_\mu(\mathbb{R}^2_+)} + \|\partial_y\tilde{w}_\varepsilon\|^2_{H^{m,m-1}_\mu(\mathbb{R}^2_+)}\\&\quad\leq C\left(\|\tilde{w}_\varepsilon\|^2_{H^4_\mu(\mathbb{R}^2_+)} + \|\tilde{w}_\varepsilon\|^4_{H^4_\mu(\mathbb{R}^2_+)}\right),\end{aligned} \tag{3.2.15}$$

where positive constant C *is independent of* ε.

Proof For $m = \alpha = \beta = 0$, multiplying (3.2.13) by $e^{2\mu}\tilde{w}_\varepsilon$ and integrating it by parts over $\mathbb{R}^2_+$, we can conclude that

$$\begin{aligned}&\frac{d}{dt}\|\tilde{w}_\varepsilon\|^2_{L^2_\mu} + \varepsilon\|\partial_x\tilde{w}_\varepsilon\|^2_{L^2_\mu} + \|\partial_y\tilde{w}_\varepsilon\|^2_{L^2_\mu}\\&\leq \frac{1}{8}\|\partial_y\tilde{w}_\varepsilon\|^2_{L^2_\mu} + C\|\tilde{w}_\varepsilon\|^2_{L^2_\mu} + \|\partial_x\tilde{u}_\varepsilon\|_{L^2_\mu}\|\tilde{w}_\varepsilon\|^2_{L^2_\mu} + \|\partial_x\tilde{u}_\varepsilon\|_{L^2_\mu}\|\tilde{w}_\varepsilon\|_{L^2_\mu}\\&\quad+\|\partial_x\tilde{u}_\varepsilon\|_{L^2_\mu}\|\partial_y\tilde{w}_\varepsilon\|_{L^2_\mu}\|\tilde{w}_\varepsilon\|_{L^2_\mu}\\&\leq \frac{1}{2}\|\partial_y\tilde{w}_\varepsilon\|^2_{L^2_\mu} + C(\|\tilde{w}_\varepsilon\|^2_{L^2_\mu} + \|\tilde{w}_\varepsilon\|^4_{L^2_\mu} + \|\partial_x\tilde{w}_\varepsilon\|^2_{L^2_\mu}\\&\quad+\|\partial_x^2\tilde{w}_\varepsilon\|^2_{L^2_\mu} + \|\partial_x^2\tilde{w}_\varepsilon\|^4_{L^2_\mu} + \|\partial_x^2\tilde{w}_\varepsilon\|^4_{L^2_\mu}),\end{aligned}$$

which implies

$$\frac{d}{dt}\|\tilde{w}_\varepsilon\|^2_{L^2_\mu} + \varepsilon\|\partial_x\tilde{w}_\varepsilon\|^2_{L^2_\mu} + \|\partial_y\tilde{w}_\varepsilon\|^2_{L^2_\mu} \leq C\left(\|\tilde{w}_\varepsilon\|^2_{H^2_\mu} + \|\tilde{w}_\varepsilon\|^4_{H^2_\mu}\right). \tag{3.2.16}$$

For $\alpha+\beta=m$ and $\alpha\le m-1$, respectively, applying the operator ∂^m on (3.2.13), multiplying the resulting equation by $e^{2\mu}\partial^m\tilde{w}$ and integrating it by parts over $\mathbb{R}^2_+$, we conclude that

$$\begin{aligned}
&\frac{d}{dt}\|\partial^m\tilde{w}_\varepsilon\|^2_{L^2_\mu}+\varepsilon\|\partial_x\partial^m\tilde{w}_\varepsilon\|^2_{L^2_\mu}+\|\partial_y\partial^m\tilde{w}_\varepsilon\|^2_{L^2_\mu}=\int_{\mathbb{R}}\partial_y\partial^m\tilde{w}_\varepsilon\partial^m\tilde{w}_\varepsilon|_{y=0}dx\\
&\quad-\int_{\mathbb{R}^2_+}e^{2\mu}\partial_y\partial^m\tilde{w}_\varepsilon\partial^m\tilde{w}_\varepsilon dxdy\\
&\quad-\int_{\mathbb{R}^2_+}e^{2\mu}\partial^m\Big((u^s+\tilde{u}_\epsilon)\partial_x\tilde{w}_\epsilon+\tilde{v}_\epsilon(u^s_{yy}+\partial_y\tilde{w}_\epsilon)\Big)\partial^m\tilde{w}_\varepsilon dxdy\\
&=j_1+j_2+j_3+j_4.
\end{aligned}\tag{3.2.17}$$

Although α and β have many various cases, there is a same estimate on j_2 by the Cauchy-Schwarz inequality, that is,

$$|j_2|\le\frac{1}{8}\|\partial_y\partial^m\tilde{w}_\varepsilon\|^2_{L^2_\mu}+C\|\partial^m\tilde{w}_\varepsilon\|^2_{L^2_\mu}.$$

Firstly, we use the boundary conditions of $\partial_y\tilde{w}_\varepsilon$ and $\tilde{v}_\varepsilon$ to deal with j_1. After a complicated analysis and calculation, we can get the following cases which do not vanish on the boundary of $y=0$.

Case 1: $\alpha=1$, $\beta=2$. We use the boundary value of $\partial_y^3\tilde{w}$ on $y=0$ and Lemma 3.2.4 to conclude that

$$\begin{aligned}
\Big|\int_{\mathbb{R}}\partial_x\partial_y^3\tilde{w}_\varepsilon\partial_x\partial_y^2\tilde{w}_\varepsilon|_{y=0}dx\Big|&\le\|\partial_x((u^s_y+\tilde{w}_\varepsilon)\partial_x\tilde{w}_\varepsilon)|_{y=0}\|_{L^2_\mu}\|\partial_x\partial_y^2\tilde{w}_\varepsilon|_{y=0}\|_{L^2_\mu}\\
&\le\|\partial_x\tilde{w}_\varepsilon|_{y=0}\|^2_{L^2_\mu}\|\partial_x\partial_y^2\tilde{w}_\varepsilon|_{y=0}\|_{L^2_\mu}\\
&\le C\|\partial_x\partial_y\tilde{w}_\varepsilon\|^2_{L^2_\mu}\|\partial_x\partial_y^3\tilde{w}_\varepsilon\|_{L^2_\mu}\\
&\le C\|\tilde{w}_\varepsilon\|^4_{H^4_\mu}+C\|\tilde{w}_\varepsilon\|^2_{H^4_\mu}.
\end{aligned}\tag{3.2.18}$$

Case 2: $\alpha=0$, $\beta=4$. We use the boundary value of $\partial_y^5\tilde{w}$ on $y=0$ and Lemma 3.2.4 to deduce that

$$\begin{aligned}
\Big|\int_{\mathbb{R}}\partial_y^5\tilde{w}_\varepsilon\partial_y^4\tilde{w}_\varepsilon|_{y=0}dx\Big|&\le\|\partial_y^5\tilde{w}_\varepsilon|_{y=0}\|_{L^2_\mu}\|\partial_y^5\tilde{w}_\varepsilon\|_{L^2_\mu}\\
&\le C\|\tilde{w}_\varepsilon\|^2_{H^4_\mu}+C\|\tilde{w}_\varepsilon\|^4_{H^4_\mu}+\frac{1}{8}\|\partial_y^5\tilde{w}_\varepsilon\|^2_{L^2_\mu}.
\end{aligned}\tag{3.2.19}$$

Case 3: $\alpha = 1,\ \beta = 3$. We use the boundary value of $\partial_y^3 \tilde{w}$ on $y = 0$, Lemma 3.2.4 and integrate it by part with respect to x variable to obtain that

$$\begin{aligned}\Big|\int_{\mathbb{R}} \partial_x \partial_y^4 \tilde{w}_\varepsilon \partial_x \partial_y^3 \tilde{w}_\varepsilon|_{y=0} dx\Big| &= \Big|\int_{\mathbb{R}} \partial_y^4 \tilde{w}_\varepsilon \partial_x^2 \partial_y^3 \tilde{w}_\varepsilon|_{y=0} dx\Big| \\ &\le \|\partial_x^2((u_y^s + \tilde{w}_\varepsilon)\partial_x \tilde{w}_\varepsilon)|_{y=0}\|_{L^2_\mu} \|\partial_y^5 \tilde{w}_\varepsilon\|_{L^2_\mu} \\ &\le C\|\tilde{w}_\varepsilon\|^4_{H^4_\mu} + \frac{1}{8}\|\partial_y^5 \tilde{w}_\varepsilon\|^2_{L^2_\mu}. \end{aligned} \tag{3.2.20}$$

Case 4: $\alpha = 2,\ \beta = 2$. We use the boundary value of $\partial_y^3 \tilde{w}$ on $y = 0$, Lemma 3.2.4 and integrate it by part with respect to x variable to derive that

$$\begin{aligned}\Big|\int_{\mathbb{R}} \partial_x^2 \partial_y^3 \tilde{w}_\varepsilon \partial_x^2 \partial_y^2 \tilde{w}_\varepsilon|_{y=0} dx\Big| &\le \|\partial_x^2((u_y^s + \tilde{w}_\varepsilon)\partial_x \tilde{w}_\varepsilon)|_{y=0}\|_{L^2_\mu} \|\partial_x^2 \partial_y^3 \tilde{w}_\varepsilon\|_{L^2_\mu} \\ &\le C\|\tilde{w}_\varepsilon\|^4_{H^4_\mu} + \frac{1}{8}\|\partial_x^2 \partial_y^3 \tilde{w}_\varepsilon\|^2_{L^2_\mu}. \end{aligned} \tag{3.2.21}$$

We can use the estimates of commutator operator to deal with remaining terms j_3 and j_4. For the third term j_3, we can write it as the following pattern and use integration by part with respect to x variable and use Lemma 3.2.2 to conclude that

$$\begin{aligned} &-\int_{\mathbb{R}^2_+} e^{2\mu} \partial^m \Big((u^s + \tilde{u}_\epsilon)\partial_x \tilde{w}_\epsilon\Big)\partial^m \tilde{w}_\epsilon dxdy \\ &= -\int_{\mathbb{R}^2_+} e^{2\mu}(u^s + \tilde{u}_\epsilon)\partial_x \partial^m \tilde{w}_\epsilon \partial^m \tilde{w}_\epsilon dxdy \\ &\quad + \int_{\mathbb{R}^2_+} e^{2\mu}[(u^s + \tilde{u}_\epsilon),\ \partial^m]\partial_x \tilde{w}_\epsilon \partial^m \tilde{w}_\epsilon dxdy \\ &\le \|\partial_x \tilde{u}_\varepsilon\|_{L^\infty} \|\tilde{w}_\varepsilon\|^2_{H^4_\mu} + C\|\tilde{w}_\varepsilon\|^2_{H^m_\mu} + C\|\tilde{w}_\varepsilon\|^3_{H^4_\mu} \\ &\le C\|\tilde{w}_\varepsilon\|^2_{H^4_\mu} + C\|\tilde{w}_\varepsilon\|^3_{H^4_\mu}, \end{aligned} \tag{3.2.22}$$

where we have used the fact of the estimate of commutator operator, which has been obtained by using the Sobolev inequality,

$$\begin{aligned} &\int_{\mathbb{R}^2_+} e^{2\mu}[(u^s + \tilde{u}_\epsilon),\ \partial^m]\partial_x \tilde{w}_\epsilon \partial^m \tilde{w}_\epsilon dxdy \\ &= \int_{\mathbb{R}^2_+} e^{2\mu} \sum_{n \le m,\ 1 \le n} C_m^n \partial^n (u^s + \tilde{u}_\epsilon)\partial^{m-n}\partial_x \tilde{w}_\epsilon \partial^m \tilde{w}_\epsilon dxdy \end{aligned}$$

$$\leq C(\|\tilde{w}_\varepsilon\|_{H^4_\mu} + C\|\tilde{w}_\varepsilon\|^2_{H^4_\mu})\|\tilde{w}_\varepsilon\|_{H^4_\mu}$$

$$\leq C\|\tilde{w}_\varepsilon\|^2_{H^4_\mu} + C\|\tilde{w}_\varepsilon\|^3_{H^4_\mu},$$

where positive constant C is independent of ε.

For the last term j_4, the estimate is similar to that of (3.2.22)

$$\left|\int_{\mathbb{R}^2_+} e^{2\mu}\partial^m\Big(\tilde{v}_\epsilon(u^s_{yy} + \partial_y\tilde{w}_\epsilon)\Big)\partial^m\tilde{w}_\varepsilon dxdy\right|$$

$$= \left|\int_{\mathbb{R}^2_+} e^{2\mu}\Big(\tilde{v}_\epsilon\partial_y\partial^m\tilde{w}_\epsilon - [\tilde{v}_\epsilon, \partial^m]\partial_y\tilde{w}_\epsilon + \partial^m(\tilde{v}_\epsilon u^s_{yy})\Big)\partial^m\tilde{w}_\varepsilon dxdy\right|$$

$$\leq \|\tilde{v}_\epsilon\|_{L^\infty}\|\partial_y\partial^m\tilde{w}_\epsilon\|_{L^2_\mu}\|\partial^m\tilde{w}_\epsilon\|_{L^2_\mu} + C\|\partial^m\tilde{w}_\epsilon\|^3_{L^2_\mu} + C\|\partial^m\tilde{w}_\epsilon\|^2_{L^2_\mu}$$

$$\leq \frac{1}{8}\|\partial_y\tilde{w}_\epsilon\|^2_{H^4_\mu} + C\|\tilde{w}_\epsilon\|^4_{H^4_\mu} + C\|\tilde{w}_\epsilon\|^3_{H^4_\mu} + C\|\tilde{w}_\epsilon\|^2_{H^4_\mu}$$

$$\leq \frac{1}{8}\|\partial_y\tilde{w}_\epsilon\|^2_{H^4_\mu} + C\|\tilde{w}_\epsilon\|^4_{H^4_\mu} + C\|\tilde{w}_\epsilon\|^2_{H^4_\mu}. \tag{3.2.23}$$

Thus combining (3.2.16)–(3.2.23), we can derive the desired result (3.2.15). The proof is thus complete.

In order to ensure the completion of energy estimate, we need to establish the estimate of term $\partial_x^m\tilde{w}_\varepsilon$, which will be done in the following lemma.

Lemma 3.2.6 *Let $m = 1, 2, 3, 4$ and under the assumptions of Theorem 3.2.1, for a smooth solution $\tilde{w}_\varepsilon$ to Eqs.* (3.2.13)*, then*

$$\frac{d}{dt}\|\partial_x^m\tilde{w}_\varepsilon\|^2_{L^2_\mu(\mathbb{R}^2_+)} + \varepsilon\|\partial_x^{m+1}\tilde{w}_\varepsilon\|^2_{L^2_\mu(\mathbb{R}^2_+)} + \|\partial_y\partial_x^m\tilde{w}_\varepsilon\|^2_{L^2_\mu(\mathbb{R}^2_+)}$$
$$\leq C\left(\|\tilde{w}_\varepsilon\|^2_{H^4_\mu(\mathbb{R}^2_+)} + \|\tilde{w}_\varepsilon\|^3_{H^4_\mu(\mathbb{R}^2_+)}\right) + \frac{C}{\varepsilon}\left(\|\tilde{w}_\varepsilon\|^2_{H^4_\mu(\mathbb{R}^2_+)} + \|\tilde{w}_\varepsilon\|^4_{H^4_\mu(\mathbb{R}^2_+)}\right), \tag{3.2.24}$$

where positive constant C is independent of ε.

Proof Applying the operator ∂_x^m on (3.2.13), multiplying the resulting equation by $e^{2\mu}\partial_x^m\tilde{w}$ and integrating it by parts over $\mathbb{R}^2_+$, we derive that

$$\frac{d}{dt}\|\partial_x^m\tilde{w}_\varepsilon\|^2_{L^2_\mu} + \varepsilon\|\partial_x^{m+1}\tilde{w}_\varepsilon\|^2_{L^2_\mu} + \|\partial_y\partial_x^m\tilde{w}_\varepsilon\|^2_{L^2_\mu}$$
$$\leq C(\|\tilde{w}_\varepsilon\|^2_{H^4_\mu} + \|\tilde{w}_\varepsilon\|^3_{H^4_\mu}) + \left|\int_{\mathbb{R}^2_+} e^{2\mu}\partial_x^m\Big(\tilde{v}_\epsilon(u^s_{yy} + \partial_y\tilde{w}_\epsilon)\Big)\partial_x^m\tilde{w}_\varepsilon dxdy\right|, \tag{3.2.25}$$

where we have used the fact

$$(\partial_y \partial_x^m \tilde{w}_\varepsilon)(t, x, 0) = 0, \ (t, x) \in [0, T_\varepsilon] \times \mathbb{R}.$$

Next, we will deal with the last term on the right-hand side of (3.2.25), which is also the main difficulty to investigate Prandtl equations. Using the Leibniz formula, Sobolev inequality, and Lemma 3.2.2, we can get

$$\begin{aligned}
&\Big|\int_{\mathbb{R}_+^2} e^{2\mu}\partial_x^m\Big(\tilde{v}_\epsilon(u_{yy}^s + \partial_y \tilde{w}_\epsilon)\Big)\partial_x^m \tilde{w}_\varepsilon dxdy\Big| \le \Big|\int_{\mathbb{R}_+^2} e^{2\mu}\tilde{v}_\epsilon \partial_y \partial_x^m \tilde{w}_\epsilon \partial_x^m \tilde{w}_\varepsilon dxdy\Big| \\
&\quad + \Big|\int_{\mathbb{R}_+^2} e^{2\mu} \sum_{1\le j\le m-1} C_m^j \partial_x^j \tilde{v}_\epsilon \partial_y \partial_x^{m-j}\tilde{w}_\epsilon \partial_x^m \tilde{w}_\varepsilon dxdy\Big| \\
&\quad + \Big|\int_{\mathbb{R}_+^2} e^{2\mu}\partial_x^m \tilde{v}_\epsilon (u_{yy}^s + \partial_y \tilde{w}_\epsilon)\partial_x^m \tilde{w}_\varepsilon dxdy\Big| \\
&\le C\|\partial_x \tilde{u}\|_{L^\infty}\|\tilde{w}_\varepsilon\|_{H_\mu^4}^2 + C\|\tilde{v}\|_{L^\infty}\|\tilde{w}_\varepsilon\|_{H_\mu^4}^2 + C\|\tilde{w}_\varepsilon\|_{H_\mu^4}^3 + \|\partial_x^m \tilde{v}_\epsilon u_{yy}^s\|_{L_\mu^2}\|\tilde{w}_\varepsilon\|_{H_\mu^4} \\
&\quad + C\|\partial_x^m \tilde{v}_\epsilon\|_{L^2(\mathbb{R}_x; L^\infty(\mathbb{R}^+))}\|\partial_y \tilde{w}_\epsilon\|_{L^\infty(\mathbb{R}_x; L_\mu^2)}\|\tilde{w}_\varepsilon\|_{H_\mu^4} \\
&\le C\|\tilde{w}_\varepsilon\|_{H_\mu^4}^3 + \|\partial_x^{m+1}\tilde{w}_\varepsilon\|_{L_\mu^2}\|\tilde{w}_\varepsilon\|_{H_\mu^4} + \|\partial_x^{m+1}\tilde{w}_\varepsilon\|_{L_\mu^2}\|\tilde{w}_\varepsilon\|_{H_\mu^4}^2 \\
&\le C\|\tilde{w}_\varepsilon\|_{H_\mu^4}^3 + \frac{C}{\varepsilon}(\|\tilde{w}_\varepsilon\|_{H_\mu^4}^2 + \|\tilde{w}_\varepsilon\|_{H_\mu^4}^4) + \frac{\varepsilon}{4}\|\partial_x^{m+1}\tilde{w}_\varepsilon\|_{L_\mu^2}^2. \qquad (3.2.26)
\end{aligned}$$

Inserting (3.2.26) into (3.2.25), we can derive the desired result (3.2.24). The proof is thus complete. □

Proof of Theorem 3.2.1 Combining (3.2.15) and (3.2.24), we have for any $\varepsilon \in (0, 1]$

$$\frac{d}{dt}\|\tilde{w}_\varepsilon\|_{H_\mu^4(\mathbb{R}_+^2)}^2 \le \frac{C}{\varepsilon}\left(\|\tilde{w}_\varepsilon\|_{H_\mu^4(\mathbb{R}_+^2)}^2 + \|\tilde{w}_\varepsilon\|_{H_\mu^4(\mathbb{R}_+^2)}^4\right), \qquad (3.2.27)$$

where a positive constant C is independent of ε.

Let $y = \|\tilde{w}_\varepsilon\|_{H_\mu^4(\mathbb{R}_+^2)}^2$. We can derive from (3.2.27),

$$\frac{dy}{dt} \le \frac{C}{\varepsilon}(y + y^2),$$

i.e.,

$$\Big(-\frac{1}{y}\Big)_t \le \frac{C}{\varepsilon}\Big(-\big(-\frac{1}{y}\big) + 1\Big).$$

Integrating above inequality over $[0, t]$, we arrive at

$$-\frac{1}{y} \leq -\frac{1}{y(0)} e^{-\frac{C}{\varepsilon}t} + (1 - e^{-\frac{C}{\varepsilon}t}),$$

which implies

$$\begin{aligned} y &\leq \frac{y_0}{e^{-\frac{C}{\varepsilon}t} - (1 - e^{-\frac{C}{\varepsilon}t}) y_0} \\ &\leq \frac{y_0}{e^{-\frac{C}{\varepsilon}t} - \frac{C}{\varepsilon} t y_0}, \end{aligned}$$

where we used the fact $e^{-\frac{C}{\varepsilon}t} - 1 \geq -\frac{C}{\varepsilon}t$ as $0 \leq t \leq \frac{\varepsilon}{2C}$.

Indeed, let $f(x) = e^x - 1 - x$ for $x < 0$, thus $f'(x) = e^x - 1 < 0$ for $[-\frac{1}{2}, 0]$. We have $f(0) \leq f(x) \leq f(-\frac{1}{2})$, i.e., $0 \leq e^x - 1 - x \leq e^{-\frac{1}{2}} - \frac{1}{2}$. Taking $x = -\frac{C}{\varepsilon}t$, namely, $-\frac{1}{2} \leq -\frac{C}{\varepsilon}t \leq 0$, i.e., $0 \leq t \leq \frac{\varepsilon}{2C}$, we get $f(-\frac{C}{\varepsilon}t) = e^{-\frac{C}{\varepsilon}t} - 1 - (-\frac{C}{\varepsilon}t) \geq 0$ for $0 \leq t \leq \frac{\varepsilon}{2C}$. Therefore, we can obtain

$$\|\tilde{w}_\varepsilon\|^2_{H^4_\mu(\mathbb{R}^2_+)} \leq \frac{\|\tilde{w}_\varepsilon(0)\|^2_{H^4_\mu(\mathbb{R}^2_+)}}{e^{-\frac{C}{\varepsilon}t} - \frac{C}{\varepsilon}t\|\tilde{w}_\varepsilon(0)\|^2_{H^4_\mu(\mathbb{R}^2_+)}}, \quad 0 < t \leq \min\left\{T_\varepsilon, \frac{\varepsilon}{2C}\right\}, \tag{3.2.28}$$

where we have chosen $T_\varepsilon > 0$ so small that

$$\left(e^{-\frac{C}{\varepsilon}T_\varepsilon} - \frac{C}{\varepsilon}T_\varepsilon \eta^2\right)^{-1} \leq 2, \text{ when } 0 \leq T_\varepsilon \leq \frac{\varepsilon}{2C}. \tag{3.2.29}$$

In fact, since $(e^{-\frac{C}{\varepsilon}T_\varepsilon} - \frac{C}{\varepsilon}T_\varepsilon\eta^2)^{-1}$ for $0 \leq T_\varepsilon \leq \frac{\varepsilon}{2C}$ is a monotonically increasing function in T_ε and $\eta \in [0, \sqrt{2}(e^{-\frac{1}{2}} - \frac{1}{2})^{\frac{1}{2}}]$, (3.2.29) follows immediately.

Thus we can conclude from (3.2.28) and (3.2.29) that for any $\|\tilde{w}_\varepsilon(0)\|_{H^4_\mu(\mathbb{R}^2_+)} \leq \eta$ and $0 < \varepsilon \leq \varepsilon_0$,

$$\|\tilde{w}_\varepsilon(t)\|^2_{H^4_\mu(\mathbb{R}^2_+)} \leq 2\|\tilde{w}_\varepsilon(0)\|^2_{H^4_\mu(\mathbb{R}^2_+)} \leq 4\|\tilde{w}(0)\|^2_{H^4_\mu(\mathbb{R}^2_+)}, \quad 0 < t \leq T_\varepsilon.$$

The proof is thus complete. □

We know that the solutions of the Prandtl equation depend on ε from (3.2.24), we need to improve the results of Lemma 3.2.6 and establish an uniform estimate with respect to ε. We now follow the approach in [151] to obtain the new estimate (3.2.44) below.

3.2.3 Formal Transformations and Uniform Estimates

As for the estimate (3.2.24) depending on ε, we have to choose a new way to obtain an estimate which is independent of ε instead of (3.2.24). To simplify the notations, from now on, we abandon the notation tilde and sub-index ε, namely,

$$u = \tilde{u}_\varepsilon, \ v = \tilde{v}_\varepsilon, \ w = \tilde{w}_\varepsilon.$$

Let $w \in L^\infty([0,T]; H^4_\mu(\mathbb{R}^2_+))$ be a classical solution of (3.2.13) which satisfies the following a priori estimate

$$\|w\|_{L^\infty([0,T];H^4_\mu(\mathbb{R}^2_+))} \le 2\eta. \tag{3.2.30}$$

We know that

$$\begin{aligned}
&\|e^{\frac{y}{4}} w\|_{L^\infty([0,T]\times\mathbb{R}^2_+)} \\
&= \Big|\int_0^y (e^{\frac{y}{4}} w)_y dy\Big| \le \Big(\int_0^y e^{-\frac{y}{2}} dy\Big)^{\frac{1}{2}} \Big(\int_0^y e^{\frac{y}{2}}((e^{\frac{y}{4}} w)_y)^2 dy\Big)^{\frac{1}{2}} \\
&\le C\Big(\int_0^y e^{\frac{y}{2}}((e^{\frac{y}{4}} w)_y)^2 dy\Big)^{\frac{1}{2}}_{L^\infty(\mathbb{R}^+_x)} \\
&\le C\Big(\int_0^\infty \int_0^y e^{\frac{y}{2}}((e^{\frac{y}{4}} w)_y)^2 dxdy\Big)^{\frac{1}{2}} \Big(\int_0^\infty \int_0^y e^{\frac{y}{2}}((e^{\frac{y}{4}} w)_{xy})^2 dxdy\Big)^{\frac{1}{2}} \\
&\le C\|w\|_{H^4_\mu},
\end{aligned} \tag{3.2.31}$$

which implies

$$|\partial_y u(t,x,y)| = |w(t,x,y)| \le C\eta e^{-\frac{y}{4}}, \ \forall (t,x,y) \in [0,T] \times \mathbb{R}^2_+.$$

Now choose $\eta > 0$ is so small that

$$C\eta \le \frac{C^{-1}}{4}.$$

Then we have

$$\frac{C^{-1}}{4} e^{-\frac{y}{4}} \le |u^s_y + u_y| \le 4Ce^{-\frac{y}{4}}, (t,x,y) \in [0,T] \times \mathbb{R}^2_+. \tag{3.2.32}$$

Under the conditions (3.2.31)–(3.2.32), system (3.2.11) reads as follows, for $0 \le m \le 4$,

$$\begin{cases} \partial_t g_m + (u^s + u)\partial_x g_m - \partial_y^2 g_m - \varepsilon\partial_x^2 g_m - 2\varepsilon(\partial_x\partial_y^{-1} g_m)\partial_y\eta_1 \triangleq M_m, \\ \partial_y g_m = 0, \\ g_m|_{t=0} = g_{m,0}, \end{cases} \tag{3.2.33}$$

with $M_m = \sum_{i=1}^6 M_i^m$,

$$\begin{cases} M_1^m = -(u^s + u)(g_m\eta_1 + (\partial_y^{-1} g_m)\partial_y\eta_1), \\ M_2^m = 2(\partial_y g_m)\eta_2 + 2g_m(\partial_y\eta_2 - 2\eta_2^2) - 8(\partial_y^{-1} g_m)\eta_2\partial_y\eta_2, \\ M_3^m = \varepsilon\big(2(\partial_x g_m)\eta_1 - 2g_m\eta_1^2 - 4(\partial_y^{-1} g_m)\eta_1\partial_y\eta_1\big), \\ M_4^m = \partial_y\Big(\partial_y^{-1} g_m \frac{(u^s+u)w_x + v(w_y + u^s_{yy}))}{u^s_y + u_y}\Big), \\ M_5^m = -\partial_y\Big(\frac{\sum_j^4 C_4^j \partial_x^j u \cdot \partial_x^{5-j} u}{u^s_y + u_y}\Big), \\ M_6^m = -\partial_y\Big(\frac{\sum_j^4 C_4^j \partial_x^j w \cdot \partial_x^{4-j} v}{u^s_y + u_y}\Big), \end{cases}$$

where

$$g_m = \Big(\frac{\partial_x^m u}{u^s_y + u_y}\Big)_y, \quad \eta_1 = \frac{u_{xy}}{u^s_y + \tilde{u}_y}, \quad \eta_2 = \frac{u^s_{yy} + u_{yy}}{u^s_y + \tilde{u}_y}, \quad \forall (t, x, y) \in [0, T] \times \mathbb{R}^2_+.$$

Lemma 3.2.7 *If $\tilde{w}_0 \in H^4_\mu(\mathbb{R}^2_+)$ and (3.2.30)–(3.2.31) hold, then we have $g_m(0) \in L^2_{\frac{\mu}{2}}(\mathbb{R}^2_+)$, and*

$$\|g_m(0)\|_{L^2_{\frac{\mu}{2}}(\mathbb{R}^2_+)} \le C\|\tilde{w}_0\|_{H^4_\mu(\mathbb{R}^2_+)}.$$

Proof Actually,

$$g_m(0) = \Big(\frac{\partial_x^m \tilde{u}_0}{u^s_{0,y} + \tilde{u}_{0,y}}\Big)_y = \frac{\partial_y\partial_x^m \tilde{u}_0}{u^s_{0,y} + \tilde{u}_{0,y}} - \frac{\partial_x^n \tilde{u}_0}{u^s_{0,y} + \tilde{u}_{0,y}}\eta_2(0).$$

From (3.2.32), we can derive

$$e^{\frac{y}{4}} g_m(0) \le e^{\frac{y}{2}}\partial_x^m \tilde{w}_0 + (1 + e^{\frac{y}{4}})\partial_x^m \tilde{u}_0,$$

implying

$$\|g_m(0)\|_{L^2_{\frac{\mu}{2}}(\mathbb{R}^2_+)} \le C\|\partial_x^m \tilde{w}_0\|_{L^2_\mu(\mathbb{R}^2_+)} \le C\|\tilde{w}_0\|_{H^4_\mu(\mathbb{R}^2_+)}.$$

This completes the proof of the lemma. □

Lemma 3.2.8 *Let $w \in L^\infty([0,T]; H^4_\mu(\mathbb{R}^2_+))$ and (3.2.30)–(3.2.31) hold. Assume $u^s_0(y)$ satisfies (3.2.2) and g_m satisfies the Eq. (3.2.33) for $1 \le m \le 4$, then we have the following estimates, for $t \in [0,T]$,*

$$\frac{d}{dt}\sum_{m=1}^{4}\|g_m\|^2_{L^2_{\frac{\mu}{2}}(\mathbb{R}^2_+)} + \sum_{m=1}^{4}\|\partial_y g_m\|^2_{L^2_{\frac{\mu}{2}}(\mathbb{R}^2_+)} + \varepsilon\sum_{m=1}^{4}\|\partial_x g_m\|^2_{L^2_{\frac{\mu}{2}}(\mathbb{R}^2_+)} \le C\Big(\sum_{m=1}^{4}\|g_m\|^2_{L^2_{\frac{\mu}{2}}(\mathbb{R}^2_+)} + \|w\|^2_{H^4_\mu}\Big), \quad (3.2.34)$$

where a constant $C > 0$ is independent of ε.

Proof Multiplying (3.2.33) by $e^{\frac{y}{4}}g_m$, integrating it by part over $\mathbb{R}^2_+$, and applying the boundary values and Lemma 3.2.2, we have

$$\begin{aligned}
&\frac{1}{2}\frac{d}{dt}\|g_m\|^2_{L^2_{\frac{\mu}{2}}(\mathbb{R}^2_+)} + \|\partial_y g_m\|^2_{L^2_{\frac{\mu}{2}}(\mathbb{R}^2_+)} + \varepsilon\|\partial_x g_m\|^2_{L^2_{\frac{\mu}{2}}(\mathbb{R}^2_+)}\\
&= -\int_{\mathbb{R}^2_+}(u^s+u)\partial_x g_m e^{\frac{y}{4}} g_m dxdy - \frac{1}{4}\int_{\mathbb{R}^2_+}\partial_y g_m e^{\frac{y}{4}} g_m dxdy\\
&\quad -2\varepsilon\int_{\mathbb{R}^2_+}(\partial_x\partial_y^{-1} g_n)\partial_y\eta_1 e^{\frac{y}{4}} g_m dxdy + \sum_{j=1}^{6}\Big|\int_{\mathbb{R}^2_+} M^m_j e^{\frac{y}{4}} g_m dxdy\Big|\\
&\le C\|u_x\|_{L^\infty(\mathbb{R}^2_+)}\|g_m\|^2_{L^2_{\frac{\mu}{2}}(\mathbb{R}^2_+)} + \frac{1}{4}\|\partial_y g_m\|^2_{L^2_{\frac{\mu}{2}}(\mathbb{R}^2_+)} + C\|g_m\|^2_{L^2_{\frac{\mu}{2}}(\mathbb{R}^2_+)}\\
&\quad +2\varepsilon\int_{\mathbb{R}^2_+}(\partial_y^{-1} g_n)\partial_y\eta_1 e^{\frac{y}{4}}\partial_x g_m dxdy + 2\varepsilon\int_{\mathbb{R}^2_+}(\partial_y^{-1} g_n)\partial_y\partial_x\eta_1 e^{\frac{y}{4}} g_m dxdy\\
&\quad +\sum_{j=1}^{6}\Big|\int_{\mathbb{R}^2_+} M^m_j e^{\frac{y}{4}} g_m dxdy\Big|\\
&\le C\|w\|_{H^2_\mu(\mathbb{R}^2_+)}\|g_m\|^2_{L^2_{\frac{\mu}{2}}(\mathbb{R}^2_+)} + \frac{1}{4}\|\partial_y g_m\|^2_{L^2_{\frac{\mu}{2}}(\mathbb{R}^2_+)} + C\|g_m\|^2_{L^2_{\frac{\mu}{2}}(\mathbb{R}^2_+)}\\
&\quad +\varepsilon\|\partial_y^{-1} g_m\partial_y\eta_1\|^2_{L^2_{\frac{\mu}{2}}(\mathbb{R}^2_+)} + \frac{\varepsilon}{8}\|\partial_x g_m\|^2_{L^2_{\frac{\mu}{2}}(\mathbb{R}^2_+)} + \varepsilon\|\partial_y^{-1} g_m\partial_y\partial_x\eta_1\|^2_{L^2_{\frac{\mu}{2}}(\mathbb{R}^2_+)}\\
&\quad +\varepsilon\|g_m\|^2_{L^2_{\frac{\mu}{2}}(\mathbb{R}^2_+)} + \sum_{j=1}^{6}\Big|\int_{\mathbb{R}^2_+} M^m_j e^{\frac{y}{4}} g_m dxdy\Big|,
\end{aligned}$$

for $0 < \varepsilon \le 1$, which implies

$$\begin{aligned}
&\frac{d}{dt}\|g_m\|^2_{L^2_{\frac{\mu}{2}}(\mathbb{R}^2_+)} + \|\partial_y g_m\|^2_{L^2_{\frac{\mu}{2}}(\mathbb{R}^2_+)} + \varepsilon\|\partial_x g_m\|^2_{L^2_{\frac{\mu}{2}}(\mathbb{R}^2_+)}\\
&\le C\|g_m\|^2_{L^2_{\frac{\mu}{2}}(\mathbb{R}^2_+)} + \|\partial_y^{-1} g_m \partial_y \eta_1\|^2_{L^2_{\frac{\mu}{2}}(\mathbb{R}^2_+)} + \|\partial_y^{-1} g_m \partial_y\partial_x \eta_1\|^2_{L^2_{\frac{\mu}{2}}(\mathbb{R}^2_+)}\\
&\quad +2\sum_{j=1}^{6}\left|\int_{\mathbb{R}^2_+} M_j^m e^{\frac{y}{4}} g_m dxdy\right|.
\end{aligned}$$

We use the similar method as in [151] to deal with right-hand side of terms in above inequality. Firstly, we derive the estimate of second and third terms from (3.2.30) and (3.2.32):

$$\begin{aligned}
&|\eta_1| = \left|\frac{u_{xy}}{u^s_y + \tilde{u}_y}\right| \le Ce^{-\frac{y}{4}},\ |\partial_x \eta_1| = \left|\frac{u_{xxy}}{u^s_y + \tilde{u}_y} - \frac{u_{xy}\tilde{u}_{xy}}{(u^s_y + \tilde{u}_y)^2}\right| \le Ce^{-\frac{y}{4}},\\
&|\partial_y \eta_1| = \left|\frac{u_{xyy}}{u^s_y + \tilde{u}_y} - \frac{u_{xy}\tilde{u}_{yy}}{(u^s_y + \tilde{u}_y)^2}\right| \le Ce^{-\frac{y}{4}},\ |\partial_x\partial_y \eta_1| \le Ce^{-\frac{y}{4}}.
\end{aligned}$$

Then we have

$$\|\partial_y^{-1} g_m \partial_y \eta_1\|^2_{L^2_{\frac{\mu}{2}}(\mathbb{R}^2_+)} \le C\int_{\mathbb{R}^2_+} e^{-\frac{y}{2}}\Big(\int_0^y g_m(t,x,\tilde{y})d\tilde{y}\Big)^2 dxdy \le C\|g_m\|^2_{L^2_{\frac{\mu}{2}}},$$

and

$$\|\partial_y^{-1} g_m \partial_y\partial_x \eta_1\|^2_{L^2_{\frac{\mu}{2}}(\mathbb{R}^2_+)} \le C\int_{\mathbb{R}^2_+} e^{-\frac{y}{2}}\Big(\int_0^y g_m(t,x,\tilde{y})d\tilde{y}\Big)^2 dxdy \le C\|g_m\|^2_{L^2_{\frac{\mu}{2}}}.$$

Then we estimate the terms of M_j^m, $j = 1, 2, \cdots, 6$. For the term M_1^m, we have

$$\begin{aligned}
\left|\int_{\mathbb{R}^2_+} M_1^m e^{\frac{y}{4}} g_m dxdy\right| &= \left|\int_{\mathbb{R}^2_+} (u^s + u)(g_m\eta_1 + (\partial_y^{-1} g_m)\partial_y\eta_1)e^{\frac{y}{4}} g_m dxdy\right|\\
&\le C\|g_m\|^2_{L^2_{\frac{\mu}{2}}} + \|\partial_y^{-1} g_m \partial_y \eta_1\|^2_{L^2_{\frac{\mu}{2}}(\mathbb{R}^2_+)}\\
&\le C\|g_m\|^2_{L^2_{\frac{\mu}{2}}}.
\end{aligned} \tag{3.2.35}$$

We use the similar method to that of η_1 above to get decay rate of η_2,

$$\begin{aligned}
&|\eta_2| \le C(1 + e^{-\frac{y}{4}}),\ |\partial_x \eta_2| \le Ce^{-\frac{y}{4}},\\
&|\partial_y \eta_2| \le Ce^{-\frac{y}{4}},\ |\partial_x\partial_y \eta_2| \le Ce^{-\frac{y}{4}}.
\end{aligned}$$

Thus we get

$$\begin{aligned}
&\Big|\int_{\mathbb{R}_+^2} M_2^m e^{\frac{y}{4}} g_m dxdy\Big| \\
&\le C\Big|\int_{\mathbb{R}_+^2}\Big((\partial_y g_m)\eta_2 + 2g_m(\partial_y\eta_2 - 2\eta_2^2) - 8(\partial_y^{-1}g_m)\eta_2\partial_y\eta_2\Big)e^{\frac{y}{4}} g_m dxdy\Big| \\
&\le C\|g_m\|^2_{L^2_{\frac{\mu}{2}}} + \frac{1}{4}\|\partial_y g_m\|^2_{L^2_{\frac{\mu}{2}}} + \|\partial_y^{-1} g_m\partial_y\eta_2\|^2_{L^2_{\frac{\mu}{2}}(\mathbb{R}_+^2)} \\
&\le C\|g_m\|^2_{L^2_{\frac{\mu}{2}}} + \frac{1}{4}\|\partial_y g_m\|^2_{L^2_{\frac{\mu}{2}}}.
\end{aligned} \tag{3.2.36}$$

Similarly as above for M_2^m, we get for any $\varepsilon \in (0, 1]$,

$$\Big|\int_{\mathbb{R}_+^2} M_3^m e^{\frac{y}{4}} g_m dxdy\Big| \le C\|g_m\|^2_{L^2_{\frac{\mu}{2}}} + \frac{\varepsilon}{4}\|\partial_x g_m\|^2_{L^2_{\frac{\mu}{2}}}. \tag{3.2.37}$$

For M_4^m, using a direct computation and (3.2.30)–(3.2.32), we have

$$\Big|\int_{\mathbb{R}_+^2} M_4^m e^{\frac{y}{4}} g_m dxdy\Big| \le C\|g_m\|^2_{L^2_{\frac{\mu}{2}}}. \tag{3.2.38}$$

Noting that M_5^m can be written as

$$M_5^m = g_4\partial_x u + \sum_{1\le m\le 3} C_4^m \partial_x u g_{5-m} + \partial_y^{-1} g_4 \partial_x w + \sum_{1\le m\le 3} C_4^m \partial_x^m w \partial_y^{-1} g_{5-m},$$

we have

$$\begin{aligned}
\Big|\int_{\mathbb{R}_+^2} M_5^m e^{\frac{y}{4}} g_m dxdy\Big| &\le C\|g_m\|_{L^2_{\frac{\mu}{2}}}(\|g_4\partial_x u\|_{L^2_{\frac{\mu}{2}}} + \|\partial_y^{-1} g_4\partial_x w\|_{L^2_{\frac{\mu}{2}}}) \\
&\quad + C\|g_m\|_{L^2_{\frac{\mu}{2}}}\sum_{1\le m\le 3}(\|\partial_x u g_{5-m}\|_{L^2_{\frac{\mu}{2}}} + \|\partial_x^m w\partial_y^{-1} g_{5-m}\|_{L^2_{\frac{\mu}{2}}}) \\
&\le C\|g_m\|_{L^2_{\frac{\mu}{2}}}(\|g_4\|_{L^2_{\frac{\mu}{2}}}\|\partial_x u\|_{L^\infty} + \|g_4\|_{L^2_{\frac{\mu}{2}}}) \\
&\quad + C\|g_m\|_{L^2_{\frac{\mu}{2}}}\sum_{1\le m\le 3}(\|g_{5-m}\|_{L^2_{\frac{\mu}{2}}}\|\partial_x u\|_{L^\infty} + \|g_{5-m}\|_{L^2_{\frac{\mu}{2}}}) \\
&\le C\|g_m\|_{L^2_{\frac{\mu}{2}}}(\|g_4\|_{L^2_{\frac{\mu}{2}}}\|\partial_x w\|_{L^2_\mu(\mathbb{R}_+^2)} + \|g_4\|_{L^2_{\frac{\mu}{2}}}) \\
&\quad + C\|g_m\|_{L^2_{\frac{\mu}{2}}}\sum_{1\le m\le 3}(\|g_{5-m}\|_{L^2_{\frac{\mu}{2}}}\|\partial_x w\|_{L^2_\mu(\mathbb{R}_+^2)} + \|g_{5-m}\|_{L^2_{\frac{\mu}{2}}})
\end{aligned}$$

$$\leq C\sum_{m=1}^{4}\|g_m\|^2_{L^2_{\frac{\mu}{2}}}, \tag{3.2.39}$$

where we have used the fact

$$\begin{aligned}
&\|e^{\frac{y}{4}}\partial_x^m w\|_{L^\infty([0,T]\times\mathbb{R}^2_+)}\\
&=\int_0^y (e^{\frac{y}{4}}\partial_x^m w)_y dy \leq \Big(\int_0^y e^{-\frac{y}{2}}dy\Big)^{\frac{1}{2}}\Big(\int_0^y e^{\frac{y}{2}}((e^{\frac{y}{4}}\partial_x^m w)_y)^2 dy\Big)^{\frac{1}{2}}\\
&\leq C\Big(\int_0^y e^{\frac{y}{2}}((e^{\frac{y}{4}} w)_y)^2 dy\Big)^{\frac{1}{2}}_{L^\infty(\mathbb{R}^+_x)}\\
&\leq C\Big(\int_0^\infty\int_0^y e^{\frac{y}{2}}((e^{\frac{y}{4}}\partial_x^m w)_y)^2 dxdy\Big)^{\frac{1}{2}}\Big(\int_0^\infty\int_0^y e^{\frac{y}{2}}((e^{\frac{y}{4}}\partial_x^m w)_{xy})^2 dxdy\Big)^{\frac{1}{2}}\\
&\leq C\|w\|_{H^4_\mu}.
\end{aligned}$$

Therefore it follows

$$|\partial_x^m w(x,y,t)|\leq Ce^{-\frac{y}{4}}, \forall (x,y,t)\in\mathbb{R}^2_+\times[0,T].$$

Last, we deal with the term M_6^m by using integration by part and boundary condition

$$\begin{aligned}
&-\int_{\mathbb{R}^2_+} M_6^m e^{\frac{y}{4}} g_m dxdy\\
&=\frac{1}{4}\int_{\mathbb{R}^2_+}\Big(\frac{\sum_j^4 C_4^j\partial_x^j w\cdot\partial_x^{4-j}v}{u_y^s+u_y}\Big)e^{\frac{y}{4}} g_m dxdy\\
&\quad+\int_{\mathbb{R}^2_+}\Big(\frac{\sum_j^4 C_4^j\partial_x^j w\cdot\partial_x^{4-j}v}{u_y^s+u_y}\Big)e^{\frac{y}{4}}\partial_y g_m dxdy\\
&\leq C\int_{\mathbb{R}^2_+}\sum_{j=1}^{4}\partial_x^j w\partial_x^{4-i} v g_m dxdy + C\int_{\mathbb{R}^2_+}\sum_{j=1}^{4}\partial_x^j w\partial_x^{4-i} v\partial_y g_m dxdy\\
&\leq C(\|\partial_x^4 w\|_{L^2_\mu}\|v\|_{L^\infty}+\sum_{j=1}^{3}\|\partial_x^j w\|_{L^\infty(R_x;L^2(R_y))}\|\partial_x^{4-j}v\|_{L^\infty(R_y;L^2(R_x))})\\
&\quad\times(\|g_m\|_{L^2_\mu}+\|\partial_y g_m\|_{L^2_\mu})\\
&\leq C\|w\|_{H^4_\mu}(\|g_m\|_{L^2_\mu}+\|\partial_y g_m\|_{L^2_\mu})\\
&\leq C\|w\|^2_{H^4_\mu}+C\|g_m\|^2_{L^2_\mu}+\frac{1}{4}\|\partial_y g_m\|^2_{L^2_\mu}.
\end{aligned} \tag{3.2.40}$$

Thus, inserting (3.2.35)–(3.2.40) into (3.2.35) completes the proof of (3.2.34). □

In this position, we give the energy estimate for the sequence of approximate solutions. Let's go back to the notations with tilde and the sub-index ε and g_m^ε is the function defined by $\tilde{u}_\varepsilon$.

We now collect some established estimates (3.2.15) and (3.2.34) to be used,

$$\frac{d}{dt}\sum_{m=1}^{4}\|\tilde{w}_\varepsilon\|^2_{H_\mu^{m,m-1}(\mathbb{R}_+^2)}+\varepsilon\sum_{m=1}^{4}\|\partial_x\tilde{w}_\varepsilon\|^2_{H_\mu^{m,m-1}(\mathbb{R}_+^2)}+\sum_{m=1}^{4}\|\partial_y\tilde{w}_\varepsilon\|^2_{H_\mu^{m,m-1}(\mathbb{R}_+^2)}$$
$$\le C(\|\tilde{w}_\varepsilon\|^2_{H_\mu^4(\mathbb{R}_+^2)}+\|\tilde{w}_\varepsilon\|^4_{H_\mu^4(\mathbb{R}_+^2)}), \tag{3.2.41}$$

$$\frac{d}{dt}\sum_{m=1}^{4}\|g_m^\varepsilon\|^2_{L^2_{\frac{\mu}{2}}(\mathbb{R}_+^2)}+\sum_{m=1}^{4}\|\partial_y g_m^\varepsilon\|^2_{L^2_{\frac{\mu}{2}}(\mathbb{R}_+^2)}+\varepsilon\sum_{m=1}^{4}\|\partial_x g_m^\varepsilon\|^2_{L^2_{\frac{\mu}{2}}(\mathbb{R}_+^2)}$$
$$\le C\Big(\sum_{m=1}^{4}\|g_m^\varepsilon\|^2_{L^2_{\frac{\mu}{2}}(\mathbb{R}_+^2)}+\|\tilde{w}_\varepsilon\|^2_{H_\mu^4}\Big). \tag{3.2.42}$$

Lemma 3.2.9 *Let $m=1,2,3,4$. Then we have*

$$h_m^\varepsilon(g,w)(0)=\sum_{m=1}^{4}\|g_m^\varepsilon\|^2_{L^2_{\frac{\mu}{2}}(\mathbb{R}_+^2)}+\|\tilde{w}_\varepsilon(0)\|^2_{H_\mu^{m,m-1}(\mathbb{R}_+^2)}\le C\|\tilde{u}_0\|^2_{D_\mu^4(\mathbb{R}_+^2)}, \tag{3.2.43}$$

where a constant $C>0$ is independent of ε.

Proof For any $1\le m\le 4$, applying Lemma 3.2.7 and $\tilde{u}_\varepsilon(0)=\tilde{u}(0)$, we conclude that

$$h_m^\varepsilon(g,w)(0)\le C\|\tilde{w}_0\|^2_{H_\mu^4(\mathbb{R}_+^2)}\le C\|\tilde{u}_0\|^2_{D_\mu^4(\mathbb{R}_+^2)},$$

which is just (3.2.43). □

Lemma 3.2.10 *Let $m=1,2,3,4$, the following estimate holds:*

$$\|\partial_x^m\tilde{w}_\varepsilon\|^2_{L^2_\mu(\mathbb{R}_+^2)}\le C\|g_m^\varepsilon\|^2_{L^2_{\frac{\mu}{2}}(\mathbb{R}_+^2)}, \tag{3.2.44}$$

where a constant $C>0$ is independent of ε.

Proof By the definition, we know that

$$\partial_x^m\tilde{u}_\varepsilon(t,x,y)=(u_y^s+\tilde{w}_\varepsilon)\int_0^y g_m^\varepsilon(t,x,z)dz,\ \ y\ge 0.$$

Thus,

$$\partial_x^m \tilde{w}_\varepsilon(t,x,y) = (u^s_{yy} + (\tilde{w}_\varepsilon)_y)\int_0^y g^\varepsilon_m(t,x,z)dz + (u^s_y + \tilde{w}_\varepsilon)g^\varepsilon_m(t,x,y),\ y \ge 0,$$

which gives

$$\|\partial_x^m \tilde{w}_\varepsilon\|^2_{L^2_\mu(\mathbb{R}^2_+)} \le C\|g^\varepsilon_m\|^2_{L^2(\mathbb{R}^2_+)} + C\|g^\varepsilon_m\|^2_{L^2_{\frac{\mu}{2}}(\mathbb{R}^2_+)} \le C\|g^\varepsilon_m\|^2_{L^2_{\frac{\mu}{2}}(\mathbb{R}^2_+)}.$$

□

Theorem 3.2.11 *Assume* $\tilde{u}_0 \in D^4_\mu(\mathbb{R}^2_+)$ *satisfies the compatibility condition* (3.2.10). *Suppose that* $\tilde{w}_\varepsilon \in L^\infty([0,T]; H^4_\mu(\mathbb{R}^2_+))$ *is a solution to* (3.2.13) *such that*

$$\|\tilde{w}_\varepsilon\|_{L^\infty([0,T];H^4_\mu(\mathbb{R}^2_+))} \le 2\eta, \tag{3.2.45}$$

where $0 \le T \le T_{\max} = \sup\{t \in [0,T],\ 1 - C\|\tilde{u}_0\|_{D^4_\mu(\mathbb{R}^2_+)}t - \frac{t^2}{2} > 0\}$. *Then we have*

$$\|\tilde{w}_\varepsilon\|^2_{H^4_\mu(\mathbb{R}^2_+)} \le \left(\|\tilde{u}_0\|^2_{D^4_\mu(\mathbb{R}^2_+)} + 1\right)\left\{1 - C\|\tilde{u}_0\|^2_{D^4_\mu(\mathbb{R}^2_+)}t - \frac{t^2}{2}\right\}^{-1}, \tag{3.2.46}$$

where a constant $C > 0$ *is independent of* $\varepsilon \in (0,1]$.

Proof Applying (3.2.41)–(3.2.42), for $m = 1, 2, 3, 4$, we get

$$\begin{aligned}\sum_{i=1}^4 \left(\|g^\varepsilon_m\|^2_{L^2_{\frac{\mu}{2}}(\mathbb{R}^2_+)} + \|\tilde{w}_\varepsilon\|^2_{H^{m,m-1}_\mu(\mathbb{R}^2_+)}\right) &\le (\|\tilde{u}_0\|^2_{D^4_\mu(\mathbb{R}^2_+)} + t)\\ &+C\int_0^t \left(\sum_{i=1}^4 (\|g^\varepsilon_m\|^2_{L^2_{\frac{\mu}{2}}(\mathbb{R}^2_+)} + \|\tilde{w}_\varepsilon\|^2_{H^{m,m-1}_\mu(\mathbb{R}^2_+)})\right)^2 ds.\end{aligned}$$

Applying nonlinear Gronwall inequality (Theorem 2, P362, [100]) to above inequality, we have

$$\begin{aligned}&\sum_{i=1}^4 \left(\|g^\varepsilon_m\|^2_{L^2_{\frac{\mu}{2}}(\mathbb{R}^2_+)} + \|\tilde{w}_\varepsilon\|^2_{H^{m,m-1}_\mu(\mathbb{R}^2_+)}\right)\\ &\quad\le (\|\tilde{u}_0\|^2_{D^4_\mu(\mathbb{R}^2_+)} + t)\left\{1 - C\|\tilde{u}_0\|^2_{D^4_\mu(\mathbb{R}^2_+)}t - \frac{t^2}{2}\right\}^{-1}.\end{aligned} \tag{3.2.47}$$

Combining (3.2.47) and Lemma 3.2.10, we have, for $t \in [0, T]$,

$$\|\tilde{w}_\varepsilon\|^2_{H^4_\mu(\mathbb{R}^2_+)} \leq (\|\tilde{u}_0\|^2_{D^4_\mu(\mathbb{R}^2_+)} + t)\Big\{1 - C\|\tilde{u}_0\|^2_{D^4_\mu(\mathbb{R}^2_+)} t - \frac{t^2}{2}\Big\}^{-1}, \quad (3.2.48)$$

where a constant $C > 0$ is independent of $\varepsilon \in (0, 1]$. Thus (3.2.46) follows immediately. □

3.2.4 Convergence and Consistency

Using evolution Eq. (3.2.13) and uniform H^4_μ bound in (3.2.14), we conclude that $\partial_t \tilde{w}^\varepsilon$ is uniformly (in ε) bounded in $L^\infty([0, T]; H^2_\mu)$. By the Lions-Aubin Lemma and the compact embedding of H^4_μ in $H^{4-\delta}_{\mu,loc}$, for $0 < \delta < 1$. Then taking a subsequence, as $\varepsilon_k \to 0^+$,

$$\tilde{w}^{\varepsilon_k} \overset{*}{\rightharpoonup} \tilde{w} \quad \text{in} \quad L^\infty([0, T]; H^4_\mu) \quad \text{and } \tilde{w}^{\varepsilon_k} \to \tilde{w} \quad \text{in} \quad C([0, T]; H^{4-\delta}_{\mu,loc}),$$

where $\tilde{w} = \partial_y \tilde{u} \in L^\infty([0, T]; H^4_\mu)$. Applying the local uniform convergence of $\partial^k_x u^{\varepsilon_k}$, we have following the pointwise convergence of v^{ε_k}: as $\varepsilon_k \to 0^+$,

$$v^{\varepsilon_k} = -\int_0^y \partial_x \tilde{u}^{\varepsilon_k} dy \to -\int_0^y \partial_x \tilde{u} dy =: v. \quad (3.2.49)$$

In fact, we also have

$$\lim_{y\to 0} \tilde{u}(t, x, y) = \lim_{y\to 0} \int_0^y \tilde{w}(t, x, \tilde{y}) d\tilde{y}$$

and

$$\tilde{v} = -\int_0^y \tilde{u}_x d\tilde{y} \in L^\infty([0, T_1]; L^\infty(\mathbb{R}_{+,y}; H^3(\mathbb{R}_x))).$$

We have proven that $\tilde{w}$ is a classical solution to the following vorticity Prandtl equation

$$\begin{cases} \partial_t \tilde{w} + (u^s + \tilde{u})\partial_x \tilde{w} + \tilde{v}(u^s_{yy} + \partial_y \tilde{w}) = \partial^2_y \tilde{w} \text{ in } \mathbb{R}^2_+ \times \mathbb{R}_+, \\ \partial_y \tilde{w} = 0 \text{ on } \mathbb{R} \times \mathbb{R}_+, \\ \tilde{w}|_{t=0} = \tilde{w}_0 \text{ on } \mathbb{R}^2_+, \end{cases}$$

and $(\tilde{u}, \tilde{v})$ is a classical solution to problem (3.2.3). Lastly, $(u, v) = (u^s + \tilde{u}, \tilde{v})$ is a classical solution to (3.2.1). Therefore we have proved Theorem 3.2.1. □

3.3 Bibliographic Comments

Since the energy method is suitable for the Navier-Stokes equations, there have been some results on the well-posedness of the 2D Prandtl equations (3.1.1) by using energy method under a monotonicity condition on the tangential velocity field. Recently, under a monotonicity condition on the tangential velocity field, Alexandre et al. [3] used a direct energy method instead of using the Crocco transformation to investigate the local well-posedness theory of the solution to the 2D Prandtl equations (3.1.1) in a weighted Sobolev space. Later on, under the Oleinik′s monotonicity assumption, Masmoudi and Wong [98] proved the local existence and uniqueness for the 2D Prandtl system (3.1.1) in a weighted Sobolev space by the energy method. They used a new method of nonlinear energy estimate instead of using the Crocco transformation for the Prandtl system. This new estimate is based on a cancellation property in the convection terms to overcome the loss of x-derivative in the tangential direction, which is valid under the monotonicity assumption. The difference from [98] is that the initial datum is a small perturbation of a monotonic shear flow in [3]. Based on the results of the local well-posedness for Prandtl equation (3.1.1) in [3, 98], Xu and Zhang [151] discussed the global well-posedness of solutions for nonlinear Prandtl boundary layer Eq. (3.1.1) on the half plane. More specifically, they used the energy method to obtain the existence, uniqueness, and stability of solutions in the weighted Sobolev space $H^m_\mu(\mathbb{R}^2_+)$ $(m \geq 6)$ when weighted functions μ is a polynomial function and proved the lifespan T of solutions could be any large when its initial datum is a perturbation around the monotonic shear profile of small size e^{-T}, which is the first result of global existence of solutions of 2D Prandtl equations (3.1.1) in a polynomial weighted Sobolev space.

Using the paralinearization technique, Chen, Wang, and Zhang [12] proved the local well-posedness of Prandtl equation (3.1.1) for monotonic data in the anisotropic Sobolev space with an exponential weighted and low regularity and provided a new way for the zero viscosity limit problem of the Navier-Stokes equations with the non-slip boundary condition, who obtained the local well-posedness of Prandtl equation (3.1.1) when weighted function μ is an exponential function in $H^{3,1}_\mu(\mathbb{R}^2_+) \cap H^{1,2}_\mu(\mathbb{R}^2_+)$. To be more precise, our motivation in this chapter is to use the energy method to prove the existence of solutions of Prandtl equation (3.1.1) with the non-slip boundary condition when the weighted function μ is an exponential function in $H^4_\mu(\mathbb{R}^2_+)$, which is still an open problem in Sobolev spaces before the present paper [115] solved it.

To deal with our problem (3.1.1), we have encountered some difficulties. The first difficulty arises from the loss of horizontal direction x-derivative by the term $v\partial_y u$. With the loss of x-derivative and the absence of any horizontal diffusion, it is necessary to use a type of cancellation mechanism in the analysis. To overcome this difficulty, inspired by recent results in [3, 98, 151], we construct a new unknown function $g_n = \left(\frac{\partial_x^n u}{u_y^s + u_y}\right)_y$ to avoid the loss of x derivative and establish the estimates of g_n in L^2-norm and also give the relationship between the quantities g_n and $\partial_x^n \tilde{w}$

in L^2-norm by Lemma 3.2.7. Using the boundary condition of vertical velocity and the divergence free, we obtain the representation formula of vertical velocity:

$$v(x, y, t) = -\int_0^y \partial_x u(x, \tilde{y}, t) d\tilde{y}.$$

The second difficulty is that vertical velocity v creates a loss of x-variable, so that the standard energy estimates fail. To overcome this one, we first use the inequality (3.2.1) to take the L^∞ norm of normal velocity $v(t, x, y)$, which plays a crucial role in the estimates of this chapter since the estimate of normal velocity $v(t, x, y)$ cannot be controlled directly by the norm of the vorticity $w = \partial_y u$, then we use incompressible equation $(3.1.1)_2$.

In order to prove our desired result (Theorem 3.2.1) on the local-in-time existence of solutions to 2D Prandtl equation (3.1.1), inspired by the ideas of [98] and [151], we investigate the parabolic regularized Prandtl equations (3.2.11) by constructing an approximate scheme, which keeps the nonlinear term of the original Prandtl equation (3.1.1) and nonlinear cancellation properties. Therefore, the local-in-time existence of solutions of the original Eq. (3.1.1) follows by uniform energy estimates of the approximate solutions. The first duty of this chapter is to derive uniform energy estimates of the approximates solutions. The second one is to construct a new unknown function g_n and to establish the relationship between the norms of g_n and $\partial_x^m \tilde{w}$ from Lemma 3.2.7, and then we can obtain a priori estimate of solutions independent of ε.

Chapter 4
Local Well-Posedness of Solutions to the 2D Mixed Prandtl Equations in a Sobolev Space Without Monotonicity and Lower Bound

In this chapter, we shall investigate the 2D Prandtl-Shercliff regime equations on the half plane and prove the local existence and uniqueness of solutions for any initial datum by using the classical energy methods in a Sobolev space. Compared to the existence and uniqueness of solutions to the classical Prandtl equations where the monotonicity condition of the tangential velocity plays a key role, this monotonicity condition is not needed for the 2D mixed Prandtl equations. Besides, compared with the existence and uniqueness of solutions to the 2D MHD boundary layer where the initial tangential magnetic field has a lower bound that plays an important role, this lower bound condition is also not needed for the 2D mixed Prandtl equations. In other words, we need neither the monotonicity condition of the tangential velocity nor the initial tangential magnetic field has a lower bound and for any initial datum in this chapter. As far as we have learned, this is the first result of the 2D mixed Prandtl-Shercliff regime equations in a Sobolev space. The content of this chapter is chosen from [116].

4.1 Introduction

The following Prandtl-Shercliff regime equations were derived in [45] by the classical 2D incompressible MHD system when the physical parameters such as Reynolds number, magnetic Reynolds number, and Hartmann number satisfy some constraints in the high Reynolds numbers limit. In this chapter, we shall investigate the local existence and uniqueness of solutions to the following initial boundary

Y. Qin et al., *Prandtl Equations and Related Boundary Layer Equations*,
https://doi.org/10.1007/978-981-97-4565-4_4

value problem for the 2D Prandtl-Shercliff regime system in the upper half plane $\mathbb{R}_+^2 = \{(x, y) : x \in \mathbb{R},\ y \in \mathbb{R}_+\}$, which reads as

$$\begin{cases} \partial_t u + u\partial_x u + v\partial_y u - \partial_y^2 u = \partial_x b, \\ \partial_x u + \partial_y^2 b = 0, \\ \partial_x u + \partial_y v = 0, \end{cases} \tag{4.1.1}$$

where (u, v) denotes the velocity field of the boundary layer and b stands for the corresponding tangential magnetic component, respectively.

The initial datum of (4.1.1) is given by

$$u|_{t=0} = u_0. \tag{4.1.2}$$

The no-slip boundary conditions are imposed on the velocity field and magnetic field

$$(u, v, b)|_{y=0} = \mathbf{0}. \tag{4.1.3}$$

Far fields boundary conditions take the form

$$\lim_{y \to +\infty} (u, b) = (\bar{u}, \bar{b}), \tag{4.1.4}$$

where $(\bar{u}, \bar{b})$ is a pair of positive constants.

4.2 Local Well-Posedness of Solutions

First, we define some Sobolev spaces which will be used throughout the chapter as follows:

$$\|f\|_{H^{m,m-1}(\mathbb{R}_+^2)}^2 = \sum_{0 \le \alpha+\beta \le m,\ \alpha \le m-1} \|\partial_x^\alpha \partial_y^\beta f\|_{L^2(\mathbb{R}_+^2)}^2,$$

$$\|f\|_{H^m(\mathbb{R}_+^2)}^2 = \|f\|_{H^{m,m-1}(\mathbb{R}_+^2)}^2 + \|\partial_x^m f\|_{L^2(\mathbb{R}_+^2)}^2.$$

Next, we will state the main result as follows.

Theorem 4.2.1 *Let $|\gamma| \le m = 4$ and $\partial_y u_0 \in H^4(\mathbb{R}_+^2)$. Then there exists a constant $0 < T = \min\left\{\|w(0)\|_{L^2(\mathbb{R}_+^2)}^2, \frac{3}{32C\|w(0)\|_{H^4(\mathbb{R}_+^2)}^4}\right\}$ such that if*

$$\|\partial_y u_0\|_{H^4(\mathbb{R}_+^2)} \le C,$$

then for $0 \leq t \leq T$, *the system* (4.1.1) *admits a unique solution* (u, v, b) *in* $[0, T]$ *such that*

$$u \in L^\infty([0,T]; H^4(\mathbb{R}_+^2)), \ (v, b) \in L^\infty([0,T]; L^\infty(\mathbb{R}_+ \times H^3(\mathbb{R}))),$$

$$u \in L^2([0,T]; H^5(\mathbb{R}_+^2)),$$

where C *is a positive constant.*

Before proving Theorem 4.2.1, we first list an important inequality as follows.

Lemma 4.2.2 (Agmon Inequality [59]) *Let* $f \in H^1(\mathbb{R}_+^2)$, *then*

$$\|f\|_{L_x^\infty L_y^2} \leq C\|f\|_{L^2(\mathbb{R}_+^2)}^{\frac{1}{2}} \|f\|_{H^1(\mathbb{R}_+^2)}^{\frac{1}{2}}.$$

Lemma 4.2.3 ([100]) *Let* $\eta(t)$ *be nonnegative, absolutely continues function on* $[0, T]$, *which satisfies for a.e.* $t \in (0, T)$, *the integral inequality*

$$\eta(t) \leq a(t) + b(t) \int_0^t k(s)\eta^m(s)ds,$$

where $m \geq 2$ *is a positive integer and* $a(t)$, $b(t)$, $k(t)$ *are positive, continuous functions on* $[0, T]$, *and* $\frac{a(t)}{b(t)}$ *is a nondecreasing function. Then for* $0 \leq t \leq \alpha_m$,

$$\eta(t) \leq a(t)\Big\{1 - (m-1)\int_0^t k(s)b(s)a^{m-1}(s)ds\Big\}^{\frac{1}{1-m}},$$

where α_m *is defined to be the supreme of those* $t \in [0, T]$ *satisfying*

$$\alpha_m = \sup\Big\{t \in [0,T] : 1 - (m-1)\int_0^t k(s)b(s)a^{m-1}(s)ds > 0\Big\}.$$

Next, we will prove Theorem 4.2.1 by the following a series of lemmas. First, we show the boundary value of the (u, b) of higher order partial derivatives in y on $y = 0$ for systems (4.1.1).

Lemma 4.2.4 *Assume that* (u, b) *is a smooth solution of the system* (4.1.1), *then the boundary values of* (u, b) *satisfy*

$$\begin{cases} u(t,x,0) = 0, \ b(t,x,0) = 0, \\ \partial_y^2 u(t,x,0) = 0, \ \ \partial_y^2 b(t,x,0) = 0, \\ \partial_y^4 u(t,x,0) = \partial_x\partial_y u\partial_y u(t,x,0), \\ \partial_y^4 b(t,x,0) = 0, \\ \partial_y^6 u(t,x,0) = (4\partial_x\partial_y^3 u + \partial_x^2\partial_y b)\partial_y u(t,x,0) + (\partial_x\partial_y b - \partial_y^3 u)\partial_x\partial_y u(t,x,0), \\ \partial_y^6 b(t,x,0) = -\Big(\partial_x^2\partial_y u\partial_y u + (\partial_x\partial_y u)^2\Big)(t,x,0). \end{cases} \tag{4.2.1}$$

Proof Looking back the boundary conditions in (4.1.3),

$$u(t,x,0)=0,\ v(t,x,0)=0,\ b(t,x,0)=0,\ (t,x)\in[0,T]\times\mathbb{R}, \tag{4.2.2}$$

By use of the equations $(4.1.1)_3$, we have

$$u(t,x,y)=-\int_{-\infty}^{x}\partial_y v(t,\tilde{x},y)d\tilde{x} \text{ and } \partial_x^n v(t,x,y)=-\int_0^y \partial_x^{n+1}u(t,x,\tilde{y})d\tilde{y},$$

thus, by (4.1.3), we get $\partial_y v(t,x,0)=0$, $\partial_x u(t,x,0)=0$ and $\partial_x^n v(t,x,0)=0$.

We know that

$$u_x(t,x,y)=\int_{-\infty}^{x}\partial_x^2 u(t,\tilde{x},y)d\tilde{x}$$

which, combined with $\partial_x u(t,x,0)=0$, we have $\partial_x^2 u(t,x,0)=0$. Similarly, we can derive $\partial_x^3 u(t,x,0)=0$ and $\partial_x^4 u(t,x,0)=0$.

By $(4.1.1)_2$, we give

$$\begin{aligned} b(t,x,y) &= \int_0^y\int_0^{\tilde{y}}\partial_y^2 b(t,x,\bar{y})d\bar{y}d\tilde{y}-\int_0^y \partial_y b(t,x,0)d\tilde{y} \\ &= -\int_0^y\int_0^{\tilde{y}}\partial_x u(t,x,\bar{y})d\bar{y}d\tilde{y}-\int_0^y \partial_y b(t,x,0)d\tilde{y}, \end{aligned}$$

which gives $\partial_x^n b(t,x,0)=0$. Therefore, we can deduce that

$$\partial_x^n(u,v,b)(t,x,0)=0,\quad (t,x)\in[0,T]\times\mathbb{R}, 0\le n\le 4. \tag{4.2.3}$$

Applying (4.1.1) and the boundary conditions (4.2.3), we have

$$\partial_y^2 u|_{y=0}=0,\quad \partial_y^2 b|_{y=0}=0. \tag{4.2.4}$$

Besides, we also have

$$\partial_x^n\partial_y^2 u|_{y=0}=0,\quad \partial_x^n\partial_y^2 b|_{y=0}=0,\quad (t,x)\in[0,T]\times\mathbb{R},\quad 0\le n\le 4. \tag{4.2.5}$$

Differentiating the equation of $(4.1.1)_1$ with respect to y,

$$\partial_t\partial_y u-\partial_y^3 u+\partial_y\big(u\partial_x u+v\partial_y u\big)=\partial_x\partial_y b, \tag{4.2.6}$$

and using the boundary conditions (4.2.2)–(4.2.4), then we have

$$\partial_t\partial_y u|_{y=0}=\partial_y^3 u|_{y=0}+\partial_x\partial_y b|_{y=0}. \tag{4.2.7}$$

Differentiating (4.1.1) with respect to y twice, we arrive at

$$\partial_t\partial_y^2 u - \partial_y^4 u + \partial_y^2\big(u\partial_x u + v\partial_y u\big) = \partial_x\partial_y^2 b. \tag{4.2.8}$$

Now appealing to the Leibniz formula

$$\begin{aligned}\partial_y^2\big(u\partial_x u + v\partial_y u\big) = {} & \partial_y^2 u\partial_x u + \partial_y^2 v\partial_y u + u\partial_x\partial_y^2 u + v\partial_y^3 u \\ & +2\partial_y u\partial_x\partial_y u + 2\partial_y v\partial_y^2 u,\end{aligned} \tag{4.2.9}$$

we conclude

$$\partial_y^4 u(t,x,0) = \partial_x\partial_y u\partial_y u(t,x,0). \tag{4.2.10}$$

Differentiating (4.2.10) with respect to t and using (4.2.7), we get

$$\begin{aligned}\partial_t\partial_y^4 u(t,x,0) &= \partial_t\partial_y u\partial_x\partial_y u(t,x,0) + \partial_y u\partial_t\partial_{xy}^2 u(t,x,0) \\ &= (\partial_y^3 u + \partial_x\partial_y b)\partial_x\partial_y u(t,x,0) + (\partial_x\partial_y^3 u + \partial_x^2\partial_y b)\partial_y u(t,x,0).\end{aligned} \tag{4.2.11}$$

Similarly, differentiating (4.1.1) four times with respect to y, we get

$$\partial_t\partial_y^4 u - \partial_y^6 u + \partial_y^4\big(u\partial_x u + v\partial_y u\big) = \partial_x\partial_y^4 b, \tag{4.2.12}$$

and invoking the Leibniz formula, we deduce

$$\begin{aligned}\partial_y^4\big(u\partial_x u + v\partial_y u\big) = {} & \partial_y^4 u\partial_x u + \partial_y^4 v\partial_y u + 4\partial_y^3 u\partial_x\partial_y u + 4\partial_y^3 v\partial_y^2 u \\ & +6\partial_y^2 u\partial_x\partial_y^2 u + 6\partial_y^2 v\partial_y^3 u + 4\partial_y u\partial_x\partial_y^3 u + 4\partial_y v\partial_y^4 u \\ & +u\partial_x\partial_y^4 u + v\partial_y^5 u,\end{aligned} \tag{4.2.13}$$

which, combined with (4.2.11), gives

$$\partial_y^6 u|_{y=0} = (4\partial_x\partial_y^3 u + \partial_x^2\partial_y b)\partial_y u(t,x,0) + (\partial_x\partial_y b - \partial_y^3 u)\partial_x\partial_y u(t,x,0). \tag{4.2.14}$$

Analogously, we can arrive at

$$\partial_y^6 b(t,x,0) = -\Big(\partial_x^2\partial_y u\partial_y u + (\partial_x\partial_y u)^2\Big)(t,x,0). \tag{4.2.15}$$

The proof is therefore complete. □

4.2.1 Existence of Solutions

In this subsection, we shall investigate the existence of solutions of the 2D mixed Prandtl equation by using classical energy method. We are going to investigate the existence of solutions to the Eqs. (4.1.1) by using the following system of vorticity $w = \partial_y u$, which reads as:

$$\begin{cases} \partial_t w + u\partial_x w + v\partial_y w - \partial_y^2 w = \partial_x \partial_y b, \quad (x, y, t) \in \mathbb{R}_+^2 \times T, \\ \partial_x u + \partial_y^2 b = 0, \quad (x, y, t) \in \mathbb{R}_+^2 \times T, \\ \partial_x u + \partial_y v = 0, \quad (x, y, t) \in \mathbb{R}_+^2 \times T, \\ w|_{t=0} = w(0) = \partial_y u_0, (x, y) \in \mathbb{R}_+^2, \\ \partial_y w|_{y=0} = v|_{y=0} = b|_{y=0} = 0, \quad (x, t) \in \mathbb{R} \times T. \end{cases} \tag{4.2.16}$$

First, we will use the following lemma, which is helpful to deal with the boundary value.

Lemma 4.2.5 ([2]) *Let $1 < p < +\infty$. If $U \in W^{m,p}(\mathbb{R}^{n+1})$, then its trace u belongs to the space $B = B^{m-\frac{1}{p};p,p}(\mathbb{R}^n)$ and*

$$\|u\|_B \leq K\|U\|_{W^{m,p}(\mathbb{R}^{n+1})},$$

with the constant $K > 0$ independent of U.

Proof See Lemma 3.2.4. □

Corollary 4.2.1 *Let $1 < p < +\infty$. If $U \in W^{m,p}(\mathbb{R}^{n+1})$, then its trace u belongs to the space $W^{m-1,p}(\mathbb{R}^n)$ and*

$$\|u\|_{W^{m-1,p}(\mathbb{R}^n)} \leq K\|U\|_{W^{m,p}(\mathbb{R}^{n+1})},$$

with the constant $K > 0$ independent of U.

Proof See Corollary 3.2.1. □

We will prove the existence of solutions to problem (4.2.16) by the standard energy method and complete the proof by the following two lemmas, where the first one is devoted to the estimate with loss of x-derivative of solutions w to problem (4.2.16).

Lemma 4.2.6 *Let $|\gamma| \leq m = 4$ and under the assumptions of Theorem 4.2.1, for a smooth solution w to system* (4.2.16), *then*

$$\begin{aligned} &\frac{d}{dt}\|w\|^2_{H^{m,m-1}(\mathbb{R}_+^2)} + \|\partial_x u\|^2_{H^{m,m-1}(\mathbb{R}_+^2)} + \|\partial_y w\|^2_{H^{m,m-1}(\mathbb{R}_+^2)} \\ &\quad \leq C(1 + \|w\|^6_{H^4(\mathbb{R}_+^2)}), \end{aligned} \tag{4.2.17}$$

where C is a positive constant.

Proof For $|\gamma| = \alpha = \beta = 0$, multiplying (4.2.16) by w and integrating it by parts over $\mathbb{R}^2_+$, we conclude that

$$\frac{d}{dt}\|w\|^2_{L^2(\mathbb{R}^2_+)} + \|\partial_y w\|^2_{L^2(\mathbb{R}^2_+)} = \int_{\mathbb{R}^2_+} \partial_x \partial_y b w dy dx = \int_{\mathbb{R}^2_+} \partial_x \partial_y^2 b u dy dx$$
$$= -\|\partial_x u\|^2_{L^2(\mathbb{R}^2_+)},$$

which implies

$$\frac{d}{dt}\|w\|^2_{L^2(\mathbb{R}^2_+)} + \|\partial_x u\|^2_{L^2(\mathbb{R}^2_+)} + \|\partial_y w\|^2_{L^2(\mathbb{R}^2_+)} = 0. \tag{4.2.18}$$

For $1 \leq \alpha + \beta = m$ and $\alpha \leq m-1$, respectively, applying the operator $\partial^\gamma = \partial_x^\alpha \partial_y^\beta$ on $(4.2.16)_1$, multiplying the resulting equation by $\partial^\gamma w$, and integrating it by parts over $\mathbb{R}^2_+$, we arrive at

$$\begin{aligned}
&\frac{d}{dt}\sum_{|\gamma|\leq m,\ \alpha\leq m-1} \|\partial^\gamma w\|^2_{L^2(\mathbb{R}^2_+)} + \sum_{|\gamma|\leq m,\ \alpha\leq m-1} \|\partial_y \partial^\gamma w\|^2_{L^2(\mathbb{R}^2_+)} \\
&= \sum_{|\gamma|\leq m,\ \alpha\leq m-1} \Big(\int_{\mathbb{R}} \partial_y \partial^\gamma w \partial^\gamma w|_{y=0} dx \\
&\quad + \int_{\mathbb{R}^2_+} \Big([\partial^\gamma, u]\partial_x w + [\partial^\gamma, v]\partial_y w \Big) \partial^\gamma w dx dy \\
&\quad + \int_{\mathbb{R}^2_+} \partial_x \partial_y \partial^\gamma b \partial^\gamma \partial_y u dx dy \Big) \\
&= \sum_{i=1}^4 J_i.
\end{aligned} \tag{4.2.19}$$

Actually, for the commutator, we can write it as follows:

$$[\partial^\gamma, u]\partial_x w = \sum_{\gamma_1\leq\gamma,\ 1\leq|\gamma_1|} C_\gamma^{\gamma_1} \partial^{\gamma_1} u \partial^{\gamma-\gamma_1} \partial_x w.$$

Then, for $|\gamma| \leq 4$, applying Lemma 4.2.2 (the Agmon inequality), we deduce

$$\begin{aligned}
&\|[\partial^\gamma, u]\partial_x w\|_{L^2(\mathbb{R}^2_+)} \\
&\leq C \sum_{\gamma_1\leq\gamma,\ 1\leq|\gamma_1|} \|\partial^{\gamma_1} u\|_{L^\infty(\mathbb{R}\times L^2(\mathbb{R}_+))} \|\partial^{\gamma-\gamma_1+1} w\|_{L^2(\mathbb{R}\times L^\infty(\mathbb{R}_+))}
\end{aligned}$$

$$
\begin{aligned}
&\le C \sum_{\gamma_1\le\gamma,\ 1\le|\gamma_1|} \|\partial^{\gamma_1}u\|^{\frac12}_{L^2(\mathbb{R}^2_+)}\|\partial_x\partial^{\gamma_1}u\|^{\frac12}_{L^2(\mathbb{R}^2_+)}\|\partial^{\gamma}w\|^{\frac12}_{L^2(\mathbb{R}^2_+)}\|\partial_y\partial^{\gamma}w\|^{\frac12}_{L^2(\mathbb{R}^2_+)}\\
&\le C \sum_{\gamma_1\le\gamma,\ 1\le|\gamma_1|} \|\partial^{\gamma_1}u\|_{L^2(\mathbb{R}^2_+)}\|\partial_x\partial^{\gamma_1}u\|_{L^2(\mathbb{R}^2_+)}\\
&\quad+\|\partial^{\gamma}w\|_{L^2(\mathbb{R}^2_+)}\|\partial_y\partial^{\gamma}w\|_{L^2(\mathbb{R}^2_+)}.
\end{aligned}
\tag{4.2.20}
$$

Similarly, we also have

$$
\begin{aligned}
&\|[\partial^{\gamma},v]\partial_y w\|_{L^2(\mathbb{R}^2_+)}\|\partial^{\gamma}w\|_{L^2(\mathbb{R}^2_+)}\\
&\le \sum_{\gamma_1\le\gamma,\ 1\le|\gamma_1|} \|\partial^{\gamma_1}v\|_{L^2(\mathbb{R}\times L^\infty(\mathbb{R}_+))}\|\partial^{\gamma-\gamma_1}\partial_y w\|_{L^\infty(\mathbb{R}\times L^2(\mathbb{R}_+))}\|\partial^{\gamma}w\|_{L^2(\mathbb{R}^2_+)}\\
&\le C \sum_{\gamma_1\le\gamma,\ 1\le|\gamma_1|} \|\partial_x\partial^{\gamma_1}u\|_{L^2(\mathbb{R}^2_+)}\|\partial^{\gamma}w\|^{\frac32}_{L^2(\mathbb{R}^2_+)}\|\partial_x\partial_y\partial^{\gamma-\gamma_1}w\|^{\frac12}_{L^2(\mathbb{R}^2_+)}\\
&\le \frac14 \sum_{\gamma_1\le\gamma,\ 1\le|\gamma_1|} \Big(\|\partial_x\partial^{\gamma_1}u\|^2_{L^2(\mathbb{R}^2_+)}+\frac14\|\partial_y\partial^{\gamma}w\|^2_{L^2(\mathbb{R}^2_+)}\Big)+C\|\partial^{\gamma}w\|^6_{L^2(\mathbb{R}^2_+)}.
\end{aligned}
\tag{4.2.21}
$$

Firstly, although α and β have many various cases, there are same estimates on j_2 and j_3 by using (4.2.20)–(4.2.21) and the Cauchy-Schwarz inequality, that is,

$$
\begin{aligned}
&|j_2|+|j_3|\\
&\le \sum_{|\gamma|\le m,\ \alpha\le m-1} \Big(C\|\partial^{\gamma}w\|^6_{L^2(\mathbb{R}^2_+)}+\frac14\|\partial_x\partial^{\gamma}u\|^2_{L^2(\mathbb{R}^2_+)}+\frac14\|\partial_y\partial^{\gamma}w\|^2_{L^2(\mathbb{R}^2_+)}\Big).
\end{aligned}
\tag{4.2.22}
$$

Next, we use the boundary conditions of $\partial_y w$ and v to deal with j_1. After a complicated analysis and calculation, we can get following cases that do not vanish on the boundary $y=0$.

Case 1: $\alpha=1,\ \beta=2$. We use the boundary value of $\partial_y^3 w$ on $y=0$, the Young inequality, and Lemma 4.2.5 to conclude that

$$
\begin{aligned}
\Big|\int_{\mathbb{R}}\partial_x\partial_y^3 w\,\partial_x\partial_y^2 w|_{y=0}dx\Big| &\le \|\partial_x(\partial_x\partial_y u\partial_y u))|_{y=0}\|_{L^2(\mathbb{R})}\|\partial_x\partial_y^2 w|_{y=0}\|_{L^2(\mathbb{R})}\\
&\le \|\partial_x w|_{y=0}\|^2_{L^2(\mathbb{R})}\|\partial_x\partial_y^2 w|_{y=0}\|_{L^2(\mathbb{R})}\\
&\le C\|\partial_x\partial_y w\|^2_{L^2(\mathbb{R}^2_+)}\|\partial_x\partial_y^3 w\|_{L^2(\mathbb{R}^2_+)}\\
&\le C\|w\|^6_{H^4(\mathbb{R}^2_+)}+C.
\end{aligned}
$$

Case 2: $\alpha = 0,\ \beta = 4$. We use the boundary value of $\partial_y^5 w$ on $y = 0$, the Young inequality, and Lemma 4.2.5 to deduce that

$$
\begin{aligned}
&\Big|\int_{\mathbb{R}} \partial_y^5 w \partial_y^4 w|_{y=0} dx\Big| \\
&\le \|\Big((\partial_x\partial_y^3 u + \partial_x^2\partial_y b)\partial_y u + (\partial_x\partial_y b + \partial_y^3 u)\partial_x\partial_y u\Big)|_{y=0}\|_{L^2(\mathbb{R})}\|\partial_y^5 w\|_{L^2(\mathbb{R}_+^2)} \\
&\le C\|w\|_{H^4(\mathbb{R}_+^2)}^4 + \frac{1}{12}\|\partial_y^5 w\|_{L^2(\mathbb{R}_+^2)}^2.
\end{aligned}
$$

Case 3: $\alpha = 1,\ \beta = 3$. We use the boundary value of $\partial_y^3 w$ on $y = 0$, the Young inequality, and Lemma 4.2.5 and integrate it by part with respect to x variable to obtain that

$$
\begin{aligned}
\Big|\int_{\mathbb{R}} \partial_x\partial_y^4 w \partial_x\partial_y^3 w|_{y=0} dx\Big| &= \Big|\int_{\mathbb{R}} \partial_y^4 w \partial_x^2\partial_y^3 w|_{y=0} dx\Big| \\
&\le \|\partial_x^2(\partial_x\partial_y u\partial_y u)|_{y=0}\|_{L^2(\mathbb{R})}\|\partial_y^5 w\|_{L^2(\mathbb{R}_+^2)} \\
&\le C\|w\|_{H^4(\mathbb{R}_+^2)}^4 + \frac{1}{12}\|\partial_y^5 w\|_{L^2(\mathbb{R}_+^2)}^2.
\end{aligned}
$$

Case 4: $\alpha = 2,\ \beta = 2$. We use the boundary value of $\partial_y^3 w$ on $y = 0$, the Young inequality, Lemma 4.2.5 and integrate it by part with respect to x variable to derive that

$$
\begin{aligned}
\Big|\int_{\mathbb{R}} \partial_x^2\partial_y^3 w \partial_x^2\partial_y^2 w|_{y=0} dx\Big| &\le \|\partial_x^2(\partial_x\partial_y u\partial_y u)|_{y=0}\|_{L^2(\mathbb{R})}\|\partial_x^2\partial_y^3 w\|_{L^2(\mathbb{R}_+^2)} \\
&\le C\|w\|_{H^4(\mathbb{R}_+^2)}^4 + \frac{1}{12}\|\partial_x^2\partial_y^3 w\|_{L^2(\mathbb{R}_+^2)}^2.
\end{aligned}
$$

Summing up, we can derive that

$$
|j_1| \le C + C\|w\|_{H^4(\mathbb{R}_+^2)}^6 + \frac{1}{4}\|\partial_y w\|_{H^4(\mathbb{R}_+^2)}^2. \tag{4.2.23}
$$

Finally, we deal with the term j_4 by integrating it by parts over $\mathbb{R}_+^2$:

$$
\begin{aligned}
\int_{\mathbb{R}_+^2} \partial_x\partial_y\partial^\gamma b\partial^\gamma\partial_y u dxdy &= \int_{\mathbb{R}} \partial_x\partial_y\partial^\gamma b\partial^\gamma u|_{y=0} dx - \int_{\mathbb{R}_+^2} \partial_x\partial_y^2\partial^\gamma b\partial^\gamma u dxdy \\
&= \int_{\mathbb{R}} \partial_x\partial_y\partial^\gamma b\partial^\gamma u|_{y=0} dx - \|\partial_x\partial^\gamma u\|_{L^2(\mathbb{R}_+^2)}^2.
\end{aligned} \tag{4.2.24}
$$

Similarly to the estimate on j_1, we need to discuss the term $\int_{\mathbb{R}} \partial_x \partial_y \partial^\gamma b \partial^\gamma u|_{y=0} dx$ for the different types of α and β; we can get the following case not vanishing on the boundary of $y = 0$ by a complicated analysis and calculation:

$$\begin{aligned}\Big|\int_{\mathbb{R}} \partial_x \partial_y^5 b \partial_y^4 u|_{y=0} dx\Big| &= \Big|\int_{\mathbb{R}} \partial_x \partial_y^3 u \partial_x \partial_y^4 u|_{y=0} dx\Big| \\ &\le C\|\partial_x \partial_y^3 u|_{y=0}\|_{L^2(\mathbb{R})} \|\partial_x \partial_y^4 u|_{y=0}\|_{L^2(\mathbb{R})} \\ &\le C\|w\|^2_{H^4(\mathbb{R}^2_+)} + \frac{1}{4}\|\partial_y w\|^2_{H^4(\mathbb{R}^2_+)}, \end{aligned} \tag{4.2.25}$$

where we have used Lemma 4.2.5.

Therefore, we attain the following inequality:

$$|j_4| \le C + C\|w\|^6_{H^4(\mathbb{R}^2_+)} + \frac{1}{4}\|\partial_y w\|^2_{H^4(\mathbb{R}^2_+)} - \sum_{|\gamma|\le m,\ \alpha\le m-1} \|\partial_x \partial^\gamma u\|^2_{L^2(\mathbb{R}^2_+)}. \tag{4.2.26}$$

Inserting above inequalities (4.2.22)–(4.2.23) and (4.2.26) into (4.2.19), we conclude

$$\begin{aligned}&\sum_{|\gamma|\le m,\ \alpha\le m-1} \Big(\frac{d}{dt}\|\partial^\gamma w\|^2_{L^2(\mathbb{R}^2_+)} + \|\partial_x \partial^\gamma u\|^2_{L^2(\mathbb{R}^2_+)} + \|\partial_y \partial^\gamma w\|^2_{L^2(\mathbb{R}^2_+)}\Big) \\ &\quad \le C + C\|w\|^6_{H^4(\mathbb{R}^2_+)}.\end{aligned}$$

The proof is thus complete. □

We have completed the estimate with loss of x-derivative of solutions w to problem (4.2.16) in previous section. In order to ensure the completion of energy estimate, we need to establish the estimate of term $\partial_x^\alpha w$, which will be done in the following lemma.

Lemma 4.2.7 *Let $\alpha \le m = 4$ and under the assumptions of Theorem 4.2.1, for a smooth solution w to problem* (4.2.16), *then*

$$\begin{aligned}&\frac{d}{dt}\sum_{\alpha\le m} \|\partial_x^\alpha w\|^2_{L^2(\mathbb{R}^2_+)} + \sum_{\alpha\le m} \|\partial_x \partial_x^\alpha u\|^2_{L^2(\mathbb{R}^2_+)} \\ &\qquad + \sum_{\alpha\le m} \|\partial_y \partial_x^\alpha w\|^2_{L^2(\mathbb{R}^2_+)} \le C\|w\|^6_{H^4(\mathbb{R}^2_+)},\end{aligned} \tag{4.2.27}$$

where C is a positive constant.

Proof Applying the operator ∂_x^α on $(4.2.16)_1$, multiplying the resulting equation by $\partial_x^\alpha w$, and integrating it by parts over $\mathbb{R}_+^2$, we arrive at

$$\begin{aligned}&\frac{d}{dt}\|\partial_x^\alpha w\|_{L^2(\mathbb{R}_+^2)}^2+\|\partial_x\partial_x^\alpha u\|_{L^2(\mathbb{R}_+^2)}^2+\|\partial_y\partial_x^\alpha w\|_{L^2(\mathbb{R}_+^2)}^2\\&=\int_{\mathbb{R}_+^2}\Big([\partial_x^\alpha,u]\partial_x w+[\partial_x^\alpha,v]\partial_y w\Big)\partial_x^\alpha w dxdy,\end{aligned}\tag{4.2.28}$$

where we have used the fact

$$(\partial_x^m\partial_y w)(t,x,0)=0,\ (t,x)\in[0,T]\times\mathbb{R}.$$

In fact, for the commutator, we can write it as follows:

$$[\partial_x^\alpha,u]\partial_x w=\sum_{\alpha_1\le\alpha,\ 1\le|\alpha_1|}C_\alpha^{\alpha_1}\partial_x^{\alpha_1}u\partial_x^{\alpha-\alpha_1}\partial_x w.$$

Then, for $|\alpha|\le 4$, applying Lemma 4.2.2 (the Agmon inequality) and the Young inequality, we deduce

$$\begin{aligned}&\|[\partial_x^\alpha,u]\partial_x w\|_{L^2(\mathbb{R}_+^2)}\\&\le\sum_{\alpha_1\le\alpha,\ 1\le|\alpha_1|}\|\partial_x^{\alpha_1}u\|_{L^\infty(\mathbb{R}\times L^2(\mathbb{R}_+))}\|\partial_x^{\alpha-\alpha_1+1}w\|_{L^2(\mathbb{R}\times L^\infty(\mathbb{R}_+))}\\&\le C\sum_{\alpha_1\le\alpha,\ 1\le|\alpha_1|}\|\partial_x^\alpha u\|_{L^2(\mathbb{R}_+^2)}^{\frac12}\|\partial_x\partial_x^\alpha u\|_{L^2(\mathbb{R}_+^2)}^{\frac12}\|\partial_x^{\alpha-\alpha_1+1}w\|_{L^2(\mathbb{R}_+^2)}^{\frac12}\\&\quad\times\|\partial_x^{\alpha-\alpha_1+1}\partial_y w\|_{L^2(\mathbb{R}_+^2)}^{\frac12}\\&\le C\sum_{\alpha_1\le\alpha,\ 1\le|\alpha_1|}\|\partial_x^\alpha u\|_{L^2(\mathbb{R}_+^2)}\|\partial_x\partial_x^\alpha u\|_{L^2(\mathbb{R}_+^2)}+C\|\partial_y\partial_x^\alpha w\|_{L^2(\mathbb{R}_+^2)}\|\partial_x^\alpha w\|_{L^2(\mathbb{R}_+^2)}.\end{aligned}$$

Hence,

$$\begin{aligned}&\|[\partial_x^\alpha,u]\partial_x w\|_{L^2(\mathbb{R}_+^2)}\|\partial_x^\alpha w\|_{L^2(\mathbb{R}_+^2)}\\&\le\frac14\|\partial_x\partial_x^\alpha u\|_{L^2(\mathbb{R}_+^2)}^2+\frac14\|\partial_y\partial_x^\alpha w\|_{L^2(\mathbb{R}_+^2)}^2+C\|w\|_{H^4(\mathbb{R}_+^2)}^6.\end{aligned}\tag{4.2.29}$$

Similarly, we also have

$$
\begin{aligned}
&\|[\partial_x^\alpha, v]\partial_y w\|_{L^2(\mathbb{R}^2_+)}\|\partial_x^\alpha w\|_{L^2(\mathbb{R}^2_+)}\\
&\le \sum_{\alpha_1\le\alpha,\ 1\le|\alpha_1|} \|\partial_x^{\alpha_1} v\|_{L^2(\mathbb{R}\times L^\infty(\mathbb{R}_+))}\|\partial_x^{\alpha-\alpha_1}\partial_y w\|_{L^\infty(\mathbb{R}\times L^2(\mathbb{R}_+))}\|\partial_x^\alpha w\|_{L^2(\mathbb{R}^2_+)}\\
&\le C \sum_{\alpha_1\le\alpha,\ 1\le|\alpha_1|} \|\partial_x\partial_x^\alpha u\|_{L^2(\mathbb{R}^2_+)}\|\partial_x^{\alpha-\alpha_1}\partial_y w\|^{\frac12}_{L^2(\mathbb{R}^2_+)}\|\partial_x^{\alpha-\alpha_1+1}\partial_y w\|^{\frac12}_{L^2(\mathbb{R}^2_+)}\\
&\quad \times \|\partial_x^\alpha w\|_{L^2(\mathbb{R}^2_+)}\\
&\le \frac14\|\partial_x\partial_x^\alpha u\|^2_{L^2(\mathbb{R}^2_+)} + \frac14\|\partial_y\partial_x^\alpha w\|^2_{L^2(\mathbb{R}^2_+)} + C\|w\|^6_{H^4(\mathbb{R}^2_+)}.
\end{aligned}
\tag{4.2.30}
$$

Inserting (4.2.29)–(4.2.30) into (4.2.28) and summing over α, we can conclude that

$$
\frac{d}{dt}\sum_{\alpha\le m}\|\partial_x^\alpha w\|^2_{L^2(\mathbb{R}^2_+)} + \sum_{\alpha\le m}\|\partial_x\partial_x^\alpha u\|^2_{L^2(\mathbb{R}^2_+)} + \sum_{\alpha\le m}\|\partial_y\partial_x^\alpha w\|^2_{L^2(\mathbb{R}^2_+)} \le C\|w\|^6_{H^4(\mathbb{R}^2_+)}.
$$

The proof is thus complete. □

Up to now, we have derived the main estimates of the solution u. Now combining Lemma 4.2.6 with Lemma 4.2.7 with $m = 4$, we have the following estimate:

$$
\frac{d}{dt}\|w\|^2_{H^4(\mathbb{R}^2_+)} + \|\partial_x u\|^2_{H^4(\mathbb{R}^2_+)} + \|\partial_y w\|^2_{H^4(\mathbb{R}^2_+)} \le C(1 + \|w\|^6_{H^4(\mathbb{R}^2_+)}). \tag{4.2.31}
$$

Applying Lemma 4.2.3 to above inequality (4.2.31), we have

$$
\|w\|^2_{H^4(\mathbb{R}^2_+)} \le \{\|w(0)\|^2_{H^4(\mathbb{R}^2_+)} + t\}\{1 - 2C(\|w(0)\|^2_{H^4(\mathbb{R}^2_+)} + t)^2 t\}^{-\frac12}, \tag{4.2.32}
$$

if $1 - 2C(\|w(0)\|^2_{H^4(\mathbb{R}^2_+)} + t)^2 t > 0$.

Taking $T = \min\left\{\|w(0)\|^2_{H^4(\mathbb{R}^2_+)}, \frac{3}{32C\|w(0)\|^4_{H^4(\mathbb{R}^2_+)}}\right\}$ such that

$$
\|w(0)\|^2_{H^4(\mathbb{R}^2_+)} + t \le 2\|w(0)\|^2_{H^4(\mathbb{R}^2_+)},\ \{1 - 2C(\|w(0)\|^2_{H^4(\mathbb{R}^2_+)} + t)^2 t\}^{-\frac12} \le 2,
$$

which, combined with (4.2.32), gives

$$
\|w\|^2_{H^4(\mathbb{R}^2_+)} \le 4\|w(0)\|^2_{H^4(\mathbb{R}^2_+)}. \tag{4.2.33}
$$

Hence

$$\|w\|^2_{H^4(\mathbb{R}^2_+)} + \int_0^t (\|\partial_x u\|^2_{H^4(\mathbb{R}^2_+)} + \|\partial_y w\|^2_{H^4(\mathbb{R}^2_+)})ds \le C_1,\ 0 \le t \le T, \tag{4.2.34}$$

where constant $C_1 > 0$ depends on T and $\|w(0)\|^2_{H^4(\mathbb{R}^2_+)}$.

According to Eqs. (4.2.16)$_{2,3}$ and (4.2.32), we can conclude $\|(v, b)\|_{L^\infty([0,T];L^\infty(\mathbb{R}_+\times H^3(\mathbb{R})))} \le C$. We therefore have proved the existence of solutions in Theorem 4.2.1. □

4.2.2 The Uniqueness of Local Solutions

In this subsection, we will prove the uniqueness of solutions to the 2D mixed Prandtl equations. Let (w_1, b_1) and (w_2, b_2) be two solutions to problem (4.2.16) with same initial data $w_1(0, x, y) = w_2(0, x, y) \in H^4(\mathbb{R}^2_+)$. Set $w = w_1 - w_2$ and $b = b_1 - b_2$. Then

$$\begin{cases} \partial_t w + u\partial_x w_1 + v\partial_y w_1 + u_2\partial_x w + v_2\partial_y w - \partial_y^2 w = \partial_x\partial_y b, \\ (x, y, t) \in \mathbb{R}^2_+ \times T, \\ \partial_x u + \partial_y^2 b = 0, \quad (x, y, t) \in \mathbb{R}^2_+ \times T, \\ \partial_x u + \partial_y v = 0, \quad (x, y, t) \in \mathbb{R}^2_+ \times T, \\ \partial_y u(0, x, y) = 0, \quad (x, y) \in \mathbb{R}^2_+, \\ \partial_y w|_{y=0} = v|_{y=0} = b|_{y=0} = 0, \quad (x, y, t) \in \mathbb{R}^2_+ \times T. \end{cases} \tag{4.2.35}$$

Similarly to Lemma 4.2.4, we can derive the following initial data and boundary values:

$$\begin{cases} \partial_y u(0, x, y) = 0,\ (x, y) \in \mathbb{R}^2_+, \\ u(t, x, 0) = 0,\ b(t, x, 0) = 0, \\ \partial_y^2 u(t, x, 0) = 0,\ \ \partial_y^2 b(t, x, 0) = 0, \\ \partial_y^4 u(t, x, 0) = \partial_x\partial_y u_1\partial_y u(t, x, 0) + \partial_x\partial_y u\partial_y u_2(t, x, 0), \\ \partial_y^4 b(t, x, 0) = 0, \\ \partial_y^6 u|_{y=0} = (4\partial_x\partial_y^3 u_1 + \partial_x^2\partial_y b_1)\partial_y u(t, x, 0) + (\partial_x\partial_y b_1 - \partial_y^3 u_1)\partial_x\partial_y u(t, x, 0), \\ +(4\partial_x\partial_y^3 u + \partial_x^2\partial_y b)\partial_y u_2(t, x, 0) + (\partial_x\partial_y b - \partial_y^3 u)\partial_x\partial_y u_2(t, x, 0), \\ \partial_y^6 b(t, x, 0) \\ = -\Big(\partial_x^2\partial_y u_1\partial_y u + \partial_x\partial_y u_1\partial_x\partial_y u + \partial_x^2\partial_y u\partial_y u_2 + \partial_x\partial_y u\partial_x\partial_y u_2\Big)(t, x, 0). \end{cases} \tag{4.2.36}$$

Proof Looking back the boundary condition in (4.2.35)$_5$,

$$u(t, x, 0) = 0,\ v(t, x, 0) = 0,\ b(t, x, 0) = 0,\ (t, x) \in [0, T] \times \mathbb{R}, \tag{4.2.37}$$

thus, the following results are obvious:

$$\partial_x^n (u, b)(t, x, 0) = 0,\ \partial_x^n v(t, x, 0) = 0,\quad (t, x) \in [0, T] \times \mathbb{R}, 0 \le n \le 4. \tag{4.2.38}$$

Appealing to $(4.2.35)_1$ and the boundary condition (4.2.38), we have

$$\partial_y^2 u|_{y=0} = 0,\ \ \partial_y^2 b|_{y=0} = 0. \tag{4.2.39}$$

Besides, we also have

$$\partial_x^n \partial_y^2 u|_{y=0} = 0,\ \ \partial_x^n \partial_y^2 b|_{y=0} = 0,\ \ (t, x) \in [0, T] \times \mathbb{R},\quad 0 \le n \le 4. \tag{4.2.40}$$

Differentiating the equation of $(4.2.35)_1$ with y, we get

$$\partial_t \partial_y u - \partial_y^3 u + \partial_y \big(u\partial_x u_1 + v\partial_y u_1 + u_2\partial_x u + v_2\partial_y u\big) = \partial_x \partial_y b, \tag{4.2.41}$$

and using the boundary values (4.2.38), then we have

$$\partial_t \partial_y u|_{y=0} = \partial_y^3 u|_{y=0} + \partial_x \partial_y b|_{y=0}. \tag{4.2.42}$$

Differentiating $(4.2.35)_1$ with respect to y twice, it follows

$$\partial_t \partial_y^2 u - \partial_y^4 u + \partial_y^2 \big(u\partial_x u_1 + v\partial_y u_1 + u_2\partial_x u + v_2\partial_y u\big) = \partial_x \partial_y^2 b. \tag{4.2.43}$$

Applying the Leibniz formula, we arrive at

$$\begin{aligned}
&\partial_y^2 \big(u\partial_x u_1 + v\partial_y u_1 + u_2\partial_x u + v_2\partial_y u\big) \\
&\quad = \partial_y^2 u\partial_x u_1 + \partial_y^2 v\partial_y u_1 + u\partial_x \partial_y^2 u_1 + v\partial_y^3 u_1 + 2\partial_y u\partial_x \partial_y u_1 + 2\partial_y v\partial_y^2 u_1 \\
&\qquad + \partial_y^2 u_2\partial_x u + \partial_y^2 v_2\partial_y u + u_2\partial_x \partial_y^2 u + v_2\partial_y^3 u + 2\partial_y u_2\partial_x \partial_y u + 2\partial_y v_2\partial_y^2 u.
\end{aligned} \tag{4.2.44}$$

Therefore,

$$\partial_y^4 u(t, x, 0) = \partial_x \partial_y u_1 \partial_y u(t, x, 0) + \partial_x \partial_y u \partial_y u_2(t, x, 0). \tag{4.2.45}$$

Differentiating (4.2.45) with respect to t and using (4.2.42), we get

$$\begin{aligned}
\partial_t \partial_y^4 u(t, x, 0) &= \partial_t \partial_y u \partial_x \partial_y u_1(t, x, 0) + \partial_y u \partial_t \partial_x \partial_y u_1(t, x, 0) \\
&\quad + \partial_t \partial_y u_2 \partial_x \partial_y u(t, x, 0) + \partial_y u_2 \partial_t \partial_x \partial_y u(t, x, 0)
\end{aligned}$$

$$
\begin{aligned}
&= (\partial_y^3 u + \partial_x \partial_y b)\partial_x \partial_y u_1(t,x,0) + (\partial_x \partial_y^3 u_1 + \partial_x^2 \partial_y b_1)\partial_y u(t,x,0) \\
&\quad + (\partial_y^3 u_2 + \partial_x \partial_y b_2)\partial_x \partial_y u(t,x,0) + (\partial_x \partial_y^3 u + \partial_x^2 \partial_y b)\partial_y u_2(t,x,0).
\end{aligned}
\tag{4.2.46}
$$

Similarly, differentiating $(4.2.35)_1$ with respect to y four times, we have

$$
\partial_t \partial_y^4 u - \partial_y^6 u + \partial_y^4\big(u\partial_x u_1 + v\partial_y u_1 + u_2 \partial_x u + v_2 \partial_y u\big) = \partial_x \partial_y^4 b, \tag{4.2.47}
$$

and applying the Leibniz formula, we conclude

$$
\begin{aligned}
&\partial_y^4\big(u\partial_x u_1 + v\partial_y u_1 + u_2\partial_x u + v_2\partial_y u\big) \\
&= \partial_y^4 u \partial_x u_1 + \partial_y^4 v \partial_y u_1 + 4\partial_y^3 u \partial_x \partial_y u_1 + 4\partial_y^3 v \partial_y^2 u_1 + 6\partial_y^2 u \partial_x \partial_y^2 u_1 + 6\partial_y^2 v \partial_y^3 u_1 \\
&\quad + 4\partial_y u \partial_x \partial_y^3 u_1 + 4\partial_y v \partial_y^4 u_1 + u\partial_x \partial_y^4 u_1 + v\partial_y^5 u_1 + \partial_y^4 u_2 \partial_x u + \partial_y^4 v_2 \partial_y u \\
&\quad + 4\partial_y^3 u_2 \partial_x \partial_y u \\
&\quad + 4\partial_y^3 v_2 \partial_y^2 u + 6\partial_y^2 u_2 \partial_x \partial_y^2 u + 6\partial_y^2 v_2 \partial_y^3 u + 4\partial_y u_2 \partial_x \partial_y^3 u + 4\partial_y v_2 \partial_y^4 u \\
&\quad + u_2 \partial_x \partial_y^4 u + v_2 \partial_y^5 u,
\end{aligned}
\tag{4.2.48}
$$

which, combined with (4.2.46), yields to

$$
\begin{aligned}
\partial_y^6 u|_{y=0} &= (4\partial_x \partial_y^3 u_1 + \partial_x^2 \partial_y b_1)\partial_y u(t,x,0) + (\partial_x \partial_y b_1 - \partial_y^3 u_1)\partial_x \partial_y u(t,x,0) \\
&\quad + (4\partial_x \partial_y^3 u + \partial_x^2 \partial_y b)\partial_y u_2(t,x,0) + (\partial_x \partial_y b - \partial_y^3 u)\partial_x \partial_y u_2(t,x,0).
\end{aligned}
\tag{4.2.49}
$$

Analogously, we can conclude

$$
\begin{aligned}
&\partial_y^6 b(t,x,0) \\
&= -\Big(\partial_x^2 \partial_y u_1 \partial_y u + \partial_x \partial_y u_1 \partial_x \partial_y u + \partial_x^2 \partial_y u \partial_y u_2 + \partial_x \partial_y u \partial_x \partial_y u_2\Big)(t,x,0).
\end{aligned}
\tag{4.2.50}
$$

The proof is thus complete. □

Lemma 4.2.8 *Let* $|\gamma| \leq m-1=3$ *and assume* u_1 *and* u_2 *are two solutions obtained in Theorem 4.2.1 with respect to the initial data* $u_1(0,x,y)$ *and* $u_2(0,x,y)$, *then*

$$\frac{d}{dt}\|w\|^2_{H^{m-1,m-2}(\mathbb{R}^2_+)} + \|\partial_x u\|^2_{H^{m-1,m-2}(\mathbb{R}^2_+)} + \|\partial_y w\|^2_{H^{m-1,m-2}(\mathbb{R}^2_+)}$$
$$\leq C(1+\|u_2\|^2_{H^{m-1,m-2}(\mathbb{R}^2_+)} + \|\partial_y w_1\|^2_{H^4(\mathbb{R}^2_+)} + \|\partial_y w_2\|^2_{H^4(\mathbb{R}^2_+)})\|w\|^2_{H^3(\mathbb{R}^2_+)}, \tag{4.2.51}$$

where C *is a positive constant.*

Proof The proof of the lemma is similar to that of Lemma 4.2.6 for $0 \leq \alpha+\beta = m-1$ and $\alpha \leq m-2$, respectively. Applying the operator $\partial^\gamma = \partial_x^\alpha \partial_y^\beta$ on $(4.2.35)_1$, multiplying the resulting equation by $\partial^\gamma w$, and integrating it by parts over $\mathbb{R}^2_+$, we arrive at

$$\begin{aligned}
&\frac{d}{dt}\|\partial^\gamma w\|^2_{L^2(\mathbb{R}^2_+)} + \|\partial_y\partial^\gamma w\|^2_{L^2(\mathbb{R}^2_+)}\\
&= \int_{\mathbb{R}} \partial_y\partial^\gamma w \partial^\gamma w|_{y=0} dx + \int_{\mathbb{R}^2_+}\Big([\partial^\gamma, u_2]\partial_x w + [\partial^\gamma, v_2]\partial_y w\Big)\partial^\gamma w dxdy\\
&\quad + \int_{\mathbb{R}^2_+}\Big(\partial^\gamma(u\partial_x w_1) + \partial^\gamma(v\partial_y w_1)\Big)\partial^\gamma w dxdy + \int_{\mathbb{R}^2_+}\partial_x\partial_y\partial^\gamma b\partial^\gamma\partial_y u dxdy\\
&= \sum_{i=1}^6 I_i.
\end{aligned} \tag{4.2.52}$$

Actually, we can write the commutator as follows:

$$[\partial^\gamma, u_2]\partial_x w = \sum_{\gamma_1\leq\gamma,\ 1\leq|\gamma_1|} C_\gamma^{\gamma_1}\partial^{\gamma_1}u_2\partial^{\gamma-\gamma_1}\partial_x w.$$

Then, for $|\gamma|\leq 3$, applying the Sobolev inequality, we deduce

$$\begin{aligned}
&\|[\partial^\gamma, u_2]\partial_x w\|_{L^2(\mathbb{R}^2_+)}\\
&\leq C\sum_{\gamma_1\leq\gamma,\ 1\leq|\gamma_1|}\|\partial^{\gamma_1}u_2\|_{L^\infty(\mathbb{R}\times L^2(\mathbb{R}_+))}\|\partial^{\gamma-\gamma_1+1}w\|_{L^2(\mathbb{R}\times L^\infty(\mathbb{R}_+))}\\
&\leq C\sum_{\gamma_1\leq\gamma,\ 1\leq|\gamma_1|}\|\partial_x\partial^{\gamma_1}u_2\|_{L^2(\mathbb{R}^2_+)}\|\partial_y\partial^\gamma w\|_{L^2(\mathbb{R}^2_+)}.
\end{aligned} \tag{4.2.53}$$

Similarly, we also have

$$\|[\partial^\gamma, v_2]\partial_y w\|_{L^2(\mathbb{R}^2_+)} \le C \sum_{\gamma_1\le\gamma,\ 1\le|\gamma_1|} \|\partial_x\partial^{\gamma_1}u_2\|_{L^2(\mathbb{R}^2_+)}\|\partial_y\partial^\gamma w\|_{L^2(\mathbb{R}^2_+)}. \tag{4.2.54}$$

Firstly, although α and β have many various cases, there are same estimates on I_2 and I_3 by using (4.2.53)–(4.2.54) and the Cauchy-Schwarz inequality, that is,

$$|I_2| + |I_3| \le \frac{1}{4}\|\partial_y\partial^\gamma w\|^2_{L^2(\mathbb{R}^2_+)} + C\|\partial_x\partial^\gamma u_2\|^2_{L^2(\mathbb{R}^2_+)}\|w\|^2_{H^3(\mathbb{R}^2_+)}. \tag{4.2.55}$$

For the items I_4 and I_5, we can deduce the following estimates by using the Sobolev inequality:

$$\begin{aligned}|I_4| &\le C\sum_{\gamma_1\le\gamma}\|\partial^{\gamma_1}u\|_{L^\infty(\mathbb{R}\times L^2(\mathbb{R}_+))}\|\partial^{\gamma-\gamma_1}\partial_x w_1\|_{L^2(\mathbb{R}\times L^\infty(\mathbb{R}_+))}\\ &\le C\sum_{\gamma_1\le\gamma}\|\partial_x\partial^{\gamma_1}u\|_{L^2(\mathbb{R}^2_+)}\|\partial_x\partial_y\partial^{\gamma-\gamma_1}w_1\|_{L^2(\mathbb{R}^2_+)}.\end{aligned}$$

Analogously,

$$|I_5| \le C\sum_{\gamma_1\le\gamma}\|\partial_x\partial^{\gamma_1}u\|_{L^2(\mathbb{R}^2_+)}\|\partial_x\partial_y\partial^{\gamma-\gamma_1}w_1\|_{L^2(\mathbb{R}^2_+)}.$$

Hence,

$$\begin{aligned}|I_4| + |I_5| \le \frac{1}{4}\sum_{\gamma_1\le\gamma}\|\partial_x\partial^{\gamma_1}u\|^2_{L^2(\mathbb{R}^2_+)} &+ C(\|\partial_y w_1\|^2_{H^4(\mathbb{R}^2_+)}\\ &+\|\partial_y w_2\|^2_{H^4(\mathbb{R}^2_+)})\|\partial^\gamma w\|^2_{L^2(\mathbb{R}^2_+)}.\end{aligned} \tag{4.2.56}$$

Next, we use the boundary conditions of w and v to deal with I_1. After a complicated analysis and calculation, we can get following case do not vanish on the boundary $y = 0$.

Case: $\alpha = 1$, $\beta = 2$. We use the boundary value of $\partial_y^3 w$ on $y = 0$ and Lemma 4.2.5 to conclude that

$$\left|\int_{\mathbb{R}}\partial_x\partial_y^3 w\partial_x\partial_y^2 w|_{y=0}dx\right| \le C(1 + \|w_1\|^2_{H^4(\mathbb{R}^2_+)} + \|w_2\|^2_{H^4(\mathbb{R}^2_+)})\|w\|^2_{H^3(\mathbb{R}^2_+)}. \tag{4.2.57}$$

Finally, we deal with the term I_6 by integrating it by parts over $\mathbb{R}^2_+$. Similarly to the estimate on I_1, we need to discuss the term $\int_{\mathbb{R}}\partial_x\partial_y\partial^\gamma b\partial^\gamma u|_{y=0}dx$ for the different types of α and β. However, the boundary values of $y = 0$ vanish by a complicated analysis and calculation,

$$\int_{\mathbb{R}^2_+} \partial_x \partial_y \partial^\gamma b \partial^\gamma \partial_y u dx dy = \int_{\mathbb{R}} \partial_x \partial_y \partial^\gamma b \partial^\gamma u|_{y=0} dx - \int_{\mathbb{R}^2_+} \partial_x \partial_y^2 \partial^\gamma b \partial^\gamma u dx dy$$
$$= -\|\partial_x \partial^\gamma u\|^2_{L^2(\mathbb{R}^2_+)}. \tag{4.2.58}$$

Inserting (4.2.55)–(4.2.58) into (4.2.52) and summing γ, we can derive the desired result (4.2.51). Thus the proof is complete. □

Lemma 4.2.9 *Let $|\alpha| \le m - 1 = 3$ and assume u_1, u_2 be two solutions obtained in Theorem 4.2.1 with respect to the initial data $u_1(0, x, y)$, $u_2(0, x, y)$, then we have*

$$\frac{d}{dt} \sum_{\alpha \le m} \|\partial_x^\alpha w\|^2_{L^2(\mathbb{R}^2_+)} + \sum_{\alpha \le m} \|\partial_x \partial_x^\alpha u\|^2_{L^2(\mathbb{R}^2_+)} + \sum_{\alpha \le m} \|\partial_y \partial_x^\alpha w\|^2_{L^2(\mathbb{R}^2_+)}$$
$$\le C(\|\partial_y w_1\|^2_{H^4(\mathbb{R}^2_+)} + \|\partial_y w_2\|^2_{H^4(\mathbb{R}^2_+)} + \|\partial_x \partial_x^\alpha u_2\|^2_{L^2(\mathbb{R}^2_+)}) \|w\|^2_{H^3(\mathbb{R}^2_+)}, \tag{4.2.59}$$

where C is a positive constant.

Proof Acting the operator ∂_x^α on (4.2.35), multiplying the resulting equation by $\partial_x^\alpha w$, and integrating it by parts over $\mathbb{R}^2_+$, we arrive at

$$\frac{d}{dt} \|\partial_x^\alpha w\|^2_{L^2(\mathbb{R}^2_+)} + \|\partial_x \partial_x^\alpha u\|^2_{L^2(\mathbb{R}^2_+)} + \|\partial_y \partial_x^\alpha w\|^2_{L^2(\mathbb{R}^2_+)}$$
$$= \int_{\mathbb{R}^2_+} \Big([\partial_x^\alpha, u_2] \partial_x w + [\partial_x^\alpha, v_2] \partial_y w + \partial_x^\alpha (u \partial_x w_1) + \partial_x^\alpha (v \partial_y w_1) \Big) \partial_x^\alpha w dx dy, \tag{4.2.60}$$

where we have used the fact

$$(\partial_x^\alpha \partial_y w)(t, x, 0) = 0, \ \ (t, x) \in [0, T] \times \mathbb{R}.$$

In fact, for the commutator, we can write as follows:

$$[\partial_x^\alpha, u_2] \partial_x w = \sum_{\alpha_1 \le \alpha, \ 1 \le |\alpha_1|} C_\alpha^{\alpha_1} \partial_x^{\alpha_1} u_2 \partial_x^{\alpha - \alpha_1} \partial_x w.$$

Then, for $|\alpha| \leq 3$, applying the Sobolev inequality, we deduce

$$\begin{aligned}\|[\partial_x^\alpha, u_2]\partial_x w\|_{L^2(\mathbb{R}^2_+)} &\leq \sum_{\alpha_1\leq\alpha,\ 1\leq|\alpha_1|} \|\partial_x^{\alpha_1} u\|_{L^\infty(\mathbb{R}\times L^2(\mathbb{R}_+))}\|\partial_x^{\alpha-\alpha_1+1} w\|_{L^2(\mathbb{R}\times L^\infty(\mathbb{R}_+))}\\ &\leq C \sum_{\alpha_1\leq\alpha,\ 1\leq|\alpha_1|} \|\partial_x\partial_x^\alpha u_2\|_{L^2(\mathbb{R}^2_+)}\|\partial_y\partial_x^\alpha w\|_{L^2(\mathbb{R}^2_+)}.\end{aligned}$$

Similarly, we also have

$$\|[\partial_x^\alpha, v_2]\partial_y w\|_{L^2(\mathbb{R}^2_+)} \leq C\|\partial_x\partial_x^\alpha u_2\|_{L^2(\mathbb{R}^2_+)}\|\partial_y\partial_x^\alpha w\|_{L^2(\mathbb{R}^2_+)}.$$

Hence,

$$\begin{aligned}&(\|[\partial_x^\alpha, u_2]\partial_x w\|_{L^2(\mathbb{R}^2_+)} + \|[\partial_x^\alpha, v_2]\partial_y w\|_{L^2(\mathbb{R}^2_+)})\|\partial_x^\alpha w\|^2_{L^2(\mathbb{R}^2_+)}\\ &\leq \frac{1}{4}\|\partial_y\partial_x^\alpha w\|_{L^2(\mathbb{R}^2_+)} + C\|\partial_x\partial_x^\alpha u_2\|^2_{L^2(\mathbb{R}^2_+)}\|\partial_x^\alpha w\|^2_{L^2(\mathbb{R}^2_+)},\end{aligned}\tag{4.2.61}$$

and

$$\begin{aligned}&(\|\partial_x^\alpha(u\partial_x w_1)\|_{L^2(\mathbb{R}^2_+)} + \|\partial_x^\alpha(v\partial_y w_1)\|_{L^2(\mathbb{R}^2_+)})\|\partial_x^\alpha w\|^2_{L^2(\mathbb{R}^2_+)}\\ &\leq \frac{1}{4}\|\partial_x\partial^\alpha u\|_{L^2(\mathbb{R}^2_+)} + C(\|\partial_y w_1\|^2_{H^4(\mathbb{R}^2_+)} + \|\partial_y w_2\|^2_{H^4(\mathbb{R}^2_+)})\|\partial_x^\alpha w\|^2_{L^2(\mathbb{R}^2_+)}.\end{aligned}\tag{4.2.62}$$

Inserting (4.2.61)–(4.2.62) into (4.2.60) and summing α, we can conclude that

$$\begin{aligned}&\frac{d}{dt}\sum_{\alpha\leq m}\|\partial_x^\alpha w\|^2_{L^2(\mathbb{R}^2_+)} + \sum_{\alpha\leq m}\|\partial_x\partial_x^\alpha u\|^2_{L^2(\mathbb{R}^2_+)} + \sum_{\alpha\leq m}\|\partial_y\partial_x^\alpha w\|^2_{L^2(\mathbb{R}^2_+)}\\ &\leq C(\|\partial_y w_1\|^2_{H^4(\mathbb{R}^2_+)} + \|\partial_y w_2\|^2_{H^4(\mathbb{R}^2_+)} + \|\partial_x\partial_x^\alpha u_2\|^2_{L^2(\mathbb{R}^2_+)})\|w\|^2_{H^3(\mathbb{R}^2_+)}.\end{aligned}$$

The proof is thus complete. □

Finally, combining Lemma 4.2.8 with Lemma 4.2.9, we can derive the following inequality:

$$\begin{aligned}&\frac{d}{dt}\|w\|^2_{H^3(\mathbb{R}^2_+)} + \|\partial_x u\|^2_{H^3(\mathbb{R}^2_+)} + \|\partial_y w\|^2_{H^3(\mathbb{R}^2_+)}\\ &\leq C(1 + \|u_2\|^2_{H^4(\mathbb{R}^2_+)} + \|\partial_y w_1\|^2_{H^4(\mathbb{R}^2_+)} + \|\partial_y w_2\|^2_{H^4(\mathbb{R}^2_+)})\|w\|^2_{H^3(\mathbb{R}^2_+)}.\end{aligned}\tag{4.2.63}$$

Applying the Gronwall inequality to (4.2.63) and using (4.2.34), we can obtain

$$\begin{aligned}
&\|w\|^2_{H^3(\mathbb{R}^2_+)}\\
&\le \|w_0\|^2_{H^3(\mathbb{R}^2_+)} \exp\Big(\int_0^t C\big(1+\|u_2\|^2_{H^4(\mathbb{R}^2_+)}+\|\partial_y w_1\|^2_{H^4(\mathbb{R}^2_+)}+\|\partial_y w_2\|^2_{H^4(\mathbb{R}^2_+)}ds\big)\Big)\\
&=0,\ 0\le t\le T,
\end{aligned}\tag{4.2.64}$$

which yields $w_1(t,x,y)=w_2(t,x,y)$. The proof of uniqueness of solutions is thus complete. □

Combining (4.2.34) with (4.2.64), we can complete the proof of Theorem 4.2.1.

4.3 Bibliographic Comments

In this section, let us briefly review some known results to problem (4.1.1). For mixed Prandtl boundary equations, Xie and Yang [147] investigated the global existence and uniqueness of solutions with analytic regularity of the 2D mixed Prandtl boundary layer equations by using the standard energy methods in analytic space. As far as we have learned, there have been no result for the 2D mixed Prandtl-Shercliff regime equations; this is the first result of the $2D$ mixed Prandtl-Shercliff regime equations in a Sobolev space. Compared to the existence and uniqueness of solutions to the classical Prandtl equations where the monotonicity condition of the tangential velocity plays a key role, this monotonicity condition is not needed for the 2D mixed Prandtl equations. Compared with the existence and uniqueness of solutions to the 2D MHD boundary layer where the initial tangential magnetic field has a lower bound $\delta > 0$ that plays an important role, this lower bound condition is also not needed for the 2D mixed Prandtl equations. In other words, we need neither the monotonicity condition of the tangential velocity, nor the initial tangential magnetic field has a lower bound $\delta > 0$ and for any initial datum in this chapter. Besides, compared with the existence and uniqueness of solutions to the classical 2D Prandtl and MHD equations where the initial data are small enough and the vorticity satisfies decay condition, these conditions are also not needed for the 2D mixed Prandtl equations. Similarly to the classical Prandtl equations, the difficulty of solving problem (4.1.1) in the Sobolev framework is the loss of x-derivative in the term $v\partial_y u$. We can insert the Eqs. $(4.1.1)_2$ into $(4.1.1)_1$ in calculating to eliminate this difficulty. In this chapter, we investigate the existence and uniqueness of solutions to the 2D mixed Prandtl-Shercliff regime system for any initial datum by using the direct energy methods.

Chapter 5
Local Well-Posedness of Solutions to 2D Magnetic Prandtl Model in the Prandtl-Hartmann Regime

In this chapter, we shall consider the 2D magnetic Prandtl equation in the Prandtl-Hartmann regime in a periodic domain and prove the local existence and uniqueness of solutions by the energy method in a polynomial weighted Sobolev space. On the one hand, we have noted that the x-derivative of the pressure P plays a key role in all known results on the existence and uniqueness of solutions to the Prandtl-Hartmann regime equations, in which the case of favorable P ($\partial_x P < 0$) or the case of $\partial_x P = 0$ (led by constant outer flow $U =$ constant) was only considered. While in this chapter, we have no any restriction on the sign of $\partial_x P$, which has generalized all previous results and gives definitely rise to a difficulty in mathematical treatments. To overcome this difficulty, we shall use the skill of cancellation mechanism which is valid under the monotonicity assumption. One the other hand, we shall consider the general outer flow $U \neq$ constant, leading to the boundary data at $y = 0$ being much more complicated. To deal with these boundary data, some more delicate estimates and mathematical induction method will be used. Moreover, our result has given us a physical understanding that the outer flow has a stabilizing effect on the Prandtl-Hartmann regime boundary layer in mathematics. The content of this chapter is selected from [118].

5.1 Introduction

The system of magnetohydrodynamics is a fundamental system to describe the fluid under the influence of electromagnetic field. The following mixed Prandtl and Hartmann boundary layer equations arise from the incompressible MHD system when the physical parameters such as Reynolds number, magnetic Reynolds number and the Hartmann number satisfy some constraints in the high Reynolds numbers limit, which were proposed in [39, 45].

Y. Qin et al., *Prandtl Equations and Related Boundary Layer Equations*,
https://doi.org/10.1007/978-981-97-4565-4_5

In this chapter, we shall consider the Prandtl-Hartmann regime equations in a periodic domain $\mathbb{T} \times \mathbb{R}^+ := \{(t, x, y)\big|t > 0, x \in \mathbb{R}/\mathbb{Z}, y \in \mathbb{R}_+\}$:

$$\begin{cases} \partial_t u + u\partial_x u + v\partial_y u = \partial_y b + \partial_y^2 u - \partial_x P, \\ \partial_y u + \partial_y^2 b = 0, \\ \partial_x u + \partial_y v = 0, \\ \lim\limits_{y\to+\infty} u(t, x, y) = U(t, x), \qquad \lim\limits_{y\to+\infty} b(t, x, y) = B(t, x). \end{cases} \tag{5.1.1}$$

Here $(u, v) = (u(t, x, y), v(t, x, y))$ denotes the velocity field; $b(t, x, y)$ is the corresponding tangential magnetic component. The given scalar pressure $P := P(x, t)$ and the outer flow U satisfy the well-known Bernoulli's law:

$$\partial_t U + U\partial_x U + \partial_x P = 0.$$

The initial data and no-slip boundary condition are imposed by

$$\begin{cases} u(0, x, y) = u_0(x, y), \\ u(t, x, 0) = 0, \quad v(t, x, 0) = 0, \quad b(t, x, 0) = 0. \end{cases} \tag{5.1.2}$$

Noticing $(5.1.1)_4$, and integrating $(5.1.1)_2$ in y over $[y, +\infty)$, we obtain

$$\partial_y b = U - u, \qquad b = B + \int_y^{+\infty} (u - U)dy. \tag{5.1.3}$$

Inserting (5.1.3) into (5.1.1), then we arrive at the following equations:

$$\begin{cases} \partial_t u + u\partial_x u + v\partial_y u = (U - u) + \partial_y^2 u - \partial_x P, \\ \partial_x u + \partial_y v = 0. \end{cases} \tag{5.1.4}$$

Let the vorticity $w = \partial_y u$, then Eqs. (5.1.4) reduce to the following vorticity system:

$$\begin{cases} \partial_t w + u\partial_x w + v\partial_y w = -w + \partial_y^2 w, \\ w(0, x, y) = \partial_y u_0, \\ \partial_y w|_{y=0} = \partial_x P - U. \end{cases} \tag{5.1.5}$$

In this chapter, we shall consider system (5.1.1) under the Oleinik's monotonicity assumption $w = \partial_y u > 0$.

The chapter is structured as follows. Section 5.1 is the introduction, which contains many inequalities used in this chapter. In Sect. 5.2, we shall give the uniform estimates with the weighted norm: our proof is based on using estimates of regularized parabolic equation and maximal principle. The difficulty of this part is the estimates on the boundary at $y = 0$, and we shall give the regular pattern eventually by mathematical induction method. In Sect. 5.3, we shall prove local existence and uniqueness of solutions to the Prandtl system.

We now introduce the weighted Sobolev space and define the space $H^{s,\gamma}_{\sigma,\delta}$ by

$$H^{s,\gamma}_{\sigma,\delta} := \Big\{ w : \mathbb{T}\times\mathbb{R}_+ \to \mathbb{R} : \|w\|_{H^{s,\gamma}} < +\infty,\ (1+y)^{\sigma}|w| \geq \delta, \\ \times \sum_{|\alpha|\leq 2} |(1+y)^{\sigma+\alpha_2} D^{\alpha} w|^2 \leq \frac{1}{\delta^2} \Big\},$$

where $D^{\alpha} := \partial_x^{\alpha_1}\partial_y^{\alpha_2}$, $\alpha_1+\alpha_2 = s$, $s \geq 4$, $\gamma \geq 1$, $\sigma > \gamma + \frac{1}{2}$ and $\delta \in (0,1)$. We define the norm as

$$\|w\|^2_{H^{s,\gamma}} := \sum_{|\alpha|\leq s} \|(1+y)^{\gamma+\alpha_2} D^{\alpha} w\|^2_{L^2}$$

and

$$\|w\|^2_{H^{s,\gamma}_g} := \|(1+y)^{\gamma} g_s\|^2_{L^2} + \sum_{\substack{|\alpha|\leq s \\ \alpha_1 \leq s-1}} \|(1+y)^{\gamma+\alpha_2} D^{\alpha} w\|^2_{L^2},$$

here

$$g_s := \partial_x^s w - \frac{\partial_y w}{w}\partial_x^s(u-U).$$

In this chapter, for convenience, we simply write

$$\iint \cdot := \int_{\mathbb{T}}\int_{\mathbb{R}_+} \cdot dxdy.$$

Last, we state our main result as follows.

Theorem 5.1.1 *Given any integer $s \geq 4$, and real numbers γ, σ, δ satisfying $\gamma \geq 1$, $\sigma > \gamma + \frac{1}{2}$ and $\delta \in (0,1)$. Assume the following conditions hold:*

(i) suppose that the initial data $u_0 - U(0,x) \in H^{s,\gamma-1}$ and $\partial_y u_0 \in H^{s,\gamma}_{\sigma,2\delta}$ satisfy the compatibility conditions $u_0|_{y=0}$ and $\lim\limits_{y\to+\infty} u = U|_{t=0}$. In addition, when $s = 4$, assume that $\delta \geq 0$ is chosen small enough such that $\|\omega_0\|_{H^{s,\gamma}} \leq C\delta^{-1}$ with a generic constant $C > 0$.

(ii) the outer flow U satisfies

$$\sup_t \sum_{l=0}^{\frac{s}{2}+1} \|\partial_t^l U\|_{H^{s-2l+2}(\mathbb{T})} < +\infty. \tag{5.1.6}$$

Then there exists a time $T := T(s, \gamma, \sigma, \delta, \|w_0\|_{H^{s,\gamma}}, U)$ *such that the initial boundary value problems (5.1.1)–(5.1.2) have a unique local classical solution* (u, v, b) *satisfying*

$$u - U \in L^\infty([0, T]; H^{s,\gamma-1}) \cap C([0, T]; H^s - w)$$

and

$$\partial_y u \in L^\infty([0, T]; H^{s,\gamma}_{\sigma,\delta}) \cap C([0, T]; H^s - w),$$

where $H^s - w$ *is the space* H^s *endowed with its weak topology.*

5.1.1 Preliminaries

In this subsection, we will give some lemmas to our main result.

Lemma 5.1.2 ([98]) *Let* $s \geq 4$ *be an integer,* $\gamma \geq 1$, $\sigma > \gamma + \frac{1}{2}$ *and* $\delta \in (0, 1)$. *Then for any* $w \in H^{s,\gamma}_{\sigma,\delta}(\mathbb{T} \times \mathbb{R}_+)$, *we have the following inequality:*

$$C_\delta \|w\|_{H^{s,\gamma}_g} \leq \|w\|_{H^{s,\gamma}} + \|u - U\|_{H^{s,\gamma-1}} \leq C_{s,\gamma,\sigma,\delta} \left(\|w\|_{H^{s,\gamma}_g} + \|\partial_x^s U\|_{L^2(\mathbb{T})} \right), \tag{5.1.7}$$

where $C_{s,\gamma,\sigma,\delta} > 0$ *is a constant and only depends on* $s, \gamma, \sigma, \delta$.

Lemma 5.1.3 (Hardy Type Inequality, Lemma B.1 in [98]) *Let* $f : \mathbb{T} \times \mathbb{R}_+ \to \mathbb{R}$,

(i) if $\lambda > -\frac{1}{2}$ *and* $\lim\limits_{y \to +\infty} f(x, y) = 0$, *then*

$$\|(1 + y)^\lambda f\|_{L^2(\mathbb{T}\times\mathbb{R}_+)} \leq \frac{2}{2\lambda + 1} \|(1 + y)^{\lambda+1} \partial_y f\|_{L^2(\mathbb{T}\times\mathbb{R}_+)}; \tag{5.1.8}$$

(ii) if $\lambda < -\frac{1}{2}$ *and* $f(x, y)|_{y=0} = 0$, *then*

$$\|(1 + y)^\lambda f\|_{L^2(\mathbb{T}\times\mathbb{R}_+)} \leq -\frac{2}{2\lambda + 1} \|(1 + y)^{\lambda+1} \partial_y f\|_{L^2(\mathbb{T}\times\mathbb{R}_+)}. \tag{5.1.9}$$

Lemma 5.1.4 ([98]) *Let* $s \geq 4$ *be an integer,* $\gamma \geq 1$, $\sigma > \gamma + \frac{1}{2}$ *and* $\delta \in (0, 1)$. *For any* $w \in H^{s,\gamma}_{\sigma,\delta}(\mathbb{T} \times \mathbb{R}_+)$, *then for* $k = 0, 1, 2, \cdots, s$, *we have*

$$\|(1 + y)^\gamma g_k\|_{L^2} \leq \|(1 + y)^\gamma \partial_x^k w\|_{L^2} + \delta^{-2} \|(1 + y)^{\gamma-1} \partial_x^k (u - U)\|_{L^2}, \tag{5.1.10}$$

and

$$\|(1+y)^{\gamma}\partial_x^k w\|_{L^2}+\|(1+y)^{\gamma-1}\partial_x^k(u-U)\|_{L^2}$$
$$\leq C_{\gamma,\sigma,\delta}\left(\|\partial_x^k U\|_{L^2(\mathbb{T})}+\|(1+y)^{\gamma}g_k\|_{L^2}\right), \tag{5.1.11}$$

where $C_{s,\gamma,\sigma,\delta}>0$ *is a constant and only depends on* s,γ,σ,δ.

Lemma 5.1.5 (Sobolev-Type Inequality, Lemma B.2 in [98]) *Let* $f:\mathbb{T}\times\mathbb{R}_+\to\mathbb{R}$. *Then there exists a constant* $C>0$ *such that*

$$\|f(x,y)\|_{L^\infty(\mathbb{T}\times\mathbb{R}_+)}\leq C\Big(\|f(x,y)\|_{L^2(\mathbb{T}\times\mathbb{R}_+)}+\|\partial_x f(x,y)\|_{L^2(\mathbb{T}\times\mathbb{R}_+)}$$
$$+\|\partial_y^2 f(x,y)\|_{L^2(\mathbb{T}\times\mathbb{R}_+)}\Big). \tag{5.1.12}$$

Lemma 5.1.6 ([98]) *Let* $s\geq 4$ *is an integer,* $\gamma\geq 1$, $\sigma>\gamma+\frac{1}{2}$ *and* $\delta\in(0,1)$. *Then for any* $w\in H^{s,\gamma}_{\sigma,\delta}(\mathbb{T}\times\mathbb{R}_+)$, *we have the following inequalities:*

(*i*) *for* $k=0,1,2,\cdots,s-1$,

$$\|\frac{\partial_x^k v+y\partial_x^{k+1}U}{1+y}\|_{L^2}\leq C_{s,\gamma,\sigma,\delta}(\|w\|_{H^{s,\gamma}_g}+\|\partial_x^s U\|_{L^2(\mathbb{T})}); \tag{5.1.13}$$

(*ii*) *for* $k=0,1,2,\cdots,s$,

$$\|(1+y)^{\gamma-1}\partial_x^k(u-U)\|_{L^2}\leq C_{s,\gamma,\sigma,\delta}(\|w\|_{H^{s,\gamma}_g}+\|\partial_x^s U\|_{L^2(\mathbb{T})}); \tag{5.1.14}$$

(*iii*) *for* $k=0,1,2,\cdots,s-2$,

$$\|\frac{\partial_x^k v}{1+y}\|_{L^\infty}\leq C_{s,\gamma,\sigma,\delta}(\|w\|_{H^{s,\gamma}_g}+\|\partial_x^s U\|_{L^2(\mathbb{T})}); \tag{5.1.15}$$

(*iv*) *for* $k=0,1,2,\cdots,s-1$,

$$\|\partial_x^k u\|_{L^\infty}\leq C_{s,\gamma,\sigma,\delta}(\|w\|_{H^{s,\gamma}_g}+\|\partial_x^s U\|_{L^2(\mathbb{T})}); \tag{5.1.16}$$

(*v*) *for* $k=0,1,2,\cdots,s-2$,

$$\|(1+y)^{\gamma+k_{\alpha_2}}D^k w\|_{L^\infty}\leq C_{s,\gamma,\sigma,\delta}\|w\|_{H^{s,\gamma}_g}; \tag{5.1.17}$$

(vi) for all $|\alpha| \le s$,

$$\|(1+y)^{\gamma+\alpha_2} D^\alpha w\|_{L^2} \le \begin{cases} C_{s,\gamma,\sigma,\delta}(\|w\|_{H_g^{s,\gamma}} + \|\partial_x^s U\|_{L^2(\mathbb{T})}) & if\ \alpha = (s,0), \\ \|w\|_{H_g^{s,\gamma}} & if\ \alpha \ne (s,0); \end{cases} \tag{5.1.18}$$

(vii) for all $k = 0, 1, 2, \cdots, s$,

$$\|(1+y)^\gamma g_k\|_{L^2(\mathbb{T})} \le \begin{cases} C_{s,\gamma,\sigma,\delta}(\|w\|_{H_g^{s,\gamma}} + \|\partial_x^s U\|_{L^2(\mathbb{T})}) & if\ k = 0, 1, 2, \cdots, s-1, \\ \|w\|_{H_g^{s,\gamma}} & if\ k = s, \end{cases} \tag{5.1.19}$$

where $C_{s,\gamma,\sigma,\delta} > 0$ *is a constant and only depends on* $s, \gamma, \sigma, \delta$.

Lemma 5.1.7 (Maximum Principle for Parabolic Equations, Lemma E.1 in [98]) *Let* $\epsilon \ge 0$. *If* $H \in C([0,T]; C^2(\mathbb{T}\times\mathbb{R}^+) \cap C^1([0,T]; C^0(\mathbb{T}\times\mathbb{R}^+))$ *is a bounded function that satisfies the differential inequality*

$$\left\{\partial_t + b_1\partial_x + b_2\partial_y - \epsilon^2\partial_x^2 - \partial_y^2\right\} H \le fH \quad in\ [0,T]\times\mathbb{T}\times\mathbb{R}^+,$$

where the coefficients b_1, b_2, *and* f *are continuous and satisfy*

$$\left\|\frac{b_2}{1+y}\right\|_{L^\infty([0,T]\times\mathbb{T}\times\mathbb{R}^+)} < +\infty \quad and \quad \|f\|_{L^\infty([0,T]\times\mathbb{T}\times\mathbb{R}^+)} \le \lambda,$$

then for any $t \in [0,T]$,

$$\sup_{\mathbb{T}\times\mathbb{R}^+} H(t) \le \max\left\{e^{\lambda t}\|H(0)\|_{L^\infty(\mathbb{T}\times\mathbb{R}^+)}, \max_{\tau\in[0,T]}\left\{e^{\lambda(t-\tau)}\|H(\tau)\big|_{y=0}\|_{L^\infty(\mathbb{T})}\right\}\right\}.$$

Lemma 5.1.8 (Minimum Principle for Parabolic Equations, Lemma E.2 in [98]) *Let* $\epsilon \ge 0$. *If* $H \in C([0,T]; C^2(\mathbb{T}\times\mathbb{R}^+) \cap C^1([0,T]; C^0(\mathbb{T}\times\mathbb{R}^+))$ *is a bounded function with*

$$\kappa(t) := \min\left\{\min_{\mathbb{T}\times\mathbb{R}^+} H(0), \min_{[0,T]\times\mathbb{T}} H\big|_{y=0}\right\} \ge 0$$

and satisfies

$$\left\{\partial_t + b_1\partial_x + b_2\partial_y - \epsilon^2\partial_x^2 - \partial_y^2\right\} H = fH$$

where the coefficients b_1, b_2, and f are continuous and satisfy

$$\left\| \frac{b_2}{1+y} \right\|_{L^\infty([0,T]\times\mathbb{T}\times\mathbb{R}^+)} < +\infty \ \text{and} \ \|f\|_{L^\infty([0,T]\times\mathbb{T}\times\mathbb{R}^+)} \leq \lambda,$$

then for any $t \in [0, T]$

$$\sup_{\mathbb{T}\times\mathbb{R}^+} H(t) \geq (1 - \lambda t e^{\lambda t})\kappa(t).$$

5.2 Uniform Estimates on the Regularized System

In this section, we will estimate the norm $\|w\|^2_{H_g^{s,\gamma}}$ by the energy method. We consider the regularized equations to problem (5.1.4) for any ϵ

$$\begin{cases} \partial_t u^\epsilon + u^\epsilon \partial_x u^\epsilon + v^\epsilon \partial_y u^\epsilon = (U - u^\epsilon) + \epsilon^2 \partial_x^2 u^\epsilon + \partial_y^2 u^\epsilon - \partial_x P^\epsilon, \\ \partial_x u^\epsilon + \partial_y v^\epsilon = 0, \\ u^\epsilon|_{t=0} = u_0, \\ u^\epsilon|_{y=0} = v^\epsilon|_{y=0} = 0, \\ \lim\limits_{y\to+\infty} u^\epsilon(t, x, y) = U(t, x), \end{cases} \tag{5.2.1}$$

where P^ϵ and U satisfy a regularized Bernoulli's law

$$\partial_t U + U\partial_x U - \epsilon^2 \partial_x^2 U + \partial_x P^\epsilon = 0. \tag{5.2.2}$$

Then, the regularized vorticity $w^\epsilon = \partial_y u^\epsilon$ satisfies the following regularized vorticity system for any $\epsilon > 0$:

$$\begin{cases} \partial_t w^\epsilon + u^\epsilon \partial_x w^\epsilon + v^\epsilon \partial_y w^\epsilon = -w^\epsilon + \epsilon^2 \partial_x^2 w^\epsilon + \partial_y^2 w^\epsilon, \\ w^\epsilon|_{t=0} = w_0 = \partial_y u_0, \\ \partial_y w^\epsilon|_{y=0} = \partial_x P^\epsilon - U, \end{cases} \tag{5.2.3}$$

where the velocity field (u^ϵ, v^ϵ) is given

$$u^\epsilon(t, x, y) = U - \int_y^{+\infty} w(t, x, \tilde{y}) d\tilde{y},$$

$$v^\epsilon(t, x, y) = -\int_0^y \partial_x u(t, x, \tilde{y}) d\tilde{y}. \tag{5.2.4}$$

From now on, we drop the superscript ϵ for simplicity of notations.

5.2.1 Estimates on $D^\alpha w$

In this subsection, we will estimate the norm on $D^\alpha w$ with weight $(1+y)^{\gamma+\alpha_2}$, $\alpha=(\alpha_1,\alpha_2)$, $|\alpha|\le s$, and α_1 should be restricted as $\alpha_1\le s-1$. Otherwise, it is impossible to directly estimate the norm on $D^\alpha w$ with $\partial_x^s w$.

Lemma 5.2.1 (Reduction of Boundary Data) *If w solves (5.2.1) and (5.2.3), then on the boundary at $y=0$, we Have*

$$\begin{cases} \partial_y w|_{y=0} = \partial_x P - U(t,x), \\ \partial_y^3 w|_{y=0} = (\partial_t - \epsilon^2\partial_x^2)(\partial_x P - U(t,x)) + (\partial_x P - U(t,x)) + w\partial_x w|_{y=0}. \end{cases} \tag{5.2.5}$$

For any $5\le 2k+1\le s$, there are some constants C_k, $C_{\wedge_\alpha,k,l,\rho^1,\rho^2,\dots,\rho^j}$, not depending on ϵ or (u,v,w), such that

$$\begin{aligned} \partial_y^{2k+1} w|_{y=0} &= C_k\sum_{s=0}^{k}(\partial_t-\epsilon^2\partial_x^2)^s(\partial_x P - U) \\ &\quad + \sum_{l=0}^{k-1}\epsilon^{2l}\sum_{j=2}^{\max\{2,k-l\}}\sum_{\rho\in A_{k,l}^j} C_{\wedge_\alpha,k,l,\rho^1,\rho^2,\dots,\rho^j}\prod_{i=1}^{j} D^{\rho^i} w\big|_{y=0} \end{aligned} \tag{5.2.6}$$

where $A_{k,l}^j := \{\rho := (\rho^1,\rho^2,\dots,\rho^j)\in\mathbb{N}^{2j}; 3\sum_{i=1}^{j}\sum_{i=1}^{j}\rho_2^i = 2k+4l+1, \sum_{i=1}^{j}\rho_1^i \le k+2l-1, \sum_{i=1}^{j}\rho_2^j \le 2k-2l-2$, and $|\rho^i|\le 2k-l-1$ $for\ all$ $i=1,2,\dots,j\}$.

Proof According to Eq. (5.2.1) and the boundary condition $(5.1.2)_2$, we get

$$\partial_y w|_{y=0} = \partial_x P - U(t,x) = K,$$

which implies

$$\begin{aligned} \partial_y^3 w\big|_{y=0} &= \left\{\partial_t\partial_y w + \partial_y(u\partial_x w) + \partial_y(v\partial_y w) + \partial_y w - \epsilon^2\partial_y\partial_x^2 w\right\}\big|_{y=0} \\ &= (\partial_t - \epsilon^2\partial_x^2)K + K + w\partial_x w\big|_{y=0}, \end{aligned} \tag{5.2.7}$$

and

$$\partial_y^{2n+1} w\big|_{y=0} = \left\{\left(\partial_t - \epsilon^2\partial_x^2 w + 1\right)\partial_y^{2n-1} w + \partial_y^{2n-1}(u\partial_x w + v\partial_y w)\right\}\big|_{y=0}. \tag{5.2.8}$$

Hence, for $n = 2$,

$$\begin{aligned}
\partial_y^5 w\big|_{y=0} &= (\partial_t - \epsilon^2\partial_x^2)\partial_y^3 w|_{y=0} \\
&\quad + \partial_y^3 w|_{y=0} + (2\partial_y w\partial_x\partial_y w + 3w\partial_x\partial_y^2 w - 2\partial_x w\partial_y^2 w)\big|_{y=0} \\
&= (\partial_t - \epsilon^2\partial_x^2)^2 K + (\partial_t - \epsilon^2\partial_x^2)K + (\partial_t - \epsilon^2\partial_x^2)(w\partial_x w)|_{y=0} \\
&\quad + (\partial_t - \epsilon^2\partial_x^2)K \\
&\quad + K + w\partial_x w|_{y=0} + (2\partial_y w\partial_x\partial_y w + 3w\partial_x\partial_y^2 w - 2\partial_x w\partial_y^2 w)\big|_{y=0},
\end{aligned} \tag{5.2.9}$$

here

$$\begin{cases}
\partial_t w|_{y=0} = (\partial_y^2 w - u\partial_x w - v\partial_y w - w - \epsilon^2\partial_x^2 w)|_{y=0} \\
= (\partial_y^2 w - w + \epsilon^2\partial_x^2 w)\big|_{y=0}, \\
\partial_x\partial_t w|_{y=0} = (\partial_x\partial_y^2 w - \partial_x w + \epsilon^2\partial_x^3 w)\big|_{y=0},
\end{cases} \tag{5.2.10}$$

$$(\partial_t - \epsilon^2\partial_x^2)(w\partial_x w)\big|_{y=0} = (w\partial_x\partial_y^2 w + \partial_x w\partial_y^2 w - 2w\partial_x w - 2\epsilon^2\partial_x\partial_x^2 w)\big|_{y=0}. \tag{5.2.11}$$

It follows from (5.2.9), (5.2.10), and (5.2.11) that

$$\begin{aligned}
\partial_y^5 w\big|_{y=0} &= C_n\sum_{i=0}^{2}(\partial_t - \epsilon^2\partial_x^2)^i K + (2\partial_y w\partial_x\partial_y w + 4w\partial_x\partial_y^2 w - \partial_x w\partial_y^2 w)\big|_{y=0} \\
&\quad - w\partial_x w\big|_{y=0} - 2\epsilon^2\partial_x w\partial_x^2 w\big|_{y=0}.
\end{aligned} \tag{5.2.12}$$

For $n = 3$,

$$\begin{aligned}
\partial_y^7 w\big|_{y=0} &= (\partial_t - \epsilon^2\partial_x^2)\partial_y^5 w\big|_{y=0} + \partial_y^5 w\big|_{y=0} + \partial_y^5(u\partial_x w + v\partial_y w)\big|_{y=0} \\
&= C_n\sum_{i=1}^{3}(\partial_t - \epsilon^2\partial_x^2)^i K + (\partial_t - \epsilon^2\partial_x^2) \\
&\quad \times(2\partial_y w\partial_x\partial_y w + 4w\partial_x\partial_y^2 w - \partial_x w\partial_y^2 w)\big|_{y=0} \\
&\quad - (\partial_t - \epsilon^2\partial_x^2)(w\partial_x w)\big|_{y=0} - (\partial_t - \epsilon^2\partial_x^2)2\epsilon^2\partial_x w\partial_x^2 w\big|_{y=0} \\
&\quad + \partial_y^5 w\big|_{y=0} + \partial_y^5(u\partial_x w + v\partial_y w)\big|_{y=0}.
\end{aligned} \tag{5.2.13}$$

Since the first term and last four terms on the right-hand side of (5.2.13) are in the desired form, we only need to calculate the second term on the right-hand side of (5.2.13),

$$
\begin{aligned}
&(\partial_t - \epsilon^2\partial_x^2)(2\partial_y w\partial_x\partial_y w + 4w\partial_x\partial_y^2 w - \partial_x w\partial_y^2 w)\big|_{y=0}\\
&= (-\partial_x\partial_t w\partial_y^2 w - \partial_x w\partial_y^2\partial_t w\\
&\quad +2\partial_x\partial_y\partial_t w\partial_y w + 2\partial_x\partial_y w\partial_y\partial_t w + 4w\partial_x\partial_y^2\partial_t w + 4\partial_t w\partial_x\partial_y^2 w)\big|_{y=0}\\
&\quad -\epsilon^2\partial_x^2(2\partial_y w\partial_x\partial_y w + 4w\partial_x\partial_y^2 w - \partial_x w\partial_y^2 w)\big|_{y=0},
\end{aligned}
\tag{5.2.14}
$$

here

$$
-(\partial_x\partial_t w\partial_y^2 w)\big|_{y=0} = -\partial_y^2 w(\partial_x\partial_y^2 w - \partial_x w + \epsilon^2\partial_x^3 w)\big|_{y=0}, \tag{5.2.15}
$$

$$
\begin{aligned}
-\partial_y^2\partial_t w\big|_{y=0} &= -\partial_y^2(\partial_y^2 w - u\partial_x w - v\partial_y w - w + \epsilon^2\partial_x^2 w)\big|_{y=0}\\
&= -(\partial_y^4 w - \partial_y^2 w + 2w\partial_x\partial_y w + \epsilon^2\partial_y^2\partial_x^2 w)\big|_{y=0},
\end{aligned}
\tag{5.2.16}
$$

which implies

$$
\begin{aligned}
\partial_x w\partial_y^2\partial_t w\big|_{y=0} &= \sum_{j=0}^{4}\wedge_\alpha\partial_x\partial_y^j w\partial_y^{4-j} w + \sum_{j=0}^{2}\wedge_\alpha\partial_x\partial_y^j w\partial_y^{2-j} w\\
&\quad + \sum_{\rho_1+\rho_2+\rho_3=3}\wedge_\alpha D^{\rho_1} w D^{\rho_2} w D^{\rho_3} w\big|_{y=0}\\
&\quad - \epsilon^2\partial_x w\partial_y^2\partial_x^2 w\big|_{y=0}.
\end{aligned}
\tag{5.2.17}
$$

Since $v = \partial_x\partial_y^{-1} w$,

$$
\begin{aligned}
\partial_x\partial_y\partial_t w\big|_{y=0} &= \partial_x\partial_y(\partial_y^2 w - u\partial_x w - v\partial_y w - w + \epsilon^2\partial_x^2 w)\big|_{y=0}\\
&= \{\partial_x\partial_y^3 w - \partial_x\partial_y w + \epsilon^2\partial_y\partial_x^3 w\\
&\quad -\partial_x(w\partial_x w + u\partial_x\partial_y w + \partial_y v\partial_y w + v\partial_y^2 w)\}\big|_{y=0}\\
&= (\partial_x\partial_y^3 w - \partial_x\partial_y w + \epsilon^2\partial_y\partial_x^3 w\\
&\quad -\partial_x w\partial_x w - w\partial_x^2 w - w\partial_y\partial_x w)\big|_{y=0}
\end{aligned}
\tag{5.2.18}
$$

which implies

$$
\begin{aligned}
2\partial_x\partial_y\partial_t w\partial_y w|_{y=0} \in & \left(\sum_{j=0}^{4} \wedge_\alpha \partial_x \partial_y^j w \partial_y^{4-j} w + \sum_{j=0}^{2} \wedge_\alpha \partial_x \partial_y^j w \partial_y^{2-j} w \right. \\
& \left. + \sum_{\rho_1+\rho_2+\rho_3=3} \wedge_\alpha D^{\rho_1} w D^{\rho_2} w D^{\rho_3} w \right) |_{y=0} \\
& + 2\epsilon^2 \partial_y \partial_x^3 w \partial_y w|_{y=0}.
\end{aligned}
\tag{5.2.19}
$$

Similarly,

$$
\begin{aligned}
& \left(2\partial_x\partial_y w \partial_y \partial_t w + 4w\partial_x \partial_y^2 \partial_t w + 4\partial_t w \partial_x \partial_y^2 w\right)|_{y=0} \\
& \in \left(\sum_{j=0}^{4} \wedge_\alpha \partial_x \partial_y^j w \partial_y^{4-j} w + \sum_{j=0}^{2} \wedge_\alpha \partial_x \partial_y^j w \partial_y^{2-j} w \right. \\
& \left. + \sum_{\rho_1+\rho_2+\rho_3=3} \wedge_\alpha D^{\rho_1} w D^{\rho_2} w D^{\rho_3} w \right) |_{y=0} \\
& + 2\epsilon^2 \partial_x \partial_y w \partial_y \partial_x^2 w|_{y=0} + 4\epsilon^2 w \partial_x^3 \partial_y^2 w|_{y=0} + 4\epsilon^2 \partial_x^2 w \partial_x \partial_y^2 w|_{y=0}.
\end{aligned}
\tag{5.2.20}
$$

Thus it follows from (5.2.13), (5.2.14), and (5.2.15)–(5.2.20) that

$$
\begin{aligned}
\partial_y^7 w|_{y=0} = & \sum_{i=0}^{3} (\partial_t - \epsilon^2 \partial_x^2)^i K + (1 + \epsilon^2 \partial_x^2 + \epsilon^4 \partial_x^4)(w\partial_x w)|_{y=0} \\
& + \sum_{j=0}^{0} \partial_x \partial_y^0 w \partial_y^{0-j} w|_{y=0} \\
& + \sum_{j=0}^{2} \wedge_\alpha \partial_x \partial_y^j w \partial_y^{2-j} w|_{y=0} + \sum_{j=0}^{4} \wedge_\alpha \partial_x \partial_y^j w \partial_y^{4-j} w|_{y=0} \\
& + \sum_{\rho_1+\rho_2+\rho_3=3} \wedge_\alpha D^{\rho_1} w D^{\rho_2} w D^{\rho_3} w|_{y=0} \\
& + \sum_{j_x=0}^{2} \sum_{j_y=0}^{2} \wedge_\alpha \partial_x \partial_y^{j_x} \partial_y^{j_y} w \partial_y^{2-j_x} \partial_y^{2-j_y} w|_{y=0}.
\end{aligned}
\tag{5.2.21}
$$

We justify the formula (5.2.21) for $k = 3$.

Now, using the same algorithm, we are going to prove formula (5.2.6) by induction on k. For notational convenience, we denote

$$\mathcal{A}_k := \left\{ \sum_{l=0}^{k-1} \epsilon^{2l} \sum_{j=2}^{\max\{2,k-l\}} \sum_{\rho\in A^j_{k,l}} C_{\wedge_\alpha,k,l,\rho^1,\rho^2,\dots,\rho^j} \prod_{i=1}^{j} D^{\rho^i} w|_{y=0} \right\}. \tag{5.2.22}$$

Using this notation, we will prove $\partial_y^{2k+1} w|_{y=0} - (\partial_t - \epsilon^2\partial_x^2)^k(\partial_x P - U) \in \mathcal{A}_k$. Assuming that formula (5.2.6) holds for $k = n$, we will show that it also holds for $k = n+1$ as follows. Then we differentiate the vorticity equation $2n+1$ times with respect to y to obtain

$$\begin{aligned}
&\partial_y^{2n+3} w|_{y=0} \\
&= \left\{ (\partial_t - \epsilon^2\partial_x^2)\partial_y^{2n+1} w + \partial_y^{2n+1}(u\partial_x w + v\partial_y w) + \partial_y^{2n+1} w \right\}|_{y=0} \\
&= \left\{ (\partial_t - \epsilon^2\partial_x^2)\partial_y^{2n+1} w + \sum_{j=0}^{2n} \wedge_\alpha \partial_x\partial_y^j w \partial_y^{2n-j} w|_{y=0} + \partial_y^{2n+1} w \right\}|_{y=0}.
\end{aligned} \tag{5.2.23}$$

By routine checking, one may show that the last three terms of (5.2.23) belong to $\mathcal{A}_{k+1}$, so it only remains to deal with the term $(\partial_t - \epsilon^2\partial_x^2)\partial_y^{2n+1}w$. Thanks to the induction hypothesis, there exist constants $C_{\wedge_\alpha,n,l,\rho^1,\rho^2,\dots,\rho^j}$ such that

$$\begin{aligned}
\partial_y^{2n+1} w|_{y=0} &= C_n \sum_{s=0}^{n} (\partial_t - \epsilon^2\partial_x^2)^n(\partial_x P - U) \\
&\quad + \sum_{l=0}^{k-1} \epsilon^{2l} \sum_{j=2}^{\max\{2,n-l\}} \sum_{\rho\in A^j_{n,l}} C_{\wedge_\alpha,n,l,\rho^1,\rho^2,\dots,\rho^j} \prod_{i=1}^{j} D^{\rho^i} w|_{y=0},
\end{aligned} \tag{5.2.24}$$

so we have, up to a relabeling of the indices ρ^i,

$$\begin{aligned}
&(\partial_t - \epsilon^2\partial_x^2)\partial_y^{2n+1} w|_{y=0} \\
&= C_n \sum_{s=1}^{n+1} (\partial_t - \epsilon^2\partial_x^2)^s(\partial_x P - U) \\
&+ \sum_{l=0}^{n-1} \epsilon^{2l} \sum_{j=2}^{\max\{2,n-l\}} \sum_{\rho\in A^j_{n,l}} \widetilde{C}_{\wedge_\alpha,n,l,\rho^1,\rho^2,\dots,\rho^j} (\partial_t - \epsilon^2\partial_x^2) D^{\rho^1} w \prod_{i=2}^{j} D^{\rho^i} w|_{y=0}
\end{aligned}$$

$$-\sum_{l=0}^{n-1}\epsilon^{2l+2}\sum_{j=2}^{\max\{2,n-l\}}\sum_{\rho\in A_{n,l}^j}\widetilde{\widetilde{C}}_{\wedge_\alpha,n,l,\rho^1,\rho^2,\dots,\rho^j}\partial_x D^{\rho^1}w\partial_x D^{\rho^2}w\prod_{i=3}^{j}D^{\rho^i}w|_{y=0},$$

(5.2.25)

where $\widetilde{C}_{\wedge_\alpha,n,l,\rho^1,\rho^2,\dots,\rho^j}$ and $\widetilde{\widetilde{C}}_{\wedge_\alpha,n,l,\rho^1,\rho^2,\dots,\rho^j}$ are some new constants depending on $C_{\wedge_\alpha,n,l,\rho^1,\rho^2,\dots,\rho^j}$. It is worth noting that the last term on the right-hand side of (5.2.25) belongs to $\mathcal{A}_{n+1}$, so it remains to check whether the second term on right-hand side of (5.2.25) also belongs to $\mathcal{A}_{n+1}$. This completes the proof. □

Lemma 5.2.2 *Let $s \geq 4$ be an integer, $\gamma \geq 1$, $\sigma > \gamma + \frac{1}{2}$ and $\delta \in (0,1)$. If $w \in H_{\sigma,\delta}^{s,\gamma}$ solves* (5.2.1) *and* (5.2.3), *we have the following estimates:*

(i) When $|\alpha| \leq s-1$,

$$\left|\int_{\mathbb{T}} D^\alpha w\partial_y D^\alpha w dx|_{y=0}\right| \leq \frac{1}{12}\|(1+y)^{\gamma+\alpha_2+1}\partial_y^2 D^\alpha w\|_{L^2}^2 + C\|w\|_{H_g^{s,\gamma}}^2, \tag{5.2.26}$$

(ii) When $|\alpha| = s$, α_2 is even,

$$\begin{aligned}&\left|\int_{\mathbb{T}} D^\alpha w\partial_y D^\alpha w dx|_{y=0}\right|\\ &\leq \frac{1}{12}\|(1+y)^{\gamma+\alpha_2}\partial_y D^\alpha w\|_{L^2}^2\\ &\quad + C_{s,\gamma,\sigma,\delta}(1+\|w\|_{H_g^{s,\gamma}}^2)^{s-2}\|w\|_{H_g^{s,\gamma}}^2\\ &\quad + C_s\sum_{l=0}^{\frac{s}{2}}\|\partial_t^l(\partial_x P - U)\|_{H^{s-2l}(\mathbb{T})}^2,\end{aligned} \tag{5.2.27}$$

(iii) When $|\alpha| = s$, α_2 is odd,

$$\begin{aligned}\left|\int_{\mathbb{T}} D^\alpha w\partial_y D^\alpha w dx|_{y=0}\right| &\leq \frac{1}{12}\|(1+y)^{\gamma+\alpha_2+1}\partial_x^{\alpha_1-1}\partial_y^{\alpha_2+2}w\|_{L^2}^2\\ &\quad + C_{s,\gamma,\sigma,\delta}(1+\|w\|_{H_g^{s,\gamma}}^2)^{s-2}\|w\|_{H_g^{s,\gamma}}^2\\ &\quad + C_s\sum_{l=0}^{\frac{s}{2}}\|\partial_t^l(\partial_x P - U)\|_{H^{s-2l}(\mathbb{T})}^2.\end{aligned} \tag{5.2.28}$$

Proof

Case 1. When $|\alpha| \le s-1$, by using the following trace estimate:

$$\int_{\mathbb{T}} |f| dx \Big|_{y=0} \le C \Big(\int_0^1 \int_{\mathbb{T}} |f| dx dy + \int_0^1 \int_{\mathbb{T}} |\partial_y f| dx dy \Big), \quad (5.2.29)$$

we have

$$\Big| \int_{\mathbb{T}} D^\alpha w \partial_y D^\alpha w dx|_{y=0} \Big| \le \frac{1}{12} \|(1+y)^{\gamma+\alpha_2+1} \partial_y^2 D^\alpha w\|_{L^2}^2 + C\|w\|_{H_g^{s,\gamma}}^2. \quad (5.2.30)$$

Case 2. When $|\alpha| = s$, $\alpha_2 = 2k$ for some $k \in \mathbb{N}$, we can apply boundary reduction Lemma 5.2.1 to $\partial_y D^\alpha|_{y=0}$ and obtain

$$\begin{aligned} \int_{\mathbb{T}} D^\alpha w \partial_y D^\alpha w dx|_{y=0} = {} & C_k \sum_{s=0}^{k} \int_{\mathbb{T}} D^\alpha w (\partial_t - \epsilon^2 \partial_x^2)^k \partial_x^{\alpha_1} (\partial_x P - U) dx|_{y=0} \\ & + \sum_{l=0}^{k-1} \epsilon^{2l} \sum_{j=2}^{\max\{2,k-l\}} \sum_{\rho \in A_{k,l}^j} C_{\wedge_\alpha, k, l, \rho^1, \rho^2, \ldots, \rho^j} \quad (5.2.31) \\ & \times \int_{\mathbb{T}} D^\alpha w \partial_x^{\alpha_1} (\prod_{i=1}^{j} D^{\rho^i} w) dx|_{y=0}, \end{aligned}$$

then we again apply the simple trace estimate to control the boundary integral as follows:

$$\begin{aligned} \Big| \int_{\mathbb{T}} D^\alpha w \partial_y D^\alpha w dx|_{y=0} \Big| \le {} & \frac{1}{12} \|(1+y)^{\gamma+\alpha_2} \partial_y D^\alpha w\|_{L^2}^2 \\ & + C_{s,\gamma,\sigma,\delta} (1 + \|w\|_{H_g^{s,\gamma}})^{s-2} \|w\|_{H_g^{s,\gamma}}^2 \quad (5.2.32) \\ & + C_s \sum_{l=0}^{\frac{s}{2}} \|\partial_t^l (\partial_x P - U)\|_{H^{s-2l}(\mathbb{T})}^2. \end{aligned}$$

Case 3. When $|\alpha| = s$, $\alpha_2 = 2k+1$ for some $k \in \mathbb{N}$, using integration by parts in the x-variable, we have

$$\int_{\mathbb{T}} D^\alpha w \partial_y D^\alpha w dx|_{y=0} = - \int_{\mathbb{T}} \partial_x D^\alpha w \partial_x^{\alpha_1 - 1} \partial_y^{\alpha_2+1} w dx|_{y=0}. \quad (5.2.33)$$

The term $\partial_x D^\alpha w|_{y=0} = \partial_x^{\alpha_1+1}\partial_y^{2k+1} w|_{y=0}$ has an odd number of y derivatives; then we have

$$\begin{aligned}\left|\int_{\mathbb{T}} D^\alpha w \partial_y D^\alpha w dx|_{y=0}\right| \le & \frac{1}{12}\|(1+y)^{\gamma+\alpha_2+1}\partial_x^{\alpha_1-1}\partial_y^{\alpha_2+2} w\|_{L^2}^2 \\ & + C_{s,\gamma,\sigma,\delta}(1+\|w\|_{H_g^{s,\gamma}})^{s-2}\|w\|_{H_g^{s,\gamma}}^2 \\ & + C_s \sum_{l=0}^{\frac{s}{2}} \|\partial_t^l(\partial_x P - U)\|_{H^{s-2l}(\mathbb{T})}^2 .\end{aligned} \tag{5.2.34}$$

□

Proposition 5.2.1 *Let $s \ge 4$ be an integer, $\gamma \ge 1$, $\sigma > \gamma + \frac{1}{2}$ and $\delta \in (0,1)$. If $w \in H_{\sigma,\delta}^{s,\gamma}$ solves (5.2.1) and (5.2.3), we have the following uniform (in ϵ) estimate:*

$$\begin{aligned}& \frac{1}{2}\frac{d}{dt}\sum_{\substack{|\alpha|\le s \\ \alpha_1 \le s-1}} \|(1+y)^{\gamma+\alpha_2} D^\alpha w\|_{L^2}^2 + \sum_{\substack{|\alpha|\le s \\ \alpha_1 \le s-1}} \|(1+y)^{\gamma+\alpha_2} D^\alpha w\|_{L^2}^2 \\ & \le -\frac{1}{2}\sum_{\substack{|\alpha|\le s \\ \alpha_1 \le s-1}} \|(1+y)^{\gamma+\alpha_2}\partial_y D^\alpha w\|_{L^2}^2 - \epsilon^2 \sum_{\substack{|\alpha|\le s \\ \alpha_1 \le s-1}} \|(1+y)^{\gamma+\alpha_2}\partial_x D^\alpha w\|_{L^2}^2 \\ & \quad + C_{s,\gamma,\sigma,\delta}\|w\|_{H_g^{s,\gamma}}^2 (\|\partial_x^s U\|_{L^\infty(\mathbb{T})}^2 \\ & \quad + \|w\|_{H_g^{s,\gamma}}^2) + C_{s,\gamma,\sigma,\delta}(1+\|w\|_{H_g^{s,\gamma}})^{s-2}\|w\|_{H_g^{s,\gamma}}^2 \\ & \quad + C_s \sum_{l=0}^{\frac{s}{2}} \|\partial_t^l(\partial_x P - U)\|_{H^{s-2l}(\mathbb{T})}^2 .\end{aligned} \tag{5.2.35}$$

Proof Applying the operator D^α on $(5.2.1)_1$ with $\alpha = (\alpha_1, \alpha_2)$, $|\alpha| \le s$, $\alpha_1 \le s-1$,

$$\begin{aligned}& (\partial_t + u\partial_x + v\partial_y + 1 - \epsilon^2\partial_x^2 - \partial_y^2) D^\alpha w \\ & = -\sum_{0<\beta\le\alpha} \binom{\alpha}{\beta} \left\{D^\beta u \partial_x D^{\alpha-\beta} w + D^\beta v \partial_y D^{\alpha-\beta} w\right\}.\end{aligned} \tag{5.2.36}$$

Multiplying (5.2.36) by $(1+y)^{2\gamma+2\alpha_2}D^\alpha w$ and integrating it over $\mathbb{T}\times\mathbb{R}_+$ yield

$$\begin{aligned}
&\frac{1}{2}\frac{d}{dt}\|(1+y)^{\gamma+\alpha_2}D^\alpha w\|_{L^2}^2+\|(1+y)^{\gamma+\alpha_2}D^\alpha w\|_{L^2}^2+\|(1+y)^{\gamma+\alpha_2}\partial_y D^\alpha w\|_{L^2}^2\\
&+\epsilon^2\|(1+y)^{\gamma+\alpha_2}\partial_x D^\alpha w\|_{L^2}^2\\
&=-\int_{\mathbb{T}}D^\alpha w\partial_y D^\alpha w dx|_{y=0}-(2\gamma+2\alpha_2)\iint(1+y)^{2\gamma+2\alpha_2-1}D^\alpha w\partial_y D^\alpha w\\
&\quad+(\gamma+\alpha_2)\iint(1+y)^{2\gamma+2\alpha_2-1}v|D^\alpha w|^2\\
&\quad-\sum_{0<\beta\le\alpha}\binom{\alpha}{\beta}\iint(1+y)^{2\gamma+2\alpha_2}D^\alpha w\left\{D^\beta u\partial_x D^{\alpha-\beta}w+D^\beta v\partial_y D^{\alpha-\beta}w\right\}.
\end{aligned}\tag{5.2.37}$$

Indeed, (5.2.37) can be obtained by integration by parts and the boundary condition. We need to estimate the integral equation (5.2.37) term by term. Obviously, the first term on the right-hand side of (5.2.37) follows from Lemmas 5.2.1–5.2.2.

Secondly, using Hölder's inequality, we have

$$\begin{aligned}
&\left|(2\gamma+2\alpha_2)\iint(1+y)^{2\gamma+2\alpha_2-1}D^\alpha w\partial_y D^\alpha w\right|\\
&\le(2\gamma+2\alpha_2)\|\frac{1}{1+y}\|_{L^\infty}\|(1+y)^{\gamma+\alpha_2}D^\alpha w\|_{L^2}\|(1+y)^{\gamma+\alpha_2}\partial_y D^\alpha w\|_{L^2}\\
&\le C_{s,\gamma}\|w\|_{H_g^{s,\gamma}}^2+\frac{1}{4}\|(1+y)^{\gamma+\alpha_2}\partial_y D^\alpha w\|_{L^2}^2,
\end{aligned}\tag{5.2.38}$$

and using Lemma 5.1.6,

$$\begin{aligned}
&\left|(\gamma+\alpha_2)\iint(1+y)^{2\gamma+2\alpha_2-1}v|D^\alpha w|^2\right|\\
&\le\|\frac{v}{1+y}\|_{L^\infty}\|(1+y)^{\gamma+\alpha_2}D^\alpha w\|_{L^2}^2\\
&\le C_{s,\gamma,\sigma,\delta}(\|w\|_{H_g^{s,\gamma}}+\|\partial_x^s U\|_{L^2(\mathbb{T})})\|w\|_{H_g^{s,\gamma}}^2.
\end{aligned}\tag{5.2.39}$$

Lastly, noticing the fact $\partial_y v=-\partial_x u,\ \partial_y u=w$, it follows that the last term on the right-hand side of Eq. (5.2.37) has the following three cases:

(i) the first case:

$$J_1=\iint(1+y)^{2\gamma+2\alpha_2}D^\alpha w\partial_x^\eta v D^k w,\quad \eta\in[1,s-1],$$

(a) for $\eta = s-1$, we get

$$
\begin{aligned}
|J_1| &= \left|\iint (1+y)^{2\gamma+2\alpha_2} D^\alpha w \partial_x^{s-1} v \partial_y^2 w\right| \\
&\le \|(1+y)^{\gamma+\alpha_2} D^\alpha w\|_{L^2} \|\frac{\partial_x^{s-1} v + y\partial_x^s U}{1+y}\|_{L^2} \|(1+y)^{\gamma+k_2} D^k w\|_{L^\infty} \\
&\quad + \|(1+y)^{\gamma+\alpha_2} D^\alpha w\|_{L^2} \|\partial_x^s U\|_{L^\infty(\mathbb{T})} \|(1+y)^{\gamma+k_2} D^k w\|_{L^\infty} \\
&\le C_{s,\gamma,\sigma,\delta} \|w\|^2_{H_g^{s,\gamma}} (\|w\|_{H_g^{s,\gamma}} + \|\partial_x^s U\|_{L^\infty(\mathbb{T})}),
\end{aligned} \tag{5.2.40}
$$

(b) for $\eta = 1, 2, \cdots, s-2$, we derive

$$
\begin{aligned}
|J_1| &\le \|(1+y)^{\gamma+\alpha_2} D^\alpha w\|_{L^2} \|\frac{\partial_x^\eta v}{1+y}\|_{L^\infty} \|(1+y)^{\gamma+k_2} D^k w\|_{L^2} \\
&\le C_{s,\gamma,\sigma,\delta} \|w\|^2_{H_g^{s,\gamma}} (\|w\|_{H_g^{s,\gamma}} + \|\partial_x^s U\|_{L^2(\mathbb{T})}).
\end{aligned} \tag{5.2.41}
$$

(ii) the second case

$$
J_2 = \iint (1+y)^{2\gamma+2\alpha_2} D^\alpha w \partial_x^\eta u D^k w, \quad \eta \in [1, s],
$$

(a) for $\eta = s$, we get

$$
\begin{aligned}
|J_2| &= \left|\iint (1+y)^{2\gamma+2\alpha_2} D^\alpha w \partial_x^s u \partial_y w\right| \\
&\le \|(1+y)^{\gamma+\alpha_2} D^\alpha w\|_{L^2} \|\partial_x^s (u-U)\|_{L^2} \|(1+y)^{\gamma+k_2} D^k w\|_{L^\infty} \\
&\quad + \|(1+y)^{\gamma+\alpha_2} D^\alpha w\|_{L^2} \|\partial_x^s U\|_{L^\infty(\mathbb{T})} \|(1+y)^{\gamma+k_2} D^k w\|_{L^2} \\
&\le C_{s,\gamma,\sigma,\delta} \|w\|^2_{H_g^{s,\gamma}} (\|w\|_{H_g^{s,\gamma}} + \|\partial_x^s U\|_{L^\infty(\mathbb{T})}),
\end{aligned} \tag{5.2.42}
$$

(b) for $\eta = 1, 2, \cdots, s-1$, we derive

$$
\begin{aligned}
|J_2| &\le \|(1+y)^{\gamma+\alpha_2} D^\alpha w\|_{L^2} \|\partial_x^\eta u\|_{L^\infty} \|(1+y)^{\gamma+k_2} D^k w\|_{L^2} \\
&\le C_{s,\gamma,\sigma,\delta} \|w\|^2_{H_g^{s,\gamma}} (\|w\|_{H_g^{s,\gamma}} + \|\partial_x^s U\|_{L^2(\mathbb{T})}).
\end{aligned} \tag{5.2.43}
$$

(iii) the last case

$$
J_3 = \iint (1+y)^{2\gamma+2\alpha_2} D^\alpha w D^\theta w D^k w, \quad \theta \in [0, s-1], \theta + k = s,
$$

we obtain

$$
\begin{aligned}
|J_3| &\le C_{s,\gamma,\sigma,\delta}\|w\|_{H_g^{s,\gamma}}(\|w\|_{H_g^{s,\gamma}}+\|\partial_x^s U\|_{L^2(\mathbb{T})})\|(1+y)^{\gamma+\alpha_2}D^\alpha w\|_{L^2}\\
&\le C_{s,\gamma,\sigma,\delta}\|w\|^2_{H_g^{s,\gamma}}(\|w\|_{H_g^{s,\gamma}}+\|\partial_x^s U\|_{L^2(\mathbb{T})})^2\\
&\quad+\frac{1}{12}\|(1+y)^{\gamma+\alpha_2}D^\alpha w\|^2_{L^2}.
\end{aligned}
\tag{5.2.44}
$$

Hence, combining (5.2.38)–(5.2.44), (5.2.37) reduces to inequality (5.2.35). □

5.2.2 Estimates on g_s

In this subsection, we will estimate the norm of $D^\alpha w$ with weight $(1+y)^{\gamma+\alpha_2}$, here $\alpha=(s,0)$, i.e., $(1+y)^\gamma g_s$, because we have estimated the norm of $(1+y)^{\gamma+\alpha_2}D^\alpha w$, $\alpha=(\alpha_1,\alpha_2)$, $\alpha_1\le s-1$ in Sect. 2.1. The evolution equations for w and $u-U$ as follows:

$$
\begin{cases}
(\partial_t+u\partial_x+v\partial_y+1-\partial_y^2-\epsilon^2\partial_x^2)w=0,\\
(\partial_t+u\partial_x+v\partial_y+1-\partial_y^2-\epsilon^2\partial_x^2)(u-U)=-(u-U)\partial_x U.
\end{cases}
\tag{5.2.45}
$$

Proposition 5.2.2 *Under the same assumption of Proposition 5.2.1, we have the following estimate:*

$$
\begin{aligned}
&\frac{d}{dt}\|(1+y)^\gamma g_s\|^2_{L^2}+\|(1+y)^\gamma g_s\|^2_{L^2}\\
&\le-\frac{1}{2}\epsilon^2\|(1+y)^\gamma\partial_x g_s\|^2_{L^2}-\frac{1}{2}\|(1+y)^\gamma\partial_y g_s\|^2_{L^2}+\|w\|^2_{H_g^{s,\gamma}}\\
&\quad+C\|\partial_x^s(\partial_x P-U)\|^2_{L^2(\mathbb{T})}+C_{\gamma,\delta}\|\partial_x^s U\|^2_{L^2(\mathbb{T})}\|w\|^2_{H_g^{s,\gamma}}\\
&\quad+C_{s,\gamma,\sigma,\delta}\left(1+\|w\|_{H_g^{s,\gamma}}+\|\partial_x^s U\|_{L^\infty(\mathbb{T})}\right)\\
&\quad\times\left(\|w\|_{H_g^{s,\gamma}}+\|\partial_x^{s+1}U\|_{L^\infty(\mathbb{T})}\right)\|w\|_{H_g^{s,\gamma}}.
\end{aligned}
\tag{5.2.46}
$$

Proof Applying the operator ∂_x^s on $(5.2.45)_1$ and $(5.2.45)_2$, respectively, we obtain the following equations:

$$\begin{cases}
(\partial_t + u\partial_x + v\partial_y + 1 - \partial_y^2 - \epsilon^2\partial_x^2)\partial_x^s w + \partial_x^s v \partial_y w \\
= - \sum\limits_{0\le j<s} \binom{s}{j} \partial_x^{s-j} u \partial_x^{j+1} w - \sum\limits_{1\le j<s} \binom{s}{j} \partial_x^{s-j} v \partial_y \partial_x^j w, \\
(\partial_t + u\partial_x + v\partial_y + 1 - \partial_y^2 - \epsilon^2\partial_x^2)\partial_x^s (u - U) + \partial_x^s v \partial_y u \\
= - \sum\limits_{0\le j<s} \binom{s}{j} \partial_x^{s-j} u \partial_x^{j+1} (u - U) \\
- \sum\limits_{1\le j<s} \binom{s}{j} \partial_x^{s-j} v \partial_y \partial_x^j u - \sum\limits_{0\le j\le s} \binom{s}{j} \partial_x^j (u - U) \partial_x^{s-j+1} U.
\end{cases} \tag{5.2.47}$$

To overcome the difficult term $\partial_x^s v$, we use the cancellation property as in [98]. Indeed, eliminating $\frac{\partial_y w}{w} \times (5.2.47)_2$ from $(5.2.47)_1$, and letting $a(t, x, y) = \frac{\partial_y w}{w}$, it follows

$$\begin{aligned}
&(\partial_t + u\partial_x + v\partial_y + 1 - \partial_y^2) g_s + \partial_x^s (u - U)(\partial_t + u\partial_x + v\partial_y + 1 - \partial_y^2) a \\
&= 2\epsilon^2 \partial_x^{s+1} (u - U) \partial_x a + 2\partial_x^s w \partial_y a - \sum_{j=0}^{s-1} \binom{s}{j} g_{j+1} \partial_x^{s-j} u \\
&- \sum_{j=1}^{s-1} \binom{s}{j} \partial_x^{s-j} v \{\partial_y \partial_x^j w - a \partial_y \partial_x^j u\} \\
&+ a \sum_{j=0}^{s} \binom{s}{j} \partial_x^j (u - U) \partial_x^{s-j+1} U.
\end{aligned} \tag{5.2.48}$$

Applying the operator ∂_y on $(5.2.3)_1$ yields

$$(\partial_t + u\partial_x + v\partial_y + 1)\partial_y w = 2\epsilon^2 \partial_x \partial_y w + \partial_y^3 w - w\partial_x w + \partial_x u \partial_y w. \tag{5.2.49}$$

Using (5.2.49) and Eq. $(5.2.3)_1$, we get

$$\begin{aligned}
&(\partial_t + u\partial_x + v\partial_y + 1) a \\
&\quad = \frac{(\partial_t + u\partial_x + v\partial_y + 1)\partial_y w}{w} - \frac{\partial_y w (\partial_t + u\partial_x + v\partial_y + 1) w}{w^2} \\
&\quad = \epsilon^2 \frac{\partial_x^2 \partial_y w}{w} - \epsilon^2 a \frac{\partial_x^2 w}{w} + \frac{\partial_y^3 w}{w} - a \frac{\partial_y^2 w}{w} - g_1 + a\partial_x U,
\end{aligned} \tag{5.2.50}$$

and

$$\begin{cases} \partial_y a = \frac{\partial_y^2 w}{w} - \frac{\partial_y w \partial_y w}{w^2}, \\ \partial_y^2 a = \frac{\partial_y^3 w}{w} - a\frac{\partial_y^2 w}{w} - 2a\partial_y a. \end{cases} \tag{5.2.51}$$

Now inserting (5.2.50) and $(5.2.51)_2$ into (5.2.48) yields

$$\begin{aligned} &(\partial_t + u\partial_x + v\partial_y + 1 - \partial_y^2 - \epsilon^2\partial_x^2)g_s \\ &= 2\epsilon^2\big\{\partial_x^{x+1}(u-U) - \frac{\partial_x w}{w}\partial_x^2(u-U)\big\}\partial_x a + 2g_s\partial_y a - g_1\partial_x^s U \\ &\quad - \sum_{j=1}^{s-1}\binom{s}{j} g_{j+1}\partial_x^{s-j}u \\ &\quad - \sum_{j=1}^{s-1}\binom{s}{j}\partial_x^{s-j}v\{\partial_y\partial_x^j w - a\partial_y\partial_x^j u\} + a\sum_{j=0}^{s-1}\binom{s}{j}\partial_x^j(u-U)\partial_x^{s-j+1}U. \end{aligned} \tag{5.2.52}$$

Multiplying (5.2.52) by $(1+y)^{2\gamma}g_s$, and integrating it over $\mathbb{T}\times\mathbb{R}_+$,

$$\begin{aligned} &\frac{1}{2}\frac{d}{dt}\|(1+y)^\gamma g_s\|_{L^2}^2 \\ &+ \|(1+y)^\gamma g_s\|_{L^2}^2 + \|(1+y)^\gamma \partial_y g_s\|_{L^2}^2 + \epsilon^2\|(1+y)^\gamma\partial_x g_s\|_{L^2}^2 \\ &= 2\epsilon^2\iint(1+y)^{2\gamma}g_s\big\{\partial_x^{x+1}(u-U) - \frac{\partial_x w}{w}\partial_x^2(u-U)\big\}\partial_x a \\ &\quad + \int_{\mathbb{T}} g_s\partial_y g_s dx|_{y=0} - 2\gamma\iint(1+y)^{2\gamma-1}g_s\partial_y g_s \\ &\quad + \gamma\iint(1+y)^{2\gamma-1}v|g_s|^2 + 2\iint(1+y)^{2\gamma}|g_s|^2\partial_y a \\ &\quad - \sum_{j=1}^{s-1}\binom{s}{j}\iint(1+y)^{2\gamma}g_s g_{j+1}\partial_x^{s-j}u \\ &\quad - \sum_{j=1}^{s-1}\binom{s}{j}\iint(1+y)^{2\gamma}g_s\partial_x^{s-j}v\{\partial_y\partial_x^j w - a\partial_x^j w\} \\ &\quad - \iint(1+y)^{2\gamma}g_s g_1\partial_x^s U + \sum_{j=0}^{s-1}\binom{s}{j}\iint(1+y)^{2\gamma}g_s a\partial_x^j(u-U)\partial_x^{s-j+1}U, \end{aligned} \tag{5.2.53}$$

which can be obtained by integration by parts and the boundary condition. We estimate equation (5.2.53) by terms. Firstly $w \in C([0,T]; H_{\sigma,\delta}^{s+4,\gamma})$, it follows from the definition of $H^{s+4,\gamma}_{\sigma,\delta}$ that $(1+y)^\sigma w \geq \delta$ and $|(1+y)^{\sigma+\alpha} D^\alpha w| \leq \delta^{-1}$ for all $|\alpha| \leq 2$. Thus, we have $\|(1+y)\partial_x a\|_{L^\infty} \leq \delta^{-2} + \delta^{-4}$ and $\|\frac{\partial_x w}{w}\|_{L^\infty} \leq \delta^{-2}$, and hence

$$
\begin{aligned}
&2\epsilon^2 \iint (1+y)^{2\gamma} g_s \{\partial_x^{x+1}(u-U) - \frac{\partial_x w}{w} \partial_x^2 (u-U)\} \partial_x a \\
&\leq 2\epsilon^2 C_\delta \|(1+y)^\gamma g_s\|_{L^2} (\|(1+y)^{\gamma-1} \partial_x^{s+1}(u-U)\|_{L^2} \\
&\quad + \|(1+y)^{\gamma-1} \partial_x^s (u-U)\|_{L^2}).
\end{aligned} \tag{5.2.54}
$$

In addition, using Lemma 5.1.3 and the following estimate:

$$
\begin{cases}
|w|^{-1} \leq \delta^{-1}(1+y)^\sigma, \\
|\partial_y w| \leq \delta^{-1}(1+y)^{-\sigma-1}, \quad |\partial_y^2 w| \leq \delta^{-1}(1+y)^{-\sigma-2},
\end{cases} \tag{5.2.55}
$$

we have

$$
\begin{aligned}
&\|(1+y)^{\gamma-1} \partial_x^{s+1}(u-U)\|_{L^2} \\
&\quad \leq C_{\gamma,\sigma,\delta} \|(1+y)^{\gamma-\sigma-1} \frac{\partial_x^{s+1}(u-U)}{w}\|_{L^2} \\
&\quad \leq C_{\gamma,\sigma,\delta} \left(\|\partial_x^{s+1} U\|_{L^2(\mathbb{T})} + \|(1+y)^\gamma \partial_x g_s\|_{L^2} + \|(1+y)^{\gamma-1} \partial_x^s (u-U)\|_{L^2} \right),
\end{aligned} \tag{5.2.56}
$$

and

$$
\|(1+y)^{\gamma-1} \frac{\partial_x w}{w} \partial_x^s (u-U)\|_{L^2} \leq C_{\gamma,\sigma,\delta} \|(1+y)^{\gamma-1} \partial_x^s (u-U)\|_{L^2}. \tag{5.2.57}
$$

Then we can obtain

$$
\begin{aligned}
&\left| 2\epsilon^2 \iint (1+y)^{2\gamma} g_s \{\partial_x^{x+1}(u-U) - \frac{\partial_x w}{w} \partial_x^2 (u-U)\} \partial_x a \right| \\
&\leq \frac{1}{2} \epsilon^2 \|(1+y)^\gamma \partial_x g_s\|_{L^2}^2 + \epsilon^2 C_{s,\gamma,\sigma,\delta} \left(\|w\|_{H_g^{s,\gamma}} + \|\partial_x^{s+1} U\|_{L^2(\mathbb{T})} \right) \|w\|_{H_g^{s,\gamma}}^2.
\end{aligned} \tag{5.2.58}
$$

Secondly, we have the fact that

$$
\begin{aligned}
\partial_y g_s|_{y=0} &= \left(\partial_x^s \partial_y w - \frac{\partial_y w}{w} \partial_x^s w - \frac{\partial_y^2 w}{w} \partial_x^s (u-U) + \frac{\partial_y w \partial_y w}{w^2} \partial_x^s (u-U) \right) |_{y=0} \\
&= \left(\partial_x^s (\partial_x p - U) - a g_s + \frac{\partial_y^2 w}{w} \partial_x^s U \right) |_{y=0}.
\end{aligned} \tag{5.2.59}
$$

Then according to the trace theorem, we get

$$\begin{aligned}
&\left|\int_{\mathbb{T}} g_s\partial_y g_s dx|_{y=0}\right|\\
&\le \iint \left|ag_s^2\right| dxdy + \iint \left|\partial_y a g_s^2\right| dxdy + 2\iint \left|ag_s\cdot\partial_y g_s\right| dxdy\\
&\quad + \iint (\partial_y g_s + g_s)\partial_x^s(\partial_x p - U)dxdy + \iint g_s(\frac{\partial_y^2 w}{w}\\
&\quad + \frac{\partial_y^3 w}{w} - \frac{\partial_y^2 w\partial_y w}{w^2})\partial_x^s U dxdy\\
&\quad + \iint \partial_y g_s\frac{\partial_y^2 w}{w}\partial_x^s U dxdy\\
&\le C_{\gamma,\sigma}\|w\|^2_{H_g^{s,\gamma}}(1+\|\partial_x^s U\|^2_{L^\infty(\mathbb{T})}) + C\|\partial_x^s(\partial_x p - U)\|^2_{L^2(\mathbb{T})}\\
&\quad + \frac{1}{4}\|(1+y)^\gamma\partial_y g_s\|^2_{L^2},
\end{aligned} \tag{5.2.60}$$

which with (5.2.55) gives the facts that $\|a\|_{L^\infty}\le\delta^{-2}$ and $\|\partial_y a\|_{L^\infty}\le\delta^{-2}+\delta^{-4}$. Next, using Hölder inequality, we obtain

$$\begin{aligned}
\left|2\gamma\iint(1+y)^{2\gamma-1}g_s\partial_y g_s\right| &\le 2\gamma\|\frac{1}{1+y}\|_{L^\infty}\|(1+y)^\gamma g_s\|_{L^2}\|(1+y)^\gamma\partial_y g_s\|_{L^2}\\
&\le C_\gamma\|w\|^2_{H_g^{s,\gamma}} + \frac{1}{4}\|(1+y)^\gamma\partial_y g_s\|^2_{L^2},
\end{aligned} \tag{5.2.61}$$

and using Lemma 5.1.6

$$\begin{aligned}
\left|\gamma\iint(1+y)^{2\gamma-1}v|g_s|^2\right| &\le \|\frac{v}{1+y}\|_{L^\infty}\|(1+y)^\gamma g_s\|^2_{L^2}\\
&\le C_{s,\gamma,\sigma,\delta}(\|w\|_{H_g^{s,\gamma}} + \|\partial_x^s U\|_{L^2(\mathbb{T})})\|w\|^2_{H_g^{s,\gamma}}.
\end{aligned} \tag{5.2.62}$$

It follows from to (5.2.55) and (5.2.51) that

$$\left|2\iint(1+y)^{2\gamma}|g_s|^2\partial_y a\right| \le 2\|\partial_y a\|_{L^\infty}\|(1+y)^\gamma g_s\|^2_{L^2} \le C_\delta\|w\|^2_{H_g^{s,\gamma}}, \tag{5.2.63}$$

and

$$\begin{aligned}
&\left|\sum_{j=1}^{s-1}\binom{s}{j}\iint(1+y)^{2\gamma}g_s g_{j+1}\partial_x^{s-j}u\right|\\
&\quad\le\sum_{j=1}^{s-1}\binom{s}{j}\|\partial_x^{s-j}u\|_{L^\infty}\|(1+y)^\gamma g_s\|_{L^2}\|(1+y)^\gamma g_{j+1}\|_{L^2}\\
&\quad\le C_{s,\gamma,\sigma,\delta}(\|w\|_{H_g^{s,\gamma}}+\|\partial_x^s U\|_{L^2(\mathbb{T})})^2\|w\|_{H_g^{s,\gamma}}.
\end{aligned}\tag{5.2.64}$$

Lastly, for $j=2,3,\cdots,s-1$,

$$\begin{aligned}
&\left|\sum_{j=1}^{s-1}\binom{s}{j}\iint(1+y)^{2\gamma}g_s\partial_x^{s-j}v\{\partial_y\partial_x^j w-a\partial_x^j w\}\right|\\
&\le\sum_{j=1}^{s-1}\binom{s}{j}\|(1+y)^\gamma g_s\|_{L^2}\|\frac{\partial_x^{s-j}v}{1+y}\|_{L^\infty}\Big(\|(1+y)^{\gamma+1}\partial_y\partial_x^j w\|_{L^2}\\
&\qquad+\|(1+y)a\|_{L^\infty}\|(1+y)^\gamma\partial_x^j w\|_{L^2}\Big)\\
&\le C_{s,\gamma,\sigma,\delta}(\|w\|_{H_g^{s,\gamma}}+\|\partial_x^s U\|_{L^2(\mathbb{T})})\|w\|^2_{H_g^{s,\gamma}},
\end{aligned}\tag{5.2.65}$$

which with (5.2.55) gives the fact that $\|(1+y)a\|_{L^\infty}\le\delta^{-2}$. And for $j=1$,

$$\begin{aligned}
&\left|\iint(1+y)^{2\gamma}g_s\partial_x^{s-1}v\{\partial_y\partial_x w-a\partial_x w\}\right|\\
&\le\|\frac{\partial_x^{s-1}v+y\partial_x^s U}{1+y}\|_{L^2}\Big(\|(1+y)^{\gamma+1}\partial_y\partial_x w\|_{L^\infty}\\
&\quad+\|(1+y)a\|_{L^\infty}\|(1+y)^\gamma\partial_x w\|_{L^\infty}\Big)\|(1+y)^\gamma g_s\|_{L^2}\\
&\quad+\|\partial_x^s U\|_{L^\infty(\mathbb{T})}\Big(\|(1+y)^{\gamma+1}\partial_y\partial_x w\|_{L^2}+\|(1+y)a\|_{L^\infty}\|(1+y)^\gamma\partial_x w\|_{L^2}\Big)\\
&\quad\times\|(1+y)^\gamma g_s\|_{L^2}\\
&\le C_{s,\gamma,\sigma,\delta}\|w\|^2_{H_g^{s,\gamma}}(\|w\|_{H_g^{s,\gamma}}+\|\partial_x^s U\|_{L^\infty(\mathbb{T})}).
\end{aligned}\tag{5.2.66}$$

For the term $|\iint(1+y)^{2\gamma}g_s g_1\partial_x^s U|$, we have

$$\begin{aligned}
&\left|\iint (1+y)^{2\gamma} g_s g_1 \partial_x^s U\right| \\
&\le \|(1+y)^\gamma g_s\|_{L^2}\|(1+y)^\gamma g_1\|_{L^2}\|\partial_x^s U\|_{L^\infty(\mathbb{T})} \\
&\le C_{s,\gamma,\sigma,\delta}\|\partial_x^s U\|_{L^\infty(\mathbb{T})}(\|w\|_{H_g^{s,\gamma}} + \|\partial_x^s U\|_{L^2(\mathbb{T})})\|w\|_{H_g^{s,\gamma}}.
\end{aligned} \tag{5.2.67}$$

For the last term, we have

$$\begin{aligned}
&\left|\iint (1+y)^{2\gamma} g_s a \partial_x^j (u-U)\partial_x^{s-j+1} U\right| \\
&\le \|(1+y)^\gamma g_s\|_{L^2}\|(1+\gamma)a\|_{L^\infty}\|(1+y)^{\gamma-1}\partial_x^j(u-U)\|_{L^2}\|\partial_x^{s-j+1}U\|_{L^\infty(\mathbb{T})} \\
&\le C_{s,\gamma,\sigma,\delta}\|\partial_x^{s+1} U\|_{L^\infty(\mathbb{T})}(\|w\|_{H_g^{s,\gamma}} + \|\partial_x^s U\|_{L^2(\mathbb{T})})\|w\|_{H_g^{s,\gamma}}.
\end{aligned} \tag{5.2.68}$$

Hence combining (5.2.60)–(5.2.66) and (5.2.53) leads to (5.2.46). □

5.2.3 Weighted H^s Estimate on w

Now we can derive the weighted H^s estimate on w by employing Propositions 5.2.1 and 5.2.2.

Proposition 5.2.3 *Under the same assumption of Proposition 5.2.1, we have the following estimate:*

$$\begin{aligned}
\|w\|_{H_g^{s,\gamma}}^2 \le &\left\{\|w_0\|_{H_g^{s,\gamma}}^2 + \int_0^t F(\tau)d\tau\right\} \\
&\times \left\{1 - C(\frac{s}{2}-1)\left(\|w_0\|_{H_g^{s,\gamma}}^2 + \int_0^t F(\tau)d\tau\right)^{\frac{s-2}{2}} t\right\}^{-\frac{2}{s-2}},
\end{aligned} \tag{5.2.69}$$

where $C > 0$ is a constant independent of ϵ and t. The function $F(t)$ is expressed by

$$F(t) = \mathcal{P}(\|\partial_x^{s+1}U\|_{L^\infty(\mathbb{T})}) + C_s \sum_{l=0}^{\frac{s}{2}} \|\partial_t^l(\partial_x P - U)\|_{H^{s-2l}(\mathbb{T})}^2, \tag{5.2.70}$$

and $\mathcal{P}(\cdot)$ denotes a polynomial.

Proof According to Propositions 5.2.1 and 5.2.2, we know from the definition of $\|\cdot\|_{H_g^{s,\gamma}}$ that

$$\begin{aligned}
\frac{d}{dt}\|w\|^2_{H_g^{s,\gamma}} \le\ & C_{s,\gamma,\sigma,\delta}\left(1+\|w\|_{H_g^{s,\gamma}}+\|\partial_x^s U\|_{L^\infty(\mathbb{T})}\right)\\
&\times\left(\|w\|_{H_g^{s,\gamma}}+\|\partial_x^{s+1}U\|_{L^\infty(\mathbb{T})}\right)\|w\|_{H_g^{s,\gamma}}\\
&+C_{\gamma,\delta}\|\partial_x^s U\|^2_{L^\infty(\mathbb{T})}\|w\|^2_{H_g^{s,\gamma}}+C_{s,\gamma,\sigma,\delta}(1+\|w\|_{H_g^{s,\gamma}})^{s-2}\|w\|^2_{H_g^{s,\gamma}}\\
&+C_s\sum_{l=0}^{\frac{s}{2}}\|\partial_t^l(\partial_x P-U)\|^2_{H^{s-2l}(\mathbb{T})}\\
\le\ & C_{s,\gamma,\sigma,\delta}\|w\|^s_{H_g^{s,\gamma}}+\mathcal{P}(\|\partial_x^{s+1}U\|_{L^\infty(\mathbb{T})})\\
&+C_s\sum_{l=0}^{\frac{s}{2}}\|\partial_t^l(\partial_x P-U)\|^2_{H^{s-2l}(\mathbb{T})},
\end{aligned}\tag{5.2.71}$$

and hence, it follows by using the comparison principle of ordinary differential equations that

$$\begin{aligned}
\|w\|^2_{H_g^{s,\gamma}} \le\ & \left\{\|w_0\|^2_{H_g^{s,\gamma}}+\int_0^t F(\tau)d\tau\right\}\\
&\times\left\{1-C(\frac{s}{2}-1)\left(\|w_0\|^2_{H_g^{s,\gamma}}+\int_0^t F(\tau)d\tau\right)^{\frac{s-2}{2}}t\right\}^{-\frac{2}{s-2}},
\end{aligned}\tag{5.2.72}$$

provided that

$$1-C(\frac{s}{2}-1)\left(\|w_0\|^2_{H_g^{s,\gamma}}+\int_0^t F(\tau)d\tau\right)^{\frac{s-2}{2}}t>0,\tag{5.2.73}$$

this proves Proposition 5.2.3. □

5.2.4 Weighted L^∞ Estimates on Lower-Order Terms

In this subsection, we will estimate the weighted L^∞ on $D^\alpha w$ for $|\alpha|\le 2$ by using the classical maximum principles. More precisely, we will derive two parts: an L^∞-estimate on $I := \sum_{|\alpha|\le 2}|(1+y)^{\gamma+\alpha_2}D^\alpha w|_{L^2}$ and a lower bound estimate on $B_{(0,0)} := (1+y)^\sigma w$.

Lemma 5.2.3 *Under the same assumption of Proposition 5.2.1, we have the following estimate:*
For any $s \geq 4$,

$$\|I(t)\|_{L^\infty(\mathbb{T}\times\mathbb{R}^+)} \leq \max\left\{\|I(0)\|_{L^\infty(\mathbb{T}\times\mathbb{R}^+)}, 6C^2W(t)^2\right\} e^{C(1+G(t))t}, \quad (5.2.74)$$

and for any $s \geq 6$,

$$\begin{aligned}&\|I(t)\|_{L^\infty(\mathbb{T}\times\mathbb{R}^+)}\\ &\leq \left\{\|I(0)\|_{L^\infty(\mathbb{T}\times\mathbb{R}^+)} + C(1+W(t))W(t)^2 t\right\} e^{C(1+G(t))t}.\end{aligned} \quad (5.2.75)$$

In addition, if $s \geq 4$, we also have

$$\begin{aligned}\min_{\mathbb{T}\times\mathbb{R}^+}(1+y)^\sigma w(t) &\geq \max\left\{1 - C\left(1+G(t)\right) t e^{C(1+G(t))t}\right\}\\ &\quad\cdot \min\left\{\min_{\mathbb{T}\times\mathbb{R}^+}(1+y)^\sigma w_0 - CW(t)t\right\},\end{aligned} \quad (5.2.76)$$

where positive constant C depends on s, γ, σ, and δ only. The functions W and $G : [0, T] \to \mathbb{R}^+$ are respectively defined by

$$\begin{aligned}G(t) &:= \sup_{[0,t]}\|w(\tau)\|_{H_g^{s,\gamma}} + \sup_{[0,t]}\|\partial_x^s U(\tau)\|_{L^2(\mathbb{T})} \text{ and}\\ W(t) &:= \sup_{[0,t]}\|w(\tau)\|_{H_g^{s,\gamma}}.\end{aligned} \quad (5.2.77)$$

Proof For simplicity, we denote,

$$I := \sum_{|\alpha|\leq 2}\left|(1+y)^{\sigma+\alpha_2} D^\alpha w\right|^2$$

and

$$B_\alpha := (1+y)^\sigma D^\alpha w,$$

then $B_{(0,0)} = (1+y)^\gamma w$. By direct computations, we obtain

$$\left(\partial_t + u\partial_x + v\partial_y - \partial_y^2 - \epsilon^2\partial_x^2\right) B_\alpha = \sum_{i=1}^3 S_i, \quad (5.2.78)$$

where

$$S_1 = \left(\frac{\sigma+\alpha_2}{1+y}v + \frac{(\sigma+\alpha_2)(\sigma+\alpha_2-1)}{(1+y)^2} - 1\right)B_\alpha, \quad S_2 = -\frac{2(\sigma+\alpha_2)}{1+y}\partial_y B_\alpha,$$

and

$$S_3 = -\sum_{0<\beta\leq\alpha}\binom{\alpha}{\beta}\left\{(1+y)^{\beta_2}\left(D^\beta u B_{\alpha-\beta+e_1} + \frac{D^\beta v B_{\alpha-\beta+e_2}}{1+y}\right)\right\}.$$

Multiplying the Eq. (5.2.78) by $2B_\alpha$, we conclude

$$\begin{aligned}
&\left(\partial_t + u\partial_x + v\partial_y - \epsilon^2\partial_x^2 - \partial_y^2\right) I \\
&\leq 2\sum_{|\alpha|\leq 2}(|S_1 B_\alpha| + |S_2 B_\alpha| \\
&\quad + |S_3 B_\alpha| - \epsilon^2|\partial_x B_\alpha|^2 - |\partial_y B_\alpha|^2\Big) \\
&\leq C_{s,\gamma,\sigma,\delta}(1 + \|w(s)\|_{H_g^{s,\gamma}} + \|\partial_x^s U\|_{L^2(\mathbb{T})})I + C_\delta I \\
&\quad + \sum_{|\alpha|\leq 2}|\partial_y B_\alpha|^2 - 2\epsilon^2\sum_{|\alpha|\leq 2}|\partial_x B_\alpha|^2 \\
&\quad -2\sum_{|\alpha|\leq 2}|\partial_y B_\alpha|^2 + 2\|(1+y)^{\beta_2}\left(D^\beta u + \frac{D^\beta v}{1+y}\right)\|_{L^\infty} \\
&\quad \times \sum_{|\alpha|\leq 2}\left(\sum_{0\leq\beta<\alpha}|B_{\alpha-\beta+e_1} + B_{\alpha-\beta+e_2}|\right)B_\alpha \\
&\leq C_{s,\gamma,\sigma,\delta}(1 + \|w(s)\|_{H_g^{s,\gamma}} + \|\partial_x^s U\|_{L^2(\mathbb{T})})I,
\end{aligned} \tag{5.2.79}$$

where we have used Lemma 5.1.6 and Young inequality. Applying the classical maximum principle (see Lemma 5.1.7, for instance) for parabolic equations, we have

$$\begin{aligned}
&\|I(t)\|_{L^\infty(T\times\mathbb{R}^+)} \\
&\leq \max\Big\{e^{C(1+G(t))t}\|I(0)\|_{L^\infty(T\times\mathbb{R}^+)}, \\
&\times \max_{\tau\in[0,t]}\left(e^{C(1+G(t))(t-\tau)}\|I(\tau)|_{y=0}\|_{L^\infty(\mathbb{T})}\right)\Big\}.
\end{aligned} \tag{5.2.80}$$

To derive a lower bound estimate on $B_{(0,0)}$, we have

$$\begin{aligned}&\left(\partial_t + u\partial_x + (v + \frac{2\sigma}{1+y})\partial_y - \partial_y^2 - \epsilon^2\partial_x^2\right) B_{(0,0)}\\&= \left(\frac{\sigma}{1+y}v + \frac{\sigma(\sigma-1)}{(1+y)^2} - 1\right) B_{(0,0)}.\end{aligned} \tag{5.2.81}$$

Applying the classical minimum principle (see Lemma 5.1.8, for instance) for (5.2.81), we get

$$\begin{aligned}&\min_{\mathbb{T}\times\mathbb{R}^+} (1+y)^\sigma w(t)\\&\geq \max\left\{1 - C(1+G(t))\, te^{C(1+G(t))t}\right\}\\&\cdot \min\left\{\min_{\mathbb{T}\times\mathbb{R}^+}(1+y)^\sigma w_0,\ \min_{[0,t]\times\mathbb{T}} w|_{y=0}\right\}.\end{aligned} \tag{5.2.82}$$

Next we start estimating the boundary values, using Lemma 5.1.5, we obtain

$$\begin{aligned}\|I(\tau)|_{y=0}\|_{L^\infty(\mathbb{T})} &\leq 3C^2\sum_{|\alpha|\leq 2}\left(\|D^\alpha w\|_{L^2}^2 + \|\partial_x D^\alpha w\|_{L^2}^2 + \|\partial_y^2 D^\alpha w\|_{L^2}^2\right)\\&\leq 6C^2\|w\|_{H_g^{s,\gamma}}^2,\end{aligned} \tag{5.2.83}$$

which, together with (5.2.80), gives (5.2.74).

According to (5.1.5) and boundary condition $\partial_x^{\alpha_1}u = \partial_x^{\alpha_1}v = 0$, we have

$$\partial_t D^\alpha w|_{y=0} = D^\alpha(\epsilon^2\partial_x^2 + \partial_y^2 w - w - u\partial_x w - v\partial_y w)|_{y=0}. \tag{5.2.84}$$

When $\alpha_2 = 0$ and $\alpha_1 \leq 2$,

$$\partial_t D^\alpha w|_{y=0} = (\epsilon^2\partial_x^{\alpha_1+2}w + \partial_x^{\alpha_1}\partial_y^2 w - \partial_x^{\alpha_1}w)|_{y=0}. \tag{5.2.85}$$

When $1 \leq \alpha_2 \leq 2$ and $\alpha_1 = 0$,

$$\begin{aligned}\partial_t D^\alpha w|_{y=0} &= \left(\epsilon^2\partial_y^{\alpha_2}\partial_x^2 w + \partial_y^{\alpha_2}\partial_y^2 w - \partial_y^{\alpha_2}w - \partial_y^{\alpha_2}(u\partial_x w + v\partial_y w)\right)|_{y=0}\\&= \left(\epsilon^2\partial_y^{\alpha_2}\partial_x^2 w + \partial_y^{\alpha_2}\partial_y^2 w - \partial_y^{\alpha_2}w - \alpha_2 w\partial_x\partial_y^{\alpha_2-1}w\right)|_{y=0}.\end{aligned} \tag{5.2.86}$$

When $\alpha_2 = 1$ and $\alpha_1 = 1$,

$$\partial_t D^\alpha w|_{y=0} = (\epsilon^2\partial_y\partial_x^3 w + \partial_x\partial_y^3 w - \partial_x\partial_y w - \partial_x w\partial_x w - w\partial_x^2 w)|_{y=0}. \tag{5.2.87}$$

For $s \geq 6$, using Lemma 5.1.6, we have

$$\begin{aligned}\|\partial_t I|_{y=0}\|_{L^\infty(\mathbb{T})} &\leq C_{s,\gamma}\|D^\alpha w \partial_t D^\alpha w|_{y=0}\|_{L^\infty(\mathbb{T})}\\ &\leq C_{s,\gamma}\|w(\tau)\|_{H_g^{s,\gamma}}(\|w(\tau)\|_{H_g^{s,\gamma}} + \|w(\tau)\|^2_{H_g^{s,\gamma}}).\end{aligned} \tag{5.2.88}$$

Hence, a direct integration yields

$$\|I(t)|_{y=0}\|_{L^\infty(\mathbb{T})} \leq \|I(0)|_{y=0}\|_{L^\infty(\mathbb{T})} + C_{s,\gamma}(1 + W(t))W(t)^2 t, \tag{5.2.89}$$

which, together with (5.2.80), gives (5.2.75). For $s \geq 4$,

$$\begin{aligned}\|\partial_t w|_{y=0}\|_{L^\infty(\mathbb{T})} &= \|(\partial_y^2 w - w)|_{y=0}\|_{L^\infty(\mathbb{T})}\\ &\leq CW(t).\end{aligned} \tag{5.2.90}$$

Hence, a direct integration gives

$$\min_{\mathbb{T}} w(t)|_{y=0} \geq \min_{\mathbb{T}} w_0|_{y=0} - CW(t)t, \tag{5.2.91}$$

which together with (5.2.82) implies (5.2.76). The proof of Lemma 5.2.3 is thus complete. □

5.3 Local-in-Time Existence and Uniqueness

5.3.1 Local-in-Time Existence

In this subsection, we go back to use the symbol $(u^\epsilon, v^\epsilon, w^\epsilon)$ instead of (u, v, w) from Sects. 5.2.1–5.2.4 to denote the solution to the regularized system (5.2.1). To obtain the local-in-time solution of the initial-boundary value problem (5.1.1)–(5.1.2), we will construct the solution to the Prandtl equations (5.1.1) by passing to the limit $\epsilon \to 0^+$ in the regularized Prandtl equations (5.2.1). We only sketch the proof into five steps and more details can be found in [98].

Step 1. According to the definition of F, assumption (5.1.6), and the regularized Bernoulli's law,

$$\|F\|_{L^\infty} \leq M < +\infty. \tag{5.3.1}$$

Thus we derive the uniform estimate for any $\epsilon \in [0, 1]$ and any $t \in [0, T_1]$

$$\|w^\epsilon\|_{H_g^{s,\gamma}} \leq 4\|w_0^\epsilon\|_{H_g^{s,\gamma}}, \tag{5.3.2}$$

provided that T_1 is chosen by (5.2.69) such that

$$T_1 := \min\left\{\frac{3\|w_0\|^2_{H_g^{s,\gamma}}}{C_{s,\gamma,\sigma,\delta}M}, \frac{1-2^{2-s}}{2^{s-2}C_{s,\gamma,\sigma,\delta}\|w_0\|^{s-2}_{H_g^{s,\gamma}}}\right\}.$$

Step 2. When $s \geq 6$, we know from definition (5.2.77) of Ω and G that for any $t \in [0, T_1]$,

$$\Omega(t) \leq 4\|w_0^\epsilon\|_{H_g^{s,\gamma}} \quad and \quad G(t) \leq 4\|w_0^\epsilon\|_{H_g^{s,\gamma}} + M. \tag{5.3.3}$$

Thus, if we choose

$$T_2 := \min\left\{T_1, \frac{1}{64\delta^2 C_{s,\gamma}(1+4\|w_0\|_{H_g^{s,\gamma}})\|w_0\|^2_{H_g^{s,\gamma}}}, \right.$$
$$\left. \times \frac{\ln 2}{C_{s,\gamma,\sigma,\delta}(1+4\|w_0\|_{H_g^{s,\gamma}}+M)}\right\},$$

then using inequality (5.2.75) and initial assumption

$$\sum_{|\alpha|\leq 2} |(1+y)^{\sigma+\alpha_2} D^\alpha w|^2 \leq \frac{1}{4\delta^2}, \tag{5.3.4}$$

we have the upper bound

$$\|\sum_{|\alpha|\leq 2} |(1+y)^{\sigma+\alpha_2} D^\alpha w^\epsilon(t)|^2\|_{L^\infty(\mathbb{T}\times\mathbb{R}^+)} \leq \frac{1}{\delta^2} \tag{5.3.5}$$

for all $t \in [0, T_2]$. When $s \geq 4$, from the initial hypothesis $\|\omega_0\| \leq C\delta^{-1}$, we have the same estimate (5.3.5) for all $t \in [0, T_2]$.

Step 3. Let us choose

$$T_3 := \min\left\{T_1, \frac{\delta}{8C_{s,\gamma}\|w_0\|_{H_g^{s,\gamma}}}, \frac{1}{6C_{s,\gamma,\sigma,\delta}N}, \frac{\ln 2}{C_{s,\gamma,\sigma,\delta}N}\right\},$$

where $N := 1 + 4\|w_0\|_{H_g^{s,\gamma}} + M$. Then by using (5.2.77) and (5.3.3), we derive the uniform estimate for any $\epsilon \in [0, 1]$ and any $t \in [0, T_3]$,

$$\min_{\mathbb{T}\times\mathbb{R}^+} (1+y)^\sigma w^\epsilon(t) \geq \delta. \tag{5.3.6}$$

Step 4. In summary, the above uniform estimates hold for any $t \in [0, T]$, if T is chosen to satisfy $T = \min\{T_1, T_2, T_3\}$. Further using almost equivalence relation (5.1.7), we have

$$\sup_{0\le t\le T}(\|w^\epsilon\|_{H^{s,\gamma}} + \|u^\epsilon - U\|_{H^{s,\gamma-1}}) \le C\left(4\|w_0\|_{H_g^{s,\gamma}} + \sup_{0\le t\le T}\|\partial_x^s U\|_{L^2}\right) < +\infty. \tag{5.3.7}$$

From the Eq. (5.2.1), systems (5.2.3) and (5.3.7), and Lemma 5.1.5, we also have $\partial_t w^\epsilon$ and $\partial_t(u^\epsilon - U)$ are uniformly bounded in $L^\infty([0, T]; H^{s-2,\gamma})$ and $L^\infty([0, T]; H^{s-2,\gamma-1})$, respectively. By Lions-Aubin lemma and the compact embedding of $H^{s,\gamma}$ in $H^{s'}_{loc}$, we have after taking a subsequence, as $\epsilon_k \to 0^+$,

$$\begin{cases} \omega^{\epsilon_k} \overset{*}{\rightharpoonup} \omega, & \text{in } L^\infty([0, T]; H^{s,\gamma}), \\ \omega^{\epsilon_k} \to \omega, & \text{in } C([0, T]; H^{s'}_{loc}), \\ u^{\epsilon_k} - U \overset{*}{\rightharpoonup} u - U, & \text{in } L^\infty([0, T]; H^{s,\gamma-1}), \\ u^{\epsilon_k} \to u, & \text{in } C([0, T]; H^{s'}_{loc}), \end{cases} \tag{5.3.8}$$

for all $s' < s$, where

$$\begin{cases} w = \partial_y u \in L^\infty([0, T]; H^{s,\gamma}) \cap \bigcap_{s'<s} C([0, T]; H^{s'}_{loc}), \\ u - U \in L^\infty([0, T]; H^{s,\gamma-1}) \cap \bigcap_{s'<s} C([0, T]; H^{s'}_{loc}). \end{cases} \tag{5.3.9}$$

Step 5. Using the local uniform convergence of $\partial_x u^{\epsilon_k}$, we also have the pointwise convergence of v^{ϵ_k} : as $\epsilon \to 0^+$,

$$v^{\epsilon_k} = -\int_0^y \partial_x u^{\epsilon_k}\, dy \to -\int_0^y \partial_x u\, dy =: v. \tag{5.3.10}$$

Thus, passing the limit $\epsilon_k \to 0^+$ in the initial-boundary value problem (5.2.1), we get that the limit (u, v) solves the initial-boundary value problems (5.1.4) and (5.1.2) in the classical sense. Furthermore, we obtain (u, v, b) solves the initial-boundary value problems (5.1.1)–(5.1.2) in the classical sense. This completes the proof of the existence.

5.3.2 *Uniqueness of Solutions*

In this subsection, we are going to prove the uniqueness to 2D magnetic Prandtl model. Let us denote $(\bar{u} - \bar{v}) = (u_1, v_1) - (u_2, v_2)$, $\bar{w} = w_1 - w_2$, $\bar{b}_1 = b_{11} - b_{12}$

and $a_2 = \frac{\partial_y w_2}{w_2}$. It is easy to check that $\bar{g} = \bar{w} - a_2\bar{u} = w_2\partial_y(\frac{\bar{u}}{w_2})$ and the evolution equation on $\bar{g}$ is as follows:

$$
\begin{aligned}
(\partial_t + u_1\partial_x + v_1\partial_y - \partial_y^2 + 1)\bar{g} = &(\partial_t + u_1\partial_x + v_1\partial_y - \partial_y^2 + 1)\bar{w} \\
&- a_2(\partial_t + u_1\partial_x + v_1\partial_y - \partial_y^2 + 1)\bar{u} \\
&- \bar{u}(\partial_t + u_1\partial_x + v_1\partial_y - \partial_y^2 + 1)a_2 - 2\partial_y a_2\bar{w}.
\end{aligned}
\tag{5.3.11}
$$

Next, we calculate the values of the first three terms on the right-hand side of the equality, respectively; recalling the vorticity system (5.1.5), we have

$$
(\partial_t + u_i\partial_x + v_i\partial_y + 1)\partial_y w_i = \partial_y^3 w_i + \partial_x u_i\partial_y w_i - w_i\partial_x w_i, \tag{5.3.12}
$$

then, according to the definition of a_i, we get

$$
\begin{aligned}
(\partial_t + u_i\partial_x + v_i\partial_y + 1)a_i = &\frac{(\partial_t + u_i\partial_x + v_i\partial_y + 1)\partial_y w_i}{w_i} \\
&- \frac{\partial_y w_i(\partial_t + u_i\partial_x + v_i\partial_y + 1)w_i}{w_i^2} \\
= &\frac{\partial_y^3 w_i}{w_i} + a_i\partial_x u_i - \partial_x w_i - a_i\frac{\partial_y^2 a_i}{w_i},
\end{aligned}
\tag{5.3.13}
$$

which, combined with the fact

$$
\frac{\partial_y^3 w_i}{w_i} - a_i\frac{\partial_y^2 a_i}{w_i} = \partial_y^2 a_i + 2a_i\partial_y a_i,
$$

yields,

$$
(\partial_t + u_i\partial_x + v_i\partial_y - \partial_y^2 + 1)a_i = a_i\partial_x u_i - \partial_x w_i + 2a_i\partial_y a_i. \tag{5.3.14}
$$

Furthermore, we conclude that

$$
\begin{aligned}
(\partial_t + u_1\partial_x + v_1\partial_y - \partial_y^2 + 1)a_2 = &a_2\partial_x u_2 - \partial_x w_2 + 2a_2\partial_y a_2 \\
&+ (\bar{u}\partial_x + \bar{v}\partial_y)a_2.
\end{aligned}
\tag{5.3.15}
$$

For the estimate of $(\partial_t + u_1\partial_x + v_1\partial_y - \partial_y^2 + 1)\bar{u}$ and $(\partial_t + u_1\partial_x + v_1\partial_y - \partial_y^2 + 1)\bar{w}$, from the Eqs. (5.1.4) and (5.1.5), we can derive that

$$
(\partial_t + u_1\partial_x + v_1\partial_y - \partial_y^2 + 1)\bar{u} = -\bar{u}\partial_x u_2 - \bar{v}\partial_y u_2, \tag{5.3.16}
$$

and

$$(\partial_t + u_1\partial_x + v_1\partial_y - \partial_y^2 + 1)\bar{w} = -\bar{u}\partial_x w_2 - \bar{v}\partial_y w_2. \tag{5.3.17}$$

Using (5.3.11) and (5.3.15)–(5.3.17), we have

$$\begin{aligned}(\partial_t + u_1\partial_x + v_1\partial_y - \partial_y^2 + 1)\bar{g} &= -2\bar{w}\partial_y a_2 \\ &\quad - \bar{u}(\bar{u}\partial_x a_2 + \bar{v}\partial_y a_2 + 2a\partial_y a_2).\end{aligned} \tag{5.3.18}$$

Now we derive L^2 estimate on $\bar{g}$. For any $t \in (0, T]$, multiplying by $2\bar{g}$ and then integrating by parts over $\mathbb{T} \times \mathbb{R}_+$, we obtain

$$\begin{aligned}&\frac{d}{dt}\|\bar{g}\|_{L^2}^2 + 2\|\bar{g}\|_{L^2}^2 + 2\|\partial_y \bar{g}\|_{L^2}^2 \\ &= \int_{\mathbb{T}} \bar{g}\partial_y\bar{g}|_{y=0}dx - 2\iint \bar{g}\bar{u}(\bar{u}\partial_x a_2 + \bar{v}\partial_y a_2 + 2a\partial_y a_2) \\ &\quad - 4\iint \bar{g}\bar{w}\partial_y a_2 \\ &\quad - 2\iint \bar{g}(u_1\partial_x\bar{g} + v_1\partial_y\bar{g}).\end{aligned} \tag{5.3.19}$$

We now need to estimate the integral equation (5.3.19) term by term. Indeed, applying the simple trace theorem and Young's inequality,

$$\begin{aligned}\left|\int_{\mathbb{T}} \bar{g}\partial_y\bar{g}|_{y=0}dx\right| &\le \left|\iint a_2|\bar{g}|^2 dxdy\right| + \left|\iint \partial_y a_2|\bar{g}|^2 dxdy\right| \\ &\quad + 2\left|\iint a_2\bar{g}\partial_y\bar{g}dxdy\right| \\ &\le \frac{1}{2}\|\partial_y\bar{g}\|_{L^2}^2 + C_{\sigma,\delta}\|\bar{g}\|_{L^2}^2,\end{aligned} \tag{5.3.20}$$

where we have used the fact $\partial_y\bar{g}|_{y=0} = -a_2\bar{g}|_{y=0}$ $(\partial_y\bar{w} = 0)$. We claim $\|\frac{\bar{u}}{1+y}\|_{L^2} \le C_{\sigma,\delta}\|\bar{g}\|_{L^2}$, so by Lemma 5.1.6,

$$\begin{aligned}&-2\iint \bar{g}\bar{u}(\bar{u}\partial_x a_2 + \bar{v}\partial_y a_2 + 2a\partial_y a_2) \\ &\le 2\|(1+y)(\bar{u}\partial_x a_2 + \bar{v}\partial_y a_2 + 2a\partial_y a_2)\|_{L^\infty}\left\|\frac{\bar{u}}{1+y}\right\|_{L^2}\|\bar{g}\|_{L^2} \\ &\le C_{\gamma,\sigma,\delta}(1 + \|w_i\|_{H_g^{4,\gamma}} + \|\partial_x^4 U\|_{L^2(\mathbb{T})})\|\bar{g}\|_{L^2}^2.\end{aligned} \tag{5.3.21}$$

Below we give the fact that $\|\frac{\bar{u}}{1+y}\|_{L^2}$ can be controlled by $\|\bar{g}\|_{L^2}$, since $\delta \le (1+y)^\delta w_2 \le \delta^{-1}$,

$$\|\frac{\bar{u}}{1+y}\|_{L^2} \le \delta^{-1}\|(1+y)^{-\sigma-1}\frac{\bar{u}}{w^2}\|_{L^2} \le C_{\sigma,\delta}\|(1+y)^{-\sigma}\partial(\frac{\bar{u}}{w_2})\|_{L_2} \\ \le C_{\sigma,\delta}\|\bar{g}\|_{L^2}. \tag{5.3.22}$$

in addition, we also have

$$\|\bar{w}\|_{L^2} \le \|\bar{g}\|_{L^2} + \delta^{-2}\|\frac{\bar{u}}{1+y}\|_{L^2} \le C_{\sigma,\delta}\|\bar{g}\|_{L^2}, \tag{5.3.23}$$

and

$$-4\iint \bar{g}\bar{w}\partial_y a_2 \le C_{\sigma,\delta}\|\bar{g}\|_{L^2}^2. \tag{5.3.24}$$

For the last term in (5.3.19), using the integration by parts, boundary condition $(u_1, v_1)|_{y=0}$ and $\partial_x u_1 + \partial_y v_1 = 0$, we easily show

$$-2\iint \bar{g}(u_1\partial_x\bar{g} + v_1\partial_y\bar{g}) = 0. \tag{5.3.25}$$

Combining all of the above estimates, we have

$$\frac{d}{dt}\|\bar{g}\|_{L^2}^2 \le C_{\gamma,\sigma,\delta}(1 + \|w_i\|_{H_g^{4,\gamma}} + \|\partial_x^4 U\|_{L^2(\mathbb{T})})\|\bar{g}\|_{L^2}^2. \tag{5.3.26}$$

which, by Gronwall's inequality, gives

$$\|\bar{g}(t)\|_{L^2}^2 \le \|\bar{g}(0)\|_{L^2}^2 e^{Ct}, \tag{5.3.27}$$

here, $C = C_{\gamma,\sigma,\delta}(1 + \|w_i\|_{H_g^{4,\gamma}} + \|\partial_x^4 U\|_{L^2(\mathbb{T})})$, and this implies $\bar{g} = 0$ due to $u_1|_{t=0} = u_2|_{t=0}$. Since $w_2\partial_y(\frac{u_1-u_2}{w_2}) = \bar{g} = 0$, we have

$$u_1 - u_2 = qw_2 \tag{5.3.28}$$

for some function $q = q(t,x)$. By using the Oleinik's monotonicity assumption $w_2 > 0$ and boundary condition $u_1|_{y=0} = u_2|_{y=0} = 0$, we can get $q = 0$, and hence $u_1 = u_2$. Further, using $\partial_x u + \partial_y v = 0$ and $\partial_y u + \partial_y^2 b_1 = 0$, then v_i and b_{1i} can be uniquely determined (i.e., $v_1 = v_2$ and $b_{11} = b_{12}$).

5.4 Bibliographic Comments

Now let us briefly review the background and corresponding results about the boundary layer. Actually, Prandtl proposed the basic rules for describing a phenomenon at the Heidelberg international mathematical conference in 1904. He pointed out that there are two regions in the flow of a solid: a thin layer near an object, viscous friction plays an important role; outside this thin layer, friction is negligible. Prandtl called this thin layer as the boundary layer, which can be described by the so-called Prandtl equations. Later on, the researches on the boundary layer were appeared, such as [13, 27]. But until 1999, Oleinik and Samokhin [106] gave a systematic study of the Prandtl equations from a mathematical point of view, which is the fundamental research on boundary layer system. The Prandtl system is obtained as a simplification of the Navier-Stokes system and describes the motion of a fluid with small viscosity about a solid body in a thin layer which is formed near its surface owing to the adhesion of the viscous fluid to the solid surface. The Prandtl boundary layer theory has been extensively studied in different references; see [3, 5, 6, 11, 12, 43, 115].

The first well-known result was developed by Oleinik and Samokhin in [106], where under the monotonicity condition on tangential velocity with respect to the normal variable to the boundary, the local (in time) well-posedness of Prandtl equations was obtained by using the Crocco transformation and von Mises transformation. Since then, Crocco transformation and von Mises transformation have been used in boundary layer problems. For example, Xin and Zhang [149] established a global existence of weak solutions to the two-dimensional Prandtl system for the pressure is favorable, which had generalized the local well-posedness results of Oleinik [106]. Gong et al. [46] investigated the local spatial existence of solutions for the compressible Navier-Stokes boundary layer equations by using von Mises transformation. The coordinate transformation was also used to MHD boundary layer in [58]. Xie and Yang in [148] also obtained the local existence of solutions to the MHD boundary layer system as a general shear flow. Thereafter, Xie and Yang [89, 90] and Gao et al. [35] studied the local well-posedness of MHD boundary layer problems by the energy method.

There are not many literature about the Prandtl-Hartmann regime, only [9, 24, 37, 47, 74, 83, 147], and even fewer literature that apply coordinate transformation to the equation of Prandtl-Hartmann regime except for [47] where Gong and Wang obtained the global existence of a weak solution to the mixed Prandtl-Hartmann boundary layer problem by introducing the Crocco transformation for the case that the pressure is favorable ($\partial_x P < 0$). The authors in [9, 24, 37, 74, 83, 147] studied the well-posedness of the Prandtl-Hartmann regime with the constant outer flow U which means $\partial_x P = 0$. For example, Dong and Qin [24](see also Chap. 2) established the global well-posedness of solutions in $H^s(1 \leq s \leq 4)$ with $U =$ constant and without monotonicity condition and a lower bound, and Qin and Dong [116] (see also Chap. chap4) also proved the local well-posedness of solutions to 2D mixed Prandtl equations in Sobolev space with $U =$ constant.

From above statement, a natural question is that whether or not the existence of local solutions can be achieved for the Prandtl-Hartmann regime equations in both favorable case ($\partial_x P < 0$) and infavorable case ($\partial_x P > 0$). However, the answer to this question is still open before the paper ([118]) has been finished. This means that the paper [118] (see also this chapter) has positively answered this question. In this direction, we would like to mention some works. Masmoudi and Wong [98] proved the local existence and uniqueness of solutions for the classical Prandtl system by the energy method in a polynomial weighted Sobolev space, and Fan et al. [29] proved the local well-posedness for the compressible Prandtl boundary layer equations by the same method. Motivated by [29, 98], our purpose of this chapter is to prove the local existence of solutions to the Prandtl-Hartmann regime model under the outer flow $U \neq$ constant. Although the method used here is the same as that in [29, 98], the Prandtl-Hartmann regime model has one more tangential magnetic components than Prandtl equations have, which leads to the boundary data $y = 0$ being much more complicated. To overcome this difficulty, some more delicated estimates and mathematical induction method are employed.

Chapter 6
Local Existence of Solutions to 3D Prandtl Equations with a Special Structure

In this chapter, we shall consider the 3D Prandtl equation in a periodic domain and prove the local existence and uniqueness of solutions by the energy method in a polynomial weighted Sobolev space. Compared to the existence and uniqueness of solutions to the classical Prandtl equations where the Crocco transform has always been used with the general outer flow $U \neq$ constant, this Crocco transform is not needed here for 3D Prandtl equations. We use the skill of cancellation mechanism and construct a new unknown function to show that the existence and uniqueness of solutions to 3D Prandtl equations (cf. Masmoudi and Wong, Comm. Pure Appl. Math., 68(10)(2015), 1683–1741) which extends from the two-dimensional case in [118] to the present three-dimensional case with a special structure. The content of this chapter is adapted from [117].

6.1 Introduction

Experimental data and theoretical analysis show that in many important practical cases, for fluids whose viscosity is small, there exists a thin transition layer near the boundary, in which the behavior of flow changes dramatically, this phenomenon is called boundary layer theory in [106]. In this chapter, we shall consider the 3D Prandtl equations in the periodic domain $\{(t, x, y, z)\big|t > 0, (x, y) \in \mathbb{T}^2, z \in \mathbb{R}^+\}$:

$$\begin{cases} \partial_t u + (u\partial_x + v\partial_y + w\partial_z)u + \partial_x P = \partial_z^2 u, \\ \partial_t v + (u\partial_x + v\partial_y + w\partial_z)v + \partial_y P = \partial_z^2 v, \\ \partial_x u + \partial_y v + \partial_z w = 0, \\ (u, v)|_{t=0} = (u_0(x, y, z), v_0(x, y, z)), \\ (u, v, w)|_{z=0} = 0, \\ \lim\limits_{z\to+\infty} (u, v) = (U(t, x, y), V(t, x, y)). \end{cases} \tag{6.1.1}$$

Y. Qin et al., *Prandtl Equations and Related Boundary Layer Equations*,
https://doi.org/10.1007/978-981-97-4565-4_6

Here $\mathbb{T}^2 \subseteq \mathbb{R}^2$, $(u, v, w) = (u(t, x, y, z), v(t, x, y, z), w(t, x, y, z))$ denote the velocity field, $(U(t, x, y), V(t, x, y))$ and P are the traces of the tangential velocity field and the pressure of the Euler flow, respectively, which satisfy Bernoulli's law

$$\begin{cases} \partial_t U + (U\partial_x + V\partial_y)U + \partial_x P = 0, \\ \partial_t V + (U\partial_x + V\partial_y)V + \partial_y P = 0. \end{cases} \tag{6.1.2}$$

Despite its importance in physics, there are very few mathematical results on the Prandtl equations in three space variables, and the three-dimensional results are based on special structures, such as [85, 86], where authors constructed a solution of the three-dimensional Prandtl equations (6.1.1) with the structure

$$(u(t, x, y, z), K(t, x, y)u(t, x, y, z), w(t, x, y, z)), \tag{6.1.3}$$

and the outer Euler flow takes the following form on the boundary: $\{z = 0\}$,

$$(U(t, x, y), K(t, x, y)U(t, x, y), 0).$$

But in what follows, we shall consider the following the equivalent system of system (6.1.1) (see Proposition 6.1.1 for the specific proof):

$$\begin{cases} \partial_t u + (u\partial_x + Ku\partial_y + w\partial_z)u = \partial_z^2 u^\epsilon - \partial_x P, \\ \partial_x u + \partial_y(Ku) + \partial_z w = 0, \\ u\big|_{t=0} = u_0(x, y, z), \\ (u, w)\big|_{z=0} = 0, \quad \lim_{z\to+\infty} u = U(t, x, y), \end{cases} \tag{6.1.4}$$

where U satisfies

$$\partial_t U + U\partial_x U + KU\partial_y U + \partial_x P = 0. \tag{6.1.5}$$

Hence, the regularized equivalent system of equations (6.1.4) can be expressed as follows:

$$\begin{cases} \partial_t u^\epsilon + (u^\epsilon\partial_x + K^\epsilon u^\epsilon\partial_y + w^\epsilon\partial_z)u^\epsilon - \epsilon^2\partial_x^2 u^\epsilon - \epsilon^2\partial_y^2 u^\epsilon = \partial_z^2 u^\epsilon - \partial_x P^\epsilon, \\ \partial_x u^\epsilon + \partial_y(K^\epsilon u^\varepsilon) + \partial_z w^\epsilon = 0, \\ u^\epsilon\big|_{t=0} = u_0(x, y, z), \\ (u^\epsilon, K^\epsilon u^\epsilon, w^\epsilon)\big|_{z=0} = 0, \quad \lim_{z\to+\infty} u^\epsilon = U(t, x, y) \end{cases} \tag{6.1.6}$$

with regularized Bernoulli's law

$$\partial_t U + U\partial_x U + K^\epsilon U\partial_y U - \epsilon^2\partial_x^2 U - \epsilon^2\partial_y^2 U + \partial_x P^\epsilon = 0. \tag{6.1.7}$$

Let the vorticity $\varphi^\epsilon = \partial_z u^\epsilon$, then equations (6.1.6) satisfy the following vorticity system: for any $\epsilon > 0$,

$$\begin{cases} \partial_t \varphi^\epsilon + (u^\epsilon \partial_x + K^\epsilon u^\epsilon \partial_y + w^\epsilon \partial_z)\varphi^\epsilon - \varepsilon \partial_x^2 \varphi^\epsilon - \varepsilon \partial_y^2 \varphi^\epsilon = \frac{1}{2}\partial_y K^\epsilon \partial_z |u^\epsilon|^2 + \partial_z^2 \varphi^\epsilon, \\ \partial_x u^\epsilon + \partial_y (K^\epsilon u^\epsilon) + \partial_z w^\epsilon = 0, \\ \varphi^\epsilon(0, x, y, z) = \partial_z u_0, \\ \partial_z \varphi^\epsilon|_{z=0} = \partial_x P^\epsilon, \end{cases} \tag{6.1.8}$$

where the velocity field (u^ϵ, w^ϵ) is given by

$$u^\epsilon(t, x, y, z) = U - \int_z^{+\infty} \varphi^\epsilon(t, x, y, \tilde{z}) d\tilde{z}, \tag{6.1.9}$$

and

$$w^\epsilon(t, x, y, z) = -\int_0^z \partial_x u^\epsilon(t, x, y, \tilde{z}) d\tilde{z} - \int_0^z \partial_y (K^\epsilon u^\epsilon(t, x, y, \tilde{z})) d\tilde{z}. \tag{6.1.10}$$

The three-dimensional Prandtl equation describes the relationship between the velocity and boundary shear stress in the boundary layer and is used to study the behavior and characteristics of fluids within the boundary layer. However, it is challenging to obtain solutions for the boundary layer due to the presence of complex three-dimensional boundary layer structures and secondary flows. Therefore, a specific structure $v(t, x, y, z) = K(t, x, y)u(t, x, y, z)$ is employed to study the three-dimensional Prandtl equation. This structure simplifies the problem by reducing it to a set of equations involving only u and w. In Proposition 6.1.1, the Burgers equation $\partial_t K + 2u(\partial_x + K\partial_y)K = 0$ is derived, where the coefficient K is shown to be independent of z. The classical Burgers equation, a nonlinear partial differential equation, describes the velocity field of a fluid with both viscous and nonlinear transport terms. It finds wide applications in studying shock waves, shock wave interactions, and nonlinear phenomena. The introduction of the Burgers equation simplifies the problem and provides a better mathematical formulation to describe the behavior of the boundary layer. This simplification aids in understanding the characteristics of the boundary layer and offers a potential approach to explore solutions of the three-dimensional Prandtl equation.

The main idea behind this approximation scheme (6.1.4)→ (6.1.6) or (6.1.8) is to introduce viscous terms, namely, $\epsilon^2\partial_x^2 u^\epsilon$, $\epsilon^2\partial_y^2 u^\epsilon$, $\epsilon^2\partial_x^2 \varphi^\epsilon$ and $\epsilon^2\partial_y^2 \varphi^\epsilon$, in order to prevent the loss of xy-derivatives. This regularization offers the advantage that our new weighted H^s and L^∞ a priori estimates also remain valid for φ^ϵ. It is this key aspect that allows us to derive the uniform (in ϵ) estimates in the subsequent section. However, a consequence of this regularization is the emergence of additional terms, such as $\frac{\partial_{xy}\varphi}{\varphi}$, $\frac{\partial_{xy}^2\varphi}{\varphi}$, $\frac{\partial_{xy}\partial_z\varphi}{\varphi}$, during the estimation process. Nevertheless, these terms can be controlled within the function space $C([0,T]; H^{s,\gamma}_{\sigma,\delta})$. In addition, it is worth emphasizing that the replacement of Bernoulli's law with the regularized Bernoulli's law is crucial. Without this regularization, it would not be possible to simultaneously satisfy the conditions $u|_{z=0} = 0$ and $\lim_{z\to+\infty} u = U$.

Next, we introduce the weighted Sobolev space and define the space $H^{s,\gamma}_{\sigma,\delta}$ by

$$H^{s,\gamma}_{\sigma,\delta} :=$$

$$\Big\{\varphi : \mathbb{T}^2 \times \mathbb{R}^+ \to \mathbb{R}, \|\varphi\|_{H^{s,\gamma}} < +\infty, (1+z)^\sigma |\varphi| \ge \delta,$$

$$\sum_{|\alpha|\le 2} |(1+z)^{\sigma+\alpha_3} D^\alpha \varphi|^2 \le \frac{1}{\delta^2}\Big\},$$

where $D^\alpha := \partial_x^{\alpha_1}\partial_y^{\alpha_2}\partial_z^{\alpha_3}, \alpha_1+\alpha_2+\alpha_3 = s, s \ge 5, \gamma \ge 1, \sigma > \gamma + \frac{1}{2}$ and $\delta \in (0,1)$. We define the weighted norm as

$$\|\varphi\|^2_{H^{s,\gamma}} := \sum_{|\alpha|\le s} \|(1+z)^{\gamma+\alpha_3} D^\alpha \varphi\|^2_{L^2} \tag{6.1.11}$$

and

$$\|\varphi\|^2_{H^{s,\gamma}_g} := \|(1+z)^\gamma g_s\|^2_{L^2} + \sum_{\substack{|\alpha|\le s \\ \alpha_1+\alpha_2 \le s-1}} \|(1+z)^{\gamma+\alpha_3} D^\alpha \varphi\|^2_{L^2}. \tag{6.1.12}$$

Here

$$g_s := \partial^s_{xy}\varphi - \frac{\partial_z\varphi}{\varphi}\partial^s_{xy}(u-U) = \sum_{i=0}^{s}\left(\partial_x^i\partial_y^{s-i}\varphi - \frac{\partial_z\varphi}{\varphi}\partial_x^i\partial_y^{s-i}(u-U)\right),$$

and

$$g_s = (g_s)_x + (g_s)_y = \partial_x\partial^{s-1}_{xy}\varphi - \frac{\partial_z\varphi}{\varphi}\partial_x\partial^{s-1}_{xy}(u-U) + \partial_y\partial^{s-1}_{xy}\varphi - \frac{\partial_z\varphi}{\varphi}\partial_y\partial^{s-1}_{xy}(u-U),$$

provided that $\varphi = \partial_z u > 0$. Hence, we can calculate that

$$g_{j+1} = \partial_x g_j + \partial_y g_j + \partial^j_{xy} u \partial_x a + \partial^j_{xy} u \partial_y a, \quad a = \frac{\partial_z\varphi}{\varphi}.$$

As pointed out that, the difference between norm (6.1.11) and norm (6.1.12) are that the weighted L^2-norm of $\partial_{xy}\varphi$ is replaced by that of g_s, which avoids the loss of xy-derivative by the nonlinear cancellation. On the other hand, these two weighted norms are the almost equivalent, that is, there exists a positive constant C such that the following inequality holds for any $\varphi \in H^{s,\gamma}_{\sigma,\delta}(\mathbb{T}^2 \times \mathbb{R}^+)$, $\gamma \geq 1$, $\sigma > \gamma + \frac{1}{2}$ and $\delta \in (0, 1)$:

$$C_\delta \|\varphi\|_{H^{s,\gamma}_g} \leq \|\varphi\|_{H^{s,\gamma}} + \|u - U\|_{H^{s,\gamma-1}} \leq C_{s,\gamma,\sigma,\delta} \left(\|\varphi\|_{H^{s,\gamma}_g} + \|\partial^s_{xy} U\|_{L^2(\mathbb{T}^2)} \right), \tag{6.1.13}$$

In this chapter, for convenience, we simply write

$$\partial^s_{xy} := \sum_{i=0}^{s} \partial^i_x \partial^{s-i}_y, \quad \iiint \cdot := \iint_{\mathbb{T}^2} \int_{\mathbb{R}^+} \cdot dxdydz.$$

Firstly, we state our main result as follows.

Theorem 6.1.1 *Given any integer $s \geq 5$, and real numbers γ, σ, δ satisfying $\gamma \geq 1$, $\sigma > \gamma + \frac{1}{2}$ and $\delta \in (0, 1)$. Assume the following conditions on the initial data, the outer flow U and $K(x, y)$:*

(i) Suppose that the initial data $u_0 - U(0, x) \in H^{s,\gamma-1}$ and $\partial_z u_0 \in H^{s,\gamma}_{\sigma,2\delta}$ satisfy the compatibility conditions $u_0|_{z=0}$ and $\lim_{z \to +\infty} u_0 = U|_{t=0}$. In addition, when $s = 5$, it is further assumed that $\delta \geq 0$ is chosen small enough such that

$$\|\omega_0\|_{H^{s,\gamma}} \leq C\delta^{-1} \tag{6.1.14}$$

with a generic constant C.

(ii) The outer flow U is supposed to satisfy

$$\sup_t \sum_{l=0}^{\frac{s}{2}+1} \|\partial^l_t U\|_{H^{s-2l+2}(\mathbb{T}^2)} < +\infty. \tag{6.1.15}$$

(iii) The $K(x, y)$ is supposed to satisfy

$$\|\partial^s_{xy} K\|_{L^\infty(\mathbb{T}^2)} < +\infty. \tag{6.1.16}$$

Then there exist a time $T := T(K, s, \gamma, \sigma, \delta, \|\varphi_0\|_{H^{s,\gamma}}, U)$ such that the initial boundary value problem (6.1.1)–(6.1.3) has a unique classical solution (u, v, w) (v=Ku) satisfying

$$u - U \in L^\infty([0, T]; H^{s,\gamma-1}) \cap C([0, T]; H^s - \varphi)$$

and

$$\partial_z u \in L^\infty([0, T]; H^{s,\gamma}_{\sigma,\delta}) \cap C([0, T]; H^s - \varphi),$$

where $H^s - \varphi$ is the space H^s endowed with its weak topology.

As we have known, Crocco transformation and the outer flow $(U, V) = \text{constant}$ have been used in three-dimensional Prandtl system, such as [70, 73, 79, 84–86, 108, 109]. But, we need neither the Crocco transformation nor the outer flow $(U, V) = \text{constant}$ in this chapter. In fact, we first use the energy method to prove the local existence of three-dimensional Prandtl system based on a cancellation property. Our result is an improvement of the existence and uniqueness of solutions in Masmoudi and Wong [98] from the two-dimensional case to the three dimensional with special structure case.

The chapter is arranged as follows. Section 6.1 is an introduction, which contains many inequalities used in this paper. In Sect. 6.2, we shall give the uniform estimates with the weighted norm. The difficulty of this paper is the estimates on the boundary at $z = 0$, and we give the regular pattern at $z = 0$ for $s \geq 5$ eventually through a series of precise calculations; see Lemmas 6.2.1–6.2.2. In Sect. 6.3, we shall prove local existence and uniqueness of solutions to the Prandtl system to problem (6.1.1).

6.1.1 Preliminaries

Proposition 6.1.1 *Under the assumption* $\partial_z u > 0$ *and special structures* $v(t, x, y, z) = K(t, x, y)u(t, x, y, z)$, *to study problem (6.1.1) is equivalent to studying the following reduced problem:*

$$\begin{cases} \partial_t u + (u\partial_x + Ku\partial_y + w\partial_z)u = \partial_z^2 u^\epsilon - \partial_x P, \\ \partial_x u + \partial_y (Ku) + \partial_z w = 0, \\ u\big|_{t=0} = u_0(x, y, z), \\ (u, w)\big|_{z=0} = 0, \quad \lim\limits_{z\to+\infty} u = U(t, x, y), \end{cases} \tag{6.1.17}$$

where K *is independent of* t *and* U *satisfies*

$$\begin{cases} \partial_t U + U\partial_x U + KU\partial_y U + \partial_x P = 0, \\ \partial_y P - K\partial_x P = 0. \end{cases} \tag{6.1.18}$$

Proof Using special structures $v(k, x, y, z) = K(t, x, y)u(k, x, y, z)$ for the equation $(6.1.1)_2$, we have

$$\partial_t (Ku) + (u\partial_x + Ku\partial_y + w\partial_z)(Ku) + \partial_y P - K\partial_z^2 u = 0, \tag{6.1.19}$$

which, by the equation $(6.1.1)_1$, yields

$$u\left[\partial_t K + u(\partial_x + \partial_y)K\right] - K\partial_x P + \partial_y P = 0. \tag{6.1.20}$$

Noting that $K(t, x, y)$ is independent of z, by differentiating (6.1.20) with respect to z, it follows that

$$\partial_z u \partial_t K + 2u\partial_z u(\partial_x + K\partial_y)K = 0. \tag{6.1.21}$$

Due to the monotonicity assumption, we can further get

$$\partial_t K + 2u(\partial_x + K\partial_y)K = 0. \tag{6.1.22}$$

Differentiating (6.1.22) with respect to z gives the *Burgers* equation

$$(\partial_x + K\partial_y)K = 0. \tag{6.1.23}$$

Plugging (6.1.23) into (6.1.22), it follows

$$\partial_t K = 0, \tag{6.1.24}$$

which, combined (6.1.20) with (6.1.23), gives

$$-K\partial_x P + \partial_y P = 0. \tag{6.1.25}$$

The above equation means that $(\partial_x P, \partial_y P)$ is parallel to both the velocity field of the outer Euler flow and the tangential velocity field in the boundary layer. Now, we assume that the classical solution (u, v, w) to problem (6.1.1) satisfies $v(t, x, y, z) = K(x, y)u(t, x, y, z)$, then by using (6.1.23) and (6.1.25), $W(t, x, y, z) :\equiv v(t, x, y, z) - K(x, y)u(t, x, y, z)$ satisfies the following problem:

$$\begin{cases} \partial_t W + (u\partial_x + v\partial_y + w\partial_z)W + (\partial_y v - K\partial_y u)W = \partial_z^2 W, \\ W|_{t=0} = 0, \\ W|_{z=0} = 0, \\ \lim\limits_{z\to+\infty} W = 0. \end{cases} \tag{6.1.26}$$

Equation (6.1.26) has only trivial solution $W \equiv 0$ based on the energy argument. Therefore, studying problem (6.1.1) is equivalent to studying the following reduced problem for only two unknown functions u and w:

$$\begin{cases} \partial_t u + (u\partial_x + Ku\partial_y + w\partial_z)u = \partial_z^2 u^\epsilon - \partial_x P, \\ \partial_x u + \partial_y(Ku) + \partial_z w = 0, \\ u\big|_{t=0} = u_0(x, y, z), \\ (u, w)\big|_{z=0} = 0, \quad \lim\limits_{z\to+\infty} u = U(t, x, y). \end{cases} \tag{6.1.27}$$

Hence, the proof of Proposition 6.1.1 is complete. □

Proposition 6.1.2 *Let $s \geq 5$ be an integer, $\gamma \geq 1$, $\sigma > \gamma + \frac{1}{2}$ and $\delta \in (0, 1)$. Then for any $\varphi \in H^{s,\gamma}_{\sigma,\delta}(\mathbb{T}^2 \times \mathbb{R}^+)$, we have the following inequality:*

$$C_\delta \|\varphi\|_{H^{s,\gamma}_g} \leq \|\varphi\|_{H^{s,\gamma}} + \|u - U\|_{H^{s,\gamma-1}} \leq C_{s,\gamma,\sigma,\delta} \left(\|\varphi\|_{H^{s,\gamma}_g} + \|\partial^s_{xy} U\|_{L^2(\mathbb{T}^2)} \right), \tag{6.1.28}$$

where $C_{s,\gamma,\sigma,\delta} > 0$ is a constant and only depends on $s, \gamma, \sigma, \delta$.

Proof Firstly, by the definition of $\|\varphi\|_{H^{s,\gamma}}$ and $\|u - U\|_{H^{s,\gamma-1}}$ that

$$\begin{aligned} \|\varphi\|_{H^{s,\gamma}} + \sum_{k=0}^{s} \|(1+z)^{\gamma-1} \partial^k_{xy}(u - U)\|_{L^2} &\leq \|\varphi\|_{H^{s,\gamma}} + \|u - U\|_{H^{s,\gamma-1}} \\ &\leq 2 \left(\|\varphi\|_{H^{s,\gamma}} + \sum_{k=0}^{s} \|(1+z)^{\gamma-1} \partial^k_{xy}(u - U)\|_{L^2} \right). \end{aligned} \tag{6.1.29}$$

Furthermore, applying Wirtinger's inequality and Lemma 6.1.2, we get

$$\begin{cases} \displaystyle\sum_{k=1}^{s} \|(1+z)^{\gamma-1} \partial^k_{xy}(u - U)\|_{L^2} \leq \frac{1 + \pi^{-2s}}{1 - \pi^{-2}} \|(1+z)^{\gamma-1} \partial^s_{xy}(u - U)\|_{L^2} \\ \|(1+z)^{\gamma-1}(u - U)\|_{L^2} \leq \dfrac{2}{2\gamma - 1} \|(1+z)^{\gamma} \varphi\|_{L^2} \leq \dfrac{2}{2\gamma - 1} \|\varphi\|_{H^{s,\gamma}}, \end{cases}$$

and hence, there exists a constant $C_{s,\gamma} > 0$ such that

$$\begin{aligned} \|\varphi\|_{H^{s,\gamma}} + \|(1+z)^{\gamma-1} \partial^s_{xy}(u - U)\|_{L^2} &\leq \|\varphi\|_{H^{s,\gamma}} + \sum_{k=0}^{s} \|(1+z)^{\gamma-1} \partial^k_{xy}(u - U)\|_{L^2} \\ &\leq C_{s,\gamma} \left(\|\varphi\|_{H^{s,\gamma}} + \|(1+z)^{\gamma-1} \partial^s_{xy}(u - U)\|_{L^2} \right). \end{aligned} \tag{6.1.30}$$

Therefore, according to Lemma 6.1.3 and inequalities (6.1.29)–(6.1.30), it suffices to prove

$$\begin{aligned} C_\delta \|\varphi\|_{H^{s,\gamma}_g} &\leq \|\varphi\|_{H^{s,\gamma}} + \|(1+z)^{\gamma-1} \partial^s_{xy}(u - U)\|_{L^2} \\ &\leq C_{s,\gamma,\sigma,\delta} \left(\|\varphi\|_{H^{s,\gamma}_g} + \|\partial^s_{xy} U\|_{L^2} \right) \end{aligned} \tag{6.1.31}$$

for some constants C_δ and $C_{s,\gamma,\sigma,\delta} > 0$. □

Lemma 6.1.2 *Let* $f : \mathbb{T}^2 \times \mathbb{R}^+ \to \mathbb{R}$,

(i) if $\lambda > -\frac{1}{2}$ *and* $\lim_{z\to+\infty} f(x,y,z) = 0$, *then*

$$\|(1+z)^\lambda f\|_{L^2(\mathbb{T}^2\times\mathbb{R}_+)} \le \frac{2}{2\lambda+1}\|(1+z)^{\lambda+1}\partial_z f\|_{L^2(\mathbb{T}^2\times\mathbb{R}_+)}; \quad (6.1.32)$$

(ii) if $\lambda < -\frac{1}{2}$ *and* $f(x,y,z)|_{z=0} = 0$, *then*

$$\|(1+z)^\lambda f\|_{L^2(\mathbb{T}^2\times\mathbb{R}^+)} \le -\frac{2}{2\lambda+1}\|(1+z)^{\lambda+1}\partial_z f\|_{L^2(\mathbb{T}^2\times\mathbb{R}^+)}. \quad (6.1.33)$$

Proof The proof is elementary. Noting that the term

$$\begin{aligned}
\|(1+z)^\lambda f\|^2_{L^2(\mathbb{T}^2\times\mathbb{R}_+)} &= \frac{1}{2\lambda+1}\int_{\mathbb{T}^2}\int_{\mathbb{R}_+} f^2(x,y,z)d(1+z)^{2\lambda+1}dxdy \\
&= \frac{1}{2\lambda+1}\int_{\mathbb{T}^2}(1+z)^{2\lambda+1}f^2(x,y,z)\Big|_{z=0}^{z=+\infty}dxdy \\
&\quad -\frac{2}{2\lambda+1}\int_{\mathbb{T}^2}\int_{\mathbb{R}_+}(1+z)^{2\lambda+1}f(x,y,z)\partial_z f(x,y,z)dxdydz \\
&\le \left|-\frac{2}{2\lambda+1}\right| \|(1+z)^\lambda f\|_{L^2(\mathbb{T}^2\times\mathbb{R}_+)}\|(1+z)^{\lambda+1}\partial_z f\|_{L^2(\mathbb{T}^2\times\mathbb{R}_+)},
\end{aligned}$$

thus we prove (6.1.32)–(6.1.33). □

Lemma 6.1.3 *Let* $s \ge 5$ *be an integer,* $\gamma \ge 1$, $\sigma > \gamma + \frac{1}{2}$ *and* $\delta \in (0,1)$. *If* $\varphi \in H^{s,\gamma}_{\sigma,\delta}(\mathbb{T}^2\times\mathbb{R}^+)$, *then for any* $k = 0, 1, 2, \cdots, s$,

$$\|(1+z)^\gamma g_k\|_{L^2} \le \|(1+z)^\gamma \partial^k_{xy}\varphi\|_{L^2} + \delta^{-2}\|(1+z)^{\gamma-1}\partial^k_{xy}(u-U)\|_{L^2}, \quad (6.1.34)$$

where $g_k := \partial^k_{xy}\varphi - \frac{\partial_z\varphi}{\varphi}\partial^k_{xy}(u-U)$. *In addition, if* $u\big|_{z=0} = 0$, *then for any* $k = 1, 2, \ldots, s$,

$$\begin{aligned}
&\|(1+z)^\gamma \partial^k_{xy}\varphi\|_{L^2} + \|(1+z)^{\gamma-1}\partial^k_{xy}(u-U)\|_{L^2} \\
&\le C_{s,\gamma,\sigma,\delta}\left(\|(1+z)^\gamma g_k\|_{L^2} + \|\partial^k_{xy}U\|_{L^2(\mathbb{T}^2)}\right),
\end{aligned} \quad (6.1.35)$$

where $C_{s,\gamma,\sigma,\delta} > 0$ *is a constant and only depends on* $s, \gamma, \sigma, \delta$.

Proof The proof is similar to that given in [98]; it is only related to the defined weighted Sobolev space and norms, not to the condition $(6.1.6)_2$, so we outline it here for brevity.

Firstly, according to the definition of $H^{s,\gamma}_{\sigma,\delta}$, we have

$$\delta(1+z)^{-\sigma} \le |\varphi| \le \delta^{-1}(1+z)^{-\sigma}, \quad |\partial_z\varphi| \le \delta^{-1}(1+z)^{-\sigma-1}, \tag{6.1.36}$$

so $\frac{|\partial_z\varphi|}{\varphi} \le \delta^{-2}(1+z)^{-1}$, for any $k = 0, 1, 2, \cdots, s$,

$$\begin{aligned}
\|(1+z)^\gamma g_k\|_{L^2} &\le \|(1+z)^\gamma \partial^k_{xy}\varphi\|_{L^2} + \|(1+z)^\gamma \frac{\partial_z\varphi}{\varphi}\partial^k_{xy}(u-U)\|_{L^2} \\
&\le \|(1+z)^\gamma \partial^k_{xy}\varphi\|_{L^2} + \|(1+z)\frac{\partial_z\varphi}{\varphi}\|_{L^\infty}\|(1+z)^{\gamma-1}\partial^k_{xy}(u-U)\|_{L^2} \\
&\le \|(1+z)^\gamma \partial^k_{xy}\varphi\|_{L^2} + \delta^{-2}\|(1+z)^{\gamma-1}\partial^k_{xy}(u-U)\|_{L^2}.
\end{aligned}$$

Next, using (6.1.33) in Lemma 6.1.2 with the index $\gamma - \sigma - 1 < -\frac{3}{2} < -\frac{1}{2}$, we get

$$\begin{aligned}
&\|(1+z)^{\gamma-1}\partial^k_{xy}(u-U)\|_{L^2} \le \delta^{-1}\|(1+z)^{\gamma-\sigma-1}\frac{\partial^k_{xy}(u-U)}{\varphi}\|_{L^2} \\
&\le C_{\gamma,\sigma}\delta^{-1}\left(\left\|\frac{\partial^k_{xy}(u-U)}{\varphi}|_{z=0}\right\|_{L^2} + \left\|(1+z)^{\gamma-\sigma}\partial_z\left(\frac{\partial^k_{xy}(u-U)}{\varphi}\right)\right\|_{L^2}\right) \\
&\le C_{\gamma,\sigma}\delta^{-2}\left(\|\partial^k_{xy}U\|_{L^2(\mathbb{T}^2)} + \|(1+z)^\gamma g_k\|_{L^2}\right).
\end{aligned} \tag{6.1.37}$$

Finally, for $k = 0, 1, 2, \cdots, s$, according to (6.1.37) and the triangle inequality, we have

$$\begin{aligned}
\|(1+z)^\gamma \partial^k_{xy}\varphi\|_{L^2} &\le \|(1+z)^\gamma g_k\|_{L^2} + \|(1+z)^\gamma \frac{\partial_z\varphi}{\varphi}\partial^k_{xy}(u-U)\|_{L^2} \\
&\le C_{\gamma,\sigma,\delta}\left(\|\partial^k_{xy}U\|_{L^2(\mathbb{T}^2)} + \|(1+z)^\gamma g_k\|_{L^2}\right).
\end{aligned} \tag{6.1.38}$$

The proof is thus complete. □

Lemma 6.1.4 ([35]) *Let $f : \mathbb{T}^2 \times \mathbb{R}^+ \to \mathbb{R}$ and $f(x, y, z)|_{z=0} = 0$. Then there exists a constant $C > 0$ such that*

$$\begin{aligned}
&\|f(x,y,z)\|_{L^\infty(\mathbb{T}^2\times\mathbb{R}^+)} \\
&\le C(\|f(x,y,z)\|_{L^2(\mathbb{T}^2\times\mathbb{R}^+)} + \|\partial_z f(x,y,z)\|_{L^2(\mathbb{T}^2\times\mathbb{R}^+)} \\
&\quad + \partial^1_{xy}\|f(x,y,z)\|_{L^2(\mathbb{T}^2\times\mathbb{R}^+)} + \|\partial_z\partial^1_{xy}f(x,y,z)\|_{L^2(\mathbb{T}^2\times\mathbb{R}^+)} \\
&\quad + \|\partial^2_{xy}f(x,y,z)\|_{L^2(\mathbb{T}^2\times\mathbb{R}^+)} + \|\partial_z\partial^2_{xy}f(x,y,z)\|_{L^2(\mathbb{T}^2\times\mathbb{R}^+)}).
\end{aligned} \tag{6.1.39}$$

Proof Firstly, we have

$$\begin{aligned}f^2(x,y,z)&=\int_0^z \partial_\tau[f^2(x,y,\tau)]d\tau=2\int_0^z f(x,y,\tau)\partial_\tau f(x,y,\tau)d\tau\\&\le C\left(\int_0^z f^2(x,y,\tau)d\tau\right)^{\frac{1}{2}}\left(\int_0^z |\partial_\tau f(x,y,\tau)|^2d\tau\right)^{\frac{1}{2}}\\&\le C\left(\int_0^{+\infty}|f(x,y,z)|^2dz+\int_0^{+\infty}|\partial_z f(x,y,z)|^2dz\right).\end{aligned}\tag{6.1.40}$$

Secondly, we apply the two-dimensional Sobolev inequality for x, y variables to get

$$|f(x,y,z)|^2\le C(\|f(z)\|^2_{L^2(\mathbb{T}^2)}+\|\partial^1_{xy}f(z)\|^2_{L^2(\mathbb{T}^2)}+\|\partial^2_{xy}f(z)\|^2_{L^2(\mathbb{T}^2)})\tag{6.1.41}$$

and

$$|\partial_z f(x,y,z)|^2\le C(\|\partial_z f(z)\|^2_{L^2(\mathbb{T}^2)}+\|\partial_z\partial^1_{xy}f(z)\|^2_{L^2(\mathbb{T}^2)}+\|\partial_z\partial^2_{xy}f(z)\|^2_{L^2(\mathbb{T}^2)}).\tag{6.1.42}$$

At last, substituting (6.1.41) and (6.1.42) into (6.1.40), we complete the proof of estimate (6.1.39). □

Lemma 6.1.5 *Let the vector field* (u, Ku, w) *defined on* $\mathbb{T}^2\times\mathbb{R}^+$ *satisfy the condition* $\partial_x u+\partial_y(Ku)+\partial_z w=0$. *If* $s\ge 5$ *is an integer,* $\gamma\ge 1$, $\sigma>\gamma+\frac{1}{2}$ *and* $\delta\in(0,1)$, *then for any* $\varphi\in H^{s,\gamma}_{\sigma,\delta}(\mathbb{T}^2\times\mathbb{R}^+)$, *we have the following inequalities:*

(*i*) *For* $k=0,1,2,\cdots,s$,

$$\|(1+z)^{\gamma-1}\partial^k_{xy}(u-U)\|_{L^2}\le C_{s,\gamma,\sigma,\delta}\left(\|\varphi\|_{H^{s,\gamma}_g}+\|\partial^s_{xy}U\|_{L^2(\mathbb{T}^2)}\right).\tag{6.1.43}$$

(*ii*) *For* $k=0,1,2,\cdots,s-1$, *if* $K\in C^{s+1}(\mathbb{T}^2)$, *then*

$$\left\|\frac{\partial^k_{xy}w+z\tilde{U}}{1+z}\right\|_{L^2}\le C_{s,\gamma,\sigma,\delta}\left(\|\varphi\|_{H^{s,\gamma}_g}+\|\partial^s_{xy}U\|_{L^2(\mathbb{T}^2)}\right).\tag{6.1.44}$$

where $\tilde{U}=\partial_x\partial^k_{xy}U+\partial_y\partial^k_{xy}(KU)$

(iii) For all $|\alpha| \le s$,

$$\|(1+z)^{\gamma+\alpha_3} D^\alpha \varphi\|_{L^2} \le \begin{cases} C_{s,\gamma,\sigma,\delta}\left(\|\varphi\|_{H_g^{s,\gamma}} + \|\partial_{xy}^s U\|_{L^2(\mathbb{T}^2)}\right) & \text{if } \alpha = (\alpha_1, \alpha_2, 0), \\ \|\varphi\|_{H_g^{s,\gamma}} \ \text{if } \alpha \ne (\alpha_1, \alpha_2, 0). \end{cases} \tag{6.1.45}$$

(iv) For $k = 0, 1, 2, \cdots, s$,

$$\|(1+z)^\gamma g_k\|_{L^2(\mathbb{T}^2)} \le \begin{cases} C_{s,\gamma,\sigma,\delta}\left(\|\varphi\|_{H_g^{s,\gamma}} + \|\partial_{xy}^s U\|_{L^2(\mathbb{T}^2)}\right) & \text{if } k \ne s, \\ \|\varphi\|_{H_g^{s,\gamma}} \ \text{if } k = s. \end{cases} \tag{6.1.46}$$

(v) For $k = 0, 1, 2, \cdots, s-2$,

$$\|\partial_{xy}^k u\|_{L^\infty} \le C_{s,\gamma,\sigma,\delta}\left(\|\varphi\|_{H_g^{s,\gamma}} + \|\partial_{xy}^s U\|_{L^2(\mathbb{T}^2)}\right). \tag{6.1.47}$$

(vi) For $k = 0, 1, 2, \cdots, s-3$, *if* $K \in C^{s+1}(\mathbb{T}^2)$, *then*

$$\left\|\frac{\partial_{xy}^k w}{1+z}\right\|_{L^\infty} \le C_{s,\gamma,\sigma,\delta}\left(\|\varphi\|_{H_g^{s,\gamma}} + \|\partial_{xy}^s U\|_{L^2(\mathbb{T}^2)} + 1\right). \tag{6.1.48}$$

(vii) For all $|\alpha| \le s-3$,

$$\|(1+z)^{\gamma+\alpha_3} D^\alpha \varphi\|_{L^\infty} \le C_{s,\gamma,\sigma,\delta}\|\varphi\|_{H_g^{s,\gamma}} \tag{6.1.49}$$

where $C_{s,\gamma,\sigma,\delta} > 0$ *is a constant and only depends on* $s, \gamma, \sigma, \delta$.

Proof First, by the definition of $\|\cdot\|_{H^{s,\gamma-1}}$ and Lemma 6.1.3, we can directly derive the inequality (6.1.43). Using Lemma 6.1.2, (6.1.43) and the fact that $\partial_x u + \partial_y(Ku) + \partial_z w = 0$, we have

$$\begin{aligned} \left\|\frac{\partial_{xy}^k w + z\tilde{U}}{1+z}\right\|_{L^2} &\le 2\|\partial_{xy}^k(-\partial_x u - \partial_y(Ku)) + \tilde{U}\|_{L^2} \\ &\le 2\|\partial_{xy}^{k+1}(u-U)\|_{L^2} + C\sum_{i=0}^k \|\partial_y\{\partial_{xy}^i K \partial_{xy}^{k-i}(u-U)\}\|_{L^2} \\ &\le C_{s,\gamma,\sigma,\delta}\left(\|\varphi\|_{H_g^{s,\gamma}} + \|\partial_{xy}^s U\|_{L^2(\mathbb{T}^2)}\right). \end{aligned} \tag{6.1.50}$$

The proofs of (6.1.45) and (6.1.46) are obvious, so we omit it here for brevity. Second, for $k = 0, 1, 2, \cdots, s-2$, applying Lemma 6.1.4 and (6.1.43), and (6.1.45), we have

$$\begin{aligned}\|\partial_{xy}^k(u-U)\|_{L^\infty} &\le C(\|\partial_{xy}^k(u-U)\|_{L^2} + \|\partial_{xy}^{k+1}(u-U)\|_{L^2} + \|\partial_{xy}^{k+2}(u-U)\|_{L^2} \\ &\quad + \|\partial_{xy}^k\varphi\|_{L^2} \\ &\quad + \|\partial_{xy}^{k+1}\varphi\|_{L^2} + \|\partial_{xy}^{k+2}\varphi\|_{L^2}) \\ &\le C_{s,\gamma,\sigma,\delta}\left(\|\varphi\|_{H_g^{s,\gamma}} + \|\partial_{xy}^s U\|_{L^2(\mathbb{T}^2)}\right). \end{aligned} \tag{6.1.51}$$

Hence, by the triangle inequality and $\|\partial_{xy}^k U\|_{L^2(\mathbb{T}^2)} \le \|\partial_{xy}^s U\|_{L^2(\mathbb{T}^2)}$, we can justify (6.1.47). Next, for $k = 0, 1, 2, \cdots, s-3$, using (6.1.43)–(6.1.45) and Lemma 6.1.4, we conclude that

$$\begin{aligned}&\left\|\frac{\partial_{xy}^k w}{1+z}\right\|_{L^\infty} \le \left\|\frac{z\tilde{U}}{1+z}\right\|_{L^\infty} + \left\|\frac{\partial_{xy}^k w + z\tilde{U}}{1+z}\right\|_{L^\infty} \\ &\le C\Big(\|\tilde{U}\|_{L^2(\mathbb{T}^2)} + \|\partial_{xy}\tilde{U}\|_{L^2(\mathbb{T}^2)} + \|\partial_{xy}^2\tilde{U}\|_{L^2(\mathbb{T}^2)} + \|(1+z)^{-1}(\partial_{xy}^k w + z\tilde{U})\|_{L^2} \\ &\quad + \|(1+z)^{-1}(\partial_{xy}^k\partial_z w + \tilde{U})\|_{L^2} + \|(1+z)^{-2}(\partial_{xy}^k w + z\tilde{U})\|_{L^2} \\ &\quad + \|(1+z)^{-1}(\partial_{xy}^{k+1} w + z\partial_{xy}\tilde{U})\|_{L^2} + \|(1+z)^{-1}(\partial_{xy}^{k+2} w + z\partial_{xy}^2\tilde{U})\|_{L^2} \\ &\quad + \|(1+z)^{-1}(\partial_{xy}^{k+1}\partial_z w + \partial_{xy}\tilde{U})\|_{L^2} + \|(1+z)^{-2}(\partial_{xy}^{k+1} w + z\partial_{xy}\tilde{U})\|_{L^2} \\ &\quad + \|(1+z)^{-1}(\partial_{xy}^{k+2}\partial_z w + \partial_{xy}^2\tilde{U})\|_{L^2} + \|(1+z)^{-2}(\partial_{xy}^{k+2} w + z\partial_{xy}^2\tilde{U})\|_{L^2}\Big) \\ &\le C_{s,\gamma,\sigma,\delta}\left(\|\varphi\|_{H_g^{s,\gamma}} + \|\partial_{xy}^s U\|_{L^2(\mathbb{T}^2)} + 1\right). \end{aligned} \tag{6.1.52}$$

Last, inequality (6.1.49) follows directly from inequality (6.1.45) in Lemma 6.1.4. □

Lemma 6.1.6 ([28] p.369 Theorem 9) *Denote the operator*

$$\mathcal{L}u = -\sum_{i,j=1}^{n} a^{ij}u_{x_ix_j} + \sum_{i=1}^{n} b^i u_{x_i} + cu,$$

where a^{ij}, b^i and c are continuous. Assume $u \in C_1^2([0,T]\times\mathbb{T}^2\times\mathbb{R}^+)\cap C([0,T]\times\mathbb{T}^2\times\mathbb{R}^+)$, and $c \ge 0$ in $[0,T]\times\mathbb{T}^2\times\mathbb{R}^+$.

(i) If

$$u_t + \mathcal{L}u \le 0$$

in $[0, T] \times \mathbb{T}^2 \times \mathbb{R}^+$, *then*

$$\max_{\overline{[0,T]\times\mathbb{T}^2\times\mathbb{R}^+}} \leq \max_{\partial([0,T]\times\mathbb{T}^2\times\mathbb{R}^+)} u^+.$$

(ii) If

$$u_t + \mathcal{L}u \geq 0$$

in $[0, T] \times \mathbb{T}^2 \times \mathbb{R}^+$, *then*

$$\min_{\overline{[0,T]\times\mathbb{T}^2\times\mathbb{R}^+}} \geq - \max_{\partial([0,T]\times\mathbb{T}^2\times\mathbb{R}^+)} u^-,$$

where $u^+ = \max\{u, 0\}$ *and* $u^- = -\min\{u, 0\}$.

Lemma 6.1.7 is the maximum principle for bounded solutions to parabolic equations.

Lemma 6.1.7 (Maximum Principle for Parabolic Equations) *Let* $\epsilon \geq 0$. *If* $H \in C([0, T]; C^2(\mathbb{T}^2 \times \mathbb{R}^+) \cap C^1([0, T]; C^0(\mathbb{T}^2 \times \mathbb{R}^+))$ *is a bounded function that satisfies the differential inequality*

$$\left\{\partial_t + b_1\partial_x + b_2\partial_y + b_3\partial_z - \epsilon^2\partial_x^2 - \epsilon^2\partial_y^2 - \partial_z^2\right\} H \leq fH \quad \text{in } [0, T] \times \mathbb{T}^2 \times \mathbb{R}^+,$$

where the coefficients b_1, b_2, b_3 *and* f *are continuous and satisfy*

$$\left\| \frac{b_3}{1+z} \right\|_{L^\infty([0,T]\times\mathbb{T}^2\times\mathbb{R}^+)} < +\infty, \tag{6.1.53}$$

and

$$\|f\|_{L^\infty([0,T]\times\mathbb{T}^2\times\mathbb{R}^+)} \leq \lambda, \tag{6.1.54}$$

then for any $t \in [0, T]$,

$$\sup_{\mathbb{T}^2\times\mathbb{R}^+} H(t) \leq \max\left\{e^{\lambda t}\|H(0)\|_{L^\infty(\mathbb{T}^2\times\mathbb{R}^+)}, \max_{\tau\in[0,T]}\left\{e^{\lambda(t-\tau)}\|H(\tau)\big|_{z=0}\|_{L^\infty(\mathbb{T}^2)}\right\}\right\}. \tag{6.1.55}$$

Proof For any $\mu > 0$, let us define

$$\mathcal{H} := e^{-\lambda t} H - \mu\|\frac{b_3}{1+z}\|_{L^\infty} \cdot t - \mu \ln(1+z). \tag{6.1.56}$$

Then for any $\tilde{t} \in (0, T]$, we can get

$$
\begin{aligned}
&\left\{\partial_t + b_1\partial_x + b_2\partial_y + b_3\partial_z - \epsilon^2\partial_x^2 - \epsilon^2\partial_y^2 - \partial_z^2 + (\lambda - f)\right\}\mathcal{H} \\
&= \left\{\partial_t + b_1\partial_x + b_2\partial_y + b_3\partial_z - \epsilon^2\partial_x^2 - \epsilon^2\partial_y^2 - \partial_z^2 + (\lambda - f)\right\} e^{-\lambda t} H \\
&\quad - \left\{\partial_t + b_1\partial_x + b_2\partial_y + b_3\partial_z - \epsilon^2\partial_x^2 - \epsilon^2\partial_y^2 - \partial_z^2 + (\lambda - f)\right\} \mu \|\frac{b_3}{1+z}\|_{L^\infty} \cdot t \\
&\quad - \left\{\partial_t + b_1\partial_x + b_2\partial_y + b_3\partial_z - \epsilon^2\partial_x^2 - \epsilon^2\partial_y^2 - \partial_z^2 + (\lambda - f)\right\} \mu \ln(1+z) \\
&= e^{-\lambda t}\left\{\partial_t + b_1\partial_x + b_2\partial_y + b_3\partial_z - \epsilon^2\partial_x^2 - \epsilon^2\partial_y^2 - \partial_z^2 - f\right\} H - \mu \|\frac{b_3}{1+z}\|_{L^\infty} \\
&\quad - (\lambda - f)\mu \|\frac{b_3}{1+z}\|_{L^\infty} \cdot t - \frac{\mu b_3}{1+z} - \frac{\mu}{(1+z)^2} \\
&< 0, \quad \text{in} \quad [0, \tilde{t}] \times \mathbb{T}^2 \times \mathbb{R}^+,
\end{aligned}
\tag{6.1.57}
$$

where we have used $-\mu \left\| \frac{b_3}{1+z} \right\|_{L^\infty([0,T]\times\mathbb{T}^2\times\mathbb{R}^+)} - \frac{\mu b_3}{1+z} < 0$, $f - \lambda \le 0$, and the following fact:

$$
\left\{\partial_t + b_1\partial_x + b_2\partial_y + b_3\partial_z - \epsilon^2\partial_x^2 - \epsilon^2\partial_y^2 - \partial_z^2\right\} H \le f H \quad in\ [0, T] \times \mathbb{T}^2 \times \mathbb{R}^+.
$$

Now, we want to use the classical maximum principle (Lemma 6.1.6) for parabolic equations on $\mathcal{H}$. To apply this maximum principle, we need the function $\mathcal{H}$ to be defined and continuous on the closure of the domain $[0, \tilde{t}] \times \mathbb{T}^2 \times \mathbb{R}^+$. Since $\ln(1+z)$ can blow up to infinity as z approaches -1, we need to make sure that R is chosen to be large enough such that $\ln(1+z)$ is well-defined and finite for all $(x, y, z) \in \mathbb{T}^2 \times \mathbb{R}^+$. And we also need to ensure that the maximum inequality itself is established.

So by the classical maximum principle for parabolic equations, we have

$$
\begin{aligned}
\max_{[0,\tilde{t}]\times\mathbb{T}^2\times[0,R]} \mathcal{H} &\le \max\left\{\| \max \mathcal{H}(0)\|_{L^\infty(\mathbb{T}^2\times\mathbb{R}^+)}, \max_{\tau\in[0,\tilde{t}]} \|\mathcal{H}(\tau)|_{z=0}\|_{L^\infty(\mathbb{T}^2)}\right\} \\
&\le \max\left\{\|H(0)\|_{L^\infty(\mathbb{T}^2\times\mathbb{R}^+)}, \max_{\tau\in[0,\tilde{t}]} \left\{e^{-\lambda\tau}\|H(\tau)|_{z=0}\|_{L^\infty(\mathbb{T}^2)}\right\}\right\}
\end{aligned}
\tag{6.1.58}
$$

provided that $R \geq \exp\left(\frac{1}{\mu}\|H\|_{L^\infty(\mathbb{T}^2\times\mathbb{R}^+)}\right) - 1$. Therefore, for any $(x, y, z) \in \mathbb{T}^2 \times \mathbb{R}^+$, we have

$$H(\tilde{t}, x, y) - \mu \left\|\frac{b_3}{1+z}\right\|_{L^\infty} e^{\lambda\tilde{t}}\tilde{t} - \mu e^{\lambda\tilde{t}} \ln(1+z)$$
$$\leq \max\left\{e^{\lambda\tilde{t}}\|H(0)\|_{L^\infty(\mathbb{T}^2\times\mathbb{R}^+)}, \max_{\tau\in[0,\tilde{t}]}\left\{e^{\lambda(\tilde{t}-\tau)}\|H(\tau)|_{z=0}\|_{L^\infty(\mathbb{T}^2)}\right\}\right\} \tag{6.1.59}$$

which is just inequality (6.1.55), after taking the limit $\mu \to 0^+$ and replace the arbitrary time $\tilde{t}$ by t. Hence, this completes the proof of the lemma. □

Lemma 6.1.8 is the lower bound estimate on bounded solutions for parabolic equations.

Lemma 6.1.8 (Minimum Principle for Parabolic Equations) *Let $\epsilon \geq 0$. If $H \in C([0, T]; C^2(\mathbb{T}^2 \times \mathbb{R}^+) \cap C^1([0, T]; C^0(\mathbb{T}^2 \times \mathbb{R}^+))$ is a bounded function with*

$$\kappa(t) := \min\left\{\min_{\mathbb{T}^2\times\mathbb{R}^+} H(0), \min_{[0,T]\times\mathbb{T}^2} H\big|_{z=0}\right\} \geq 0$$

and satisfies

$$\left\{\partial_t + b_1\partial_x + b_2\partial_y + b_3\partial_z - \epsilon^2\partial_x^2 - \epsilon^2\partial_y^2 - \partial_z^2\right\} H = fH$$

where the coefficients b_1, b_2, b_3 and f are continuous and satisfy

$$\left\|\frac{b_3}{1+z}\right\|_{L^\infty([0,T]\times\mathbb{T}^2\times\mathbb{R}^+)} < +\infty, \tag{6.1.60}$$

and

$$\|f\|_{L^\infty([0,T]\times\mathbb{T}^2\times\mathbb{R}^+)} \leq \lambda, \tag{6.1.61}$$

then for any $t \in [0, T]$,

$$\sup_{\mathbb{T}^2\times\mathbb{R}^+} H(t) \geq (1 - \lambda t e^{\lambda t})\kappa(t). \tag{6.1.62}$$

Proof For any fixed $\tilde{t} \in [0, T]$ and $\mu > 0$, let us define

$$h := e^{-\lambda t}\left(H - \kappa(\tilde{t})\right) + \left(\lambda\kappa(\tilde{t}) + \mu\left\|\frac{b_3}{1+z}\right\|_{L^\infty}\right) \cdot t + \mu\ln(1+z). \tag{6.1.63}$$

Then for any $\tilde{t} \in (0, T]$, we have

$$
\begin{aligned}
&\left\{\partial_t + b_1\partial_x + b_2\partial_y + b_3\partial_z - \epsilon^2\partial_x^2 - \epsilon^2\partial_y^2 - \partial_z^2 + (\lambda - f)\right\} h \\
&= \left\{\partial_t + b_1\partial_x + b_2\partial_y + b_3\partial_z - \epsilon^2\partial_x^2 - \epsilon^2\partial_y^2 - \partial_z^2 + (\lambda - f)\right\} e^{-\lambda t} H \\
&- \left\{\partial_t + b_1\partial_x + b_2\partial_y + b_3\partial_z - \epsilon^2\partial_x^2 - \epsilon^2\partial_y^2 - \partial_z^2 + (\lambda - f)\right\} e^{-\lambda t}\kappa(\tilde{t}) \\
&+ \left\{\partial_t + b_1\partial_x + b_2\partial_y + b_3\partial_z - \epsilon^2\partial_x^2 - \epsilon^2\partial_y^2 - \partial_z^2 + (\lambda - f)\right\} \\
&\left(\lambda\kappa(\tilde{t}) + \mu\left\|\frac{b_3}{1+z}\right\|_{L^\infty}\right)\cdot t \\
&+ \left\{\partial_t + b_1\partial_x + b_2\partial_y + b_3\partial_z - \epsilon^2\partial_x^2 - \epsilon^2\partial_y^2 - \partial_z^2 + (\lambda - f)\right\} \mu\ln(1+z) \\
&= 0 - \{\partial_t + (\lambda - f)\} e^{-\lambda t}\kappa(\tilde{t}) + \{\partial_t + (\lambda - f)\}\lambda\kappa(\tilde{t})\cdot t + \mu\|\frac{b_3}{1+z}\|_{L^\infty} \\
&+ (\lambda - f)\mu\|\frac{b_3}{1+z}\|_{L^\infty}\cdot t + \frac{\mu b_3}{1+z} + \frac{\mu}{(1+z)^2} \\
&= \lambda\kappa(\tilde{t})(e^{-\lambda t} + 1) + (\lambda t - e^{-\lambda t})(\lambda - f)\kappa(\tilde{t}) + (\lambda - f)\mu\|\frac{b_3}{1+z}\|_{L^\infty}\cdot t \\
&+ \frac{\mu b_3}{1+z} + \frac{\mu}{(1+z)^2} \\
&> 0, \quad \text{in} \quad [0, \tilde{t}] \times \mathbb{T}^2 \times \mathbb{R}^+,
\end{aligned}
\tag{6.1.64}
$$

where we have used $\lambda t - e^{-\lambda t} > 0$ for $t > 0$, $\mu\left\|\frac{b_3}{1+z}\right\|_{L^\infty([0,T]\times\mathbb{T}^2\times\mathbb{R}^+)} + \frac{\mu b_3}{1+z} \geq 0$, $\lambda - f \geq 0$, and the following fact:

$$
\left\{\partial_t + b_1\partial_x + b_2\partial_y + b_3\partial_z - \epsilon^2\partial_x^2 - \epsilon^2\partial_y^2 - \partial_z^2\right\} H = fH \quad in\ [0, T] \times \mathbb{T}^2 \times \mathbb{R}^+,
$$

so by the classical maximum principle (Lemma 6.1.6) for parabolic equations, we have

$$
\min_{[0,\tilde{t}]\times\mathbb{T}^2\times[0,R]} h \geq -\max_{\partial([0,T]\times\mathbb{T}^2\times\mathbb{R}^+)} (-\min\{0, h\}) \geq 0,
\tag{6.1.65}
$$

provided that

$$
R \geq \exp\left(\frac{1}{\mu}(\|H\|_{L^\infty(\mathbb{T}^2\times\mathbb{R}^+)} + \kappa(\tilde{t}))\right) - 1.
\tag{6.1.66}
$$

Here, we have ensured that R is chosen to be large enough such that $\ln(1+z)$ is well-defined and finite for all $(x, y, z) \in \mathbb{T}^2 \times \mathbb{R}^+$. And we also ensured that the above inequality itself is established, that is

$$\min\left\{e^{-\lambda t}\left(H - \kappa(\tilde{t})\right) + \left(\lambda\kappa(\tilde{t}) + \mu\left\|\frac{b_3}{1+z}\right\|_{L^\infty}\right)t + \mu\ln(1+z)\right\} \geq 0. \tag{6.1.67}$$

Taking the limit $R \to +\infty$, and then $\mu \to 0^+$, we obtain

$$H(\tilde{t}) \geq (1 - \lambda\tilde{t}e^{\lambda\tilde{t}})\kappa(\tilde{t}) \tag{6.1.68}$$

which implies inequality (6.1.62) if we replace the arbitrary time $\tilde{t}$ by t. Hence, this completes the proof of the lemma.

□

6.2 Uniform Estimates

In this section, we will estimate the norm $\|\varphi\|^2_{H^{s,\gamma}_g}$ to problem (6.1.8) by the energy method, whose idea comes from [98].

6.2.1 Estimates on $D^\alpha\varphi$

We will estimate the norm with weight $(1+z)^{\gamma+\alpha_3}$ on $D^\alpha\varphi$, $\alpha = (\alpha_1, \alpha_2, \alpha_3)$, $|\alpha| \leq s$, and $\alpha_1 + \alpha_2$ should be restricted as $\alpha_1 + \alpha_2 \leq s - 1$. Otherwise, it is impossible to directly estimate the norm with $\partial^s_{xy}\varphi$.

Lemma 6.2.1 (Reduction of Boundary Data) *If φ solves (6.1.6) and (6.1.8), then on the boundary at $z = 0$, we have*

$$\begin{cases} \partial_z\varphi|_{z=0} = \partial_x P, \\ \partial_z^3\varphi|_{z=0} = (\partial_t - \epsilon^2\partial_x^2 - \epsilon^2\partial_y^2)\partial_x P + (\varphi\partial_x\varphi + K\varphi\partial_y\varphi)\big|_{z=0} - \partial_y K|\varphi|^2\big|_{z=0}. \end{cases} \tag{6.2.1}$$

For any $5 \le 2k+1 \le s$, there are some constants $C_{K,k,l,\rho^1,\rho^2,\ldots,\rho^j}$, not depending on ϵ (in this section, the superscript ϵ is removed for simplification) or (u, Ku, w), such that

$$\partial_z^{2k+1}\varphi|_{z=0} = (\partial_t - \epsilon^2\partial_x^2 - \epsilon^2\partial_y^2)^k \partial_x P + \sum_{h=0}^{2k-1} \partial_{xy}^h K \sum_{l=0}^{k-1} \epsilon^{2l} \sum_{j=2}^{\max\{2,k-l\}} \sum_{\rho \in A_{k,l}^j} C_{K,k,l,\rho^1,\rho^2,\ldots,\rho^j} \prod_{i=1}^{j} D^{\rho^i}\varphi\Big|_{z=0}, \tag{6.2.2}$$

where $A_{k,l}^j := \{\rho := (\rho^1, \rho^2, \ldots, \rho^j) \in \mathbb{N}^{2j};\ \sum_{i=1}^{j} \rho_1^i \le k+2l-1,\ \sum_{i=1}^{j} \rho_2^i \le k+2l-1,\ \sum_{i=1}^{j} \rho_{12}^i \le k+2l-1,\ \sum_{i=1}^{j} \rho_3^i \le 2k-2l-2,$ *and* $|\rho^i| \le 2k-l-1$ *for all* $i = 1, 2, \ldots, j\}$.

Proof According to the equation (6.1.8) and the boundary condition $\partial_z\varphi|_{z=0} = \partial_x P$, we get

$$\begin{aligned}\partial_z^3\varphi\big|_{z=0} &= \partial_z\left\{\partial_t\varphi + (u\partial_x + Ku\partial_y + w\partial_z)\varphi - \epsilon^2\partial_x^2\varphi - \epsilon^2\partial_y^2\varphi - \frac{1}{2}\partial_y K\partial_z|u|^2\right\}\Big|_{z=0}\\ &= (\partial_t - \epsilon^2\partial_x^2 - \epsilon^2\partial_y^2)\partial_x P + (\varphi\partial_x\varphi + K\varphi\partial_y\varphi)\big|_{z=0} - \partial_y K|\varphi|^2\big|_{z=0},\end{aligned} \tag{6.2.3}$$

and

$$\begin{aligned}&\varphi_z^{2n+1}\varphi\big|_{z=0}\\ &= \left\{\left(\partial_t - \epsilon^2\partial_x^2\varphi - \epsilon^2\partial_y^2\varphi\right)\partial_z^{2n-1}\varphi + \partial_z^{2n-1}(u\partial_x\varphi + Ku\partial_y\varphi + w\partial_z\varphi\right.\\ &\quad \left. - \frac{1}{2}\partial_y K\partial_z|u|^2)\right\}\Big|_{z=0}.\end{aligned} \tag{6.2.4}$$

Hence, for $n = 2$, we arrive at

$$\begin{aligned}\partial_z^5\varphi\big|_{z=0} &= \left\{(\partial_t - \epsilon^2\partial_x^2 - \epsilon^2\partial_y^2)\partial_z^3\varphi + \partial_z^3(u\partial_x + Ku\partial_y + w\partial_z)\varphi\right.\\ &\qquad \left. - \partial_z^3(\frac{1}{2}\partial_y k\partial_z|u|^2)\right\}\Big|_{z=0}\\ &= (\partial_t - \epsilon^2\partial_x^2 - \epsilon^2\partial_y^2)^2\partial_x P + (\partial_t - \epsilon^2\partial_x^2 - \epsilon^2\partial_y^2)\\ &\quad \times \left\{(\varphi\partial_x\varphi + K\varphi\partial_y\varphi) - \partial_y K|\varphi|^2\right\}\Big|_{z=0}\end{aligned}$$

$$
\begin{aligned}
&+ (3\varphi\partial_x\partial_z^2\varphi + 2\partial_z\varphi\partial_x\partial_z\varphi - 2\partial_x\varphi\partial_z^2\varphi)|_{z=0}\\
&+ K(3\varphi\partial_y\partial_z^2\varphi + 2\partial_z\varphi\partial_y\partial_z\varphi - 2\partial_y\varphi\partial_z^2\varphi)|_{z=0}\\
&- \partial_y K(4\varphi\partial_z^2\varphi + 3\partial_z\varphi\partial_z\varphi)\big|_{z=0}.
\end{aligned} \tag{6.2.5}
$$

Since the right-hand side of (6.2.5) except for the second term in the formula is in the desired form, we only need to deal with the terms $(\partial_t - \epsilon^2\partial_x^2 - \epsilon^2\partial_y^2)\left\{(\varphi\partial_x\varphi + K\varphi\partial_y\varphi) - \partial_y K|\varphi|^2\right\}\big|_{z=0}$. Using the evolution equations for φ, $\partial_x\varphi$ and $\partial_y\varphi$ as well as $(u, w)\big|_{z=0} = 0$, we may check that

$$
\begin{aligned}
&(\partial_t - \epsilon^2\partial_x^2 - \epsilon^2\partial_y^2)\left\{(\varphi\partial_x\varphi + K\varphi\partial_y\varphi) - \partial_y K|\varphi|^2\right\}\big|_{z=0}\\
&= \left\{\partial_t\varphi(\partial_x\varphi + K\partial_y\varphi - 2\partial_y K\varphi) + \varphi(\partial_t\partial_x\varphi + K\partial_t\partial_y\varphi)\right\}\big|_{z=0}\\
&\quad - (\epsilon^2\partial_x^2 + \epsilon^2\partial_y^2)\left\{(\varphi\partial_x\varphi + K\varphi\partial_y\varphi) - \partial_y K|\varphi|^2\right\}\big|_{z=0}\\
&= \left(\epsilon^2\partial_x^2\varphi + \epsilon^2\partial_y^2\varphi + \partial_z^2\varphi\right)(\partial_x\varphi + K\partial_y\varphi - 2\partial_y K\varphi)\big|_{z=0}\\
&\quad + \varphi\left\{\left(\epsilon^2\partial_x^3\varphi + \epsilon^2\partial_x\partial_y^2\varphi + \partial_x\partial_z^2\varphi\right) + K\left(\epsilon^2\partial_y\partial_x^2\varphi + \epsilon^2\partial_y^3\varphi + \partial_y\partial_z^2\varphi\right)\right\}\big|_{z=0}\\
&\quad - (\epsilon^2\partial_x^2 + \epsilon^2\partial_y^2)\left\{(\varphi\partial_x\varphi + K\varphi\partial_y\varphi) - \partial_y K|\varphi|^2\right\}\big|_{z=0},
\end{aligned} \tag{6.2.6}
$$

where we have used the following facts:

$$
\begin{aligned}
&\partial_t\varphi|_{z=0}\\
&= \left\{-(u\partial_x + Ku\partial_y + w\partial_z)\varphi + \epsilon^2\partial_x^2\varphi + \epsilon^2\partial_y^2\varphi + \partial_z^2\varphi + \frac{1}{2}\partial_y K\partial_z|u|^2\right\}\big|_{z=0}\\
&= \left(\epsilon^2\partial_x^2\varphi + \epsilon^2\partial_y^2\varphi + \partial_z^2\varphi\right)\big|_{z=0},\\
&\partial_t\partial_x\varphi|_{z=0}\\
&= \partial_x\left\{-(u\partial_x + Ku\partial_y + w\partial_z)\varphi + \epsilon^2\partial_x^2\varphi + \epsilon^2\partial_y^2\varphi + \partial_z^2\varphi + \frac{1}{2}\partial_y K\partial_z|u|^2\right\}\big|_{z=0}\\
&= \left(\epsilon^2\partial_x^3\varphi + \epsilon^2\partial_x\partial_y^2\varphi + \partial_x\partial_z^2\varphi\right)\big|_{z=0},\\
&\partial_t\partial_y\varphi|_{z=0} = \left(\epsilon^2\partial_y\partial_x^2\varphi + \epsilon^2\partial_y^3\varphi + \partial_y\partial_z^2\varphi\right)\big|_{z=0}.
\end{aligned}
$$

Combining (6.2.6) and (6.2.5), we have

$$
\begin{aligned}
\partial_z^5\varphi\big|_{z=0} &= (\partial_t - \epsilon^2\partial_x^2 - \epsilon^2\partial_y^2)^2\partial_x P + (3\varphi\partial_x\partial_z^2\varphi + 2\partial_z\varphi\partial_x\partial_z\varphi - 2\partial_x\varphi\partial_z^2\varphi)|_{z=0}\\
&\quad + K(3\varphi\partial_y\partial_z^2\varphi + 2\partial_z\varphi\partial_y\partial_z\varphi - 2\partial_y\varphi\partial_z^2\varphi)|_{z=0} - \partial_y K(4\varphi\partial_z^2\varphi + 3\partial_z\varphi\partial_z\varphi)\big|_{z=0}
\end{aligned}
$$

$$+\left(\epsilon^2\partial_x^2\varphi+\epsilon^2\partial_y^2\varphi+\partial_z^2\varphi\right)(\partial_x\varphi+K\partial_y\varphi-2\partial_yK\varphi)\big|_{z=0}$$
$$+\varphi\left\{\left(\epsilon^2\partial_x^3\varphi+\epsilon^2\partial_x\partial_y^2\varphi+\partial_x\partial_z^2\varphi\right)+K\left(\epsilon^2\partial_y\partial_x^2\varphi+\epsilon^2\partial_y^3\varphi+\partial_y\partial_z^2\varphi\right)\right\}\big|_{z=0}$$
$$-(\epsilon^2\partial_x^2+\epsilon^2\partial_y^2)\left\{(\varphi\partial_x\varphi+K\varphi\partial_y\varphi)-\partial_yK|\varphi|^2\right\}\big|_{z=0}, \tag{6.2.7}$$

which implies that

$$\partial_z^5\varphi\big|_{z=0}=(\partial_t-\epsilon^2\partial_x^2-\epsilon^2\partial_y^2)^2\partial_xP$$
$$+(4\varphi\partial_x\partial_z^2\varphi+2\partial_z\varphi\partial_x\partial_z\varphi-\partial_x\varphi\partial_z^2\varphi)|_{z=0}$$
$$+K(4\varphi\partial_y\partial_z^2\varphi+2\partial_z\varphi\partial_y\partial_z\varphi-\partial_y\varphi\partial_z^2\varphi)|_{z=0}$$
$$-\partial_yK(6\varphi\partial_z^2\varphi+3\partial_z\varphi\partial_z\varphi)\big|_{z=0}$$
$$+\left(\epsilon^2\partial_x^2\varphi+\epsilon^2\partial_y^2\varphi\right)(\partial_x\varphi+K\partial_y\varphi-2\partial_yK\varphi)\big|_{z=0}+\varphi\left(\epsilon^2\partial_x^3\varphi+\epsilon^2\partial_x\partial_y^2\varphi\right)\big|_{z=0}$$
$$+K\left(\epsilon^2\partial_y\partial_x^2\varphi+\epsilon^2\partial_y^3\varphi\right)\big|_{z=0}$$
$$-(\epsilon^2\partial_x^2+\epsilon^2\partial_y^2)\left\{(\varphi\partial_x\varphi+K\varphi\partial_y\varphi)-\partial_yK|\varphi|^2\right\}\big|_{z=0}. \tag{6.2.8}$$

We start completing the verification of formula (6.2.2) for $k=2$. When $k=2$: formula (6.2.2) becomes the following the equality

$$\partial_z^5\varphi|_{z=0}=(\partial_t-\epsilon^2\partial_x^2-\epsilon^2\partial_y^2)^2\partial_xP$$
$$+\sum_{h=0}^{3}\partial_{xy}^hK\sum_{l=0}^{1}\epsilon^{2l}\sum_{\rho\in A_{2,l}^2}C_{K,k,l,\rho^1,\rho^2}\prod_{i=1}^{2}D^{\rho^i}\varphi\big|_{z=0}, \tag{6.2.9}$$

where $\sum\limits_{i=1}^{2}\rho_1^i\le 1+2l$, $\sum\limits_{i=1}^{2}\rho_2^i\le 1+2l$, $\sum\limits_{i=1}^{2}\rho_{12}^i\le 1+2l$, $\sum\limits_{i=1}^{2}\rho_3^j\le 2-2l$, *and* $|\rho^i|\le 3-l$ *for all* $i=1,2$.

When $l=0$, then ϵ^{2l}=1, by the second, third, and fourth terms on the right-hand side of the equation (6.2.8), it holds that

$$\sum_{i=1}^{2}\rho_1^i\le 1,\ \sum_{i=1}^{2}\rho_2^i\le 1,\ \sum_{i=1}^{2}\rho_{12}^i\le 1,\ \sum_{i=1}^{2}\rho_3^j\le 2,\ \textit{and}\ |\rho^i|\le 3\ \textit{for all}\ i=1,2.$$

When $l = 1$, then $\epsilon^{2l}=\epsilon^2$, by the fifth, sixth, and seventh terms on the right-hand side of the equation (6.2.8), it holds that

$$\sum_{i=1}^{2} \rho_1^i \leq 3, \sum_{i=1}^{2} \rho_2^i \leq 3, \sum_{i=1}^{2} \rho_{12}^i \leq 3, \sum_{i=1}^{2} \rho_3^j \leq 0, \ and\ |\rho^i| \leq 2\ for\ all\ i = 1, 2.$$

Note that although $\epsilon^2\partial_x^3\varphi$ and $\epsilon^2\partial_y^3\varphi$ appears in the seventh term of the right-hand side of the equation (6.2.8), it can be canceled out by $-(\epsilon^2\partial_x^2 + \epsilon^2\partial_y^2)\left(\varphi\partial_x\varphi + K\varphi\partial_y\varphi\right)$ in the eighth term. Therefore, $|\rho^i| \leq 2$ is indeed true.

Moreover, $\partial_{xy}^h K$ is not only related to k in (6.2.2) and (6.2.9) in fact, but also related to other quantities. However, due to assumption (6.1.16), we do not need to accurately give the relationship between $\partial_{xy}^h K$ and other quantities.

According to the above argument, the equality (6.2.8) satisfies (6.2.9); we have completed the verification of formula (6.2.2) for $k = 2$.

Now, using the same algorithm, we are going to prove formula (6.2.2) by induction on k. For notational convenience, we denote

$$\mathcal{A}_k := \left\{ \sum_{h=0}^{2k-1} \partial_{xy}^h K \sum_{l=0}^{k-1} \epsilon^{2l} \sum_{j=2}^{\max\{2,k-l\}} \sum_{\rho\in A_{k,l}^j} C_{K,k,l,\rho^1,\rho^2,\ldots,\rho^j} \prod_{i=1}^{j} D^{\rho^i}\varphi\big|_{z=0} \right\}. \tag{6.2.10}$$

Using this notation, we will prove $\partial_z^{2k+1}\varphi\big|_{z=0} - (\partial_t - \epsilon^2\partial_x^2 - \epsilon^2\partial_y^2)^k\partial_x P \in \mathcal{A}_k$. Assuming that formula (6.2.2) holds for $k = n$, we will show that it also holds for $k = n+1$ as follows. Then we differentiate the vorticity equation $2n+1$ times with respect to z to obtain

$$\begin{aligned}
&\partial_z^{2n+3}\varphi\big|_{z=0}\\
&= \Big\{ \left(\partial_t - \epsilon^2\partial_x^2\varphi - \epsilon^2\partial_y^2\varphi\right)\partial_z^{2n+1}\varphi + \partial_z^{2n+1}\\
&\quad\times (u\partial_x\varphi + Ku\partial_y\varphi + w\partial_z\varphi - \partial_y K\varphi u)\Big\}\big|_{z=0}\\
&= \left(\partial_t - \epsilon^2\partial_x^2\varphi - \epsilon^2\partial_y^2\varphi\right)\partial_z^{2n+1}\varphi\big|_{z=0}\\
&\quad + \sum_{m=1}^{2n+1}\binom{2n+1}{m}\partial_z^{m-1}\varphi(\partial_x + K\partial_y - \partial_y K)\partial_z^{2n-m+1}\varphi\big|_{z=0}\\
&\quad - \sum_{m=2}^{2n+1}\binom{2n+1}{m}(\partial_x + K\partial_y)\partial_z^{m-2}\varphi\partial_z^{2n-m+2}\varphi\big|_{z=0}.
\end{aligned} \tag{6.2.11}$$

The last two terms of (6.2.11) satisfy the situation where $k = n+1$, $j = 2$ and $l = 0$ in (6.2.2). So $\sum_{i=1}^{2} \rho_3^i$ and $|\rho^i|$ need to satisfy $\sum_{i=1}^{j} \rho_3^i \le 2n$ and $|\rho^i| \le 2n+1$ for all $i = 1, 2, \dots, j$ respectively. By routine checking, it is easy to obtain $\sum_{i=1}^{j} \rho_3^i = (m-1)+2n-m+1$ or $(m-2)+2n-m+2 = 2n$ and $\max|\rho^i| \le 2n+1$ (i.e., $m = 1$ in $(\partial_x + K\partial_y)\partial_z^{2n-m+1}\varphi$); one may show that the last two terms of (6.2.11) belong to $\mathcal{A}_{k+1}$, so it only remains to deal with the term $(\partial_t - \epsilon^2\partial_x^2 - \epsilon^2\partial_y^2)\partial_z^{2n+1}\varphi\big|_{z=0}$.

Thanks to the induction hypothesis, there exist constants $C_{K,n,l,\rho^1,\rho^2,\dots,\rho^j}$ such that

$$\partial_z^{2n+1}\varphi\big|_{z=0} = (\partial_t - \epsilon^2\partial_x^2 - \epsilon^2\partial_y^2)^n\partial_x P + \sum_{h=0}^{2k-1} \partial_{xy}^h K \sum_{l=0}^{k-1} \epsilon^{2l} \sum_{j=2}^{\max\{2,n-l\}} \sum_{\rho\in A_{n,l}^j} C_{K,n,l,\rho^1,\rho^2,\dots,\rho^j} \prod_{i=1}^{j} D^{\rho^i}\varphi\big|_{z=0}. \tag{6.2.12}$$

Thus we have, up to a relabeling of the indices ρ^i,

$$\begin{aligned}
&(\partial_t - \epsilon^2\partial_x^2 - \epsilon^2\partial_y^2)\partial_z^{2n+1}\varphi\big|_{z=0} \\
&= (\partial_t - \epsilon^2\partial_x^2 - \epsilon^2\partial_y^2)^{n+1}\partial_x P \\
&+ \sum_{h=0}^{2n-1} \partial_{xy}^h K \sum_{l=0}^{n-1} \epsilon^{2l} \sum_{j=2}^{\max\{2,n-l\}} \sum_{\rho\in A_{n,l}^j} \widetilde{C}_{K,n,l,\rho^1,\rho^2,\dots,\rho^j} (\partial_t - \epsilon^2\partial_x^2 - \epsilon^2\partial_y^2) D^{\rho^1}\varphi \prod_{i=2}^{j} \\
&\quad D^{\rho^i}\varphi\big|_{z=0} \\
&- \sum_{h=0}^{2n-1} \partial_{xy}^h K \sum_{l=0}^{n-1} \epsilon^{2l+2} \sum_{j=2}^{\max\{2,n-l\}} \sum_{\rho\in A_{n,l}^j} \widetilde{\widetilde{C}}_{K,n,l,\rho^1,\rho^2,\dots,\rho^j} (\partial_x + \partial_y) D^{\rho^1}\varphi(\partial_x + \partial_y) \\
&\quad D^{\rho^2}\varphi \prod_{i=3}^{j} D^{\rho^i}\varphi\big|_{z=0},
\end{aligned} \tag{6.2.13}$$

where $\widetilde{C}_{K,n,l,\rho^1,\rho^2,\dots,\rho^j}$ and $\widetilde{\widetilde{C}}_{K,n,l,\rho^1,\rho^2,\dots,\rho^j}$ are some new constants depending on $C_{K,n,l,\rho^1,\rho^2,\dots,\rho^j}$.

For $\mathcal{A}_n$, $\rho^i \leq 2n - l - 1$; for $\mathcal{A}_{n+1}$, $\rho^i \leq 2n - l + 1$. It can be seen that every time the number of n increases by 1, the maximum value of the order ρ^i will increase by 2. Due to

$$-\sum_{h=0}^{2n-1} \partial_{xy}^h K \sum_{l=0}^{n-1} \epsilon^{2l} \sum_{j=2}^{\max\{2,n-l\}} \sum_{\rho \in A_{n,l}^j} \tilde{\tilde{C}}_{K,n,l,\rho^1,\rho^2,\ldots,\rho^j} \prod_{i=2}^{j} D^{\rho^i} \varphi\big|_{z=0} \in \mathcal{A}_{n+1}, \tag{6.2.14}$$

the last term on the right-hand side of (6.2.13) belongs to $\mathcal{A}_{n+1}$. So it remains to check whether the second term on right-hand side of (6.2.13) also belongs to $\mathcal{A}_{n+1}$.

Indeed, differentiating the vorticity equation $(6.1.8)_1$ ρ_1^1 times, ρ_2^1 times, ρ_3^1 times with respect to x, y, z respectively, and then calculating at $z = 0$, we have, by denoting $e_3 = (0, 0, 1)$,

$$\begin{aligned}
&(\partial_t - \epsilon^2\partial_x^2 - \epsilon^2\partial_y^2) D^{\rho^1}\varphi\big|_{z=0} \\
&= -\sum_{\substack{\beta \leq \rho^1 \\ \beta_3 \geq 1}} \binom{\rho^1}{\beta} D^{\beta - e_3}\varphi \partial_x D^{\rho_1 - \beta}\varphi\big|_{z=0} - \sum_{\substack{\beta \leq \rho^1 \\ \beta_3 \geq 1}} \binom{\rho^1}{\beta} D^{\beta - e_3}(K\varphi)\partial_y D^{\rho_1 - \beta}\varphi\big|_{z=0} \\
&\quad + \sum_{\substack{\beta \leq \rho^1 \\ \beta_3 \geq 2}} \binom{\rho^1}{\beta} D^{\beta - 2e_3}(\partial_x\varphi + \partial_y(K\varphi))\partial_z D^{\rho_1 - \beta} w\big|_{z=0} + \partial_z^2 D^{\rho^1}\varphi\big|_{z=0} \\
&\quad + \sum_{\substack{\beta \leq \rho^1 \\ \beta_3 \geq 1}} \binom{\rho^1}{\beta} D^{\beta - e_3}(\partial_y K\varphi) D^{\rho_1 - \beta}\varphi\big|_{z=0}.
\end{aligned} \tag{6.2.15}$$

Using (6.2.15), one may justify by a routine counting of indices that the second term on the right-hand side of (6.2.13) belongs to $\mathcal{A}_{n+1}$. This thus completes the proof. □

Lemma 6.2.2 *Let $s \geq 5$ be an integer, $\gamma \geq 1$, $\sigma > \gamma + \frac{1}{2}$ and $\delta \in (0, 1)$. If $\varphi \in H_{\sigma,\delta}^{s,\gamma}$ solves* (6.1.6) *and* (6.1.8), *we have the following estimates:*

(i) when $|\alpha| \leq s - 1$,

$$\left| \iint_{\mathbb{T}^2} D^\alpha \varphi \partial_z D^\alpha \varphi dxdy\big|_{z=0} \right| \leq \frac{1}{12} \|(1+z)^{\gamma + \alpha_3 + 1}\partial_z^2 D^\alpha \varphi\|_{L^2}^2 + C\|\varphi\|_{H_g^{s,\gamma}}^2, \tag{6.2.16}$$

(ii) when $|\alpha| = s$, α_3 *is even,*

$$\left|\iint_{\mathbb{T}^2} D^\alpha \varphi \partial_z D^\alpha \varphi dxdy\big|_{z=0}\right|$$

$$\leq \frac{1}{12}\|(1+z)^{\gamma+\alpha_2}\partial_z D^\alpha \varphi\|_{L^2}^2 + \sum_{l=0}^{\frac{s}{2}} \|\partial_t^l \partial_x P\|_{H^{s-2l}(\mathbb{T}^2)}^2$$

$$+ C_{s,\gamma,\sigma,\delta}\|\partial_{xy}^s K\|_{L^\infty(\mathbb{T}^2)}(1+\|\varphi\|_{H_g^{s,\gamma}}^2)^{s-2}\|\varphi\|_{H_g^{s,\gamma}}^2, \tag{6.2.17}$$

(iii) when $|\alpha| = s$, α_3 *is odd,*

$$\left|\iint_{\mathbb{T}^2} D^\alpha \varphi \partial_z D^\alpha \varphi dxdy\big|_{z=0}\right|$$

$$\leq \frac{1}{12}\|(1+z)^{\gamma+\alpha_3+1}\partial_x^{\alpha_1-1}\partial_y^{\alpha_2}\partial_z^{\alpha_3+2}\varphi\|_{L^2}^2 + \sum_{l=0}^{\frac{s}{2}} \|\partial_t^l \partial_x P\|_{H^{s-2l}(\mathbb{T})^2}^2$$

$$+ C_{s,\gamma,\sigma,\delta}\|\partial_{xy}^s K\|_{L^\infty(\mathbb{T}^2)}(1+\|\varphi\|_{H_g^{s,\gamma}}^2)^{s-2}\|\varphi\|_{H_g^{s,\gamma}}^2. \tag{6.2.18}$$

Proof

Case 1. When $|\alpha| \leq s-1$, using the following trace estimate:

$$\iint_{\mathbb{T}^2} |f| dxdy\big|_{z=0} \leq C\left(\int_0^1 \int_{\mathbb{T}^2} |f| dxdydz + \int_0^1 \int_{\mathbb{T}^2} |\partial_z f| dxdydz\right), \tag{6.2.19}$$

we have

$$\left|\int_{\mathbb{T}^2} D^\alpha \varphi \partial_z D^\alpha \varphi dxdy\big|_{z=0}\right| \leq \frac{1}{12}\|(1+z)^{\gamma+\alpha_3+1}\partial_z^2 D^\alpha \varphi\|_{L^2}^2 + C\|\varphi\|_{H_g^{s,\gamma}}^2. \tag{6.2.20}$$

Case 2. When $|\alpha| = s$, $\alpha_3 = 2k$ for some $k \in \mathbb{N}$, we can apply Lemma 6.2.1 to $\partial_z D^\alpha \varphi|_{z=0}$ to obtain

$$\int_{\mathbb{T}} D^\alpha \varphi \partial_z D^\alpha \varphi dxdy\big|_{z=0}$$

$$= \int_{\mathbb{T}^2} D^\alpha \varphi(\partial_t - \epsilon^2\partial_x^2 - \epsilon^2\partial_y^2)^k \partial_x^{\alpha_1}\partial_y^{\alpha_2}\partial_x P dxdy\big|_{z=0}$$

$$+\sum_{l=0}^{2k-1}\partial_{xy}^{l}K\sum_{l=0}^{k-1}\epsilon^{2l}\sum_{j=2}^{\max\{2,k-l\}}\sum_{\rho\in A_{k,l}^{j}}C_{K,k,l,\rho^1,\rho^2,\dots,\rho^j}\int_{\mathbb{T}^2}$$

$$D^{\alpha}\varphi\partial_x^{\alpha_1}\partial_y^{\alpha_2}\left(\prod_{i=1}^{j}D^{\rho^i}\varphi\right)dx\big|_{z=0}. \tag{6.2.21}$$

Then we again apply the simple trace estimate to control the boundary integral as follows:

$$\left|\int_{\mathbb{T}^2}D^{\alpha}\varphi\partial_z D^{\alpha}\varphi dxdy\big|_{z=0}\right|\le\frac{1}{12}\|(1+z)^{\gamma+\alpha_3}\partial_z D^{\alpha}\varphi\|_{L^2}^2+\sum_{l=0}^{\frac{s}{2}}\|\partial_t^l\partial_x P\|_{H^{s-2l}(\mathbb{T}^2)}^2$$
$$+C_{s,\gamma,\sigma,\delta}\|\partial_{xy}^{s}K\|_{L^\infty(\mathbb{T}^2)}(1+\|\varphi\|_{H_g^{s,\gamma}})^{s-2}\|\varphi\|_{H_g^{s,\gamma}}^2. \tag{6.2.22}$$

Case 3. When $|\alpha|=s$, $\alpha_3=2k+1$ for some $k\in\mathbb{N}$, using integration by parts in the x, y variables, we have

$$\int_{\mathbb{T}^2}D^{\alpha}\varphi\partial_z D^{\alpha}\varphi dxdy\big|_{z=0}=-\int_{\mathbb{T}^2}\partial_x D^{\alpha}\varphi\partial_x^{\alpha_1-1}\partial_y^{\alpha_2}\partial_z^{\alpha_3+1}\varphi dxdy\big|_{z=0}. \tag{6.2.23}$$

The term $\partial_x D^{\alpha}\varphi\big|_{z=0}=\partial_x^{\alpha_1-1}\partial_y^{\alpha_2}\partial_z^{\alpha_3+1}\varphi\big|_{z=0}$ has an odd number of derivatives in z; then we get

$$\left|\int_{\mathbb{T}^2}D^{\alpha}\varphi\partial_y D^{\alpha}\varphi dxdy\big|_{z=0}\right|$$
$$\le\frac{1}{12}\|(1+z)^{\gamma+\alpha_3+1}\partial_x^{\alpha_1-1}\partial_y^{\alpha_2}\partial_z^{\alpha_3+2}\varphi\|_{L^2}^2+\sum_{l=0}^{\frac{s}{2}}\|\partial_t^l\partial_x P\|_{H^{s-2l}(\mathbb{T}^2)}^2$$
$$+C_{s,\gamma,\sigma,\delta}\|\partial_{xy}^{s}K\|_{L^\infty(\mathbb{T}^2)}(1+\|\varphi\|_{H_g^{s,\gamma}})^{s-2}\|\varphi\|_{H_g^{s,\gamma}}^2. \tag{6.2.24}$$

□

Proposition 6.2.1 *Let $s \geq 5$ be an integer, $\gamma \geq 1$, $\sigma > \gamma + \frac{1}{2}$ and $\delta \in (0, 1)$. If $\varphi \in H^{s,\gamma}_{\sigma,\delta}$ solves* (6.1.6) *and* (6.1.8), *then we have the following uniform (in ϵ) estimate:*

$$\begin{aligned}
&\frac{d}{dt}\|(1+z)^{\gamma+\alpha_3}D^\alpha\varphi\|^2_{L^2}\\
&\quad\leq C_{s,\gamma,\sigma,\delta}\left(1+\|\partial^s_{xy}K\|_{L^\infty(\mathbb{T}^2)}\right)\left(\|\varphi\|_{H^{s,\gamma}_g}+\|\partial^s_{xy}U\|_{L^2(\mathbb{T}^2)}+1\right)\|\varphi\|^2_{H^{s,\gamma}_g}\\
&\quad+C_{s,\gamma,\sigma,\delta}\|\partial^s_{xy}K\|_{L^\infty(\mathbb{T}^2)}(1+\|w\|_{H^{s,\gamma}_g})^{s-2}\|\varphi\|^2_{H^{s,\gamma}_g}\\
&\quad+\sum_{l=0}^{\frac{s}{2}}\|\partial^l_t\partial_x P\|^2_{H^{s-2l}(\mathbb{T}^2)}.
\end{aligned}\tag{6.2.25}$$

Proof Applying the operator D^α on $(6.1.8)_1$ with $\alpha = (\alpha_1, \alpha_2, \alpha_3)$, $|\alpha| \leq s$, $\alpha_1 + \alpha_2 \leq s - 1$,

$$\begin{aligned}
&(\partial_t+u\partial_x+Ku\partial_y+w\partial_z-\varepsilon^2\partial_x^2\varphi-\varepsilon^2\partial_y^2\varphi-\partial_z^2)D^\alpha\varphi\\
&=-\sum_{0<\beta\leq\alpha}\binom{\alpha}{\beta}\left\{D^\beta u\partial_xD^{\alpha-\beta}\varphi+D^\beta(Ku)\partial_yD^{\alpha-\beta}\varphi+D^\beta w\partial_zD^{\alpha-\beta}\varphi\right\}\\
&\quad+\sum_{0\leq\beta\leq\alpha}\binom{\alpha}{\beta}D^\beta(u\partial_yK)D^{\alpha-\beta}\varphi.
\end{aligned}\tag{6.2.26}$$

Multiplying (6.2.26) by $(1+z)^{2\gamma+2\alpha_3}D^\alpha\varphi$ and integrating it over $\mathbb{T}^2\times\mathbb{R}^+$ yield

$$\begin{aligned}
&\frac{1}{2}\frac{d}{dt}\|(1+z)^{\gamma+\alpha_3}D^\alpha\varphi\|^2_{L^2}+\|(1+z)^{\gamma+\alpha_3}\partial_zD^\alpha\varphi\|^2_{L^2}+\varepsilon^2\|(1+z)^{\gamma+\alpha_3}\partial_zD^\alpha\varphi\|^2_{L^2}\\
&+\varepsilon^2\|(1+z)^{\gamma+\alpha_3}\partial_zD^\alpha\varphi\|^2_{L^2}\\
&=-\iint_{\mathbb{T}^2}D^\alpha\varphi\partial_zD^\alpha\varphi dxdy|_{z=0}-(2\gamma+2\alpha_3)\iiint(1+z)^{2\gamma+2\alpha_3-1}D^\alpha\varphi\partial_zD^\alpha\varphi\\
&\quad+(\gamma+\alpha_3)\iiint(1+z)^{2\gamma+2\alpha_3-1}w|D^\alpha\varphi|^2\\
&\quad-\sum_{0<\beta\leq\alpha}\binom{\alpha}{\beta}\iiint(1+z)^{2\gamma+2\alpha_3}D^\alpha\varphi\\
&\left\{D^\beta u\partial_xD^{\alpha-\beta}\varphi+D^\beta(Ku)\partial_yD^{\alpha-\beta}\varphi+D^\beta w\partial_zD^{\alpha-\beta}\varphi\right\}\\
&\quad+\sum_{0<\beta\leq\alpha}\binom{\alpha}{\beta}\iiint(1+z)^{2\gamma+2\alpha_3}D^\alpha\varphi\left\{D^\beta(u\partial_yK)D^{\alpha-\beta}\varphi\right\}.
\end{aligned}\tag{6.2.27}$$

Indeed, (6.2.27) can be obtained by integration by parts and by using the boundary condition, i.e.,

$$\begin{aligned}&\iiint (1+z)^{2\gamma+2\alpha_3} D^\alpha\varphi(u\partial_x + ku\partial_y + w\partial_z)D^\alpha\varphi\\&= -(\gamma+\alpha_3)\iiint (1+z)^{2\gamma+2\alpha_3-1} w|D^\alpha\varphi|^2,\end{aligned} \tag{6.2.28}$$

and

$$\begin{aligned}&-\iiint (1+z)^{2\gamma+2\alpha_3} D^\alpha\varphi\partial_z^2 D^\alpha\varphi\\&= \|(1+z)^{\gamma+\alpha_3}\partial_z D^\alpha\varphi\|_{L^2} + \iint_{\mathbb{T}^2} D^\alpha\varphi\partial_z D^\alpha\varphi dxdy|_{z=0}\\&+ (2\gamma+2\alpha_3)\iiint (1+z)^{2\gamma+2\alpha_3-1} D^\alpha\varphi\partial_z D^\alpha\varphi.\end{aligned} \tag{6.2.29}$$

We need to estimate equation (6.2.27) term by term. Obviously, the first term on the right-hand side of (6.2.27) follows from Lemmas 6.2.1–6.2.2.

Secondly, using the Hölder inequality, we have

$$\begin{aligned}&\left|(2\gamma+2\alpha_3)\iiint (1+z)^{2\gamma+2\alpha_3-1} D^\alpha\varphi\partial_z D^\alpha\varphi\right|\\&\le (2\gamma+2\alpha_3)\|\frac{1}{1+z}\|_{L^\infty}\|(1+z)^{\gamma+\alpha_3}D^\alpha\varphi\|_{L^2}\|(1+z)^{\gamma+\alpha_3}\partial_z D^\alpha\varphi\|_{L^2}\\&\le C_{\gamma,\alpha}\|(1+z)^{\gamma+\alpha_3}D^\alpha\varphi\|_{L^2}^2 + \frac{1}{4}\|(1+z)^{\gamma+\alpha_3}\partial_z D^\alpha\varphi\|_{L^2}^2\\&\le C_{\gamma,\alpha}\|\varphi\|_{H_g^{s,\gamma}}^2 + \frac{1}{4}\|(1+z)^{\gamma+\alpha_3}\partial_z D^\alpha\varphi\|_{L^2}^2,\end{aligned} \tag{6.2.30}$$

which, together with Lemma 6.1.5, yields

$$\begin{aligned}&\left|(\gamma+\alpha_3)\iiint (1+z)^{2\gamma+2\alpha_3-1} w|D^\alpha\varphi|^2\right| \le \|\frac{w}{1+z}\|_{L^\infty}\|(1+z)^{\gamma+\alpha_3}D^\alpha\varphi\|_{L^2}^2\\&\le C_{s,\gamma,\sigma,\delta}\left(\|\varphi\|_{H_g^{s,\gamma}} + \|\partial_{xy}^s U\|_{L^2(\mathbb{T}^2)} + 1\right)\|\varphi\|_{H_g^{s,\gamma}}^2.\end{aligned} \tag{6.2.31}$$

Last, it follows that the first term of the third line on the right-hand side of equation (6.2.27) has the following three cases:

$$J_1 = -\sum_{0<\beta\le\alpha}\binom{\alpha}{\beta}\iiint (1+z)^{2\gamma+2\alpha_3} D^\alpha\varphi D^\beta u\partial_x D^{\alpha-\beta}\varphi, \tag{6.2.32}$$

$$J_2 = -\sum_{0<\beta\le\alpha}\binom{\alpha}{\beta}\iiint(1+z)^{2\gamma+2\alpha_3}D^\alpha\varphi D^\beta(Ku)\partial_y D^{\alpha-\beta}\varphi, \quad (6.2.33)$$

and

$$J_3 = -\sum_{0<\beta\le\alpha}\binom{\alpha}{\beta}\iiint(1+z)^{2\gamma+2\alpha_3}D^\alpha\varphi D^\beta w\partial_z D^{\alpha-\beta}\varphi. \quad (6.2.34)$$

Here J_1, J_2 and J_3 can be estimated in the following.

For J_1, when $\beta_3>0$, we have

$$\begin{aligned}
|J_1| &\le \left|\sum_{0<\beta\le\alpha}\binom{\alpha}{\beta}\iiint(1+z)^{2\gamma+2\alpha_3}D^\alpha\varphi D^{\beta-e_3}\varphi D^{\alpha-\beta+e_1}\varphi\right| \\
&\le C\|(1+z)^{\gamma+\beta_3-1}D^{\beta-e_3}\varphi\cdot(1+z)^{\gamma+\alpha_3-\beta_3}D^{\alpha-\beta+e_1}\varphi\|_{L^2}\|(1+z)^{\gamma+\alpha_3}D^\alpha\|_{L^2} \\
&\le C\|(1+z)^{\gamma+\alpha_3}D^\alpha\|_{L^2} \\
&\quad\times\begin{cases}\|(1+z)^{\gamma+\beta_3-1}D^{\beta-e_3}\varphi\|_{L^\infty}\|(1+z)^{\gamma+\alpha_3-\beta_3}D^{\alpha-\beta+e_1}\varphi\|_{L^2}, & \beta\le s-2,\\ \|(1+z)^{\gamma+\beta_3-1}D^{\beta-e_3}\varphi\|_{L^2}\|(1+z)^{\gamma+\alpha_3-\beta_3}D^{\alpha-\beta+e_1}\varphi\|_{L^\infty}, & \beta\ge s-3,\end{cases} \\
&\le C_{s,\gamma,\sigma,\delta}\|\varphi\|^3_{H_g^{s,\gamma}}. \quad (6.2.35)
\end{aligned}$$

When $\beta_3=0$, $\beta_1+\beta_2=s-1$, we get

$$\begin{aligned}
|J_1| &\le \left|\sum_{0<\beta\le\alpha}\binom{\alpha}{\beta}\iiint(1+z)^{2\gamma+2\alpha_3}D^\alpha\varphi\partial_{xy}^{s-1}u\partial_x D^{\alpha+1-s}\varphi\right| \\
&\le C\left(\|\partial_{xy}^{s-1}(u-U)\|_{L^2}+\|\partial_{xy}^{s-1}U\|_{L^2}\right)\|(1+z)^{\gamma+\alpha_3}\partial_x D^{\alpha+1-s}\varphi\|_{L^\infty} \\
&\qquad\times\|(1+z)^{\gamma+\alpha_3}D^\alpha\|_{L^2} \\
&\le C\left(\left\|\frac{1}{(1+z)^{\gamma-1}}\right\|_{L^\infty}\|(1+z)^{\gamma-1}\partial_{xy}^{s-1}(u-U)\|_{L^2}+\|\partial_{xy}^{s-1}U\|_{L^2}\right)\|\varphi\|^2_{H_g^{s,\gamma}} \\
&\le C_{s,\gamma,\sigma,\delta}\left(\|\varphi\|_{H_g^{s,\gamma}}+\|\partial_{xy}^s U\|_{L^2(\mathbb{T}^2)}\right)\|\varphi\|^2_{H_g^{s,\gamma}}. \quad (6.2.36)
\end{aligned}$$

When $\beta_3=0$, $\beta_1+\beta_2\le s-2$, we derive

$$\begin{aligned}
|J_1| &\le \left|\sum_{0<\beta\le\alpha}\binom{\alpha}{\beta}\iiint(1+z)^{2\gamma+2\alpha_3}D^\alpha\varphi\partial_x^{\beta_1}\partial_y^{\beta_2}u\partial_x^{\alpha_1-\beta_1+1}\partial_y^{\alpha_2-\beta_2}\partial_z^{\alpha_3}\varphi\right| \\
&\le C\|\partial_x^{\beta_1}\partial_y^{\beta_2}u\|_{L^\infty}\|(1+z)^{\gamma+\alpha_3}\partial_x^{\alpha_1-\beta_1+1}\partial_y^{\alpha_2-\beta_2}\partial_z^{\alpha_3}\varphi\|_{L^2}\|(1+z)^{\gamma+\alpha_3}D^\alpha\|_{L^2}
\end{aligned}$$

$$\leq C_{s,\gamma,\sigma,\delta}\left(\|\varphi\|_{H_g^{s,\gamma}} + \|\partial_{xy}^s U\|_{L^2(\mathbb{T}^2)}\right)\|\varphi\|_{H_g^{s,\gamma}}^2. \tag{6.2.37}$$

For J_2, when $\beta_3 > 0$, we obtain

$$\begin{aligned}
|J_2| &\leq \left|\sum_{0<\beta\leq\alpha}\binom{\alpha}{\beta}\iiint(1+z)^{2\gamma+2\alpha_3}D^\alpha\varphi D^\beta(Ku)\partial_y D^{\alpha-\beta}\varphi\right| \\
&\leq C\|(1+z)^{\gamma+\beta_3-1}D^{\beta-e_3}(K\varphi)\cdot(1+z)^{\gamma+\alpha_3-\beta_3}D^{\alpha-\beta+e_2}\varphi\|_{L^2}\|(1+z)^{\gamma+\alpha_3}D^\alpha\|_{L^2} \\
&\leq C\|(1+z)^{\gamma+\alpha_3}D^\alpha\|_{L^2} \\
&\times\begin{cases}\|(1+z)^{\gamma+\beta_3-1}D^{\beta-e_3}(K\varphi)\|_{L^\infty}\|(1+z)^{\gamma+\alpha_3-\beta_3}D^{\alpha-\beta+e_2}\varphi\|_{L^2}, & \beta\leq s-2,\\ \|(1+z)^{\gamma+\beta_3-1}D^{\beta-e_3}(K\varphi)\|_{L^2}\|(1+z)^{\gamma+\alpha_3-\beta_3}D^{\alpha-\beta+e_2}\varphi\|_{L^\infty}, & \beta\geq s-3,\end{cases} \\
&\leq C_{s,\gamma,\sigma,\delta}\|\partial_{xy}^s K\|_{L^\infty(\mathbb{T}^2)}\|\varphi\|_{H_g^{s,\gamma}}^3.
\end{aligned} \tag{6.2.38}$$

When $\beta_3 = 0$, $\beta_1+\beta_2 = s-1$, we conclude

$$\begin{aligned}
|J_2| &\leq \left|\sum_{0<\beta\leq\alpha}\binom{\alpha}{\beta}\iiint(1+z)^{2\gamma+2\alpha_3}D^\alpha\varphi\partial^{s-1}(Ku)\partial_y D^{\alpha+1-s}\varphi\right| \\
&\leq C\left(\|\partial_{xy}^{s-1}K(u-U)\|_{L^2} + \|\partial_{xy}^{s-1}KU\|_{L^2}\right)\|(1+z)^{\gamma+\alpha_3}\partial_y D^{\alpha+1-s}\varphi\|_{L^\infty} \\
&\|(1+z)^{\gamma+\alpha_3}D^\alpha\|_{L^2} \\
&\leq C\left(\left\|\frac{1}{(1+z)^{\gamma-1}}\right\|_{L^\infty}\|(1+z)^{\gamma-1}\partial_{xy}^{s-1}K(u-U)\|_{L^2} + \|\partial_{xy}^{s-1}KU\|_{L^2}\right)\|\varphi\|_{H_g^{s,\gamma}}^2 \\
&\leq C_{s,\gamma,\sigma,\delta}\|\partial_{xy}^s K\|_{L^\infty(\mathbb{T}^2)}\left(\|\varphi\|_{H_g^{s,\gamma}} + \|\partial_{xy}^s U\|_{L^2(\mathbb{T}^2)}\right)\|\varphi\|_{H_g^{s,\gamma}}^2.
\end{aligned} \tag{6.2.39}$$

When $\beta_3 = 0$, $\beta_1+\beta_2 \leq s-2$, we get

$$\begin{aligned}
|J_2| &\leq \left|\sum_{0<\beta\leq\alpha}\binom{\alpha}{\beta}\iiint(1+z)^{2\gamma+2\alpha_3}D^\alpha\varphi\partial_x^{\beta_1}\partial_y^{\beta_2}(Ku)\partial_x^{\alpha_1-\beta_1}\partial_y^{\alpha_2-\beta_2+1}\partial_z^{\alpha_3}\varphi\right| \\
&\leq C\|\partial_x^{\beta_1}\partial_y^{\beta_2}(Ku)\|_{L^\infty}\|(1+z)^{\gamma+\alpha_3}\partial_x^{\alpha_1-\beta_1}\partial_y^{\alpha_2-\beta_2+1}\partial_z^{\alpha_3}\varphi\|_{L^2}\|(1+z)^{\gamma+\alpha_3}D^\alpha\|_{L^2} \\
&\leq C_{s,\gamma,\sigma,\delta}\|\partial_{xy}^s K\|_{L^\infty(\mathbb{T}^2)}\left(\|\varphi\|_{H_g^{s,\gamma}} + \|\partial_{xy}^s U\|_{L^2(\mathbb{T}^2)}\right)\|\varphi\|_{H_g^{s,\gamma}}^2.
\end{aligned} \tag{6.2.40}$$

For J_3, when $\beta_3 = 0$, $\beta_1 + \beta_2 = s - 2$ or $s - 1$, we achieve

$$
\begin{aligned}
|J_3| &\le \left| \sum_{0<\beta\le\alpha} \binom{\alpha}{\beta} \iiint (1+z)^{2\gamma+2\alpha_3} D^\alpha \varphi \partial_x^{\beta_1} \partial_y^{\beta_2} w \partial_x^{\alpha_1-\beta_1} \partial_y^{\alpha_2-\beta_2} \partial_z^{\alpha_3+1} \varphi \right| \\
&\le C\|(1+z)^{\gamma+\alpha_3} D^\alpha \varphi\|_{L^2} \|(1+z)^{-1}(\partial_x^{\beta_1}\partial_y^{\beta_2} w + z\tilde{U})\|_{L^2} \\
&\quad \|(1+z)^{\gamma+\alpha_3+1} \partial_x^{\alpha_1-\beta_1} \partial_y^{\alpha_2-\beta_2} \partial_z^{\alpha_3+1} \varphi\|_{L^\infty} \\
&\quad + C\|(1+z)^{\gamma+\alpha_3} D^\alpha \varphi\|_{L^2} \|(1+z)^{-1} z\tilde{U}\|_{L^\infty} \|(1+z)^{\gamma+\alpha_3+1} \partial_x^{\alpha_1-\beta_1} \partial_y^{\alpha_2-\beta_2} \partial_z^{\alpha_3+1} \varphi\|_{L^2} \\
&\le C_{s,\gamma,\sigma,\delta} \left(\|\varphi\|_{H_g^{s,\gamma}} + \|\partial_{xy}^s U\|_{L^2(\mathbb{T}^2)} \right) \|\varphi\|_{H_g^{s,\gamma}}^2 .
\end{aligned} \tag{6.2.41}
$$

When $\beta_3 = 0$, $\beta_1 + \beta_2 \le s - 3$, we deduce

$$
\begin{aligned}
|J_3| &\le \left| \sum_{0<\beta\le\alpha} \binom{\alpha}{\beta} \iiint (1+z)^{2\gamma+2\alpha_3} D^\alpha \varphi \partial_x^{\beta_1} \partial_y^{\beta_2} w \partial_x^{\alpha_1-\beta_1} \partial_y^{\alpha_2-\beta_2} \partial_z^{\alpha_3+1} \varphi \right| \\
&\le C\|(1+z)^{\gamma+\alpha_3} D^\alpha \varphi\|_{L^2} \|(1+z)^{-1} \partial_x^{\beta_1}\partial_y^{\beta_2} w\|_{L^\infty} \\
&\quad \|(1+z)^{\gamma+\alpha_3+1} \partial_x^{\alpha_1-\beta_1} \partial_y^{\alpha_2-\beta_2} \partial_z^{\alpha_3+1} \varphi\|_{L^2} \\
&\le C_{s,\gamma,\sigma,\delta} \left(\|\varphi\|_{H_g^{s,\gamma}} + \|\partial_{xy}^s U\|_{L^2(\mathbb{T}^2)} + 1 \right) \|\varphi\|_{H_g^{s,\gamma}}^2 .
\end{aligned} \tag{6.2.42}
$$

When $\beta_3 = 1$, we obtain

$$
\begin{aligned}
&|J_3| \\
&\le \left| \sum_{0<\beta\le\alpha} \binom{\alpha}{\beta} \iiint (1+z)^{2\gamma+2\alpha_3} D^\alpha \varphi \partial_x^{\beta_1} \partial_y^{\beta_2} [\partial_x u + \partial_y (Ku)] \partial_x^{\alpha_1-\beta_1} \partial_y^{\alpha_2-\beta_2} \partial_z^{\alpha_3} \varphi \right| \\
&\le C\|(1+z)^{\gamma+\alpha_3} D^\alpha \varphi\|_{L^2} \\
&\quad \times \|(1+z)^{\gamma} \partial_x^{\beta_1} \partial_y^{\beta_2} [\partial_x u + \partial_y (Ku)] \cdot (1+z)^{\gamma+\alpha_3} \partial_x^{\alpha_1-\beta_1} \partial_y^{\alpha_2-\beta_2} \partial_z^{\alpha_3} \varphi\|_{L^2} \\
&\le C\|(1+z)^{\gamma+\alpha_3} D^\alpha\|_{L^2} \\
&\quad \times \begin{cases} \|(1+z)^{\gamma} \partial_x^{\beta_1} \partial_y^{\beta_2} [\partial_x u + \partial_y (Ku)]\|_{L^\infty} \|(1+z)^{\gamma+\alpha_3} \partial_x^{\alpha_1-\beta_1} \partial_y^{\alpha_2-\beta_2} \partial_z^{\alpha_3} \varphi\|_{L^2}, \\ \qquad \beta_1 + \beta_2 \le s - 4, \\ \|(1+z)^{\gamma} \partial_x^{\beta_1} \partial_y^{\beta_2} [\partial_x u + \partial_y (Ku)]\|_{L^2} \|(1+z)^{\gamma+\alpha_3} \partial_x^{\alpha_1-\beta_1} \partial_y^{\alpha_2-\beta_2} \partial_z^{\alpha_3} \varphi\|_{L^\infty}, \\ \qquad s - 3 \le \beta_1 + \beta_2 \le s - 1, \end{cases}
\end{aligned}
$$

$$\le C_{s,\gamma,\sigma,\delta}(1+\|\partial_y\partial_{xy}^{s-1}K\|_{L^\infty(\mathbb{T}^2)})\left(\|\varphi\|_{H_g^{s,\gamma}}+\|\partial_{xy}^{s}U\|_{L^2(\mathbb{T}^2)}+1\right)\|\varphi\|_{H_g^{s,\gamma}}^2. \tag{6.2.43}$$

When $\beta_3 \ge 2$, we get

$$\begin{aligned}
&|J_3|\\
&\le \left|\sum_{0<\beta\le\alpha}\binom{\alpha}{\beta}\iiint(1+z)^{2\gamma+2\alpha_3}D^\alpha\varphi\partial_x^{\beta_1}\partial_y^{\beta_2}[\partial_x\varphi+\partial_yK\varphi\right.\\
&\left.+K\partial_y\varphi]\partial_x^{\alpha_1-\beta_1}\partial_y^{\alpha_2-\beta_2}\partial_z^{\alpha_3-1}\varphi\right|\\
&\le C\|(1+z)^{\gamma+\alpha_3}D^\alpha\|_{L^2}\\
&\times\|(1+z)^\gamma\partial_x^{\beta_1}\partial_y^{\beta_2}[\partial_x\varphi+\partial_yK\varphi+K\partial_y\varphi]\cdot(1+z)^{\gamma+\alpha_3-1}\partial_x^{\alpha_1-\beta_1}\partial_y^{\alpha_2-\beta_2}\partial_z^{\alpha_3-1}\varphi\|_{L^2}\\
&\le C\|(1+z)^{\gamma+\alpha_3}D^\alpha\|_{L^2}\\
&\times\begin{cases}\|(1+z)^\gamma\partial_x^{\beta_1}\partial_y^{\beta_2}[\partial_x\varphi+\partial_yK\varphi+K\partial_y\varphi]\|_{L^2}\\
\quad\times\|(1+z)^{\gamma+\alpha_3-1}\partial_x^{\alpha_1-\beta_1}\partial_y^{\alpha_2-\beta_2}\partial_z^{\alpha_3-1}\varphi\|_{L^\infty},\ \beta_1+\beta_2\ge 2,\\
\|(1+z)^\gamma\partial_x^{\beta_1}\partial_y^{\beta_2}[\partial_x\varphi+\partial_yK\varphi+K\partial_y\varphi]\|_{L^\infty}\\
\quad\times\|(1+z)^{\gamma+\alpha_3-1}\partial_x^{\alpha_1-\beta_1}\partial_y^{\alpha_2-\beta_2}\partial_z^{\alpha_3-1}\varphi\|_{L^2},\ \beta_1+\beta_2=1,\end{cases}\\
&\le C_{s,\gamma,\sigma,\delta}(1+\|\partial_y\partial_{xy}^{s-1}K\|_{L^\infty(\mathbb{T}^2)})\left(\|\varphi\|_{H_g^{s,\gamma}}+\|\partial_{xy}^{s}U\|_{L^2(\mathbb{T}^2)}+1\right)\|\varphi\|_{H_g^{s,\gamma}}^2.
\end{aligned} \tag{6.2.44}$$

When $\beta_3 \ge 3$, we get

$$\begin{aligned}
&|J_3|\\
&\le \left|\sum_{0<\beta\le\alpha}\binom{\alpha}{\beta}\iiint(1+z)^{2\gamma+2\alpha_3}D^\alpha\varphi\partial_x^{\beta_1}\partial_y^{\beta_2}\partial_z^{\beta_3-2}\right.\\
&\left.[\partial_x\varphi+\partial_yK\varphi+K\partial_y\varphi]\partial_x^{\alpha_1-\beta_1}\partial_y^{\alpha_2-\beta_2}\partial_z^{\alpha_3-\beta_3+1}\varphi\right|\\
&\le C\|(1+z)^{\gamma+\alpha_3}D^\alpha\|_{L^2}\\
&\times\|(1+z)^{\gamma+\beta_3-2}\partial_x^{\beta_1}\partial_y^{\beta_2}\partial_z^{\beta_3-2}[\partial_x\varphi+\partial_yK\varphi+K\partial_y\varphi]\\
&\cdot(1+z)^{\gamma+\alpha_3-\beta_3+1}\partial_x^{\alpha_1-\beta_1}\partial_y^{\alpha_2-\beta_2}\partial_z^{\alpha_3-\beta_3+1}\varphi\|_{L^2}\\
&\le C\|(1+z)^{\gamma+\alpha_3}D^\alpha\|_{L^2}
\end{aligned}$$

$$
\times \begin{cases} \|(1+z)^{\gamma+\beta_3-2}\partial_x^{\beta_1}\partial_y^{\beta_2}\partial_z^{\beta_3-2}[\partial_x\varphi+\partial_y K\varphi+K\partial_y\varphi]\|_{L^\infty} \\ \quad \cdot \|(1+z)^{\gamma+\alpha_3-\beta_3+1}\partial_x^{\alpha_1-\beta_1}\partial_y^{\alpha_2-\beta_2}\partial_z^{\alpha_3-\beta_3+1}\varphi\|_{L^2}, \ \beta_1+\beta_2 \le s-5, \\ \|(1+z)^{\gamma+\beta_3-2}\partial_x^{\beta_1}\partial_y^{\beta_2}\partial_z^{\beta_3-2}[\partial_x\varphi+\partial_y K\varphi+K\partial_y\varphi]\|_{L^2} \\ \quad \cdot \|(1+z)^{\gamma+\alpha_3-\beta_3+1}\partial_x^{\alpha_1-\beta_1}\partial_y^{\alpha_2-\beta_2}\partial_z^{\alpha_3-\beta_3+1}\varphi\|_{L^\infty}, \ \beta_1+\beta_2 = s-4, s-3, \end{cases}
$$

$$
\le C_{s,\gamma,\sigma,\delta}(1+\|\partial_y\partial_{xy}^{s-1}K\|_{L^\infty(\mathbb{T}^2)})\|\varphi\|_{H_g^{s,\gamma}}^3. \tag{6.2.45}
$$

The estimate of J_4 is similar to that of J_1, then, when $\beta > 0$, we conclude

$$
|J_4| \le C_{s,\gamma,\sigma,\delta}\|\partial_y K\|_{L^\infty(\mathbb{T}^2)})\|\varphi\|_{H_g^{s,\gamma}}^3. \tag{6.2.46}
$$

When $\beta_3 = 0$, $\beta_1+\beta_2 = s-1$, we get

$$
|J_4| \le C_{s,\gamma,\sigma,\delta}\|\partial_y\partial_{xy}^{s-1}K\|_{L^\infty(\mathbb{T}^2)}\left(\|\varphi\|_{H_g^{s,\gamma}}+\|\partial_{xy}^s U\|_{L^2(\mathbb{T}^2)}+1\right)\|\varphi\|_{H_g^{s,\gamma}}^2. \tag{6.2.47}
$$

When $\beta_3 = 0$, $\beta_1+\beta_2 \le s-1$, we achieve

$$
|J_4| \le C_{s,\gamma,\sigma,\delta}\|\partial_y\partial_{xy}^{s-1}K\|_{L^\infty(\mathbb{T}^2)}\left(\|\varphi\|_{H_g^{s,\gamma}}+\|\partial_{xy}^s U\|_{L^2(\mathbb{T}^2)}+1\right)\|\varphi\|_{H_g^{s,\gamma}}^2. \tag{6.2.48}
$$

Hence, combining (6.2.27) and (6.2.30)–(6.2.31) with (6.2.35)–(6.2.48), we conclude

$$
\begin{aligned}
&\frac{d}{dt}\|(1+z)^{\gamma+\alpha_3}D^\alpha\varphi\|_{L^2}^2 \\
&\le C_{s,\gamma,\sigma,\delta}\left(1+\|\partial_{xy}^s K\|_{L^\infty(\mathbb{T}^2)}\right)\left(\|\varphi\|_{H_g^{s,\gamma}}+\|\partial_{xy}^s U\|_{L^2(\mathbb{T}^2)}+1\right)\|\varphi\|_{H_g^{s,\gamma}}^2 \\
&\quad + C_{s,\gamma,\sigma,\delta}\|\partial_{xy}^s K\|_{L^\infty(\mathbb{T}^2)}(1+\|w\|_{H_g^{s,\gamma}})^{s-2}\|\varphi\|_{H_g^{s,\gamma}}^2 \\
&\quad + \sum_{l=0}^{\frac{s}{2}}\|\partial_t^l\partial_x P\|_{H^{s-2l}(\mathbb{T}^2)}^2,
\end{aligned} \tag{6.2.49}
$$

which completes the proof. □

6.2.2 Estimates on g_s

In this subsection, we will estimate the norm of $\partial_x^i \partial_y^{s-i}\varphi$ with weight $(1+z)^\gamma$. To overcome the difficult term $\partial_z^s w$, we use the usual cancellation property, i.e., we estimate the norm of $(1+z)^\gamma g_s$.

Proposition 6.2.2 *Under the same assumption of Proposition 6.2.1, we have the following estimate:*

$$\begin{aligned}\frac{d}{dt}\|(1+z)^\gamma g_s\|_{L^2}^2 \le\; & C_{s,\gamma,\sigma,\delta}\left(1+\|\partial_{xy}^s K\|_{L^\infty}\right)\left(1+\|\varphi\|_{H_g^{s,\gamma}}+\|\partial_{xy}^s U\|_{L^\infty(\mathbb{T}^2)}\right)\\ &\times\left(\|\varphi\|_{H_g^{s,\gamma}}+\|\partial_{xy}^{s+1}U\|_{L^\infty(\mathbb{T}^2)}\right)\|\varphi\|_{H_g^{s,\gamma}} \qquad (6.2.50)\\ &+C_{\gamma,\delta}\left\{1+\|\partial_{xy}U\|_{L^\infty(\mathbb{T}^2)}^2\right\}\|\varphi\|_{H_g^{s,\gamma}}^2+C\|\partial_{xy}^s\partial_x P\|_{L^2(\mathbb{T}^2)}.\end{aligned}$$

Proof Applying the operator ∂_{xy}^s on $(6.1.8)_1$ and $(6.1.6)_1$, respectively, we obtain the following equations:

$$\begin{cases}(\partial_t+u\partial_x+Ku\partial_y+w\partial_z-\epsilon^2\partial_x^2-\epsilon^2\partial_y^2-\partial_z^2)\partial_{xy}^s\varphi+\partial_{xy}^s w\partial_z\varphi\\ =-\sum\limits_{0\le j<s}\binom{s}{j}\partial_{xy}^{s-j}u\partial_x\partial_{xy}^j\varphi-\sum\limits_{0\le j<s}\binom{s}{j}\partial_{xy}^{s-j}(Ku)\partial_y\partial_{xy}^j\varphi\\ \quad-\sum\limits_{1\le j<s}\binom{s}{j}\partial_x^{s-j}w\partial_z\partial_{xy}^j\varphi+\sum\limits_{0\le j\le s}\binom{s}{j}\partial_x^{s-j}(\partial_y Ku)\partial_{xy}^j\varphi,\\ (\partial_t+u\partial_x+Ku\partial_y+w\partial_z-\epsilon^2\partial_x^2-\epsilon^2\partial_y^2-\partial_z^2)\partial_{xy}^s(u-U)+\partial_{xy}^s w\varphi\\ =-\sum\limits_{0\le j<s}\binom{s}{j}\partial_{xy}^{s-j}u\partial_x\partial_{xy}^j(u-U)-\sum\limits_{0\le j<s}\binom{s}{j}\partial_{xy}^{s-j}(Ku)\partial_y\partial_{xy}^j(u-U)\\ \quad-\sum\limits_{1\le j<s}\binom{s}{j}\partial_{xy}^{s-j}w\partial_{xy}^j\varphi-\sum\limits_{0\le j\le s}\binom{s}{j}\partial_{xy}^j(u-U)\partial_{xy}^{s-j}\partial_x U\\ \quad-\sum\limits_{0\le j\le s}\binom{s}{j}\partial_{xy}^j(u-U)\partial_{xy}^{s-j}(K\partial_y U).\end{cases} \qquad (6.2.51)$$

Eliminating $\frac{\partial_z\varphi}{\varphi}\times(6.2.51)_2$ from $(6.2.51)_1$, and letting $a(t,x,y,z)=\frac{\partial_z\varphi}{\varphi}$, it follows

$$\begin{aligned}&(\partial_t+u\partial_x+Ku\partial_y+w\partial_z-\epsilon^2\partial_x^2-\epsilon^2\partial_y^2-\partial_z^2)g_s\\ &\quad+\partial_{xy}^s(u-U)(\partial_t+u\partial_x+Ku\partial_y+w\partial_z-\epsilon^2\partial_x^2-\epsilon^2\partial_y^2-\partial_z^2)a\\ &=2\partial_{xy}^s\varphi\partial_z a+2\epsilon^2\partial_x\partial_{xy}^s(u-U)\partial_x a+2\epsilon^2\partial_y\partial_{xy}^s(u-U)\partial_y a\\ &-\sum_{j=0}^{s-1}\binom{s}{j}(\partial_{xy}^{s-j}u(\partial_x\partial_{xy}^j\varphi-a\partial_x\partial_{xy}^j(u-U))\\ &+\partial_{xy}^{s-j}(Ku)(\partial_y\partial_{xy}^j\varphi-a\partial_y\partial_{xy}^j(u-U)))\end{aligned}$$

$$-\sum_{j=1}^{s-1}\binom{s}{j}\partial_{xy}^{s-j}w(\partial_z\partial_{xy}^j\varphi-a\partial_{xy}^j\varphi)+\sum_{0\le j\le s}\binom{s}{j}\partial_{xy}^j(u-U)\partial_{xy}^{s-j}\partial_xU$$
$$+a\sum_{0\le j\le s}\binom{s}{j}\partial_{xy}^j(u-U)\partial_{xy}^{s-j}(K\partial_yU)+\sum_{0\le j\le s}\binom{s}{j}\partial_x^{s-j}(\partial_yKu)\partial_{xy}^j\varphi. \quad (6.2.52)$$

Applying the operator ∂_z on $(6.1.8)_1$ yields

$$\begin{aligned}&(\partial_t+u\partial_x+Ku\partial_y+w\partial_z)\partial_z\varphi\\&=\epsilon^2(\partial_x^2+\partial_y^2)\partial_z\varphi+\partial_z^3\varphi-\varphi\partial_x\varphi-K\varphi\partial_y\varphi+\partial_xu\partial_z\varphi+\partial_y(Ku)\partial_z\varphi\\&\quad+\partial_yK(\varphi^2+u\partial_z\varphi).\end{aligned}\quad (6.2.53)$$

Using (6.2.53) and equation $(6.1.8)_1$, we get

$$\begin{aligned}&(\partial_t+u\partial_x+Ku\partial_y+w\partial_z)a\\&=\frac{(\partial_t+u\partial_x+Ku\partial_y+w\partial_z)\partial_z\varphi}{\varphi}-\frac{\partial_z\varphi(\partial_t+u\partial_x+Ku\partial_y+w\partial_z)\varphi}{\varphi^2}\\&=\epsilon^2\left\{\frac{(\partial_x^2+\partial_y^2)\partial_z\varphi}{\varphi}-a\frac{(\partial_x^2+\partial_y^2)\varphi}{\varphi}\right\}+\left\{\frac{\partial_z^3\varphi}{\varphi}-a\frac{\partial_z^2\varphi}{\varphi}\right\}\\&\quad-\partial_x\varphi-K\partial_y\varphi+a\partial_xu+a\partial_y(Ku)+\partial_yK\varphi,\end{aligned}\quad (6.2.54)$$

and

$$\begin{cases}\partial_za=\frac{\partial_z^2\varphi}{\varphi}-\frac{\partial_zw\partial_z\varphi}{\varphi^2},\\ \partial_z^2a=\frac{\partial_z^3\varphi}{\varphi}-a\frac{\partial_z^2\varphi}{\varphi}-2a\partial_za,\\ \partial_x^2a=\frac{\partial_x^2\partial_z\varphi}{\varphi}-a\frac{\partial_x^2\varphi}{\varphi}-2\frac{\partial_x\varphi}{\varphi}\partial_xa,\\ \partial_y^2a=\frac{\partial_y^2\partial_z\varphi}{\varphi}-a\frac{\partial_y^2\varphi}{\varphi}-2\frac{\partial_y\varphi}{\varphi}\partial_ya.\end{cases}\quad (6.2.55)$$

Substituting (6.2.55) into (6.2.54), we obtain an equation for a:

$$\begin{aligned}&(\partial_t+u\partial_x+Ku\partial_y+w\partial_z-\epsilon^2\partial_x^2-\epsilon^2\partial_y^2-\partial_z^2)a\\&=2\epsilon^2(\frac{\partial_x\varphi}{\varphi}\partial_xa+\frac{\partial_y\varphi}{\varphi}\partial_ya)+2a\partial_za-g_{1x}+a\partial_xU-Kg_{1y}+Ka\partial_yU\quad (6.2.56)\\&\quad+a\partial_yKu+\partial_yK\varphi.\end{aligned}$$

Now inserting (6.2.56) and $(6.2.55)_2$ into (6.2.52) yields

$$
\begin{aligned}
&(\partial_t + u\partial_x + Ku\partial_y + w\partial_z - \epsilon^2\partial_x^2 - \epsilon^2\partial_y^2 - \partial_z^2)g_s \\
&= 2g_s\partial_z a + 2\epsilon^2\partial_x a\left(\partial_x\partial_{xy}^s(u-U) - \frac{\partial_x\varphi}{\varphi}\partial_{xy}^s(u-U)\right) \\
&+ 2\epsilon^2\partial_y a\left(\partial_y\partial_{xy}^s(u-U) - \frac{\partial_y\varphi}{\varphi}\partial_{xy}^s(u-U)\right) \\
&- \sum_{j=1}^{s-1}\binom{s}{j} g_{j+1}\partial_{xy}^{s-j}u - g_{x1}\partial_{xy}^s U - \sum_{j=1}^{s-1}\binom{s}{j} g_{j+1}\partial_{xy}^{s-j}(Ku) \\
&- g_{1y}\partial_{xy}^s(Ku) + g_{1y}(K\partial_{xy}^s(u-U)) \\
&- \sum_{j=1}^{s-1}\binom{s}{j}\partial_{xy}^{s-j}w(\partial_z\partial_{xy}^j\varphi - a\partial_{xy}^j\varphi) + a\sum_{0\le j\le s-1}\binom{s}{j}\partial_{xy}^j(u-U)\partial_{xy}^{s-j}\partial_x U \\
&+ a\sum_{0\le j\le s-1}\binom{s}{j}\partial_{xy}^j(u-U)\partial_{xy}^{s-j}(K\partial_y U) + \sum_{0\le j\le s}\binom{s}{j}\partial_x^{s-j}(\partial_y Ku)\partial_{xy}^j\varphi \\
&- (a\partial_y Ku + \partial_y K\varphi)\partial_{xy}^s(u-U).
\end{aligned}
\tag{6.2.57}
$$

Multiplying (6.2.57) by $(1+z)^{2\gamma}g_s$, and integrating it over $\mathbb{T}^2\times\mathbb{R}^+$, we arrive at

$$
\begin{aligned}
&\frac{1}{2}\frac{d}{dt}\|(1+z)^\gamma g_s\|_{L^2}^2 + \|(1+z)^\gamma\partial_z g_s\|_{L^2}^2 + \epsilon^2\|(1+z)^\gamma\partial_x g_s\|_{L^2}^2 \\
&+ \epsilon^2\|(1+z)^\gamma\partial_y g_s\|_{L^2}^2 \\
&= \iint_{\mathbb{T}^2} g_s\partial_z g_s dxdy|_{z=0} - 2\gamma\iiint(1+z)^{2\gamma-1}g_s\partial_z g_s + \gamma\iiint(1+z)^{2\gamma-1}w|g_s|^2 \\
&+ 2\iiint(1+z)^{2\gamma}|g_s|^2\partial_z a \\
&+ 2\epsilon^2\iiint(1+z)^{2\gamma}g_s\partial_x a\left(\partial_x\partial_{xy}^s(u-U) - \frac{\partial_x\varphi}{\varphi}\partial_{xy}^s(u-U)\right) \\
&+ 2\epsilon^2\iiint(1+z)^{2\gamma}g_s\partial_y a\left(\partial_y\partial_{xy}^s(u-U) - \frac{\partial_y\varphi}{\varphi}\partial_{xy}^s(u-U)\right) \\
&- \sum_{j=1}^{s-1}\binom{s}{j}\iiint(1+z)^{2\gamma}g_s g_{j+1}\partial_{xy}^{s-j}u - \iiint(1+z)^{2\gamma}g_s g_{x1}\partial_{xy}^s U \\
&- \sum_{j=1}^{s-1}\binom{s}{j}\iiint(1+z)^{2\gamma}g_s g_{j+1}\partial_{xy}^{s-j}(Ku) - \iiint(1+z)^{2\gamma}g_s g_{1y}\partial_{xy}^s(Ku)
\end{aligned}
$$

$$+\iiint (1+z)^{2\gamma} g_s g_{1y}(K\partial_{xy}^s(u-U))$$
$$-\sum_{j=1}^{s-1}\binom{s}{j}\iiint (1+z)^{2\gamma} g_s \partial_{xy}^{s-j} w(\partial_z\partial_{xy}^j\varphi - a\partial_{xy}^j\varphi)$$
$$+\sum_{0\le j\le s-1}\binom{s}{j}\iiint (1+z)^{2\gamma} g_s a\partial_{xy}^j(u-U)\partial_{xy}^{s-j}\partial_x U$$
$$+\sum_{0\le j\le s-1}\binom{s}{j}\iiint (1+z)^{2\gamma} g_s a\partial_{xy}^j(u-U)\partial_{xy}^{s-j}(K\partial_y U)$$
$$+\sum_{0\le j\le s}\binom{s}{j}\iiint (1+z)^{2\gamma} g_s \partial_x^{s-j}(\partial_y K u)\partial_{xy}^j\varphi$$
$$-\iiint (1+z)^{2\gamma} g_s(a\partial_y K u + \partial_y K\varphi)\partial_{xy}^s(u-U), \tag{6.2.58}$$

which can be obtained by integration and by using the boundary condition, i.e.,

$$\begin{aligned}
&\iiint (1+z)^{2\gamma} g_s(u\partial_x + Ku\partial_y + w\partial_z)g_s\\
&= \frac{1}{2}\iiint (1+z)^{2\gamma}\left\{u(|g_s|^2)_x + u(|g_s|^2)_y + w(|g_s|^2)_z\right\}\\
&= -\gamma\iiint (1+z)^{2\gamma-1} w|g_s|^2,
\end{aligned} \tag{6.2.59}$$

and

$$\begin{aligned}
\iiint (1+z)^{2\gamma} g_s\partial_z^2 g_s &= -\|(1+z)^\gamma\partial_z g_s\|_{L^2} - \iint_{\mathbb{T}^2} g_s\partial_z g_s dxdy|_{z=0}\\
&- 2\gamma\iiint (1+z)^{2\gamma-1} g_s\partial_z g_s.
\end{aligned} \tag{6.2.60}$$

By definition of $H_{\sigma,\delta}^{s,\gamma}$, we have

$$\begin{cases}
|\varphi|^{-1} \le \delta^{-1}(1+z)^\sigma,\\
|\partial_x\varphi|, |\partial_y\varphi| \le \delta^{-1}(1+z)^{-\sigma},\\
|\partial_z\varphi|, |\partial_x\partial_z\varphi|, |\partial_y\partial_z\varphi| \le \delta^{-1}(1+z)^{-\sigma-1},\\
|\partial_z^2\varphi| \le \delta^{-1}(1+z)^{-\sigma-2}.
\end{cases} \tag{6.2.61}$$

We estimate equation (6.2.58) by terms. Firstly, we have the fact that

$$\partial_z g_s\big|_{z=0} = \left(\partial_{xy}^s \partial_z \varphi - \frac{\partial_z \varphi}{\varphi}\partial_{xy}^s \varphi - \frac{\partial_z^2 \varphi}{\varphi}\partial_{xy}^s (u-U) + \frac{\partial_z \varphi \partial_z \varphi}{\varphi^2}\partial_{xy}^s (u-U)\right)\big|_{z=0}$$

$$= \partial_{xy}^s \partial_x P - a g_s\big|_{z=0} + \frac{\partial_z^2 \varphi}{\varphi}\partial_{xy}^s U\big|_{z=0}. \tag{6.2.62}$$

Then according to the trace theorem, together with the facts that $\|a\|_{L^\infty} \le \delta^{-2}$ and $\|\partial_z a\|_{L^\infty} \le \delta^{-2} + \delta^{-4}$, we get

$$\begin{aligned}
&\left|\iint_{\mathbb{T}^2} g_s \partial_z g_s dxdy|_{z=0}\right| \\
&= \left|\iint_{\mathbb{T}^2} g_s|_{z=0}\partial_{xy}^s \partial_x P dxdy\right| + \left|\iint_{\mathbb{T}^2} a g_s^2|_{z=0} dxdy\right| + \left|\iint_{\mathbb{T}^2} \frac{\partial_z^2 \varphi}{\varphi}\partial_{xy}^s U g_s|_{z=0} dxdy\right| \\
&\le C\iint_{\mathbb{T}^2} (g_s + \partial_z g_s)\partial_{xy}^s \partial_x P dxdydz + C\iint_{\mathbb{T}^2} (g_s + \partial_z g_s) g_s dxdydz \\
&+ C\iint_{\mathbb{T}^2} \frac{\partial_z^2 \varphi}{\varphi}\partial_{xy}^s U g_s dxdydz \\
&+ C\iint_{\mathbb{T}^2} \partial_z\left(\frac{\partial_z^2 \varphi}{\varphi}\right)\partial_{xy}^s U g_s dxdydz + C\iint_{\mathbb{T}^2} \frac{\partial_z^2 \varphi}{\varphi}\partial_{xy}^s U \partial_z g_s dxdydz \\
&\le \frac{1}{4}\|(1+z)^\gamma \partial_z g_s\|_{L^2}^2 + C_\delta\left\{1 + \|\partial_{xy} U\|_{L^\infty(\mathbb{T}^2)}^2\right\}\|\varphi\|_{H_g^{s,\gamma}}^2 + C\|\partial_{xy}^s \partial_x P\|_{L^2(\mathbb{T}^2)}^2.
\end{aligned} \tag{6.2.63}$$

Using the Hölder inequality, we obtain

$$\begin{aligned}
\left|2\gamma\iiint (1+z)^{2\gamma-1} g_s \partial_z g_s\right| &\le 2\gamma\left\|\frac{1}{1+z}\right\|_{L^\infty}\|(1+z)^\gamma g_s\|_{L^2}\|(1+z)^\gamma \partial_z g_s\|_{L^2} \\
&\le C_\gamma \|\varphi\|_{H_g^{s,\gamma}}^2 + \frac{1}{4}\|(1+z)^\gamma \partial_z g_s\|_{L^2}^2,
\end{aligned} \tag{6.2.64}$$

and using Lemma 6.1.5,

$$\begin{aligned}
\left|\gamma\iiint (1+z)^{2\gamma-1} w|g_s|^2\right| &\le \left\|\frac{w}{1+z}\right\|_{L^\infty}\|(1+z)^\gamma g_s\|_{L^2}^2 \\
&\le C_{s,\gamma,\sigma,\delta}\left(1 + \|\varphi\|_{H_g^{s,\gamma}} + \|\partial_{xy} U\|_{L^\infty(\mathbb{T}^2)}^2\right)\|\varphi\|_{H_g^{s,\gamma}}^2.
\end{aligned} \tag{6.2.65}$$

Thus, it follows from to (6.2.61) and (6.2.55) that

$$\left|2\iiint(1+z)^{2\gamma}|g_s|^2\partial_z a\right| \le 2\|\partial_z a\|_{L^\infty}\|(1+z)^\gamma g_s\|_{L^2}^2 \le C_\delta\|\varphi\|_{H_g^{s,\gamma}}^2, \tag{6.2.66}$$

and by Lemma 6.1.3,

$$\begin{aligned}&\left|2\epsilon^2\iiint(1+z)^{2\gamma}g_s\partial_x a\left(\partial_x\partial_{xy}^s(u-U)-\frac{\partial_x\varphi}{\varphi}\partial_{xy}^s(u-U)\right)\right|\\&\le 2\epsilon^2 C_\delta\|(1+z)^\gamma g_s\|_{L^2}\\&\left(\|(1+z)^{\gamma-1}\partial_{xy}^s\partial_x(u-U)\|_{L^2}+\|(1+z)^{\gamma-1}\partial_x^s(u-U)\|_{L^2}\right),\end{aligned} \tag{6.2.67}$$

which, together with (6.2.61), implies the fact that $\|(1+z)(\partial_x a+\partial_y a)\|_{L^\infty}\le \delta^{-2}+\delta^{-4}$. In addition, by using Lemma 6.1.2 and the following relation:

$$\varphi\partial_z\left(\frac{\partial_x\partial_{xy}^s(u-U)}{\varphi}\right)=g_{(s+1)x}=\partial_x g_s+\partial_x a\partial_{xy}^s(u-U), \tag{6.2.68}$$

we get

$$\begin{aligned}&\|(1+z)^{\gamma-1}\,\partial_x\partial_{xy}^s(u-U)\|_{L^2}\\&\le\delta^{-1}\|(1+y)^{\gamma-\sigma-1}\frac{\partial_x\partial_{xy}^s(u-U)}{\varphi}\|_{L^2}\\&\le C_{\gamma,\sigma,\delta}\left(\|\partial_x\partial_{xy}^s U\|_{L^2(\mathbb{T})}+\left\|(1+z)^\gamma\varphi\partial_z\left(\frac{\partial_x\partial_{xy}^s(u-U)}{\varphi}\right)\right\|_{L^2}\right)\\&\le C_{\gamma,\sigma,\delta}\left(\|\partial_x\partial_{xy}^s U\|_{L^2(\mathbb{T}^2)}+\|(1+z)^\gamma\partial_x g_s\|_{L^2}\right.\\&\left.+\|(1+z)\partial_x a\|_{L^\infty}\|(1+z)^{\gamma-1}\partial_{xy}^s(u-U)\|_{L^2}\right).\end{aligned} \tag{6.2.69}$$

Then, we can obtain

$$\begin{aligned}&\left|2\epsilon^2\iiint(1+z)^{2\gamma}g_s\partial_x a\left(\partial_x\partial_{xy}^s(u-U)-\frac{\partial_x\varphi}{\varphi}\partial_{xy}^s(u-U)\right)\right|\\&\le\epsilon^2 C_{\gamma,\sigma,\delta}\left(\|\varphi\|_{H_g^{s,\gamma}}+\|\partial_{xy}^s U\|_{L^2(\mathbb{T}^2)}\right.\\&\left.+\|\partial_x\partial_{xy}^s U\|_{L^2(\mathbb{T}^2)}+\|(1+z)^\gamma\partial_x g_s\|_{L^2}\right)\|\varphi\|_{H_g^{s,\gamma}}.\end{aligned} \tag{6.2.70}$$

Similarly, we also have

$$\begin{aligned}
&\left|2\epsilon^2 \iiint (1+z)^{2\gamma} g_s \partial_y a \left(\partial_y \partial_{xy}^s (u-U) - \frac{\partial_y \varphi}{\varphi} \partial_{xy}^s (u-U)\right)\right| \\
&\le \epsilon^2 C_{\gamma,\sigma,\delta} \Big(\|\varphi\|_{H_g^{s,\gamma}} + \|\partial_{xy}^s U\|_{L^2(\mathbb{T}^2)} + \|\partial_y \partial_{xy}^s U\|_{L^2(\mathbb{T}^2)} \\
&+\|(1+z)^\gamma \partial_y g_s\|_{L^2}\Big) \|\varphi\|_{H_g^{s,\gamma}}. \qquad (6.2.71)
\end{aligned}$$

Now we estimate some of the remaining complex terms in (6.2.58). Thus we need to discuss the classification to complete the estimation. For $i \le s-3$, we use the usual inequality to derive

$$\begin{aligned}
\|(1+z)^\gamma g_i\|_{L^\infty} &\le \|(1+z)^\gamma \partial_{xy}^i \varphi\|_{L^\infty} + \|(1+z)^{\gamma-1} a\|_{L^\infty} \|\partial_{xy}^i (u-U)\|_{L^\infty} \\
&\le C_\delta \left(\|(1+z)^{\gamma+\alpha_3} \partial_{xy}^i \varphi\|_{L^\infty} + \|\partial_{xy}^i u\|_{L^\infty} + \|\partial_{xy}^i U\|_{L^\infty}\right) \\
&\le C_{s,\gamma,\sigma,\delta} \left(\|\varphi\|_{H_g^{s,\gamma}} + \|\partial_{xy}^s U\|_{L^2(\mathbb{T}^2)}\right). \qquad (6.2.72)
\end{aligned}$$

By Lemma 6.1.5, we obtain

$$\begin{aligned}
&\left|\iiint (1+z)^{2\gamma} g_s g_{j+1} \partial_{xy}^{s-j} u\right| \\
&\le \begin{cases} \|(1+z)^\gamma g_s\|_{L^2} \|(1+z)^\gamma g_{j+1}\|_{L^2} \|\partial_{xy}^{s-j} u\|_{L^\infty}, \ j = [2, s-1], \\ \|(1+z)^\gamma g_s\|_{L^2} \|(1+z)^\gamma g_2\|_{L^\infty} \|\partial_{xy}^{s-1}(u-U) + \partial_{xy}^{s-1} U\|_{L^2}, \ j = 1. \end{cases} \\
&\le C_{s,\gamma,\sigma,\delta} \left(\|\varphi\|_{H_g^{s,\gamma}} + \|\partial_{xy}^s U\|_{L^2(\mathbb{T}^2)}\right)^2 \|\varphi\|_{H_g^{s,\gamma}}, \qquad (6.2.73)
\end{aligned}$$

for $j = 1, 2, \cdots, s-1$, and

$$\begin{aligned}
&\left|\iiint (1+z)^{2\gamma} g_s g_{x1} \partial_{xy}^s U\right| \\
&\le C_{s,\gamma,\sigma,\delta} \left(\|\varphi\|_{H_g^{s,\gamma}} + \|\partial_{xy}^s U\|_{L^2(\mathbb{T}^2)}\right) \|\partial_{xy}^s U\|_{L^2(\mathbb{T}^\infty)} \|\varphi\|_{H_g^{s,\gamma}}. \qquad (6.2.74)
\end{aligned}$$

Similarly,

$$\begin{aligned}
&\left|\sum_{j=1}^{s-1} \binom{s}{j} \iiint (1+z)^{2\gamma} g_s g_{j+1} \partial_{xy}^{s-j} (Ku)\right| \\
&\le C_{s,\gamma,\sigma,\delta} \|\partial_{xy}^{s-1} K\|_{L^\infty} \left(\|\varphi\|_{H_g^{s,\gamma}} + \|\partial_{xy}^s U\|_{L^2(\mathbb{T}^2)}\right)^2 \|\varphi\|_{H_g^{s,\gamma}} \qquad (6.2.75)
\end{aligned}$$

and

$$
\begin{aligned}
&\left|-\iiint (1+z)^{2\gamma} g_s g_{1y} \partial_{xy}^s (Ku) + \iiint (1+z)^{2\gamma} g_s g_{1y} (K\partial_{xy}^s (u-U))\right| \\
&\le C_{s,\gamma,\sigma,\delta} \|\partial_{xy}^{s-1} K\|_{L^\infty} \|(1+z)^\gamma g_s\|_{L^2} \|(1+z)^\gamma g_{1y}\|_{L^\infty} \| -\partial_{xy}^s (Ku) \\
&+ (K\partial_{xy}^s (u-U))\|_{L^2} \\
&\le C_{s,\gamma,\sigma,\delta} \|\partial_{xy}^{s-1} K\|_{L^\infty} \left(\|\varphi\|_{H_g^{s,\gamma}} + \|\partial_{xy}^s U\|_{L^2(\mathbb{T}^2)} \right)^2 \|\varphi\|_{H_g^{s,\gamma}}.
\end{aligned}
\tag{6.2.76}
$$

It should be noted that the highest order of u will not reach s, due to cancelling each other out, in the inequality (6.2.76).

Last, for $j = 3, \cdots, s-1$,

$$
\begin{aligned}
&\left|\iiint (1+z)^{2\gamma} g_s \partial_{xy}^{s-j} w (\partial_z \partial_{xy}^j \varphi - a\partial_{xy}^j \varphi)\right| \\
&\le \left\| \frac{\partial_{xy}^{s-j} w}{1+z} \right\|_{L^\infty} \left(\|(1+z)^{\gamma+1} \partial_{xy}^j \partial_z \varphi\|_{L^2} + \|(1+z)a\|_{L^2} \|(1+z)^\gamma \partial_{xy}^j \varphi\|_{L^2} \right) \\
&\quad \|(1+z)^\gamma g_s\|_{L^2} \\
&\le C_{s,\gamma,\sigma,\delta} \left(\|\varphi\|_{H_g^{s,\gamma}} + \|\partial_{xy}^s U\|_{L^2(\mathbb{T}^2)} \right) \|\varphi\|_{H_g^{s,\gamma}}^2
\end{aligned}
\tag{6.2.77}
$$

and for $j = 1, 2$,

$$
\begin{aligned}
&\left|\iiint (1+z)^{2\gamma} g_s \partial_{xy}^{s-j} w (\partial_z \partial_{xy}^j \varphi - a\partial_{xy}^j \varphi)\right| \\
&\le \left\| \frac{\partial_{xy}^{s-j} w + z\tilde{U}}{1+z} \right\|_{L^2} \left(\|(1+z)^{\gamma+1} \partial_{xy} \partial_z \varphi\|_{L^\infty} + \|(1+z)a\|_{L^\infty} \|(1+z)^\gamma \partial_{xy}^j \varphi\|_{L^\infty} \right) \\
&\quad \|(1+z)^\gamma g_s\|_{L^2} \\
&\quad + \|\tilde{U}\|_{L^\infty(\mathbb{T}^2)} \left(\|(1+z)^{\gamma+1} \partial_{xy} \partial_z \varphi\|_{L^2} + \|(1+z)a\|_{L^\infty} \|(1+z)^\gamma \partial_{xy}^j \varphi\|_{L^2} \right) \\
&\quad \|(1+z)^\gamma g_s\|_{L^2} \\
&\le C_{s,\gamma,\sigma,\delta} \left(\|\varphi\|_{H_g^{s,\gamma}} + \|\partial_{xy}^s U\|_{L^2(\mathbb{T}^2)} \right) \|\varphi\|_{H_g^{s,\gamma}}^2.
\end{aligned}
\tag{6.2.78}
$$

For $j = 1, 2, \cdots, s-1$, by Lemma 6.1.5 and using the fact $\|(1+z)a\|_{L^\infty} \le \delta^{-2}$, it follows

$$
\left|\iiint (1+z)^{2\gamma} g_s a \partial_{xy}^j (u-U) \partial_{xy}^{s-j} \partial_x U\right|
$$

$$
\begin{aligned}
&\le C\|(1+z)^\gamma g_s\|_{L^2}\|(1+z)a\|_{L^\infty}\|(1+z)^{\gamma-1}\partial_{xy}^j(u-U)\|_{L^2}\|\partial_x\partial_{xy}^{s-j}U\|_{L^\infty(\mathbb{T}^2)}\\
&\le C_{s,\gamma,\sigma,\delta}\|\partial_x\partial_{xy}^s U\|_{L^\infty(\mathbb{T}^2)}\left(\|\varphi\|_{H_g^{s,\gamma}}+\|\partial_{xy}^s U\|_{L^2(\mathbb{T}^2)}\right)\|\varphi\|_{H_g^{s,\gamma}}, \qquad (6.2.79)
\end{aligned}
$$

and

$$
\begin{aligned}
&\left|\iiint (1+z)^{2\gamma} g_s a\partial_{xy}^j(u-U)\partial_{xy}^{s-j}(K\partial_y U)\right|\\
&\le C_{s,\gamma,\sigma,\delta}\|\partial_{xy}^s(K\partial_y U)\|_{L^\infty(\mathbb{T}^2)}\left(\|\varphi\|_{H_g^{s,\gamma}}+\|\partial_{xy}^s U\|_{L^2(\mathbb{T}^2)}\right)\|\varphi\|_{H_g^{s,\gamma}}.
\end{aligned}
\qquad (6.2.80)
$$

For $j=1,2,\cdots,s$, we achieve

$$
\begin{aligned}
&\left|\iiint (1+z)^{2\gamma} g_s\partial_x^{s-j}(\partial_y K u)\partial_{xy}^j\varphi\right|\\
&\le C_\delta\|\varphi\|_{H_g^{s,\gamma}}\begin{cases}\|(1+z)^\gamma\partial_{xy}^j\varphi\|_{L^2}\|\partial_x^{s-j}(\partial_y K u)\|_{L^\infty}, & j\ge 2,\\ \|(1+z)^\gamma\partial_{xy}^j\varphi\|_{L^\infty}\|\partial_x^{s-j}(\partial_y K u)\|_{L^2}, & j=0,1.\end{cases}\\
&\le C_{s,\gamma,\sigma,\delta}\|\partial_{xy}^{s-2}K\|_{L^\infty(\mathbb{T}^2)}\left(\|\varphi\|_{H_g^{s,\gamma}}+\|\partial_{xy}^s U\|_{L^2(\mathbb{T}^2)}\right)^2\|\varphi\|_{H_g^{s,\gamma}}, \qquad (6.2.81)
\end{aligned}
$$

and

$$
\begin{aligned}
&-\iiint (1+z)^{2\gamma} g_s(a\partial_y K u+\partial_y K\varphi)\partial_{xy}^s(u-U)\\
&\quad\le C\|\partial_y K\|_{L^\infty}\|(1+z)^\gamma g_s\|_{L^2}\|(1+z)^{\gamma-1}\partial_{xy}^s(u-U)\|_{L^2}\\
&\quad\times(\|(1+z)a\|_{L^\infty}\|u\|_{L^\infty}+\|(1+z)^{1-\gamma}\|_{L^\infty}\|(1+z)^\gamma\varphi\|_{L^\infty})\\
&\quad\le C_{s,\gamma,\sigma,\delta}\|\partial_y K\|_{L^\infty}\left(\|\varphi\|_{H_g^{s,\gamma}}+\|\partial_{xy}^s U\|_{L^2(\mathbb{T}^2)}\right)^2\|\varphi\|_{H_g^{s,\gamma}}. \qquad (6.2.82)
\end{aligned}
$$

Hence combining (6.2.63)–(6.2.66), (6.2.70)–(6.2.71), and (6.2.73)–(6.2.82) with (6.2.58) leads to

$$
\begin{aligned}
\frac{d}{dt}\|(1+z)^\gamma g_s\|_{L^2}^2 &\le C_{s,\gamma,\sigma,\delta}\left(1+\|\partial_{xy}^s K\|_{L^\infty}\right)\left(1+\|\varphi\|_{H_g^{s,\gamma}}+\|\partial_{xy}^s U\|_{L^\infty(\mathbb{T}^2)}\right)\\
&\quad\times\left(\|\varphi\|_{H_g^{s,\gamma}}+\|\partial_{xy}^{s+1}U\|_{L^\infty(\mathbb{T}^2)}\right)\|\varphi\|_{H_g^{s,\gamma}} \qquad (6.2.83)\\
&\quad+C_{\gamma,\delta}\left\{1+\|\partial_{xy}U\|_{L^\infty(\mathbb{T}^2)}^2\right\}\|\varphi\|_{H_g^{s,\gamma}}^2+C\|\partial_{xy}^s\partial_x P\|_{L^2(\mathbb{T}^2)}.
\end{aligned}
$$

□

6.2.3 Weighted H^s Estimate on φ

Now we can derive the weighted H^s estimate on φ by employing Propositions 6.2.1–6.2.2.

Proposition 6.2.3 *Under the same assumption of Proposition 6.2.1, we have the following estimate:*

$$\|\varphi\|^2_{H^{s,\gamma}_g} \le \left\{\|\varphi_0\|^2_{H^{s,\gamma}_g} + \int_0^t F(\tau)d\tau\right\} \times \left\{1 - C(s-1)\left(\|\varphi_0\|^2_{H^{s,\gamma}_g} + \int_0^t F(\tau)d\tau\right)^{s-1} t\right\}^{-\frac{1}{s-1}}, \tag{6.2.84}$$

where $C > 0$ is a constant independent of ϵ and t. The function $F(t)$ is expressed by

$$F(t) = \mathcal{P}(\|\partial_{xy}^{s+1}U\|_{L^\infty(\mathbb{T}^2)}, \|\partial_{xy}^{s}K\|_{L^\infty(\mathbb{T}^2)}) + C\sum_{l=0}^{\frac{s}{2}} \|\partial_t^l\partial_x P\|^2_{H^{s-2l}(\mathbb{T}^2)}, \tag{6.2.85}$$

and $\mathcal{P}(\cdot,\cdot)$ denotes a polynomial.

Proof According to Propositions 6.2.1–6.2.2, we deduce from the definition of $\|\cdot\|_{H^{s,\gamma}_g}$ that

$$\frac{d}{dt}\|\varphi\|^2_{H^{s,\gamma}_g} \le C_{s,\gamma,\sigma,\delta}\|\varphi\|^{2s}_{H^{s,\gamma}_g} + \sum_{l=0}^{\frac{s}{2}} \|\partial_t^l\partial_x P\|^2_{H^{s-2l}(\mathbb{T}^2)} + \mathcal{P}(\|\partial_{xy}^{s+1}U\|_{L^\infty(\mathbb{T}^2)}, \|\partial_{xy}^{s}K\|_{L^\infty(\mathbb{T}^2)}) \tag{6.2.86}$$

and hence, it follows by using the comparison principle of ordinary differential equations that

$$\|\varphi\|^2_{H^{s,\gamma}_g} \le \left\{\|\varphi_0\|^2_{H^{s,\gamma}_g} + \int_0^t F(\tau)d\tau\right\} \times \left\{1 - C(s-1)\left(\|\varphi_0\|^2_{H^{s,\gamma}_g} + \int_0^t F(\tau)d\tau\right)^{s-1} t\right\}^{-\frac{1}{s-1}}, \tag{6.2.87}$$

provided that

$$1 - C(s-1)\left(\|\varphi_0\|^2_{H^{s,\gamma}_g} + \int_0^t F(\tau)d\tau\right)^{s-1} t > 0. \tag{6.2.88}$$

This proves Proposition 6.2.3. □

6.2.4 Weighted L^∞ Estimates on Lower-Order Terms

In this subsection, we will estimate the weighted L^∞ on $D^\alpha\varphi$ for $|\alpha| \leq 2$ by using the classical maximum principles. More precisely, we will derive two parts: an L^∞-estimate on $I := \sum_{|\alpha|\leq 2} \left|(1+z)^{\sigma+\alpha_3} D^\alpha\varphi\right|^2$ and a lower bound estimate on $B_{(0,0,0)} := (1+z)^\sigma\varphi$. Here

$$I := \sum_{|\alpha|\leq 2} \left|(1+z)^{\sigma+\alpha_3} D^\alpha\varphi\right|^2, \quad B_\alpha := (1+z)^{\sigma+\alpha_3} D^\alpha\varphi,$$

and $B_{(0,0,0)} = (1+z)^\sigma\varphi$.

Lemma 6.2.3 *Under the same assumption of Proposition 6.2.1, we have the following estimates.*
For any $s \geq 5$,

$$\|I(t)\|_{L^\infty(\mathbb{T}^2\times\mathbb{R}^+)} \leq \max\left\{\|I(0)\|_{L^\infty(\mathbb{T}^2\times\mathbb{R}^+)}, 6C^2W(t)^2\right\} e^{C(1+G_1(t))t}, \tag{6.2.89}$$

and for any $s \geq 7$,

$$\|I(t)\|_{L^\infty(\mathbb{T}^2\times\mathbb{R}^+)} \leq \left\{\|I(0)\|_{L^\infty(\mathbb{T}^2\times\mathbb{R}^+)} + C[1+Y(t)W(t)]W(t)^2t\right\} e^{C(1+G_1(t))t}. \tag{6.2.90}$$

In addition, if $s \geq 5$, we also have

$$\begin{aligned}\min_{\mathbb{T}^2\times\mathbb{R}^+} (1+z)^\sigma\varphi(t) \geq{} & \max\left\{1 - C(1+G_2(t))\, te^{C(1+G_2(t))t}\right\} \\ & \cdot \min\left\{\min_{\mathbb{T}^2\times\mathbb{R}^+} (1+z)^\sigma\varphi_0 - CW(t)t\right\},\end{aligned} \tag{6.2.91}$$

where constant $C > 0$ depends on s, γ, σ, and δ only. The functions G_1, G_2, W, and Y: $[0, T] \to \mathbb{R}^+$ are respectively defined by

$$
\begin{aligned}
G_1(t) : = & \sup_{[0,t]} \|\partial_{xy}^2 K(\tau)\|_{L^\infty(\mathbb{T}^2)} + \sup_{[0,t]} \|w(\tau)\|_{H_g^{s,\gamma}} + \sup_{[0,t]} \|\partial_{xy}^s U(\tau)\|_{L^2(\mathbb{T}^2)} \\
& + \sup_{[0,t]} \|\partial_{xy}^2 K(\tau)\|_{L^\infty(\mathbb{T}^2)} \left(\sup_{[0,t]} \|w(\tau)\|_{H_g^{s,\gamma}} + \sup_{[0,t]} \|\partial_{xy}^s U(\tau)\|_{L^2(\mathbb{T}^2)} \right), \\
G_2(t) : = & \sup_{[0,t]} \|w(\tau)\|_{H_g^{s,\gamma}} + \sup_{[0,t]} \|\partial_{xy}^s U(\tau)\|_{L^2(\mathbb{T}^2)},
\end{aligned}
\tag{6.2.92}
$$

and

$$
W(t) := \sup_{[0,t]} \|w(\tau)\|_{H_g^{s,\gamma}}, \quad Y(t) := \sup_{[0,t]} \|\partial_{xy} K(\tau)\|_{L^\infty(\mathbb{T}^2)}. \tag{6.2.93}
$$

Proof This proof will be presented in two distinct steps, making use of the classical maximum principle for parabolic equations. In the first step, we will establish weighted L^∞ controls on $D^\alpha \varphi$, employing the maximum principle outlined in Lemma 6.1.7. These controls will depend on the boundary values of $D^\alpha \varphi$ at $z = 0$. In the second step, we will obtain estimates on the boundary values of $D^\alpha \varphi$, utilizing either Sobolev embedding or growth rate control argument.

Step 1. By direct computations, we obtain

$$
\left(\partial_t + u\partial_x + Ku\partial_y + w\partial_z - \partial_z^2 - \epsilon^2 \partial_x^2 - \epsilon^2 \partial_y^2 - \partial_y K u \right) B_\alpha = \sum_{i=1}^{3} S_i, \tag{6.2.94}
$$

where

$$
S_1 = \left(\frac{\sigma + \alpha_3}{1+z} w + \frac{(\sigma + \alpha_3)(\sigma + \alpha_3 - 1)}{(1+z)^2} \right) B_\alpha, \quad S_2 = -\frac{2(\sigma + \alpha_3)}{1+z} \partial_z B_\alpha,
$$

and

$$
\begin{aligned}
S_3 = - \sum_{0<\beta\le\alpha} \binom{\alpha}{\beta} \Big\{ & (1+z)^{\beta_3} \Big(D^\beta u B_{\alpha-\beta+e_1} + D^\beta (Ku) B_{\alpha-\beta+e_2} \\
& + \frac{D^\beta w B_{\alpha-\beta+e_3}}{1+z} + D^\beta (\partial_y K u) B_{\alpha-\beta} \Big) \Big\}.
\end{aligned}
$$

Here, $e_1 := (1, 0, 0)$, $e_2 := (0, 1, 0)$ and $e_3 := (0, 0, 1)$. Using Lemma 6.1.5, we have the following pointwise controls on Q_α, R_α and S_α: for $|\alpha| \le 2$,

$$\begin{cases} |S_1| \le C_{s,\gamma,\sigma,\delta}\left(1 + \|\varphi\|_{H_g^{s,\gamma}} + \|\partial_{xy}^s U\|_{L^2(\mathbb{T}^2)}\right), \quad |S_2| \le C_\sigma, \\ |S_3| \le C_{s,\gamma,\sigma,\delta}(1 + \|\partial_{xy}^2 K\|_{L^\infty(\mathbb{T}^2)})\left(\|\varphi\|_{H_g^{s,\gamma}} + \|\partial_{xy}^s U\|_{L^2(\mathbb{T}^2)}\right) \\ \qquad \times \sum_{0<\beta\le\alpha} |B_{\alpha-\beta+e_i}|, \quad i = 0, 1, 2, 3, \end{cases} \tag{6.2.95}$$

where C_σ and $C_{s,\gamma,\sigma,\delta}$ are some universal constants which are independent of the solution φ.

By (6.2.95), multiplying the equation (6.2.94) by $2B_\alpha$ gives

$$\begin{aligned} &\left(\partial_t + u\partial_x + Ku\partial_y + w\partial_z - \partial_z^2 - \epsilon^2\partial_x^2 - \epsilon^2\partial_y^2 - 2\partial_y Ku\right) I \\ &= 2\sum_{|\alpha|\le 2}\left(|S_1 B_\alpha| + |S_2 B_\alpha| + |S_3 B_\alpha| - \epsilon^2|\partial_x B_\alpha|^2 - \epsilon^2|\partial_y B_\alpha|^2 - |\partial_z B_\alpha|^2\right) \\ &\le C_{s,\gamma,\sigma,\delta}(1 + \|\partial_{xy}^2 K\|_{L^\infty(\mathbb{T}^2)})(1 + \|\varphi(s)\|_{H_g^{s,\gamma}} + \|\partial_{xy}^s U\|_{L^2(\mathbb{T}^2)})I, \end{aligned} \tag{6.2.96}$$

where we have used Lemma 6.1.5.

Applying the classical maximum principle Lemma 6.1.7 for parabolic equations (6.2.96), we have

$$\begin{aligned} &\|I(t)\|_{L^\infty(\mathbb{T}^2\times\mathbb{R}^+)} \\ &\le \max\left\{e^{C(1+G_1(t))t}\|I(0)\|_{L^\infty(\mathbb{T}^2\times\mathbb{R}^+)}, \right. \\ &\left. \max_{\tau\in[0,t]}\left(e^{C(1+G_1(t))(t-\tau)}\|I(\tau)\big|_{z=0}\|_{L^\infty(\mathbb{T}^2)}\right)\right\}. \end{aligned} \tag{6.2.97}$$

To derive a lower bound estimate on $B_{(0,0,0)}$, we have

$$\begin{aligned} &\left(\partial_t + u\partial_x + Ku\partial_y + (w + \frac{2\sigma}{1+z})\partial_z - \partial_z^2 - \epsilon^2\partial_x^2 - \epsilon^2\partial_y^2 - \partial_y Ku\right) B_{(0,0,0)} \\ &= \left(\frac{\sigma}{1+z}w + \frac{\sigma(\sigma-1)}{(1+z)^2}\right) B_{(0,0,0)}. \end{aligned} \tag{6.2.98}$$

Applying the classical maximum principle Lemma 6.1.8 for (6.2.98), we get

$$
\begin{aligned}
&\min_{\mathbb{T}^2\times\mathbb{R}^+} (1+z)^\sigma \varphi(t) \\
&\geq \max\left\{1 - C\left(1+G_2(t)\right) te^{C(1+G_2(t))t}\right\} \\
&\times \min\left\{\min_{\mathbb{T}^2\times\mathbb{R}^+} (1+z)^\sigma \varphi_0,\ \min_{[0,t]\times\mathbb{T}^2} \varphi\big|_{z=0}\right\}.
\end{aligned}
\tag{6.2.99}
$$

Step 2. Based on inequalities (6.2.99) and (6.2.97), we have successfully managed to control the underlying quantities I and $B_{(0,0,0)}$ by considering their initial and boundary values. Nonetheless, the problem does not provide specific boundary values for I and $B_{(0,0,0)}$; thus it becomes necessary to estimate them in this step.
In order to control $I|_{z=0}$, we will make use of the Sobolev embedding argument when $s \geq 5$ and the growth rate control argument when $s \geq 7$. By combining these boundary estimates on I with (6.2.99), we can eventually derive the desired inequalities (6.2.89) and (6.2.90).

By Lemma 6.1.4, we obtain

$$
\begin{aligned}
&\|I(\tau)|_{z=0}\|_{L^\infty(\mathbb{T})} \\
&\leq C^2 \sum_{|\alpha|\leq 2} \Big(\|D^\alpha\varphi\|_{L^2}^2 + \|\partial_z D^\alpha\varphi\|_{L^2}^2 + \|\partial_{xy}^1 D^\alpha\varphi\|_{L^2}^2 + \|\partial_z\partial_{xy}^1 D^\alpha\varphi\|_{L^2}^2 \\
&\quad + \|\partial_{xy}^2 D^\alpha\varphi\|_{L^2}^2 + \|\partial_z\partial_{xy}^2 D^\alpha\varphi\|_{L^2}^2\Big) \\
&\leq 6C^2\|\varphi\|_{H_g^{s,\gamma}}^2,
\end{aligned}
\tag{6.2.100}
$$

which, together with (6.2.97), gives (6.2.89).

Now, we derive inequality (6.2.90), according to (6.1.8) and boundary condition $\partial_x^{\alpha_1}\partial_y^{\alpha_2} u = \partial_x^{\alpha_1}\partial_y^{\alpha_2} w = 0$, we have

$$
\partial_t D^\alpha\varphi\big|_{z=0} = D^\alpha(\epsilon^2\partial_x^2 + \epsilon^2\partial_y^2 + \partial_y^2 + \partial_y Ku - u\partial_x - Ku\partial_y - w\partial_z)\varphi\big|_{z=0}.
\tag{6.2.101}
$$

When $\alpha_3 = 0$ and $\alpha_1 + \alpha_2 \leq 2$,

$$
\partial_t D^\alpha\varphi\big|_{z=0} = (\epsilon^2\partial_x^{\alpha_1+2}\partial_y^{\alpha_2} + \epsilon^2\partial_x^{\alpha_1}\partial_y^{\alpha_2+2} + \partial_x^{\alpha_1}\partial_y^{\alpha_2}\partial_z^2)\varphi\big|_{z=0}. \tag{6.2.102}
$$

When $1 \le \alpha_3 \le 2$ and $\alpha_1 = \alpha_2 = 0$,

$$\partial_t D^\alpha \varphi\big|_{z=0} = \big(\epsilon^2 \partial_z^{\alpha_3}\partial_x^2 + \epsilon^2 \partial_z^{\alpha_3}\partial_y^2 + \partial_z^{\alpha_3+2} + (3\alpha_3 - 2)\partial_y K\varphi \partial_z^{\alpha-1} \\ - \alpha_3 \varphi \partial_z^{\alpha_3-1}\partial_x - K\alpha_3\varphi\partial_z^{\alpha_3-1}\partial_y\big)\varphi\big|_{z=0}. \tag{6.2.103}$$

When $\alpha_3 = 1$ and $\alpha_1 + \alpha_2 = 1$,

$$\partial_t D^\alpha \varphi|_{z=0} = (\epsilon^2 \partial_z \partial_{xy}^1 \partial_x^2 + \epsilon^2 \partial_z \partial_{xy}^1 \partial_y^2 + \partial_{xy}^1 \partial_z^3 + \partial_{xy}^1 \partial_y K\varphi + 2\varphi\partial_{xy} + \varphi\partial_{xy}^1\partial_x \\ + \partial_{xy}^1 \varphi \partial_x + K\varphi\partial_{xy}^1\partial_y + \partial_{xy}^1(K\varphi)\partial_y)\varphi\big|_{z=0}. \tag{6.2.104}$$

For $s \ge 7$, using Lemma 6.1.5, we have

$$\|\partial_t I\big|_{z=0}\|_{L^\infty(\mathbb{T}^2)} \le C_{s,\gamma}\|D^\alpha\varphi\partial_t D^\alpha\varphi\big|_{z=0}\|_{L^\infty(\mathbb{T}^2)} \\ \le C_{s,\gamma}\|\varphi(\tau)\|_{H_g^{s,\gamma}}(\|\varphi(\tau)\|_{H_g^{s,\gamma}} + \|\partial_{xy}K(\tau)\|_{L^\infty(\mathbb{T}^2)}\|\varphi(\tau)\|^2_{H_g^{s,\gamma}}). \tag{6.2.105}$$

Hence, a direct integration yields

$$\|I(t)\big|_{z=0}\|_{L^\infty(\mathbb{T}^2)} \le \|I(0)\big|_{z=0}\|_{L^\infty(\mathbb{T}^2)} + C_{s,\gamma}[1 + Y(t)W(t)]W(t)^2 t, \tag{6.2.106}$$

which, together with (6.2.97), gives (6.2.90). For $s \ge 5$,

$$\|\partial_t\varphi|_{z=0}\|_{L^\infty(\mathbb{T}^2)} \le CW(t). \tag{6.2.107}$$

Thus, a direct integration leads to

$$\min_{\mathbb{T}^2} \varphi(t)|_{z=0} \ge \min_{\mathbb{T}^2} \varphi_0\big|_{z=0} - CW(t)t, \tag{6.2.108}$$

which, together with (6.2.99), gives (6.2.91). The proof of Lemma 6.2.3 is thus complete. □

6.3 Local-in-Time Existence and Uniqueness

6.3.1 Local-in-time Existence

In this subsection, we go back to use the symbol $(u^\epsilon, K^\epsilon u^\epsilon, w^\epsilon, \varphi^\epsilon)$ instead of (u, Ku, w, φ) from Sect. 6.2.1–6.2.4 to denote the solution to the regularized system (6.1.8). To obtain the local-in-time solution of initial-boundary value problems (6.1.1)–(6.1.2) or (6.1.6)–(6.1.7), we will construct the solution to the Prandtl equations (6.1.6) by passing to the limit $\epsilon \to 0^+$ in the regularized Prandtl equations (6.1.8). We only sketch the proof into five steps and more details can be found in [98].

Lemma 6.3.1 *Let $s \geq 5$ be an integer, $\gamma \geq 1, \sigma > \gamma + \frac{1}{2}, \delta \in (0, \frac{1}{2})$, and $\epsilon \in (0, 1]$. If $\varphi_0 \in H^{s,\gamma}_{\sigma,2\delta}$, U and P^ϵ are given and satisfy the regularized Bernoulli's law (6.1.7) and the regularity assumption (6.1.15), when $s = 5$, we further assume that $\delta > 0$ is chosen small enough such that the initial hypothesis (6.1.14) holds. Then there exists a uniform life span $T := T(s, \gamma, \sigma, \delta, \|\varphi_0\|_{H^{s,\gamma}}, U) > 0$, which is independent of ϵ, such that the regularized vorticity system (6.1.8) has a solution $\varphi^\epsilon \in C([0, T]; H^{s,\gamma}_{\sigma,\delta}) \cap C^1([0, T]; H^{s-2,\gamma})$ with the following uniform (in ϵ) estimates:*

(i) (Uniform Weighted H^s Estimate) For any $\epsilon \in [0, 1]$ and any $t \in [0, T]$,

$$\|\varphi^\epsilon\|_{H^{s,\gamma}_g} \leq 4\|\varphi_0\|_{H^{s,\gamma}_g}. \tag{6.3.1}$$

(ii) (Uniform Weighted L^∞ Bound) For any $\epsilon \in [0, 1]$ and $t \in [0, T]$,

$$\left\| \sum_{|\alpha|\leq 2} |(1+z)^{\sigma+\alpha_3} D^\alpha \varphi^\epsilon(t)|^2 \right\|_{L^\infty(\mathbb{T}^2\times\mathbb{R}^+)} \leq \frac{1}{\delta^2}. \tag{6.3.2}$$

(iii) (Uniform Weighted L^∞ Lower Bound) For any $\epsilon \in [0, 1]$ and $t \in [0, T]$,

$$\min_{\mathbb{T}^2\times\mathbb{R}^+} (1+z)^\sigma \varphi^\epsilon(t) \geq \delta. \tag{6.3.3}$$

Proof

Step 1. According to the definition of F, assumption (6.1.15) and (6.1.16), and the regularized Bernoulli's law,

$$\|F\|_{L^\infty} \leq C_{s,\gamma,\sigma,\delta}(M_U + M_K) < +\infty, \tag{6.3.4}$$

which, together with (6.2.84), implies

$$\|\varphi^\epsilon\|^2_{H^{s,\gamma}_g} \leq \left\{\|\varphi_0\|^2_{H^{s,\gamma}_g} + C_{s,\gamma,\sigma,\delta}(M_U + M_K)t_1\right\} \times \left\{1 - C_{s,\gamma,\sigma,\delta}\left(\|\varphi_0\|^2_{H^{s,\gamma}_g} + C_{s,\gamma,\sigma,\delta}(M_U + M_K)t_1\right)^{s-1} t_2\right\}^{-\frac{1}{s-1}}, \tag{6.3.5}$$

for t_1 and t_2 to be chosen later. If choosing

$$t_1 \leq \frac{3\|\varphi_0\|^2_{H^{s,\gamma}_g}}{C_{s,\gamma,\sigma,\delta}(M_U + M_K)} \tag{6.3.6}$$

in (6.3.5), then we can get

$$\|\varphi^\epsilon\|^2_{H^{s,\gamma}_g} \leq 4\|\varphi_0\|^2_{H^{s,\gamma}_g} \times \left\{1 - C_{s,\gamma,\sigma,\delta}\left(4\|\varphi_0\|^2_{H^{s,\gamma}_g}\right)^{s-1} t_2\right\}^{-\frac{1}{s-1}}. \tag{6.3.7}$$

Further, choosing

$$t_2 \leq \frac{1 - 2^{2-2s}}{2^{2s-2} C_{s,\gamma,\sigma,\delta}\|\varphi_0\|^{2s-2}_{H^{s,\gamma}_g}} \tag{6.3.8}$$

in (6.3.7) gives

$$\|\varphi^\epsilon\|^2_{H^{s,\gamma}_g} \leq 4\|\varphi_0\|^2_{H^{s,\gamma}_g} \times \left(2^{2-2s}\right)^{-\frac{1}{s-1}} \leq 16\|\varphi_0\|^2_{H^{s,\gamma}_g}. \tag{6.3.9}$$

Note that the functions in both brackets in (6.3.5) are monotonically increasing with respect to time t_1 and t_2; thus we derive the uniform estimate for any $\epsilon \in [0, 1]$ and any $t \in [0, T_1]$,

$$\|\varphi^\epsilon\|_{H^{s,\gamma}_g} \leq 4\|\varphi_0\|_{H^{s,\gamma}_g}, \tag{6.3.10}$$

provided that T_1 is chosen such that

$$T_1 := \min\left\{\frac{3\|\varphi_0\|^2_{H^{s,\gamma}_g}}{C_{s,\gamma,\sigma,\delta}(M_U + M_K)}, \frac{1 - 2^{2-2s}}{2^{2s-2} C_{s,\gamma,\sigma,\delta}\|\varphi_0\|^{2s-2}_{H^{s,\gamma}_g}}\right\}.$$

Step 2. When $s \geq 7$, we know from definition (6.2.85) of F, (6.2.93) of W, and (6.2.92) of G_1 that for any $t \in [0, T_1]$,

$$F(t) \leq M_K,\ W(t) \leq 4\|\varphi_0\|_{H^{s,\gamma}_g} and G_1(t) \leq M_K + (4\|\varphi_0\|_{H^{s,\gamma}_g} + M_U)(1 + M_K), \tag{6.3.11}$$

which, together with (6.2.90), gives

$$\begin{aligned}&\|I(t)\|_{L^\infty(\mathbb{T}^2\times\mathbb{R}^+)}\\&\le\left\{\|I(0)\|_{L^\infty(\mathbb{T}^2\times\mathbb{R}^+)}+C[1+4M_K\|\varphi_0\|_{H_g^{s,\gamma}}]16\|\varphi_0\|^2_{H_g^{s,\gamma}}t_3\right\}\\&e^{C\left(1+M_K+(4\|\varphi_0\|_{H_g^{s,\gamma}}+M_U)(1+M_K)\right)t_4}\end{aligned}\tag{6.3.12}$$

for t_3 and t_4 to be chosen later. If choosing

$$t_3\le\frac{1}{64\delta^2C(1+4M_K\|\varphi_0\|_{H_g^{s,\gamma}})\|\varphi_0\|^2_{H_g^{s,\gamma}}},\tag{6.3.13}$$

in (6.3.12), then we can get

$$\begin{aligned}\|I(t)\|_{L^\infty(\mathbb{T}^2\times\mathbb{R}^+)}\le&\left\{\|I(0)\|_{L^\infty(\mathbb{T}^2\times\mathbb{R}^+)}+\frac{1}{4\delta^2}\right\}\\&e^{C\left(1+M_K+(4\|\varphi_0\|_{H_g^{s,\gamma}}+M_U)(1+M_K)\right)t_4}.\end{aligned}\tag{6.3.14}$$

Further, choosing

$$t_4\le\frac{\ln 2}{C[1+M_K+(4\|\varphi_0\|_{H_g^{s,\gamma}}+M_U)(1+M_K)]}\tag{6.3.15}$$

in the above inequality, gives

$$\|I(t)\|_{L^\infty(\mathbb{T}^2\times\mathbb{R}^+)}\le 2\left\{\|I(0)\|_{L^\infty(\mathbb{T}^2\times\mathbb{R}^+)}+\frac{1}{4\delta^2}\right\}.\tag{6.3.16}$$

Then using initial assumption

$$\sum_{|\alpha|\le 2}|(1+z)^{\sigma+\alpha_3}D^\alpha\varphi_0|^2\le\frac{1}{4\delta^2},\tag{6.3.17}$$

we have the upper bound

$$\left\|\sum_{|\alpha|\le 2}|(1+z)^{\sigma+\alpha_3}D^\alpha\varphi^\epsilon(t)|^2\right\|_{L^\infty(\mathbb{T}^2\times\mathbb{R}^+)}\le\frac{1}{\delta^2}\tag{6.3.18}$$

provided that T_2 is chosen such that

$$T_2 := \min\left\{T_1, \frac{1}{64\delta^2 C(1+4M_K\|\varphi_0\|_{H_g^{s,\gamma}})\|\varphi_0\|^2_{H_g^{s,\gamma}}}\right.$$
$$\left., \frac{\ln 2}{C[1+M_K+(4\|\varphi_0\|_{H_g^{s,\gamma}}+M_U)(1+M_K)]}\right\},$$

for all $t \in [0, T_2]$. When $s \geq 5$, from the initial hypothesis $\|\varphi_0\| \leq C\delta^{-1}$, we have the same estimate (6.3.18) for all $t \in [0, T_2]$.

Step 3. From (6.2.91), we have

$$\min_{\mathbb{T}^2\times\mathbb{R}^+}(1+z)^\sigma\varphi(t) \geq \max\left\{1 - CNte^{CNt}\right\}$$
$$\cdot \min\left\{\min_{\mathbb{T}^2\times\mathbb{R}^+}(1+z)^\sigma\varphi_0 - 4C|\varphi_0\|_{H_g^{s,\gamma}}t\right\}, \tag{6.3.19}$$

where $N := 1 + G_2(t) \leq 1 + M_U + 4\|\varphi_0\|_{H_g^{s,\gamma}}$.
Similarly, let us choose

$$T_3 := \min\left\{T_1, \frac{\delta}{8C\|\varphi_0\|_{H_g^{s,\gamma}}}, \frac{1}{6CN}, \frac{\ln 2}{CN}\right\}.$$

Hence, we derive the uniform estimate for any $\epsilon \in [0, 1]$ and any $t \in [0, T_3]$,

$$\min_{\mathbb{T}^2\times\mathbb{R}^+}(1+z)^\sigma\varphi^\epsilon(t) \geq \delta. \tag{6.3.20}$$

Step 4. In summary, the above uniform estimates hold for any $t \in [0, T]$, if T is chosen to satisfy $T = \min\{T_1, T_2, T_3\}$. Further using almost equivalence relation (6.1.28), Lemma 6.1.5, and (6.3.10), we have

$$\sup_{0\leq t\leq T}(\|\varphi^\epsilon\|_{H^{s,\gamma}} + \|u^\epsilon - U\|_{H^{s,\gamma-1}})$$
$$\leq 4C\|\varphi_0\|_{H_g^{s,\gamma}} + C\sup_{0\leq t\leq T}\left(\sum_{k=0}^{s}\|(1+z)^{\gamma-1}\partial_{xy}^k(u^\epsilon - U)\|_{L^2}\right)$$
$$\leq C\|\varphi_0\|_{H_g^{s,\gamma}} + C_{s,\gamma,\sigma,\delta}\sup_{0\leq t\leq T}\left(\|\varphi\|_{H_g^{s,\gamma}} + \|\partial_{xy}^s U\|_{L^2(\mathbb{T}^2)}\right)$$

$$\le C_{s,\gamma,\sigma,\delta}\left(\|\varphi_0\|_{H_g^{s,\gamma}} + \sup_{0\le t\le T}\|\partial_{xy}^s U\|_{L^2(\mathbb{T}^2)}\right)$$

$$< +\infty. \tag{6.3.21}$$

From equations (6.1.6), (6.1.8) and (6.3.21) and Lemma 6.1.5, we also have $\partial_t\varphi^\epsilon$ and $\partial_t(u^\epsilon - U)$ are uniformly bounded in $L^\infty([0,T]; H^{s-2,\gamma})$ and $L^\infty([0,T]; H^{s-2,\gamma-1})$, respectively. By Lions-Aubin lemma and the compact embedding of $H^{s,\gamma}$ in $H^{s'}_{loc}$, we have after taking a subsequence, as $\epsilon_k \to 0^+$,

$$\begin{cases} \varphi^{\epsilon_k} \overset{*}{\rightharpoonup} \varphi, & \text{in} \quad L^\infty([0,T]; H^{s,\gamma}), \\ \varphi^{\epsilon_k} \to \varphi, & \text{in} \quad C([0,T]; H^{s'}_{loc}), \\ u^{\epsilon_k} - U \overset{*}{\rightharpoonup} u - U, & \text{in} \quad L^\infty([0,T]; H^{s,\gamma-1}), \\ u^{\epsilon_k} \to u, & \text{in} \quad C([0,T]; H^{s'}_{loc}), \end{cases} \tag{6.3.22}$$

for all $s' < s$, where

$$\begin{cases} \varphi = \partial_z u \in L^\infty([0,T]; H^{s,\gamma}) \cap \bigcap_{s'<s} C([0,T]; H^{s'}_{loc}), \\ u - U \in L^\infty([0,T]; H^{s,\gamma-1}) \cap \bigcap_{s'<s} C([0,T]; H^{s'}_{loc}). \end{cases} \tag{6.3.23}$$

Step 5. Using the local uniform convergence of $\partial_x u^{\epsilon_k}$ and $\partial_y(K^{\epsilon_k} u^{\epsilon_k})$, we also have the pointwise convergence of w^{ϵ_k} : as $\epsilon \to 0^+$,

$$w^{\epsilon_k} = -\int_0^z \partial_x u^{\epsilon_k}\, dz - \int_0^z \partial_y(K^{\epsilon_k} u^{\epsilon_k})\, dz \to$$

$$-\int_0^z \partial_x u\, dz - \int_0^z \partial_y(Ku)\, dz =: w. \tag{6.3.24}$$

Combining (6.3.22) and (6.3.23), one may justify the pointwise convergences of all terms in the regularized Prandtl equations (6.1.6). Thus, passing the limit $\epsilon_k \to 0^+$ in the initial-boundary value problem (6.1.6) and the regularized Bernoulli's law (6.1.7), we get that the limit (u, Ku, w) solves the initial-boundary value problem (6.1.4) in the classical sense. By the equivalent system, we get that the limit (u, Ku, w) solves the initial-boundary value problem (6.1.1) in the classical sense.

Lastly, in order to complete the proof, it remains to justify that $\varphi \in L^\infty([0,T]; H^{s,\gamma}_{\sigma,\delta})$ and matching condition $(6.1.1)_6$. Since $D^\alpha\varphi^{\epsilon_k}$ converges to $D^\alpha\varphi$ pointwisely for all $|\alpha| \le 2$ and that φ^{ϵ_k} satisfies

$$\sum_{|\alpha|\le 2} |(1+z)^{\sigma+\alpha_3} D^\alpha \varphi^{\epsilon_k}|^2 \le \frac{1}{\delta^2} \tag{6.3.25}$$

and

$$\min_{\mathbb{T}^2\times\mathbb{R}^+} (1+z)^\sigma \varphi^{\epsilon_k} \ge \delta, \tag{6.3.26}$$

we deduce that (6.3.25) and (6.3.26) still holds for φ, and hence, $\varphi \in L^\infty(0, T; H^s)$.

Also, by Lebesgue's dominated convergence theorem,

$$\int_0^{+\infty} \varphi dz = \lim_{\epsilon_k \to 0^+} \int_0^{+\infty} \varphi^{\epsilon_k} dz = U \tag{6.3.27}$$

and

$$\int_0^{+\infty} K\varphi dz = \lim_{\epsilon_k \to 0^+} \int_0^{+\infty} K\varphi^{\epsilon_k} dz = KU = V \tag{6.3.28}$$

which are equivalent to condition $(6.1.1)_6$ because $\varphi = \partial_z u > 0$. This completes the proof of the existence. □

6.3.2 Uniqueness of Solutions

The aim of this section is to prove the uniqueness of $H^{s,\gamma}_{\sigma,\delta}$ solutions to the 3D Prandtl model. To show the uniqueness, we will generalize the nonlinear cancelation to the L^2 comparison. This motivates us to consider the quantity $\bar{g}$ below.

Specifically, the uniqueness of $H^{s,\gamma}_{\sigma,\delta}$ solutions to the Prandtl equations (6.1.1) is a direct consequence of the following L^2 comparison principle.

Lemma 6.3.2 (**L^2 Comparison Principle**) *For any $s \ge 5, \gamma \ge 1, \sigma > \gamma + \frac{1}{2}$ and $\delta \in (0, 1)$, let (u_i, v_i, w_i) solve the Prandtl equations (6.1.4) with the vorticity $\varphi_i := \partial_z u_i \in C([0, T]; H^{s,\gamma}_{\sigma,\delta}) \cap C^1([0, T]; H^{s-2,\gamma})$ for $i = 1, 2$. Define $\bar{g} := \varphi_1 - \varphi_2 + \frac{\partial_z \varphi_2}{\varphi_2}(u_1 - u_2)$. Then we have*

$$\|\bar{g}\|^2_{L^2} + \int_0^t \|\partial_z \bar{g}\|^2_{L^2} \le \|\bar{g}_0\|^2_{L^2} + C_{\gamma,\sigma,\delta,\varphi,K,U} \int_0^t \|\bar{g}\|^2_{L^2}, \tag{6.3.29}$$

where the constant $C_{\gamma,\sigma,\delta,\varphi,K,U}$ depends on $\gamma, \sigma, \delta, \|w_i\|_{H^{5,\gamma}_g}, \|\partial_y K\|_{L^\infty(\mathbb{T}^2)}$ and $\|\partial^5_{xy} U\|_{L^2(\mathbb{T}^2)}$ only.

Proof Let us denote $(\bar{u}, K\bar{u}, \bar{w}) = (u_1, Ku_1, w_1) - (u_2, Ku_2, w_2)$, $\bar{\varphi} = \varphi_1 - \varphi_2$, $a_2 = \frac{\partial_z \varphi_2}{\varphi_2}$ and $\bar{g} = \bar{\varphi} - a_2\bar{u}$. It is easy to check that $\bar{g} = \bar{\varphi} - a_2\bar{u} = \varphi_2 \partial_z(\frac{\bar{u}}{\varphi_2})$ and the evolution equation on $\bar{g}$ is as follows:

$$
\begin{aligned}
&(\partial_t + u_1\partial_x + Ku_1\partial_y + w_1\partial_z - \partial_z^2)\bar{g} \\
&= (\partial_t + u_1\partial_x + Ku_1\partial_y + w_1\partial_z - \partial_z^2)\bar{\varphi} - a_2(\partial_t + u_1\partial_x + Ku_1\partial_y + w_1\partial_z - \partial_z^2)\bar{u} \\
&\quad - \bar{u}(\partial_t + u_1\partial_x + Ku_1\partial_y + w_1\partial_z - \partial_z^2)a_2 - 2\partial_z a_2 \bar{\varphi}.
\end{aligned}
\tag{6.3.30}
$$

Next, we calculate the values of the first three terms on the right-hand side of the equality (6.3.30), respectively. Recalling the vorticity system (6.1.8), we have

$$
\begin{aligned}
&(\partial_t + u_i\partial_x + Ku_i\partial_y + w_i\partial_z)\partial_z\varphi_i \\
&= \partial_z^3\varphi_i + [\partial_x u_i + \partial_y(Ku_i)]\partial_z\varphi_i - \varphi_i\partial_x\varphi_i - K\varphi_i\partial_y\varphi_i \\
&\quad + \partial_y K(\partial_z\varphi u + \varphi^2),
\end{aligned}
\tag{6.3.31}
$$

then, according to the definition of a_i, we get

$$
\begin{aligned}
&(\partial_t + u_i\partial_x + Ku_i\partial_y + w_i\partial_z)a_i \\
&= \frac{(\partial_t + u_i\partial_x + Ku_i\partial_y + w_i\partial_z)\partial_z\varphi_i}{\varphi_i} - \frac{\partial_z\varphi_i(\partial_t + u_i\partial_x + Ku_i\partial_y + w_i\partial_z)\varphi_i}{\varphi_i^2} \\
&= \frac{\partial_z^3\varphi_i}{\varphi_i} + a_i\partial_x u_i + a_i\partial_y(Ku_i) - \partial_x\varphi_i - K\partial_y\varphi_i - a_i\frac{\partial_z^2\varphi_i}{\varphi_i} + \partial_y K\varphi_i,
\end{aligned}
\tag{6.3.32}
$$

which, combined with the fact

$$
\frac{\partial_z^3\varphi_i}{\varphi_i} - a_i\frac{\partial_y^2 a_i}{\varphi_i} = \partial_z^2 a_i + 2a_i\partial_z a_i,
$$

implies

$$
\begin{aligned}
&(\partial_t + u_i\partial_x + Ku_i\partial_y + w_i\partial_z - \partial_z^2)a_i \\
&= a_i\partial_x u_i + a_i\partial_y(Ku_i) - \partial_x\varphi_i - K\partial_y\varphi_i + \partial_y K\varphi + 2a_i\partial_z a_i.
\end{aligned}
\tag{6.3.33}
$$

Furthermore, we conclude that

$$\begin{aligned}&(\partial_t + u_1\partial_x + Ku_1\partial_y + w_1\partial_z - \partial_z^2)a_2\\&= a_2\partial_x u_2 + a_2\partial_y(Ku_2) - \partial_x\varphi_2 - K\partial_y\varphi_2 + \partial_y K\varphi_2 + 2a_2\partial_z a_2\\&\quad + (\bar{u}\partial_x + K\bar{u}\partial_y + \bar{w}\partial_z)a_2.\end{aligned} \tag{6.3.34}$$

For the estimate of $(\partial_t + u_1\partial_x + Ku_1\partial_y + w_1\partial_z - \partial_z^2)\bar{u}$ and $(\partial_t + u_1\partial_x + Ku_1\partial_y + w_1\partial_z - \partial_z^2)\bar{\varphi}$, from the evolution equation on u and on φ, we can derive that

$$(\partial_t + u_1\partial_x + Ku_1\partial_y + w_1\partial_z - \partial_z^2)\bar{u} = -\bar{u}\partial_x u_2 - K\bar{u}\partial_y u_2 - \bar{w}\partial_z u_2, \tag{6.3.35}$$

and

$$\begin{aligned}&(\partial_t + u_1\partial_x + Ku_1\partial_y + w_1\partial_z - \partial_z^2)\bar{\varphi}\\&= -\bar{u}\partial_x\varphi_2 - K\bar{u}\partial_y\varphi_2 - \bar{w}\partial_z\varphi_2 + \partial_y K(\bar{u}\varphi_2 + u_1\bar{\varphi}).\end{aligned} \tag{6.3.36}$$

Using (6.3.30) and (6.3.34)–(6.3.36), we have

$$\begin{aligned}(\partial_t + u_1\partial_x + Ku_1\partial_y + w_1\partial_z - \partial_z^2)\bar{g} &= -\bar{u}(\bar{u}\partial_x a_2 + K\bar{u}\partial_y a_2 + \bar{w}\partial_z a_2 + 2a_2\partial_z a_2)\\&\quad - 2\bar{\varphi}\partial_z a_2 + \partial_y K(u_1\bar{\varphi} - \bar{u}a_2u_2).\end{aligned} \tag{6.3.37}$$

Now we derive L^2 estimate on $\bar{g}$. For any $t \in (0, T]$, multiplying by $2\bar{g}$ and then integrating by parts over $\mathbb{T}^2 \times \mathbb{R}_+$, we obtain

$$\begin{aligned}&\frac{d}{dt}\|\bar{g}\|_{L^2}^2 + 2\|\partial_z\bar{g}\|_{L^2}^2\\&= \iint_{\mathbb{T}^2} \bar{g}\partial_z\bar{g}\big|_{z=0} dxdy - 2\iiint \bar{g}\bar{u}(\bar{u}\partial_x a_2 + K\bar{u}\partial_y a_2 + \bar{w}\partial_z a_2 + 2a_2\partial_z a_2)\\&\quad - 4\iiint \bar{g}\bar{\varphi}\partial_z a_2\\&\quad + 2\iiint \bar{g}\partial_y K(u_1\bar{\varphi} - \bar{u}a_2u_2) - 2\iiint \bar{g}(u_1\partial_x\bar{g} + Ku_1\partial_y\bar{g} + w_1\partial_z\bar{g}).\end{aligned} \tag{6.3.38}$$

We need to estimate all terms on the right-hand side of equation (6.3.38). Applying the simple trace theorem and Young's inequality,

$$\left|\iint_{\mathbb{T}^2} \bar{g}\partial_z\bar{g}|_{z=0} dxdy\right| \le \left|\iint a_2|\bar{g}|^2 dxdydz\right| + \left|\iint \partial_z a_2|\bar{g}|^2 dxdydz\right|$$

$$
\begin{aligned}
&+ 2\left|\iint a_2 \bar{g} \partial_z \bar{g} dxdydz\right| \\
&\leq \frac{1}{2}\|\partial_z \bar{g}\|_{L^2}^2 + C_{\sigma,\delta}\|\bar{g}\|_{L^2}^2, \qquad (6.3.39)
\end{aligned}
$$

where we have used the fact $\partial_z \bar{g}\big|_{z=0} = -a_2 \bar{g}|_{z=0}$. We claim $\|\frac{\bar{u}}{1+z}\|_{L^2} \leq C_{\sigma,\delta}\|\bar{g}\|_{L^2}$, so by Lemma 6.1.5,

$$
\begin{aligned}
&- 2\iint \bar{g}\bar{u}(\bar{u}\partial_x a_2 + \bar{v}\partial_y a_2 + 2a\partial_y a_2) \\
&\leq C\|(1+z)(\bar{u}\partial_x a_2 + K\bar{u}\partial_y a_2 + \bar{w}\partial_z a_2 + 2a_2\partial_z a_2)\|_{L^\infty} \left\|\frac{\bar{u}}{1+z}\right\|_{L^2} \|\bar{g}\|_{L^2} \\
&\leq C_{\gamma,\sigma,\delta}\Big(1 + \|w_i\|_{H_g^{5,\gamma}} + \|\partial_{xy}^5 U\|_{L^2(\mathbb{T}^2)} + \|K\|_{L^\infty(\mathbb{T}^2)}\|w_i\|_{H_g^{5,\gamma}} \\
&\quad + \|K\|_{L^\infty(\mathbb{T}^2)}\|\partial_{xy}^5 U\|_{L^2(\mathbb{T}^2)}\Big)\|\bar{g}\|_{L^2}^2. \qquad (6.3.40)
\end{aligned}
$$

Now we shall prove that $\|\frac{\bar{u}}{1+z}\|_{L^2}$ can be controlled by $\|\bar{g}\|_{L^2}$. Indeed, since $\delta \leq (1+z)^\delta \varphi_2 \leq \delta^{-1}$ and Lemma 6.1.2,

$$
\begin{aligned}
&\left\|\frac{\bar{u}}{1+z}\right\|_{L^2} \leq \delta^{-1}\left\|(1+z)^{-\sigma-1}\frac{\bar{u}}{\varphi_2}\right\|_{L^2} \leq C_{\sigma,\delta}\left\|(1+z)^{-\sigma}\partial_z\left(\frac{\bar{u}}{\varphi_2}\right)\right\|_{L_2} \\
&\leq C_{\sigma,\delta}\|\bar{g}\|_{L^2}. \qquad (6.3.41)
\end{aligned}
$$

In addition, we also have

$$
\|\bar{\varphi}\|_{L^2} \leq \|\bar{g}\|_{L^2} + \delta^{-2}\left\|\frac{\bar{u}}{1+z}\right\|_{L^2} \leq C_{\sigma,\delta}\|\bar{g}\|_{L^2}, \qquad (6.3.42)
$$

then we can obtain that

$$
-4\iint \bar{g}\bar{\varphi}\partial_z a_2 \leq C_{\sigma,\delta}\|\bar{g}\|_{L^2}^2 \qquad (6.3.43)
$$

and

$$
\begin{aligned}
&2\iiint \bar{g}\partial_y K(u_1\bar{\varphi} - \bar{u}a_2 u_2) \\
&\leq C_{\gamma,\sigma,\delta}\|\partial_y K\|_{L^\infty(\mathbb{T}^2)}\Big(\|w_i\|_{H_g^{5,\gamma}} + \|\partial_{xy}^5 U\|_{L^2(\mathbb{T}^2)}\Big)\|\bar{g}\|_{L^2}^2. \qquad (6.3.44)
\end{aligned}
$$

For the last term in (6.3.38), using the integration by parts, boundary condition $(u_1, w_1)|_{z=0} = 0$ and $\partial_x u_1 + \partial_y(Ku_1) + \partial_z w_1 = 0$, we readily show

$$-2 \iiint \bar{g}(u_1\partial_x\bar{g} + Ku_1\partial_y\bar{g} + w_1\partial_z\bar{g}) = 0. \tag{6.3.45}$$

Combining all of the above estimates (6.3.39)–(6.3.45), we can derive from (6.3.38)

$$\frac{d}{dt}\|\bar{g}\|_{L^2}^2 \le C_{\gamma,\sigma,\delta}\left[1+\left(1+\|\partial_y K\|_{L^\infty(\mathbb{T}^2)}\right)\left(\|w_i\|_{H_g^{5,\gamma}} + \|\partial_{xy}^5 U\|_{L^2(\mathbb{T}^2)}\right)\right]\|\bar{g}\|_{L^2}^2. \tag{6.3.46}$$

Integrating the above inequality over (0, t), then the proof is complete. □

Now applying Gronwall's inequality for Lemma 6.3.2 yields

$$\|\bar{g}(t)\|_{L^2}^2 \le \|\bar{g}(0)\|_{L^2}^2 e^{Ct}, \tag{6.3.47}$$

here $C = C_{\gamma,\sigma,\delta}\left[1+\left(1+\|\partial_y K\|_{L^\infty(\mathbb{T}^2)}\right)\left(\|w_i\|_{H_g^{5,\gamma}} + \|\partial_{xy}^5 U\|_{L^2(\mathbb{T}^2)}\right)\right]$, and this implies $\bar{g} = 0$ due to $u_1|_{t=0} = u_2|_{t=0}$. Since $\varphi_2\partial_z(\frac{u_1-u_2}{\varphi_2}) = \bar{g} = 0$, we have

$$u_1 - u_2 = q\varphi_2 \tag{6.3.48}$$

for some function $q = q(t, x)$. By using the Oleinik's monotonicity assumption $\varphi_2 > 0$ and boundary condition $u_1|_{z=0} = u_2|_{z=0} = 0$, we can get $q = 0$, and hence $u_1 = u_2$. Furthermore, using $\partial_x u + \partial_y(Ku) + \partial_z w = 0$, then w_i $(i = 1, 2)$ can be uniquely determined (i.e., $w_1 = w_2$). This completes the proof of the uniqueness of solutions.

6.4 Bibliographic Comments

Let us now briefly review the background and corresponding results about the boundary layer. The mathematical studies on the Prandtl boundary layer have a very long history. Moore [101] gave an analytical framework of the three-dimensional boundary layer in 1956. In the following decades, the two-dimensional boundary layer theory developed rapidly. The first well-known result was developed by Oleinik and Samokhin in [106], where under the monotonicity condition on tangential velocity with respect to the normal variable to the boundary, the local (in time) well-posedness of Prandtl equations was obtained by using the Crocco transformation and von Mises transformation under the outer flow $U \ne$ constant. Since then, Crocco transformation and von Mises transformation have been used in boundary layer problems. For example, Xin and Zhang [149] established a global existence of weak solutions to the two-dimensional Prandtl system for the pressure is favorable (i.e., $\partial_x P < 0$), which had generalized the local well-posedness results of Oleinik [106]. But, Masmoudi and Wong [98] proved the local existence and uniqueness of solutions for classical Prandtl system by energy methods in a polynomial weighted Sobolev space under the outer flow $U \neq$ constant, and Fan, Ruan, and Yang [29] proved the local well-posedness for the compressible Prandtl boundary layer equations by the same method. The authors Qin, Wang, and Liu [118] proved the local existence and uniqueness of the Prandtl-Hartmann regime with the general outer flow $U \neq$ constant, and Dong and Qin [24] (see also Chap. 2) established the global well-posedness of solutions in $H^s(1 \leq s \leq 4)$ with $U =$ constant and without monotonicity condition and a lower bound.

Compared to the 2D case, the results of the three-dimensional boundary layer equations were very few. Until recent years, Liu, Yang, and Wang [85, 86] obtained the local and global existence of weak solutions to the three-dimensional Prandtl equations with a special structure by Crocco transformation. Liu, Yang, and Wang [84] gave an ill-posedness criterion for the Prandtl equations in three-dimensional space under the outer flow $(U, V) =$ constant, which shows that the monotonicity condition on tangential velocity fields is not sufficient for the well-posedness of the three-dimensional Prandtl equations. Based on this result, they proved the existence and uniqueness of local solutions of three-dimensional boundary layer equations without any special structure in the Gevrey function space; see [70, 73]. Lin and Zhang [79] proved the almost global existence of classical solutions to the 3D Prandtl system with the initial data which lie within ε of a stable shear flow under the outer flow $(U, V) =$ constant. Pan and Xu [108] proved the global existence of solutions to the three dimensional axially symmetric Prandtl boundary layer equations with small initial data under $(U, V) = 0$.

References

1. A. Acrivos, Solution of the laminar boundary layer energy equation at high Prandtl numbers, The Physics of Fluids, 3(4)(1960). https://doi.org/10.1063/1.1706102
2. R. A. Adams and J. J. F. Fournier, Sobolev Spaces, Vol. 140, 2003.
3. R. Alexander, Y. Wang, C. Xu and T. Yang, Well posedness of the Prandtl equation in Sobolev spaces, J. Amer. Math. Soc., 28(3)(2015), 745–784.
4. J. T. Beale and A. Majda, Rates of convergence for viscous splitting of the Navier-Stokes equations, Math. Comput., 37(1981), 243–259.
5. R. E. Caflisch and M. Sammartino, Existence and singularities for the Prandtl boundary layer equations, Z. Angew. Math. Mech., 80(11–12)(2000), 733–744.
6. M. Cannone, M. Lombardo and M. Sammartino, Existence and uniqueness for the Prandtl equations, C. R. Acad. Sci. Paris., 332(2001), 277–282.
7. M. Cannone, M. Lombardo and M. Sammartino, Well-posedness of Prandtl equations with non-compatible data, Nonlinearity, 26(2013), 3077–3100.
8. M. Cannone, M. Lombardo and M. Sammartino, On the Prandtl boundary layer equations in presence of corner singularities, Acta. Appl. Math., 132(2014), 139–149.
9. D. Chen, S. Ren, Y. Wang and Z. Zhang, Global well-posedness of the 2-D magnetic Prandtl model in the Prandtl-Hartmann regime, Asymptotic Anal., 120(3–4)(2020), 1–21.
10. D. Chen, S. Ren, Y. Wang and Z. Zhang, Long time well-posedness of the MHD boundary layer equation in Sobolev space, Anal. Theory Appl., 36(1)(2020), 1–18.
11. D. Chen, Y. Wang and Z. Zhang, Well-posedness of the linearized Prandtl equation around a non-monotonic shear flow, Ann. H. Poincaré, 35(2018), 1119–1142.
12. D. Chen, Y. Wang and Z. Zhang, Well-posedness of the Prandtl equation with monotonicity in Sobolev spaces, J. Differential. Equations, 264(2018), 5870–5893.
13. W. F. Cope and D. R. Hartree, The laminar boundary layer in compressible flow, Philos. Trans. R. Soc. A., 241(827)(1951), 1–69.
14. T. G. Cowling, Magnetohydrodynamics, Interscience Tracts on Physics and Astronomy, 4. Interscience, New York; Interscience, London, 1957.
15. A. L. Dalibard, H. Dietert, D. Gérard-Varet and F. Marbach, High frequency analysis of the unsteady interactive boundary layer model, SIAM J. Math. Anal., (2017), arXiv:1710.04510v1.
16. A. L. Dalibard and N. Masmoudi, Separation for the stationary Prandtl equation, Publications Mathématiques de l'IHéS, 130(2019), 187–297.
17. A. L. Dalibard, M. Paddick, An existence result for the steady rotating Prandtl equation, J. Math. Fluid Mech., 23(2021), Article number: 13.

Y. Qin et al., *Prandtl Equations and Related Boundary Layer Equations*,
https://doi.org/10.1007/978-981-97-4565-4

18. P. A. Davidson, An Introduction to Magnetohydrodynamics, Cambridge Texts in Applied Mathematics, Cambridge University Press, Cambridge, 2001.
19. H. Dietert and D. Gérard-Varet, Well-posedness of the Prandtl equations without any structural assumption, Ann. Partial Differential Equations, (2019), https://doi.org/10.1007/s40818-019-0063-6.
20. Y. Ding, Remarks on the Prandtl boundary layer, J. Math. Anal. Appl., 389(2012), 412–419.
21. M. Ding and S. Gong, Global existence of weak solution to the compressible Prandtl equations, J. Math. Fluid. Mech., 19(2017), 239–254.
22. S. Ding, Z. Ji and Z. Lin, Validity of Prandtl layer theory for steady magnetohydrodynamics over a moving plate with nonshear outer ideal MHD flows, J. Differential. Equations, 278(2021), 220–293.
23. S. Ding and Q. Li, Symmetrical Prandtl boundary layer expansions of steady Navier-Stokes equations on bounded domain, J. Differential Equations, 268(2020), 1771–1819.
24. X. Dong and Y. Qin, Global well-posedness of solutions to 2D Prandtl-Hartmann equations in analytic framework, J. Part. Diff. Eq., 3(35)(2022), 289–306.
25. P. Drasin, Stability of parallel flow in a parallel magnetic field at small magnetic Reynolds number, J. Fluid Mech., 8 (1960), 130–142.
26. W. E, Boundary layer theory and the zero-viscosity limit of the Navier-Stokes equation, Acta Math. Sin., 16(2000), 207–218.
27. W. E and B. Engquist, Blowup of solutions of the unsteady Prandtl's equation, Commun. Pur. Appl. Math., 50(12)(1997), 1287–1293.
28. L. C. Evans, Partial differential equations, American Mathematical Society, Vol 19, 1997.
29. L. Fan, L. Ruan and A. Yang, Local well-posedness of solutions to the boundary layer equations for 2D compressible flow, J. Math. Anal. Appl., 493(2021), 124565.
30. M. Fei, T. Tao and Z. Zhang, On the zero-viscosity limit of the Navier-Stokes equations in R^3_+ without analyticity, J. Math. Pure. Appl., 112(9)(2018), 170–229.
31. P. C. Fife, Toward the validity of Prandtl's approximation in a boundary layer, Arch. Ration. Mech. Anal., 18(1)(1965), 1–13.
32. P. C. Fife, Considerations regarding the mathematical basis for Prandtl's boundary layer theory, Arch. Ration. Mech. An., 28(1968), 184–216.
33. M. Filho, Boundary layers and the vanishing viscosity limit for incompressible 2D flow, Mathematics-Basel, (2007), arXiv:0712.0875.
34. J. Gao, B. Guo and D. Huang, Local-in-time well-posedness of boundary layer system for the full incompressible MHD equations by energy methods, (2017), arXiv:1704.06766v1.
35. J. Gao, D. Huang and Z. Yao, Boundary layer problems for the two-dimensional inhomogeneous incompressible magnetohydrodynamics equations, (2018), arXiv:1810.11258v2.
36. J. Gao, D. Huang and Z. Yao, A note on the well-posedness of Prandtl equations in dimension two, (2018), arXiv:1811.11931v1.
37. J. Gao, M. Li and Z. Yao, Higher regularity and asymptotic behavior of 2D magnetic Prandtl model in the Prandtl-Hartmann regime, math.AP., arXiv:2111.04016v1.
38. C. Gao, and L. Zhang, On the steady Prandtl boundary layer expansions, (2020), arXiv:2001.10700v1.
39. F. Gargano, M. Sammartino and V. Sciacca, Singularity formation for Prandtl's equations, Phys. D., 238(19)(2009), 1975–1991.
40. D. Gérard-Varet and E. Dormy, On the ill-posedness of the Prandtl equation, J. Amer. Math. Soc., 23(2)(2010), 591–609.
41. D. Gérard-Varet and Y. Maekawa, Sobolev stability of Prandtl expansions for the steady Navier-Stokes equations, Arch. Ration. Mech. Anal., 233(2019), 1319–1382.
42. D. Gérard-Varet, Y. Maekawa and N. Masmoudi, Gevrey stability of Prandtl expansions for 2D Navier–Stokes flows, Duke Math. J., 167(13) (2018), 2531–2631.
43. D. Gérard-Varet and N. Masmoudi, Well-posedness for the Prandtl system without analyticity or monotonicity, Ann. Sci. école. Norm. S., 48(4)(2015), 1273–1325.
44. D. Gérard-Varet and T. Nguyen, Remarks on the ill-posedness of the Prandtl equation, Asymptot. Anal., 77(1–2)(2012), 71–88.

45. D. Gérard-Varet and M. Prestipino, Formal derivation and stability analysis of boundary layer models in MHD, Z. Angew. Math. Phys., 68(3)(2017), Art., 76.
46. S. Gong, Y. Guo and Y. Wang, Boundary layer problems for the two-dimensional compressible Navier-Stokes equations, Anal. Appl., 14(1)(2016), 1–37.
47. S. Gong and X. Wang, On a global weak solution and back flow of the mixed Prandtl-Hartmann boundary layer problem, J. Math. Fluid Mech., 23(2021), Art. 11.
48. E. Greniner, On the nonlinear instability of Euler and Prandtl equations, Comm. Pure Appl. Math., 53(9)(2000), 1067–1091.
49. E. Greniner, Y. Guo and T. Nguyen, Spectral stability of Prandtl boundary layers: an overview, Analysis, 35(4)(2015), 343–355.
50. E. Greniner, Y. Guo and T. Nguyen, Spectral instability of characteristic boundary layer flows, Duke Math. J., 16(2016), 3085–3146.
51. Y. Guo and S. Lyer, Validity of steady prandtl layer expansions with external forcing, (2018), arXiv:1810.06662v1.
52. Y. Guo and T. Nguyen, A note on Prandtl boundary layers, Comm. Pure Appl. Math., 64(10)(2011), 1416–1438.
53. Y. Guo and T. Nguyen, Prandtl boundary layer expansions of steady Navier-Stokes flows over a moving plane, Ann. Partial Differential Equations, 3(10)(2017). https://doi.org/10.1007/s40818-016-0020-6
54. Y. Guo, and I. Tice, Compressible, inviscid Rayleigh-Taylor instability, Indiana Univ. Math. J., 2(2009), 677–712.
55. L. Hong and J. Hunter, Singularity formation and instability in the unsteady inviscid and viscous Prandtl equations, Comm. Math. Sci, 1(2)(2003), 293–316.
56. L. Hörmander, The Analysis of Linear Partial Differential Equations III, Vol. 257, Springer, Berlin, New York, 1983.
57. Q. Hou, C. Liu, Y. Wang and Z. Wang, Stability of boundary layers for a viscous hyperbolic system arising from chemotaxys: one dimensional case, SIAM J. Math. Anal., 50(3)(2018), 3058–3091.
58. Y. Huang, C. Liu and T. Yang, Local-in-time well-posedness for compressible MHD boundary layer, J. Differential. Equations, 266(2019), 2978–3013.
59. M. Ignatova and V. Vicol, Almost global existence for the Prandtl boundary layer equations, Arch. Ration. Mech. Anal., 220(2)(2016), 809–848.
60. S. Iyer and N. Masmoudi, Global-in-x stability of steady Prandtl expansions for 2D Navier-Stokes flows, (2020), arXiv:2008.12347v1.
61. S. Jiang and J. Zhang, On the non-resistive limit and the magnetic boundary-layer for one-dimensional compressible magnetohydrodynamics, Nonlinearity, 30(2017), 3587–3612.
62. K. T. Joseph and P. G. Lefloch, Boundary layers in weak solutions of hyperbolic conservation laws, Arch. Ration. Mech. Anal., 147(1999), 47–88.
63. T. V. Kármán and H. S. Tsien, Boundary layer in compressible fluids, J. Aeronautical Science, 5(1938), 227–232.
64. S. A. Korpela, On the viscous dissipation in the boundary layer of a high Prandtl number fluid in laminar flow over a flat plate, Z. Angew. Math. Phys., 36(1985), 624–628.
65. I. Kukavica, N. Masmoudi, V. Vicol and T. K. Wong, On the local well-posedness of the Prandtl and the hydrostatic Euler equations with multiple monotonicity regions, SIAM J. Math. Anal., 46(2014), 3865–3890.
66. I. Kukavica and V. Vicol, On the local existence of analytic solutions to the Prandtl boundary layer equations, Commun. Math. Sci., 11(2013), 269–292.
67. I. Kukavica, V. Vicol and F. Wang, The van dommelen and shen singularity in the Prandtl equations, Adv. Math., 307(2017), 288–311.
68. V. V. Kuznetsov, On solvability conditions for a boundary value problem for the prandtl boundary layer, Siberian Math. J., 35(1994), 558–563.
69. Q. Li and S. Ding, Symmetrical Prandtl boundary layer expansions of steady Navier-Stokes equations on bounded domain, J. Differential. Equations, 268(2020), 1771–1819.

70. W. Li, N. Masmoudi and T. Yang, Well-posedness in Gevrey function space for 3D Prandtl equations without structural assumption, (2020), arXiv:2001.10222v2.
71. W. Li and S. Wang, Boundary layer effects for the nonlinear evolution equations with the vanishing diffusion limit, Discrete Dyn. Nat. Soc. (2018), 1–11.
72. W. Li, D. Wu and C. Xu, Gevrey class smoothing effect for the Prandtl equation, SIAM J. Math. Anal., 48(3)(2016), 1672–1726.
73. W. Li and R. Xu, Gevrey well-posedness of the hyperbolic Prandtl equations, Commun. Math. Res., 38()(2022), 605–624.
74. W. Li, R. Xu and T. Yang, Global well-posedness of a Prandtl model from MHD in Gevrey function spaces, Acta Math. Sci., 42B(6)(2022), 2343–2366.
75. W. Li and T. Yang, Well-posedness in Gevrey function space for the three-dimensional Prandtl equations, (2019), arXiv:1708.08217v4.
76. W. Li and T. Yang, Well-posedness in Gevrey space for the Prandtl equations with non-degenerate critical points, J. Eur. Math. Soc., 22(3)(2020), 717–775.
77. W. Li and T. Yang, Well-posedness of the MHD boundary layer system in Gevrey function space without structural assumption, (2020), arXiv:2009.06513v1.
78. X. Lin and T. Zhang, Local well-posedness for 2D incompressible magneto-micropolar boundary layer system, Appl. Anal., (2019), 1–22.
79. X. Lin and T. Zhang, Almost global existence for the 3D Prandtl boundary layer equations, Acta. Appl. Math., 169(2020), 383–410.
80. C. Liu, Y. Wan, A simple exact solution of the Prandtl boundary layer equations containing a point of separation, Arch. Ration. Mech. Anal., 89(1985), 177–185.
81. C. Liu and Y. Wang, Derivation of Prandtl boundary layer equations for the incompressible Navier-Stokes equations in a curved domain, Appl. Math. Lett., 34(2014), 81–85.
82. C. Liu and Y. Wang, Stability of boundary layers for the nonisentropic compressible circularly symmetric 2D flow, SIAM J. Math. Anal., 46(1)(2014), 256–309.
83. C. Liu, D. Wang, F. Xie and T. Yang, Magnetic effects on the solvability of 2D MHD boundary layer equations without resistivity in Sobolev spaces, J. Funct. Anal., 279(2020), 108637.
84. C. Liu, Y. Wang and T. Yang, On the ill-posedness of the Prandtl equations in three-dimensional space, Arch. Ration. Mech. Anal., 220 (2016), 83–108.
85. C. Liu, Y. Wang and T. Yang, Global existence of weak solutions to the three-dimensional Prandtl equations with a special structure, Discrete Cont. Dyn. Syst., 9(6) (2016), 2011–2029.
86. C. Liu, Y. Wang and T. Yang, A well-posedness theory for the Prandtl equations in three space variables, Adv. Math., 308 (2017), 1074–1126.
87. C. Liu, F. Xie and T. Yang, A note on the ill-posedness of shear flow for the MHD boundary layer equations, Sci. China Math., 61 (2018), 2065–2078.
88. C. Liu, F. Xie and T. Yang, Justification of Prandtl ansatz for MHD boundary layer, SIAM J. Math. Anal., 51(3)(2019), 2748–2791.
89. C. Liu, F. Xie and T. Yang, MHD boundary layers theory in Sobolev spaces without monotonicity I: well-posedness theory, Commun. Pure Appl. Math., 1 (2019), 63–121.
90. C. Liu, F. Xie and T. Yang, MHD boundary layers in Sobolev spaces without monotonicity, II. convergence theory, Analysis of PDEs, (2018), arXiv:1704.00523v4.
91. C. Liu and T. Yang, Ill-posedness of the Prandtl equations in Sobolev spaces around a shear flow with general decay, J. Math. Pures Appl., 108(2)(2017), 150–162.
92. X. Lin and T. Zhang, Local well-posedness for 2D incompressible magneto-micropolar boundary layer system, Appl. Anal., (2019), 1–22.
93. X. Lin and T. Zhang, Almost global existence for the 3D Prandtl boundary layer equations, Acta. Appl. Math., 169(2020), 383–410.
94. N. Liu and P. Zhang, Global small analytic solutions of MHD boundary layer equations, J. Differ. Equations, 281(2021), 199–257.
95. M. C. Lombardo, M. Cannone and M. Sammartino, Well-posedness of the boundary layer equations, SIAM J. Math. Anal., 35(4)(2003), 987–1004.
96. T. Luo and Z. Xin, Hölder continuous solutions to the three-dimensional Prandtl system, (2018), arXiv:1804.04285.

97. Y. Maekawa, On the inviscid limit problem of the vorticity equations for viscous incompressible flows in the half-plane, Comm. Pure Appl. Math., 67(2014), 1045–1128.
98. N. Masmoudi and T. K. Wong, Local-in-time existence and uniqueness of solutions to the Prandtl equations by energy methods, Comm. Pure Appl. Math., 68(10)(2015), 1683–1741.
99. M. Medveď, A new approach to an analysis of Henry type integral inequalities and their Bihari type versions, J. Math. Anal. Appl., 214(2)(1997), 349–366.
100. D. S. Mitrinović, J. E. Pěcarić and A. M. Fink, Inequalities Involving Functions and Their Integrals and Derivatives, Mathematics and Its Applications, Vol 63, 1991.
101. F.K. Moore, Three-dimensional boundary layer theory. Adv. Appl. Mech., 4(1956), 159–228.
102. R. Narasimha, S. Vasantha, Laminar boundary layer on a flat plate at high Prandtl number, Z. Angew. Math. Phys., 17(1966), 585–592.
103. K. Nickel, Prandtl's boundary layer theory from the view-point of a mathematician, Annu. Rev. Fluid Mech., 5(1973), 405–428.
104. O. A. Oleinik, On the system of Prandtl equations in boundary-layer theory, Dokl. Akad. Nauk SSSR, 150(1)(1963), 28–31.
105. O. A. Oleinik, On the mathematical theory of boundary layer for unsteady flow of incompressible fluid, PMM-J. Appl. Math. Mec., 30(5)(1966), 951–974.
106. O. A. Oleinik and V. N. Samokhin, Mathematical Models in Boundary Layer Theory, Applied Mathematics and Mathematical Computation, Vol. 15, Chapman and Hall/CRC, Boca Raton, FL, 1999.
107. M. Paicu and P. Zhang, Global well-posdness of Prandtl system with small analytic data, (2019), arXiv: 1911.03690v1.
108. X. Pan and C. Xu, Global tangentially analytical solutions of the 3D axially symmetric Prandtl equations, https://doi.org/10.48550/arXiv.2202.05969
109. X. Pan and C. Xu, Long-time existence of Gevrey-2 solutions to the 3D Prandtl boundary layer equations, https://doi.org/10.48550/arXiv.2212.02113
110. V. E. Petrov and T.A. Suslina, Regularity of the solution of the Prandtl equation, (2020), arXiv: 2008.06715v2.
111. L. Prandtl, Uber Flussigkeitsbewegung bei sehr kleiner Reibung, in: Verhandlung des III Intern. Math. Kongresses, Heidelberg, 1904, 484–491.
112. Y. Qin, Integral and Discrete Inequalities and Their Applications: Volume I: Nonlinear Inequalities, Birkhäuser, 2016.
113. Y. Qin, Integral and Discrete Inequalities and Their Applications: Volume II: Nonlinear Inequalities, Birkhäuser, 2016.
114. Y. Qin, Analytic Inequalities and Their Applications in PDEs, Springer International Publishing, 2017.
115. Y. Qin and X. Dong, Local existence of solutions to 2D Prandtl equations in a weighted Sobolev space, Anal. Math. Phy., (2022) 12:16, https//doi.org/10.1007/s13324-021-00615-z
116. Y. Qin and X. Dong, Local well-posedness of solutions to 2D mixed Prandtl equations in Sobolev space without monotonicity and a lower bound, Nonlinear Anal. RWA, 80 (2024) 104140.
117. Y. Qin and X. Wang, Local existence of solutions to 3D Prandtl equations with a special structure, J. Math. Pure. Appl., to appear.
118. Y. Qin and X. Wang, Local well-posedness of solutions to 2D magnetic Prandtl model in the Prandtl-Hartmann regime, submitted.
119. M. Renardy, Well-posedness of the Prandtl boundary layer equations for the upper convected maxwell fluid, J. Dyn. Differ. Equ., 27(2015), 981–988.
120. M. Renardy and X. Wang, Well-posedness of boundary layer equations for time-dependent flow of non-newtonian fluids, J. Math. Fluid Mech., 16(2014), 179–191.
121. M. Sammartino, R. E. Caflisch, Zero viscosity limit for analytic solutions of the Navier-Stokes equation on a half-space, I. Existence for Euler and Prandtl Equations Comm. Math. Phys., 192(1998), 433–461.

122. M. Sammartino, R. E. Caflisch, Zero viscosity limit for analytic solutions of the Navier-Stokes equation on a half-space, II. Construction of the Navier-Stokes solution, Comm. Math. Phys., 192(1998), 463–491.
123. H. Schlichting and K. Gersten, Boundary-Layer Theory, Ninth Edition, Springer-Verlag Berlin Heidelberg, 2017.
124. M. Sermange and R. Temam, Some mathematical questions related to the MHD equations, Computer Compacts, 1(4)(1983), 635–664.
125. J. Serrin, Asymptotic behaviour of velocity profiles in the Prandtl boundary layer theory, Proc. Roy. Soc. London Ser. A., 299(1967), 491–507.
126. J. Serrin, On the mathematical basis for Prandtl's boundary layer theory: an example, Arch. Ration. Mech. An., 28(1968), 217–225.
127. W. Shen, Y. Wang and Z. Zhang, Boundary layer separation and local behavior for the steady Prandtl equation, (2019), arXiv:1904.08055v1.
128. Y. Sun, A blow-up criterion for classical solutions to the Prandtl equations, J. Math. Phys., 60(11)(2019), 111502.
129. R. Temam, Boundary layers associated with incompressible Navier-Stokes equations: The noncharacteristic boundary case, J. Differential. Equations, 179(2002), 647–686.
130. H. Triebel, Theory of Function Spaces, Modern Birkhäuser classics, Birkhäuser, 1983.
131. W. Walter, On the asymptotic behavior of solutions of the Prandtl boundary layer equations, Indiana Univ. Math. J., 20(1970), 829–841.
132. S. Wang, L. Jiang and C. Liu, Quasi-neutral limit and the boundary layer problem of Planck-Nernst-Poisson-Navier-Stokes equations for electro-hydrodynamics, J. Differential Equations, 267(2019), 3475–3523.
133. S. Wang and C. Liu, Boundary layer problem and quasineutral limit of compressible Euler-Poisson system, Comm. Pure Appl. Anal., 16(6)(2017), 2177–2199.
134. S. Wang and N. Wang, Boundary layer problem of MHD system with non-characteristic perfect conducting wall, Appl. Anal., (2017), 1–21.
135. S. Wang and G. Wang, Well-posedness and blowup of the geophysical boundary layer problem, J. Math. Fluid Mech., 22(2020), Art., 52.
136. S. Wang, B. Wang, C. Liu and N. Wang, Boundary layer problem and zero viscosity-diffusion limit of the incompressible magnetohydrodynamic system with no-slip boundary conditions, J. Differential. Equations, 263(2017), 4723–4749.
137. X. Wang, Y. Wang and Z. Xin, Boundary layers in incompressible Navier-Stokes equations with Navier friction boundary condition for vanishing viscosity limit, Comm. Math. Sci., 8(4)(2010), 965–998.
138. C. Wang, Y. Wang and Z. Zhang, Gevrey stability of hydrostatic approximate for the Navier-Stokes equations in a thin domain, arXiv:2004.05537v1.
139. C. Wang, Y. Wang and P. Zhang, On the global small solution of 2-D Prandtl system with initial data in the optimal Gevrey class, (2021), arXiv:2103.00681v1.
140. Y. Wang and F. Xie and T. Yang, Local well-posedness of Prandtl equations for compressible flow in two space variables, SIAM J. Math. Anal., 47(1)(2015), 321–346.
141. S. Wang and X. Xin, Boundary layer problems in the viscosity-diffusion vanishing limits for the incompressible MHD systems, (2017), arXiv:1706.07898v1.
142. Y. Wang and Z. Zhang, Global C^∞ regularity of the steady prandtl equation, (2019), arXiv:1912.08402v1.
143. Y. Wang and S. Zhu, Back flow of the two-dimensional unsteady Prandtl boundary layers under adverse pressure gradient, SIAM J. Math. Anal., 52(1)(2020), 954–966.
144. Y. Wang and S. Zhu, Blow-up criterion for the 2-D Prandtl equation, (2020), arXiv:2002.06327v1.
145. F. Wu, Well-posedness of the nonlinear unsteady Prandtl equations with Robin boundary condition in weighted Sobolev spaces, (2015), arXiv:1503.00682v3.
146. F. Wu, Full regularity and well-posedness of the nonlinear unsteady Prandtl equations with Robin or Dirichlet boundary condition, (2016), arXiv:1603.07725v1.

147. F. Xie and T. Yang, Global-in-time stability of 2D MHD boundary layer in the Prandtl-Hartmann regime, SIAM. J. Math. Anal. 50(6)(2018), 5749–5760.
148. F. Xie and T. Yang, Lifespan of solutions to MHD Boundary layer equations with analytic perturbation of general shear flow, Acta Math. Appl. Sin-E, 35(1)(2019), 209–229.
149. Z. Xin and L. Zhang, On the global existence of solutions to the Prandtl's system, Adv. Math., 181(2004), 88–133.
150. C. Xu and X. Zhang, Well-posedness of the Prandtl equation in Sobolev space without monotonicity, (2015), arXiv: 1511.04850v3.
151. C. Xu and X. Zhang, Long time well-posedness of Prandtl equations in Sobolev space, J. Differential. Equations, 263(2017), 8749–8803.
152. X. Xu and J. Zhao, On the global existence and uniqueness of solutions to Prandtl's system, Acta Math. Sin., 25(1)(2009), 109–132.
153. L. Xue, Well-posedness and zero microrotation viscosity limit of the 2D micropolar fluid equations, Math. Meth. Appl. Sci., 34(2011), 1760–1777.
154. T. Yang and W. Li, Well-posedness in Gevrey function spaces for the Prandtl equations with non-degenerate critical points, J. Eur. Math. Soc., (2019), 717–775.
155. X. Ye and J. Zhang, The behavior of boundary layer for the compressible planar MHD equations at small shear viscosity, J. Dyn. Differ. Equ., 32(2020), 281–311.
156. P. Zhang and Z. Zhang, Long time well-posedness of Prandtl system with small and analytic initial data, J. Funct. Anal., 270(7)(2016), 2591–2615.

MIX
Papier aus verantwortungsvollen Quellen
Paper from responsible sources
FSC® C105338

If you have any concerns about our products,
you can contact us on
ProductSafety@springernature.com

In case Publisher is established outside the EU,
the EU authorized representative is:
Springer Nature Customer Service Center GmbH
Europaplatz 3, 69115 Heidelberg, Germany

Printed by Libri Plureos GmbH
in Hamburg, Germany